Construction Dictionary Illustrated

BNi® Building News

1-800-873-6397

BNi. Building News

EDITORS
James Acret, Esq.
Richard Chylinski, FAIA
William D. Mahoney, P.E.
Arthur F. O'Leary, FAIA
Roy S. Smith

COVER DESIGN
Robert O. Wright

GRAPHICS
Reuben Chock
Eric F. Mahoney, Assoc. AIA

TECHNICAL SERVICES
Lisa Allred
Anthony Jackson
Ana Varela
Rod Yabut

BNi PUBLICATIONS, INC.

LOS ANGELES

10801 National Blvd., Ste.100
Los Angeles, CA 90064

NEW ENGLAND

172 Taunton Ave. Ste. 2
PO BOX 14527
East Providence, RI 02914

ANAHEIM

1612 S. Clementine St.
Anaheim, CA 92802

WASHINGTON, D.C.

502 Maple Ave. West
Vienna, VA 22180

ISBN 1-55701-365-9

A Jointing. See Joining.

AA. Aluminum Association.

AAA. American Arbitration Association.

AAC. Autoclaved Aerated Concrete.

AAN. American Association of Nurserymen.

AASHTO. American Association of State Highway and Transportation Officials.

Abacus. The flat slab on top of a column capital, supporting the architrave.

Abandonment. The failure of both parties to a contract to abide by its terms.

Abattoir. A slaughterhouse; a place where animals are butchered for food.

Aberration. A distortion of an image produced by a faulty lens or mirror.

Above Ground Tank. A large above ground vessel used for the storage of liquids.

Abrade. Scrape or wear away by friction.

Abrams' Law. A rule stating that with given concrete materials and conditions of test, the ratio of the amount of water to the amount of the cement in the mixture determines the strength of the concrete, provided the mixture is of a workable consistency.

Abrasion Resistance. Ability of a surface to resist being worn away by rubbing and friction.

Abrasion. Wearing away by friction.

Abrasive Coatings. In closed coating of paper no adhesive is exposed, as surface of paper is completely covered with abrasive; in open coating, surface of backing paper is covered with regulated amount of abrasive, exposing the adhesive; space between the abrasive grains reduces loading and filling when sanding gummy or soft materials.

Abrasive Paper. Paper with an abrasive surface; sandpaper, emery paper; garnet paper.

Abrasive Surface Tile. Floor tile that has been roughened to be slip-resistant.

Abrasive Surface. A surface that has been roughened for safety or for warning.

Abrasive. A substance used for wearing, grinding, cleaning, or polishing by rubbing or grinding.

ABS Pipe. A plastic pipe sold in 10 and 20 foot lengths in various diameters for plumbing stacks and drains; used primarily for drain lines.

ABS. Acrylonitrile Butadiene Styrene; a plastic used for piping; has high resistance to impact, heat, and chemicals.

Absolute Humidity. The density of water vapor per unit volume of air.

Absolute Pressure. The pressure measured by a gauge plus a correction for the effect of air pressure on the gauge (14.7 psi at sea level).

Absolute Temperature. Temperature measured from absolute zero.

Absolute Viscosity. A method of measuring viscosity using the poise as the basic measurement unit; this method utilizes a partial vacuum to induce flow in the viscometer.

Absolute Volume. The volume of an ingredient in its solid state, without voids between individual pieces or particles; in the case of fluids, the cubic content occupied; in concrete, it is the actual volume occupied by the different ingredients determined by dividing the weight of each ingredient in pounds by its specific gravity times the weight of one cubic foot of water in pounds; example. the absolute volume of one sack of cement equals 94 divided by 3.15 times 62.4 equals 0.478 cubic feet.

Absolute Zero. A theoretical lowest possible temperature, at which all molecular motion ceases, calculated to be exactly minus 273.15° Centigrade or minus 459.67° Fahrenheit.

Absorb. To swallow up or suck in, like wood absorbing a finishing material.

Absorbed Moisture. Moisture that has entered a solid material by absorption and has physical properties not substantially different from ordinary water at the same temperature and pressure; in aggregates, that water which is not available to become part of the mixing water.

Absorbent. Having the ability to suck up liquid, gas, or heat.

Absorber. The blackened surface in a solar collector that absorbs the solar radiation and converts it to heat energy.

Absorptance. The ratio of light absorbed by a material to incident light falling on it.

Absorption Chiller. A system similar to a vapor compression chiller with the exception that it does not use a compressor, but uses thermal energy (low pressure steam, hot water, or other hot liquids) to produce the cooling effect.

Absorption Coefficient. The absorption coefficient of a material or sound-absorbing device is the ratio of the sound absorbed to the sound incident on the material or device; the sound absorbed by a material or device is usually taken as the sound energy incident on the surface minus the sound energy reflected.

Absorption Rate. 1. The speed at which the real estate market can absorb new offerings of land or buildings during a specified period of time. 2. The amount of water absorbed when a brick is partially immersed for one minute; usually expressed in either grams or ounces per minute per 30 sq. in; also called suction or initial rate of absorption.

Absorption Refrigerator. Refrigerator which creates low temperatures by using the cooling effect formed when a refrigerant is absorbed by chemical substance.

Absorption, Total. The amount of water a masonry unit will absorb when immersed in water.

Absorption. The relationship of the weight of the water absorbed by a material specimen subjected to pre-scribed immersion procedure, to the weight of the dry specimen, expressed in percent.

Abstract of Title. A written summary of all transactions that could affect the ownership of a piece of real property, including deeds, leases, liens, and wills.

Abutment Piece or Member. The bearing plate or piece of a wall system to which the loads are transferred.

Abutment. The lateral supporting structure of an arch, bridge or similar pressure; that part of a pier or wall from which an arch springs, specifi-cally the support at either end of an arch, beam or bridge; that part of a structure which takes the thrust of a beam, arch, vault, truss or girder; the part of a bridge that supports the end of the span and prevents the bank from sliding under it; a foundation that carries gravity and also thrust loads.

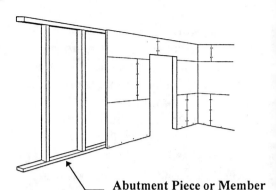

Abutment Piece or Member

AC Relay. An electromagnetic or elec-tromechanical valve or switch using small currents and voltages to control the making or breaking of electrical contacts on a circuit.

AC. 1. Air Conditioning. 2. Asphaltic Concrete. 3. Alternating Current.

Accelerate. To quicken or hasten the natural progress of certain actions or events.

Accelerated Depreciation. The de-clining balance and sum of the year's digits method which give greater de-preciation to the early years of the life of the assets.

Accelerating Admixture. Substance that increases the rate of hydration, shortens setting time, or increases strength development in concrete, mortar, grout, or plaster.

Acceleration Clause. A clause in a mortgage or trust deed that would al-low the lender to call the whole loan due at any time that certain specified events occur, such as a default in payments or sale of the property.

Acceleration. 1. Requiring change order work to be done without exten-sion of the contract time. 2. An in-crease in the rate of speed.

Accelerator. A substance which, when added to concrete, mortar, grout, or plaster, increases the rate of hydration of the hydraulic cement, shortens the time of setting, or increases the rate of hardening of strength development, or both; materials used to speed up the setting of mortar or concrete.

Accelerogram. The record from an accelerograph showing acceleration as a function of time.

Acceptance. 1. Manifestation that a party assents or agrees to a contract. 2. Approval of the work of a construction contract.

Access Control. A computerized security system designed to protect against unauthorized entry into buildings or building areas.

Access Door. A door or panel creating a means of access for the inspection or repair of concealed equipment.

Access Floor. A raised floor platform with removable panels to provide access to the area below.

Access Panel. Removable or swinging panel, usually flush with adjoining surface to provide access to concealed equipment or system components for inspection and maintenance purposes.

Access Stair. A stair system to provide specific access to roofs, mechanical equipment rooms, or as a means of emergency exit.

Access/Secure Control Unit. The controlling device of a computerized security system designed to protect against unauthorized entry into buildings or building areas.

Accessible Heremetic. Assembly of motor and compressor inside a single bolted housing unit.

Accessible Location. A location which can be reached by standing on the floor, platform, runway, or other permanent working area.

Accessible. 1. As applied to equipment, admitting close approach. 2. As applied to wiring methods, not permanently closed in by the structure or finish of the building; capable of being removed without disturbing the building structure, finish, or fixed appurtenance thereto.

Accessories. 1. Tile accessories, ceramic or non-ceramic articles, affixed to or inserted in tile work, as exemplified by towel bars, paper, soap and tumbler holders, grab bars and the like. 2. Concrete accessories, implements or devices used in the formwork, pouring, spreading, and finishing, of concrete surfaces.

Accessory, Reinforcing. Items used to facilitate the installation of masonry or concrete reinforcing.

Accessory. An object or device aiding or contributing in a secondary way.

Accord and Satisfaction. Conduct of a debtor that indicates agreement to an amount of money owed by the debtor to a creditor.

Accordion Folding Door. A folding, hinged, or creased door with rollers which run along a track.

Accordion Partition. A folded, creased, or hinged interior dividing wall.

Account Balance. The difference between the sum of the debits and credits.

Account. A statement of transactions during a fiscal period and the resulting balance in each category of income and expense.

Accounting Period. The time that elapses between the preparation of financial statements.

Accounting. The recording and auditing of financial accounts.

Accounts Payable. Money owed by the firm to vendors for services or materials.

Accounts Receivable. Money owed to the firm for services rendered or for reimbursements.

Accretion. An increase of land area by the gradual or imperceptible action of natural forces.

Accrual Accounting. A method of keeping accounting records in which income is recorded when services are rendered and expenses are recorded when incurred, rather than when cash is received or paid out.

Accruals. The recognition of income and expenses as they occur even though they are not received or paid for until a later period.

Accrue. Periodically accumulated, as interest.

Accrued Expenses. The entry into the liability accounts of expenses, incurred but not paid, at a given date.

Accrued Income. The entry into the asset accounts of income earned, but not received, at a given date.

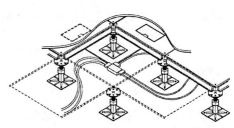

Access Floor

Accrued Interest. Interest that has been earned but not yet paid.

Accrued Liabilities. Amounts representing liabilities incurred, but not paid, by a given date.

Accumulator. Storage tank which receives liquid refrigerant from evaporator and prevents it from flowing into suction line before vaporizing.

Acetic Acid. A compound, which in the pure state is a colorless, pungent, biting liquid; vinegar contains 4 to 12 percent of acetic acid.

Acetone. A water-white volatile, highly flammable solvent with ether-like odor; made by destructive distillation of hardwood, fermentation of butyl alcohol, and from petroleum sources; used extensively in making paint removers; dimethyl ketone; see Ketones.

Acetylene. A colorless hydrocarbon gas, burning with a bright flame, used as a fuel in welding and soldering.

ACGIH. American Conference of Governmental Industrial Hygienists, Inc.

Achromatic. 1. Having no color, being black, gray, or white. 2. Being neutral in color. 3. Difficult to color. 4. Lenses practically free from light of unwanted color.

ACI. American Concrete Institute.

Acid Condition In System. Condition in which refrigerant or oil in system is mixed with fluids that are acid in nature.

Acid Demand. Amount of acid required to lower pH and total alkalinity of pool water to correct level.

Acid Etch. 1. The use of acid to cut lines into metal or glass. 2. The use of acid to remove the surface of concrete.

Acid Number. A designation of the amount of free acid in oils, flats, waxes and resins, expressed as the number of milligrams of potassium hydroxide required to neutralize one gram of the material being tested.

Acid Rain. Sulfur dioxide emissions combining with water in the atmosphere and falling to the earth.

Acid Resisting Brick. Brick suitable for use in contact with chemicals, usually in conjunction with acid-resistant mortars.

Acid. A sour substance, one which liberates hydrogen ions in water and is sour and corrosive; will turn litmus red and has a pH of less than 7; acids are generally divided into two classes: (1) strong mineral or inorganic acids such as sulfamic, sulfuric, phosphoric, hydrochloric, or nitric, (2) weak organic or natural acids such as acetic (vinegar), citric (citrus fruit juices), oxalic, and fatty acids (oleic) such as palmitic and stearic.

Acidity. A general term applying to substances on the acid side of neutral.

Acid-Proof Counter. A horizontal work surface resistant to acid spills.

Acid-Resistant Grout. A grout that resists the effect of prolonged contact with acids.

Acid-Test Radio. A calculation of a firm's liquidity position; that is the ratio of its quick assets (readily convertible to cash) to current liabilities.

ACM. Asbestos-Containing Material.

ACORD 25-S. A form of insurance certificate issued by Agency Company Organization for Research and Development which has replaced the AIA standard form.

Acoustic Paint. Paint which absorbs or deadens sound.

Acoustic. 1. Relating to sound or the sense of hearing. 2. The properties of a room or auditorium in transmitting sound.

Acoustical and Insulating Materials Association (AIMA). 205 W. Touhy Avenue, Park Ridge, Illinois 60068, (312) 692-5178.

Acoustical Block. A masonry block used for its sound-absorbing qualities.

Acoustical Materials. Those capable of absorbing sound waves.

Acoustical Panel. Ceiling and wall mounted modular units composed of sound absorbing materials.

Acoustical Plaster and Plastic. Sound absorbing finishing materials mill-formulated for application in areas where a reduction in sound reverberation or noise intensity is desired; these materials usually are applied to a minimum thickness of 1/2 inch and generally provide a noise reduction coefficient of at least .45 decibels.

Acoustical Tile. Ceiling panels in board form used for its sound absorbing properties, sometimes used on walls.

Acoustical Treatment. The act or process of applying acoustical materials to walls and ceilings.

Acoustical. Relating to sound or to the sense of hearing.

Acoustics. The science of sound including its production, transmission, and effects.

ACR Tubing. Tubing used in air conditioning and refrigeration; ends are sealed to keep tubing clean and dry.

Acre. 1. A piece of land measuring 43,560 square feet. 2. Unit for measuring land, equal to 43,560 square feet or 4840 square yards or 160 square rods.

Acre-Foot. A volume unit for measuring large quantities of water as in reservoirs and lakes; the amount of water that would cover one acre one foot deep, equal to 43,560 cubic feet.

Acre-Inch. One twelfth of an acre-foot.

ACRI. Air-Conditioning & Refrigeration Institute.

Across. The application of gypsum board where the long dimension is applied at right angles to the framing.

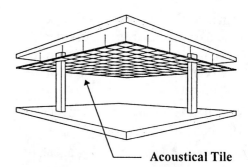

Acoustical Tile

Acrylic Carpet. A carpet made from acrylic fiber composed of synthetic polymers.

Acrylic Plastics. Plastics based on resins made by the polymerization of acrylic monomers, such as ethyl acrylate and methecrylate.

Acrylic Resins. Family of synthetic resins made by polymerizing esters of acrylic acid; synthetic resins of excellent color and clarity used in both emulsion and solvent-based paints.

Acrylic. 1. A general class of resinous polymers derived from esters, amides or other acrylic aid derivatives. 2. A transparent plastic material used in sheet form for window glass and skylights.

Acrylics. In carpeting, generic term including acrylic and modified acrylic (modacrylic) fibers; acrylic is a polymer composed of at least 85% by weight of acrylonitrile; modacrylic is a polymer composed of less than 85% but at least 35% by weight of acrylonitrile.

Acrylonitrile Butadiene-Styrene (ABS). Plastic material used in manufacturing drainage pipe and fittings.

Act of God. An unexpected event, not within the control of either party, that makes the performance of a contract impossible, unreasonable, or illegal.

Activated Carbon. Specially processed carbon used as a filter-drier; commonly used to clean air; pulverized carbon treated to be especially adsorbent.

Activated Charcoal. See Activated Carbon.

Activator. A catalyst, curing agent, or co-reactant, as for an epoxy resin.

Active Door. In a pair of doors, the leaf that opens first and the one to which the lock is applied.

Active Pressure. The pressure exerted by retained earth; such as the earth retained by a retaining wall.

Active System. A solar heating or cooling system that requires outside mechanical power to move the collected heat.

Actual Dimension. The true size of a piece of lumber after it has been milled and dried; see Nominal Dimension.

Actual Notice. The giving of notice by mailing it or handing it to the recipient.

Actuator. That portion of a regulating valve which converts mechanical fluid, thermal energy, or electrical energy into mechanical motion to open or close the valve seats.

Acute Angle. An angle of less than 90 degrees.

AD Plywood. A designation or gradation of plywood. The A and the D designate quality of surface layers.

Ad Valorem. A tax imposed at a percentage rate of the value of the property, such as property tax.

ADA. Americans with Disabilities Act.

Adapter Terminal. Electrical fitting attached to the end of a conductor or to a piece of equipment, for taking power from an outlet in a way for which it was not designed.

Adapter, Cubicle. See Cubical Adapter.

Adapter. A device for connecting two different or non-similar parts.

Adaptive reuse. Adapting an old or historical building for a new purpose.

Addenda. A revision in the contract document made prior to the execution of the owner-contractor contract.

Additive Alternate. An alternate bid that, if accepted, adds to the contract price.

Additive. A substance added to another to impart different or special qualities; an admixture.

Adhered Veneer. A veneer secured and supported through adhesion to an approved bonding material applied over an approved backing.

Adhered. Attached by adhesion, rather than mechanical anchorage, as adhered veneer.

Adherence. The properties of bodies for sticking together.

Adherend Failure. Failure of an adhesive joint when the separation is within the adherend.

Adherend. A body that is held to another body by an adhesive.

Adhesion Type Ceramic Veneer. Thinner sections of ceramic veneer, held in place by adhesion of mortar to unit and to backing; no metal anchors are required.

Adhesion, Mechanical. Adhesion between surfaces in which the adhesive holds the parts together by interlocking action.

Adhesion. 1. The state in which two surfaces are held together by interfacial forces which may consist of valence forces or interlocking action, or both; bonding strength; the attraction of a coating to the substrate, or of one coat of paint to another. 2. The soil quality of sticking to buckets, blades, and other parts of excavators.

Adhesive Application. A means of applying gypsum board utilizing adhesives and supplemental mechanical fasteners.

Adhesive Bond. A relationship between two materials in contact with each other causing them to stick or adhere together by means other than cohesion.

Adhesive Failure. Failure of an adhesive joint when the plane of separation is at the adhesive-adherend interface.

Adhesive Spreader. A notched trowel used in the application of laminating adhesives.

Adhesive Wall Cups. Special clips or nails with large perforated bases for mastic application to most firm surfaces.

Adhesive, Ceramic. Used for bonding tile to a surface; rubber solvents; rubber- and resin-based emulsions used as adhesives.

Adhesive, Pressure-Sensitive. An adhesive that will adhere to a surface at room temperature by briefly applied pressure alone.

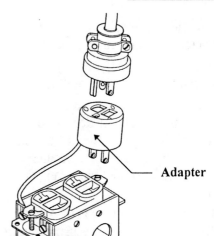

Adapter

Adhesive, Roof. A bonding agent used to cement roof materials.

Adhesive, Solvent. An adhesive having a volatile organic liquid as a vehicle, not including water-based adhesives.

Adhesive, Tile. Organic adhesive used for bonding tile to a surface; rubber solvents and resin-based and rubber emulsions can be used as adhesives.

Adhesive. 1. A material capable of holding other materials together by surface attachment; glues, cements, pastes, epoxy, and mucilage are some of the common adhesives. 2. A compound, glue, or mastic used in the application of gypsum board products to framing or for laminating one or more layers of gypsum boards.

Adiabatic Compression. Compressing refrigerant gas without removing or adding heat.

Adiabatic Curing. The maintenance of ambient conditions during the setting and hardening of concrete so that heat is neither lost nor gained.

Adiabatic. 1. Impassable to heat. 2. Occurring without gain or loss of heat.

Adjacent. Nearby or adjoining.

Adjustable Bar Hanger. A metal hanger that can be made to fit the varying distances between floor and ceiling joists or rafters to securely hold electrical outlet boxes and devices.

Adjustable Shelf Standard. Metal items to support shelves usually in the form of strips attached to vertical surfaces.

Adjustable Shelf. A shelf that can be adjusted to different heights.

Adjustable Speed Motor. One in which the speed can be varied gradually over a considerable range, but when once adjusted remains practically unaffected by the load, such as shunt motors designed for a variation of field strength.

Adjustable Triangle. A transparent plastic drafting tool that can be adjusted and set for any angle.

Adjustable Wrench. An open faced wrench which can be adjusted to different sizes.

Admixture. 1. A material other than water, aggregates, and hydraulic cement used as an ingredient of concrete or mortar, and added immediately before or during its mixing. 2. A chemical additive used to alter the normal properties of concrete. 3. Any substance added to a plaster component or to plaster mortar for the purpose of altering its properties.

Adobe Masonry. Construction that utilizes unburned (unfired) clay masonry units.

Adobe. Unburned or unfired brick, dried in the sun.

ADR. Alternative Dispute Resolution; includes mainly negotiation, mediation, and arbitration.

Adsorbed Water. Water held on surfaces in a material by either physical and/or chemical forces.

Adsorbent. Usually of a solid, having the ability to attract molecules of liquids, solutions, or gasses that adhere to its surface.

Adsorption. The process of attraction to a surface; the attachment of foreign molecules on the surface of a substance.

Adulteration. The addition of unwanted materials.

Advancing Colors. Colors that give an illusion of being closer to the observer; warm colors in which red-orange predominates.

Adverse Possession. The overt occupation of real property under some claim of right that is opposed to the claim of some other claimant.

Advertisement for Bids. Published notice for receiving of bids for a construction project.

Advisor CM. A construction manager who is an advisor to the owner and who does not guarantee the construction cost.

Adze. A tool for cutting away the surface of wood, like an axe with an arched blade at right angles to the handle.

Aerate. To introduce air into a substance, such as into water at the kitchen sink.

Aeration. Act of combining substance with air.

Aerator. Device which adds air to water; fills flowing water with bubbles.

Aerembolism. Caisson Disease.

Aerial. 1. Relating to the air or atmosphere. 2. An antenna.

Aerobic. Activities or processes that can take place only in the presence of air or oxygen.

Aerosol. 1. A colloidal suspension of fine solid or liquid particles in gas, like smoke, fog, and mist. 2. A substance dispensed from a pressurized can in aerosol form.

Aesthetic Effect. Relating to the beautiful rather than to the merely pleasing, useful, or utilitarian; artistic and in accordance with the principles of good taste.

Aesthetics. Concerned with beauty, refinement, and good taste.

AEV. Automatic Expansion Valve.

Affidavit. A written statement that is made under oath.

AFPA (Formerly NFoPA). American Forest & Paper Association.

A-Frame. A building structure where the main structural members forming the roof and floors are in an A-shape.

Aftershock. An earthquake occurring subsequent to a large earthquake, the main shock; the magnitude of an aftershock is usually smaller than the main shock.

AGA. American Gas Association.

AGC. Associated General Contractors of America.

Agent. One who acts with delegated authority for a principal.

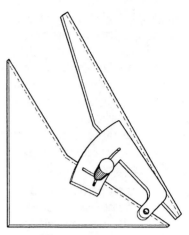

Adjustable Triangle

Agglomeration. Formation of masses or aggregates of pigments; not dispersed.

Aggregate Coated Panel. Sheet material, usually plywood, with decorative face of aggregate bonded with epoxy applied to one face.

Aggregate Storage Bins. In a concrete batching plant, the bins that store the necessary aggregate sizes and feed them to the dryer in substantially the same proportions as are required in the finished mix.

Aggregate, Coarse. One of the four ingredients of concrete, usually gravel, which is retained on a #4 sieve.

Aggregate, Fine. One of the four ingredients of concrete, usually sand, which will pass the #4 sieve and will be retained on the #200 sieve.

Aggregate, Heavyweight. Aggregate of high specific gravity such as barite, magnetite, limonite, limenite, iron, or steel used to produce heavy concrete.

Aggregate, Lightweight. Aggregate of low specific gravity, such as expanded or sintered clay, shale, slate, diatomaceous shale, perlite, vermiculite, or slag; natural pumice, scoria, volcanic cinders, tuff, and diatomite, sintered fly ash, or industrial cinders; used to produce lightweight concrete; aggregate with a dry, loose weight of 70 pounds per cubic foot or less.

Aggregate. 1. Inert particles such as sand, gravel, crushed stone, or expanded materials, in a concrete, plaster, or terrazzo mixture. 2. Granular material, such as sand, gravel, crushed stone, and iron blast-furnace slag, used with a cementing medium to form a hydraulic-cement, concrete or mortar. 3. Crushed stone, crushed slag, or water-worn gravel used for surfacing a built-up roof. 4. Any granular mineral material.

Agitator. Device used to cause motion in confined fluid.

Agreement. A mutual understanding; a meeting of the minds; a contract duly executed and legally binding.

Agreement of Sale. See Conditional Sales Contract.

Agricultural Varnishes. Varnishes designed to protect and beautify farm implements and machinery.

AHA. 1. American Hardware Association. 2. American Hardboard Association.

AHC. Architectural Hardware Consultant, a member of the Door and Hardware Institute.

AHDGA. American Hot Dipped Galvanizers Association.

AHMA. American Hardware Manufacturer's Association.

AHU. Air Handling Unit.

AI. Asphalt Institute.

AIA. American Institute of Architects.

AIMA. Acoustical and Insulating Materials Association.

AInA. American Insurance Association.

Air Adjusting Valve. Spray gun valve controlling input air.

Air Break. In a plumbing system, a physical separation between a drain outlet from a fixture and an indirect waste receptor from the fixture.

Air Bubble. Bubble in paint film caused by entrapped air.

Air Cap. Perforated housing for atomizing air at head of spray gun; also called air nozzle.

Air Carbon-Arc Cutting. An arc-cutting process in which the severing of metals is effected by melting with the heat of an arc between an electrode and the base metal and an air stream is used to facilitate cutting.

Air Chamber. A short piece of pipe about 10 long, installed above the hot and cold valves of fixtures such as sinks, lavatories, and clothes washers which traps a column of air intended to cushion the rush of water as the valve is closed and prevents water hammer.

Air Change Method. A method of calculating the quantity of infiltration air into a building.

Air Changes per Hour. The number of times the air volume of a room or building can be replaced in an hour by an air-handling, circulating, or exhaust system.

Air Cleaner. Device used for removal of airborne impurities.

Air Cleaning. A control strategy to remove various airborne particulates and/or gases from the air; the three types of air cleaning most commonly used are particulate filtration, electrostatic precipitation, and gas absorption.

Air Coil. Coil on some types of heat pumps used either as an evaporator or a condenser.

Air Compressor. A mechanism which forces air at a high pressure into a storage tank where it is released through a regulator and a hose to power small tools.

Air Conditioner. Device used to control temperature, humidity, cleanliness, and movement of air in conditioned space.

Air Content. The amount of entrained or entrapped air in concrete or mortar, exclusive of pore space in aggregate particles, usually expressed as a percentage of total volume of concrete or mortar.

Air Cooled Compressor. The condenser component of a refrigeration system placed out of the refrigerant area in a series of copper tubes; a fan blows outdoor air across the tubes which contain the refrigerant.

Air Cooler. Mechanism designed to lower temperature of air passing through it.

Air Core Solenoid. Solenoid which has a hollow core instead of a solid core.

Air Core. Coil of wire not having a metal core.

Air Diffuser. Air distribution outlet or grille designed to direct airflow into desired patterns.

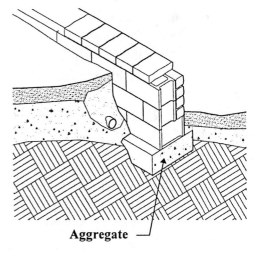
Aggregate

Air Distribution. To force air to desired locations in a building or facility.

Air Drying. Drying by oxidation or evaporating by simple exposure to air; used for drying block, brick, wood, or paint without any special equipment, simply by exposure to ambient air.

Air Eliminator. A mechanical device that expels excess air.

Air Embolism. Caisson Disease.

Air Entraining Agent. A substance added to concrete, mortar or cement that produces air bubbles during mixing, making it easier to work with and increasing its resistance to frost and freezing.

Air Entraining Cement. A portland cement with an admixture that causes a controlled quantity of stable, microscopic air bubbles to form in the concrete during mixing.

Air Entrainment. Introduction of air into a process such as in concrete mixing or in a whirlpool bath.

Air Entrapment. Inclusion of air bubbles in paint film.

Air Exchange Rate. 1. The number of times that the outdoor air replaces the volume of air in a building per unit time, typically expressed as air changes per hour. 2. The number of times that the ventilation system replaces the air within a room or area within the building.

Air Gap. 1. In a water supply system, the vertical distance from the top of the flood rim to the faucet or spout which supplies fresh water to the fixture; designed to prevent backsiphonage. 2. The space between magnetic poles or between rotating and stationary assemblies in a motor or generator.

Air Handler. Fan-blower, heat transfer coil, filter, and housing parts of a system.

Air Handling System. A system to heat, cool, humidify, dehumidify, filter, and transport air, consisting of an air handling unit, fresh air and exhaust air damper at the building exterior, ductwork, supply air, diffusers or registers, and return air grills in the conditioned space.

Air Handling Unit (AHU). Equipment that is designed to move conditioned air, containing fan(s), filter(s), heating coil(s), and/or cooling coil(s); units can be classified as either a central system or unitary; unitary equipment can be classified as rooftop unit, unitary package unit, unitary split system, or compound room unit.

Air Hose. Hose of air supply quality, usually red.

Air Infiltration. Leakage of air into rooms through cracks, windows, doors, and other openings.

Air Jet. In blast cleaning, type of blast cleaning gun in which the abrasive is conveyed to the gun by partial vacuum.

Air Lock. Air trapped within a pipe which restricts or blocks the flow of liquid through the pipe.

Air Manifold. Common air supply chamber for several lines.

Air Monitoring Test. A procedure used to determine the contents in a volume of air over a measurable period of time.

Air Nozzle. See Air Cap.

Air Plenum. Any space used to convey return air in a building or structure.

Air Powered Hoist. A hoist that is operated by compressed air.

Air Purger. A mechanical device that removes unwanted air.

Air Rights. The right to use the space above a piece of real property.

Air Space. An open space or cavity in a wall or between building materials.

Air Structure. A canvas structure supported by air produced by fans.

Air Tool. Attachments using compressed air to saw, spray-paint, sand, drill, or nail.

Air Transformer. Device for controlled reduction in air pressure.

Air Valve. Control valve in air line system.

Air Vent. 1. An opening in a building or structure for the passage of air. 2. Valve, either manual or automatic, to remove air from the highest point of a coil or piping assembly.

Air Voids. The small hollow spaces in cement paste caused by unwanted entrapped air bubbles and the smaller voids caused by air-entraining admixtures.

Air Volume. Quantity of air in cubic feet, usually per minute, at atmospheric pressure.

Air Washer. Device used to clean air while increasing or lowering its humidity.

Air, Standard. See Standard Atmosphere.

Air. 1. An invisible gaseous substance surrounding the earth, a mixture of mainly oxygen and nitrogen; the atmosphere. 2. An air conditioning system.

Airborne Sound. Sound originating in a space; airborne sound can be created from the radiation of structure-borne sound into the air.

Air-Conditioning & Refrigeration Institute (ACRI), 4301 North Fairfax Drive, #425, Arlington, Virginia 22203, (703) 524-8800.

Air-Cooled Condenser. Heat of compression is dissipated from condensing coils to surrounding air, by convection or by a fan or blower.

Air-Dried Lumber. Lumber that has dried by being stored in yards or sheds for any length of time; for North America as a whole, the minimum moisture content of thoroughly air-dried lumber is 12 to 15%, and the average is higher.

Air-Dried. Dried by exposure to air in a yard or shed without artificial heat.

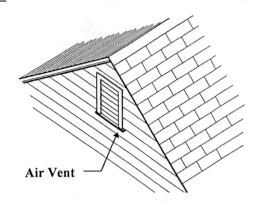

Air Vent

Air-Dry Weight. The unit weight of a light weight concrete specimen cured for seven days with neither loss nor gain of moisture at 60° F. to 80° F. and dried for 21 days in 50 plus or minus 7% relative humidity at 73.4° F. plus or minus 2° F.

Air-Entrained Concrete. Concrete containing an admixture that produces microscopic air bubbles in the concrete; used to improve workability and freeze resistance.

Air-Entertainment. The process by which air is introduced into a material while in a liquid or plastic state in the form of small isolated bubbles.

Airfield Marking. Lines, markers, or lines placed on airfield to aid in the takeoff and landing of planes.

Airless Spraying. Spraying using hydraulic pressure to atomize the paint.

Air-Sensing Thermostat. Thermostat unit in which sensing element is located in refrigerated space.

Air-Slack. A condition where soft-body clay, after absorbing moisture and being exposed to the atmosphere, will spall a piece of clay and/or glaze.

Air-To-Air Heat Exchanger. A method of heat recovery in which the intake and exhaust ducts are brought together at the heat exchanger; this system transfers only heat.

AISC. American Institute of Steel Construction, Inc.

AISE. Association of Iron and Steel Engineers.

AISI. American Iron & Steel Institute.

Aisle. A passage space between rows of seats or tables; a passage space between rows of stored goods in a warehouse.

AITC. American Institute of Timber Construction.

ALA. American Library Association.

Alabaster. A compact translucent, usually white, form of gypsum, often carved into lamps, vases, and ornaments.

Alarm. A warning sound or device, as in smoke alarm or burglar alarm.

Alclad. Trade name for an aluminum alloy coated with pure aluminum to give high corrosion resistance.

Alcohol Brine. Water and alcohol solution which remains a liquid below 32° F. (0° C.).

Alcohol Resisting. Showing no damage when in contact with alcohol.

Alcohol. A colorless volatile inflammable liquid, miscible with water, used as a solvent; in full, ethyl alcohol; the alcohols commonly used in painting are ethyl alcohol as a shellac solvent and methyl alcohol or wood alcohol in paint removers.

Alcove. A small recessed section of a room or outdoor area.

Alfresco. In the open air.

Algae (plural). A non-flowering stemless water-plant; seaweed, pond scums, and phytoplankton; growing in water in the presence of sunlight and carbon dioxide.

Algaecide. Chemical that kills algae.

Algebra. The use of letters and other symbols to represent quantities in mathematical calculations.

Algistat. Chemical that inhibits algae growth.

Alienation. A conveyance of property to another.

Align. To be or come into precise adjustment or correct relative position.

Aliphatic. Describes a major class of organic compounds, many of which are useful as solvents.

Alizarin Lake. A bright red pigment with blue undertone made from the organic coal tar dyestuff, alizarin; some purple pigments are also marketed under this name.

Alkali. A chemical substance which effectively neutralizes acid material so as to form neutral salts; a base, the opposite of acid; examples are ammonia and caustic soda; highly destructive to oil paint films.

Alkali-Aggregate Reaction. The chemical reaction in concrete or mortar between alkalis (sodium and potassium oxides) from portland cement or other sources and certain constituents of some aggregates, primarily certain phases of silica, which causes deterioration in the form of strength loss, excessive expansion, and cracks in the concrete or mortar.

Alkaline. Having the properties of an alkali; having a pH of more than 7.

Alkalinity. Amount of bicarbonate, carbonate, or hydroxide compounds in water.

Alkali-Resistant Grout. A grout that resists the effect of prolonged contact with alkalis.

Alkyd. A synthetic resin, made usually with phthalic anhydride, glycerol and fatty acids from vegetable oils.

All Risk Policy. A property insurance policy that insures against all risks of loss that are not specifically excluded.

All Stretcher Bond. A brick bond showing only stretchers on the face of the wall with each stretcher divided evenly over the stretchers under it; staggered vertical joints.

Allen Key. An Allen Wrench.

Allen Wrench. A sometimes L-shaped hexagonal rod designed to fit into a hexagonal hole in a bolt head; also called an Allen Key.

Allen Type Screw. Screw with recessed hex-shaped head.

Alley. 1. A narrow street or passageway behind buildings. 2. A building for bowling.

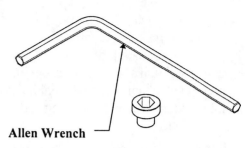

Allen Wrench

Alligatoring. 1. The cracking and crazing that occurs in asphalt roofing due to age and the effect of solar radiation and oxidation. 2. Coarse checking pattern characterized by a slipping of the new paint coating over the old coating to the extent that the old coating can be seen through the fissures, caused by improper build-up of paint. 3. In asphalt paving, interconnected cracks forming a series of small blocks resembling an alligator's skin or chicken-wire, caused by excessive deflection of the surface over unstable subgrade or lower courses of the pavement.

All-Inclusive Deed of Trust or Mortgage. A trust deed or mortgage that includes underlying financing; one payment is made to the all-inclusive mortgagee or beneficiary who then makes the payments on the underlying loans; also called a Wraparound Deed of Trust or Mortgage.

Allowable Bearing Capacity. The amount of pressure, expressed in pounds per square foot, that any particular soil will support, in the opinion of a soil mechanics engineer, taking into account acceptable settling and a safety factor.

Allowance. In the contract documents, a sum noted by the architect to be included in the contract sum, for a specific item; for example, a stipulated sum for carpeting or hardware which will be selected at a later date.

Allowances. The amount of tightness or looseness of male and female parts.

Alloy. 1. A substance composed of two or more metals. 2. Or of a metal and a nonmetallic constituent.

All-Purpose Compound. In gypsum wallboard installation, a joint treatment material that can be used as a bedding compound for tape, a finishing compound, and as a laminating adhesive or texturing product.

Alluvium. Any geologically recent deposit of fine soil, silt, sand, and gravel from an ancient river.

Alpha Gypsum. A class of specially processed calcined gypsum having properties of low consistency and high strength.

Alternate. A stipulated construction item and its cost, in addition to the original base bid, for a specific item to be included or excluded from the project; a mutually agreed upon item that is used in place of the originally specified item.

Alternating Current (AC). Electric current in which direction of flow alternates or reverses in 60-cycle (Hertz) current, direction of flow reverses every 1/120th of a second.

Alternating Current Automatic Low Voltage Secondary Network. See Secondary Network.

Alternative Dispute Resolution (ADR). Resolution of a dispute without litigation, includes negotiation, mediation, and arbitration.

Alternator. An electric generator for producing alternating current.

Alumina Cement. Used in high early strength concrete.

Alumina Porcelain. A vitreous ceramic whiteware for technical application in which Alumina ($A_{12}O_3$) is the essential crystalline phase.

Alumina. A mineral usually found in the clay used for brickmaking.

Aluminous Cement. A hydraulic cement in which the principal constituents are calcium illuminates, instead of calcium silicates which comprise the major ingredients of portland cement. (See calcium aluminate cement)

Aluminum Association (AA). 900 19th Street, NW, #300, Washington, DC 20006, (202) 862-5100.

Aluminum Extrusion. Aluminum sections formed by extrusion.

Aluminum Grid Walkway. Walkway fabricated of aluminum grid placed over roof surface to protect roofing surface from damage from traffic.

Aluminum Jacket. A watertight outer housing, fashioned from aluminum, placed around a pipe or vessel.

Aluminum Leaf. Aluminum hammered into very thin sheets.

Aluminum Oxide. 1. Corrosion on surface of an aluminum member caused by oxidation. 2. Hard and sharp abrasive made by fusing mineral Bauxite at high temperature.

Aluminum Paint. Mixture of finely divided aluminum particles in flake form combined with vehicle.

Aluminum Plate. Flat aluminum sheet material.

Aluminum Sheet Plate. Flat rolled aluminum plate.

Aluminum Silicate. White extender pigment made from China clay or feldspar, which provides very little color or opacity.

Aluminum Storefront. A facade of a building or structure which is constructed of a system of aluminum tubing and glass.

Aluminum Wire. Electrical conductors and cable manufactured from aluminum.

Aluminum, Zinc Coated. Aluminum, zinc plated for corrosion protection by hot dipping into molten zinc or by electrolysis.

Aluminum. A bluish silver-white malleable metallic element with good electrical and thermal conductivity, high reflectivity and resistance to oxidation.

Amalgam. 1. An alloy of mercury with some other metal. 2. An alloy of two or more metals.

Amalgamation. A merger or consolidation of two or more businesses.

Ambience. The feeling of the surroundings or atmosphere of a place.

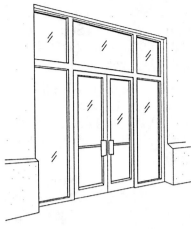

Aluminum Storefront

Ambient Background Samples. Prevalent Level Samples.

Ambient Sound. The quiet-state noise level in a room or space, which is a composite of sounds from many external sources, both near and far, over which one individual has no control.

Ambient Temperature. 1. Room temperature or the temperature of the surroundings; temperature of fluid (usually air) which surrounds object on all sides.

Ambient. Surrounding on all sides; encompassing, as the ambient temperature.

Ambiguous. Having more than one meaning, as in a contract.

Ambo. Pulpit in a church.

Ambulatory. A sheltered place for walking, as in an aisle or cloister in a church or monastery.

Amenity. Pleasant or useful features in real estate that contribute satisfaction and pleasure rather than direct financial benefit.

American Arbitration Association (AAA). 140 West 51st Street, New York, NY 10020- 1203, (212) 484-4000, Fax (212) 765-1203.

American Architectural Aluminum Manufacturers Association(AAMA). 1540 Dundee Road #310, Palatine, Illinois 60067, (708) 202-1350.

American Association of Nurserymen (AAN). 1250 I Street, NW, Washington, DC 20005, (202) 789-2900.

American Association of State Highway and Transportation Officials (AASHTO). 444 North Capitol Street, NW, #225, Washington, DC 20001, (202) 624-5800 - David Hensing.

American Bond. In masonry, a form of bond in which every sixth course is a header course and the intervening courses are stretcher courses.

American Concrete Institute (ACI). PO Box 19150, Detroit, Michigan 48219, (313) 532-2600.

American Conference of Governmental Industrial Hygienists, Inc. (ACGIH) 1330 Kemper Meadow Drive, #600, Cincinnati, Ohio 45240, (513) 742-3355

American Consulting Engineers Council (ACEC). 1015 15th St. NW, Washington, DC 20005-2605, (202) 347-7474

American Forest & Paper Association (AFPA). 111 19th St., NW, #800, Washington, DC 20036, (202) 463-2700.

American Gallon. A measure of liquid volume, 231 cubic inches.

American Gas Association (AGA). 1515 Wilson Blvd., Arlington, Virginia 22209, (703) 841-8400.

American Hardboard Association (AHA). 1210 West Northwest Highway, Palatine, Illinois 60067, (708) 934-8800.

American Hardware Association (AHA). 20 North Wacker Drive, Chicago, IL 60606.

American Hardware Manufacturer's Association (AHMA). 801 N Plaza Drive, Schaumburg, Illinois 60173, (708) 605-1025.

American Hot Dipped Galvanizers Association (AHDGA). 1000 Vermont Avenue, NW, Washington, DC 20005, (202) 628-4634.

American Institute of Architects (AIA). 1735 New York Avenue, NW, Washington, DC 20006, (202) 626-7300

American Institute of Steel Construction, Inc. (AISC). One East Wacker Drive, #3100, Chicago, Illinois 60601-2001, (312) 670-2400.

American Institute of Timber Construction (AITC). 7012 South Revere Parkway, #140, Englewood, Colorado 80112, (303) 792-9559.

American Insurance Association (AInA). 1130 Connecticut Avenue, NW, Washington, DC 20036, (202) 828-7100.

American Iron & Steel Institute (AISI). 1101 17th Street, NW #1300, Washington, DC 20036-4700, (202) 452-7100.

American Library Association (ALA). 50 E. Huron Street, Chicago, Illinois 60611, (312) 944-6780.

American Lumber Standards. Rules for softwood lumber, dealing with recognized classifications, nomenclature, sizes, descriptions, amounts, shipping provisions, basic grades, grade marking, and inspection.

American National Standards Institute, Inc. (ANSI). 11 West 42nd Street, 13th Floor, New York, NY 10036, (212) 642-4900.

American Petroleum Institute (API). 1220 L Street, NW, Washington, DC 20035, (202) 682-8000.

American Plywood Association (APA). 7011 South 19th Street, Tacoma, Washington 98411, (206) 272-2283.

American Process Zinc Oxide. Zinc oxide pigment made directly from zinc ores; also called Direct Process.

American Public Works Association (APWA), 1313 E. 60th Street, Chicago, Illinois 60637, (312) 667-2200, Fax. (312) 667-2304.

American Society for Testing and Materials (ASTM). 1916 Race Street, Philadelphia, Pennsylvania 19103, (215) 299-5420.

American Society of Architectural Hardware Consultants (ASAHC), 1815 N. Ft. Myer Drive, Suite 412, Arlington, Virginia 22209, (703) 527-2060.

American Society of Civil Engineers (ASCE), 345 East 47th St., New York, New York 10017, (212) 705-7551.

AIA Logo

American Society of Heating, Refrigeration, and Air-Conditioning Engineers, Inc (ASHRAE). 1791 Tullie Circle, NE, Atlanta, Georgia 30329, (404) 636-8400.

American Society of Mechanical Engineers (ASME). United Engineering Center, 345 East 47th St., New York, New York 10017-2392, (212) 705-8500.

American Society of Plumbing Engineers (ASPE). 3617 Thousand Oaks Boulevard., #210, Westlake, California 91362 - (805) 495-7120.

American Society of Sanitary Engineers (ASSE). P.O. Box 40362, Bay Village, Ohio 44140, (216) 835-3040.

American Standard Pipe Thread. Type of screw thread commonly used on pipe and fittings to assure a tight seal.

American Standards Association. Now known as American National Standards Institute.

American Subcontractors Association (ASA). 1004 Duke St. Alexandria, VA 22314, (703) 684-3450.

American Vermillon. Chrome orange pigment.

American Water Works Association (AWWA). 6666 West Quincy Avenue, Denver, Colorado 80235, (303) 794-7711.

American Welding Society (AWS). 550 NW LeJeune Road, P.O. Box 351040, Miami, Florida 33135, (305) 443-9353.

American Wire Gauge. A standard system for designating the diameter of wire.

American Wood Preservers Association (AWPA). P.O. Box 286, Woodstock, Maryland 21163-0286, (410) 465-3169.

American Wood Preservers Institute (AWPI). 1945 Old Gallows Road #150, Vienna, Virginia 22182, (703) 893-4005.

Americans with Disabilities Act. National legislation prohibiting discrimination against disabled individuals; the act includes detailed requirements for the planning and design of buildings.

Ames Taping Tools. Specially designed tools to mechanically apply taping compound and tape.

Amides. Curing agent combined with epoxy resins.

Amine Adduct. Amine curing agent combined with a portion of the resin.

Amines. Organic substituted ammonia; curing agent combined with epoxy resins.

Ammeter. An instrument for measuring electric current in amperes.

Ammonia. 1. Chemical combination of nitrogen and hydrogen, NH_3 2. Popular refrigerant for industrial refrigerating systems; also a popular absorption system refrigerant; ammonia refrigerant is identified as R-117. 3. Ammonia combines with free chlorine in pool water to form chloramine which causes burning eyes, skin irritation and chlorine odor.

Amortization Schedule. A chart showing the monthly, quarterly, or annual payments necessary to pay off a loan including interest over a particular period of time.

Amortization. The gradual paying off of indebtedness by regular equal payments including interest and principal over a specified period of time.

Amp. Ampere.

Ampacity. Current-carrying capacity expressed in amperes.

Amperage. The strength of a current of electricity expressed in amperes; electron or current flow of one coulomb per second past given point in circuit.

Ampere Turns. A measure of magnetic force; represents product of amperes times number of turns in coil of electromagnet.

Ampere. Unit of electric current equivalent to flow of one coulomb per second.

Amplification. An increase in earthquake motion as a result of resonance of the natural period of vibration with that of the forcing vibration.

Amplifier. An electrical device to obtain amplification of voltage, current or power.

Amplitude. Maximum deviation from mean or center line of a wave.

Amyl Acetate. Banana oil; solvent for nitrocellulose, formed by esterification of acetic acid with amyl alcohol.

Anaerobic. An organism that can live without oxygen, such as the bacteria in a septic tank.

Anaglyph. A sculptured, chased, or embossed ornament worked in low relief.

Analog. A way of expressing one quantity in terms of another quantity; an analogous or parallel thing; for example, voltage, weight, or length to represent numbers, as watch hands represent time; compare with Digital.

Analogous Harmony. Colors which are related by containing one color in common; color harmony.

Analogy. Inference from a similar case.

Analysis. 1. Separation into constituent parts. 2. In engineering, the investigative determination of the detailed aspects of a particular phenomenon; may be qualitative, meaning a general evaluation of the nature of the phenomenon, or quantitative, meaning the numerical determination of the magnitude of the phenomenon.

Anchor Bolt. A bolt embedded in concrete for the purpose of fastening a building frame to a concrete or masonry foundation.

Anchor Pattern. Profile surface roughness usually attained by blasting.

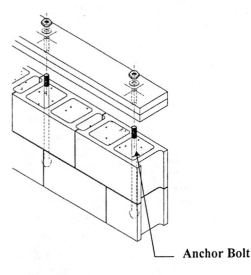

Anchor Bolt

Anchor Slot. A groove in an object into which a fastener or connector is inserted to attach objects together.

Anchor Tenant. The main tenant in a shopping center development; a large stable tenant or tenants that are expected to attract other tenants and customers to the development.

Anchor Ties. Any type of fastener used to secure wood framing, steel, or masonry to some stable object, such as a foundation or another wall; usually for tension value.

Anchor, Brick. Fasteners that are designed to attach and secure a veneer to a concrete or brick wall.

Anchor, Dovetail. A fastener with an interlocking joint that is wider at its end than at its base.

Anchor, Rafter. A bolt or fastening device which attaches the rafters to the walls or rafter plate.

Anchor, Wood. A bolt or fastening device which attaches wood to wood or wood to other materials.

Anchor. Irons or metals of special form and shapes used to fasten together and secure timbers or masonry.

Anchorage. 1. An attachment for resistance to movement; the movement can be a result of overturning, sliding or uplift; the most common anchorage for these movements are tie-downs (hold-downs) for overturning and uplift, and anchor bolts for sliding. 2. In posttensioning, a device used to anchor tendons to concrete member; in pretensioning, a device used to anchor tendons during hardening of concrete. 3. The securing of reinforcing bars in cast-in-place concrete either by hooks, bends, or embedment length.

Anchored Type Ceramic Veneer. Thicker sections of ceramic veneer held in place by grout and wire anchors connected to backing wall.

Anchoring Cement. Grout used in sleeves to anchor tubing in place.

Andalusite. A polymorph, along with sellimanite and kyanite, of composition $Al_2O_3 SiO_2$; in firing, it dissociates to yield principally mullite.

Andiron. A metal stand, usually a pair, for supporting wood fuel in a fireplace.

Anechoic Chamber. A room or building that is free from echoes and reverberations.

Anemometer. Instrument for measuring the rate of air flow or motion.

Aneroid Barometer. An instrument for measuring atmospheric air pressure by its action on the elastic lid of an evacuated box. This is the basis of a surveyor's aneroid barometer which is used to measure altitude; an altimeter.

Anesthetizing Location. Areas in hospitals in which flammable anesthetics are or may be administered to patients; such locations include operating rooms, delivery rooms, and anesthesia rooms, and will also include any corridors, utility rooms or other areas which are or may be used for administering flammable anesthetics to patients.

Angle Bar. A steel structural member in the shape of an L; classified by the thickness of the stock and the length of the legs.

Angle Blasting. Blast cleaning at angles less than 90 degrees.

Angle Block. A square of tile specially made for changing direction of the trim.

Angle Brick. Any brick shaped to an oblique angle to fit a salient corner.

Angle Divider. A tool used by tilesetters to determine the degree of an angle to cut; used for fitting trim, moldings, and floors into corners; a corner angle is measured by adjusting the divider to fit the corner.

Angle Float. A concrete or plastering finishing tool having a surface bent to form a right angle; used to finish re-entrant angles.

Angle Iron. A rolled steel structural member with an L-shaped section; used to support brickwork over doors and windows and is sometimes used as main runners in lieu of channels to support plaster.

Angle of Degree. On an airless spray cap, the orifice angle; controls width of spray and pattern angle.

Angle of Incidence. The angle that a line or light ray striking a surface makes with the perpendicular at that point, the point of incidence.

Angle of Reflection. The angle that a reflected ray makes with the perpendicular to a surface at the point of incidence.

Angle of Repose. 1. The natural angle that a pile of a material, such as earth, sand, or gravel will assume. 2. The angle at which a body will slide down an inclined plane impelled only by gravity.

Angle Steel. An L-shaped steel member.

Angle Stop. A water valve that occurs where there is a 90 degree change in direction of the piping, as under a lavatory or sink.

Angle, Shelf. A structural angle which is fastened to the structure of a building to support a wall or other component.

Angle. 1. The difference in direction of two lines which meet or tend to meet, usually measured in degrees. 2. An inside corner. 3. A structural section of steel which resembles an L in cross section. 4. In masonry, a portion of a whole brick which is used to close the bond or brickwork at corners.

Anhydrite. The mineral consisting primarily of anhydrous calcium sulfate, $CaSO_4$.

Anhydrous Calcined Gypsum. Keene's Cement.

Anhydrous Calcium Sulphate. A stable form of gypsum from which practically all of the water of crystallization has been removed; also called dead-burned gypsum; dry chemical made of calcium, sulphur and oxygen, $CaSO_4$.

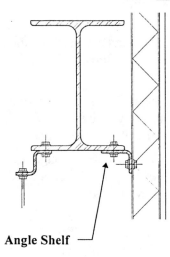

Angle Shelf

Anhydrous. Dry; free from moisture or water in any form.

Aniline Colors. Coal-tar derivatives precipitated on a colorless base.

Anisotropic. Exhibiting different properties when measured along different axes; in general, fibrous materials such as wood are anisotropic; not isotropic.

Annealed Glass. Glass whose surface has been heated to melting and then cooled to provide a toughened surface; also called heat strengthened glass.

Annealed. Cooled under controlled conditions to minimize internal stresses.

Annealing. Process of heat treating metal to get desired properties of softness and ductility.

Annex. A wing or an addition to a building.

Annual Growth Ring. The layer of wood growth put on a tree during a single growing season, comprised of springwood and summerwood.

Annuity. An investment of money entitling the investor to receive a series of equal monthly or yearly payments.

Annular Ring Nail. A nail that has grooves around the shank which prevent easy withdrawal; also called a ring shank nail.

Annunciator Panel. A panel mounted on a surface which indicates by lights which circuits have been activated.

Annunciator. A electromagnetic device used to show which of several circuits are activated.

Anode. 1. The positive electrode in an electrolytic cell. 2. The negative terminal of a primary cell such as a battery.

Anodize. Coat a metal, such as aluminum, with a protective oxide layer by electrolysis.

Anodized Door. A door which has been given an aluminum oxide coating by electrolytic action.

Anodized Tile. Tile which has been given an aluminum oxide coating by electrolytic action.

Anodized. A metal that has been subjected to electrolytic action in order to coat with a protective or decorative film.

Anodizing. An electrolytic process that forms a permanent, protective coating on aluminum.

Anomaly. An irregularity, abnormality, peculiarity, or deviation from the norm.

ANSI. American National Standards Institute, Inc.

Antenna. A metallic device used for radiating or receiving radio waves.

Anterior. Situated before or toward the front.

Anteroom. A small room leading to a main room; a waiting room.

Anticipatory Breach. A positive statement by a party to a contract that the party will not perform the terms of the contract.

Anticlastic. Saddle-shaped, or having curvature in two opposing directions.

Anti-Corrosive Paint. Metal paint designed to inhibit corrosion; applied directly to the metal.

Anti-Flooding Agent. A synthetic organic product used to reduce floating and flooding of iron blues, carbon blacks, and chrome greens.

Anti-Fouling Paint. A special coating for ship bottoms, containing poison like copper or mercury, formulated to effect the release of the poison at a controlled rate, to prevent attachment and growth of marine organisms such as barnacles and algae.

Anti-Friction Bearing. Any bearing having the capacity of effectively reducing friction.

Anti-Friction Latch Bolt. A latch bolt designed to reduce friction when the bolt starts to engage the lock strike.

Antihammer Device. An air chamber such as a closed length of pipe or a coil which is designed to absorb the shock caused by closing a valve rapidly.

Antimicrobial. Agent that kills microbial growth.

Antimony Oxide. Pure white pigment which provides about same hiding power as lithopone.

Antioxidant. Protective compound used on ends of aluminum wiring at connections to prevent corrosion that would interfere with a solid safe connection.

Antiquing. Furniture finishing technique intended to give appearance of age or wear.

Anti-Siphon Trap. A trap which is designed to prevent the siphonage of its water seal by increasing the diameter of the outlet leg of the trap so that it contains a sufficient volume of water to prevent a siphoning action.

Anti-Skinning Agent. A synthetic organic product, used to prevent forming of surface skin in packaged varnishes and paints.

APA. American Plywood Association.

Apartment Hotel. A hotel containing apartments as well as transient rooms.

Apartment House. A building containing three or more separate residential apartments; also called an Apartment Building.

Apartment. A suite of rooms, usually let as a dwelling; a dwelling unit.

Aperture. 1. Any opening. 2. An opening that varies the amount of light entering a camera.

Apex. The highest point; vertex.

API. American Petroleum Institute.

Apices. Plural of apex.

Appliance, Fixed. An appliance which is fastened or otherwise secured at a specific location.

Appliance, Portable. An appliance which is movable or can easily be moved from one place to another in normal use.

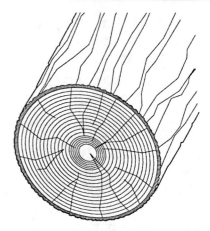

Annual Growth Ring

Appliance, Stationary. An appliance which is not easily moved from one place to another in normal use.

Appliance. An appliance is current-consuming utilization equipment, generally other than industrial, normally built in standardized sizes or types, which is installed or connected as a unit to perform one or more functions such as clothes washing, air conditioning, food mixing, or deep frying.

Application for Payment. A written document forwarded by the contractor requesting payment for work completed.

Application Rate. The quantity (mass, volume or thickness) of material applied per unit area.

Applicator. 1. One who applies. 2. Tool for applying.

Applied Preservative Treated. Applied treatment of wood or plywood to make it resistant to deterioration from moisture and insects.

Appraisal. A valuation of property by a qualified person, usually in the form of a written report.

Appraise. To give an expert judgment of the value or merit of.

Appraiser. One who, using rational systems of analysis, estimates the value of real or personal property, and usually expresses the opinion by means of a written report.

Appreciation. The increase in value of a property from any cause.

Apprentice. A person who is learning a trade, art or calling by being employed in it for an agreed period for low wages.

Approved. Acceptable to that enforcing agency having the responsibility and authority to grant the approval for the specified condition or application.

APR. Annual Percentage Rate.

Apron Wall. That part of a panel wall between the window sill and the support of the panel wall.

Apron. 1. The lower trim member under the sill of the interior casing of a window. 2. An upward or downward vertical extension of a sink or lavatory. 3. A paved area immediately adjacent to a building, structure or facility.

Apse. A projecting part of a building, such as in a church, usually semicircular and vaulted.

APWA. American Public Works Association.

Aqueduct. 1. An artificial channel or conduit for conveying large quantities of water, usually supported by a bridge-like structure. 2. A structure for conveying a canal over a river or a valley.

Aqueous. Containing water.

Aquifer. A strata of porous permeable rock or soil that is capable of holding a large quantity of water.

AR. Acid Resisting.

Arabesque. An ornament, style or design of intertwined leaves, flowers, or fruit to produce an intricate pattern of interlaced lines.

Arabic Numeral. One of the number symbols, 0, 1, 2, 3, 4, 5, 6, 7, 8, 9.

Arbitration Award. The arbitrator's judgment or decision in an arbitration.

Arbitration Demand. A written notice by one party to a contract served on the other to avail of the arbitration provisions of the contract.

Arbitration. 1. A proceeding for resolution of disputes in which a neutral person or panel, after hearing evidence presented by both sides, makes a final and binding decision that resolves the dispute. 2. A hearing used to resolve disputes.

Arbor. 1. A light open-work structure of wood or metal, covered or intended to be covered with vines, as in a park or garden. 2. An Axle or spindle on which something revolves. 3. A device holding a tool in a lathe.

Arc Blow. The deflection of an electric arc from its normal path because of magnetic forces.

Arc Voltage. The voltage (electrical potential) across the welding arc.

Arc Welding. A process of joining two pieces of metal by melting them together at their interface with a continuous electric spark and adding a controlled additional amount of molten metal from a metallic electrode.

Arc. 1. Arched or curved. 2. A portion of an ellipse or circle.

Arcade. 1. An arched covered passageway. 2. An amusement center with coin-operated games.

Arch Brick. Wedge-shaped brick for special use in an arch.

Arch Corner Bead. A job-shaped length of corner bead used to define the curved portion of arched openings.

Arch Culvert. A curved shaped drain under a roadway, canal or embankment.

Arch. A curved structural member used to span an opening or recess; also built flat. Structurally, an arch is a piece or assemblage of pieces so arranged over an opening that the supported load is resolved into pressures on the side supports, and practically normal to their faces.

Archeology. The study of human history and prehistory through the excavation of sites and the analysis of human remains.

Archetype. An original model; a Prototype.

Archimedes' Principle. A law of physics that states that a floating body displaces a weight of liquid equal to its own weight.

Archimedes' Screw. A primitive device for raising water, consisting of a pipe spirally wound around an axis.

Archipelago. A group of islands in the sea.

Architect. One who designs buildings and advises on their construction.

Arabesque

Architectonic. 1. Architectural. 2. Having an organized and unified structure or concept that suggests an architectural design.

Architectonics. The science of architecture.

Architect's Project Representative. A more or less continuous architect's representative on the jobsite; reports to the architect; formerly called Clerk of the Works.

Architectural Equipment. The implements, apparatus, or equipment used in the construction and initial outfitting of a building.

Architectural Fee. The amount of money charged by an architect for professional services such as programming, design, preparation of contract documents, and administration of the construction of a building or facility.

Architectural Terra Cota. Hard-burned, glazed or unglazed clay building units, plain or ornamental, machine-extruded or hand-molded, and generally larger in size than brick or facing tile; also see Ceramic Veneer.

Architectural Woodwork Institute (AWI). 13924 Braddlock Road #100, Centerville, Virginia 22020, (703) 222-1100.

Architectural Woodwork. Finish work using wood or composites for ornamental designs or casework.

Architectural. 1. Of or relating to or conforming to the rules of architecture. 2. Having a single, unified overall design, form, or structure.

Architrave. 1. The lowest division of an entablature. 2. The molding around a rectangular opening, such as a door or window.

Are. A metric unit of measure equal to 100 square meters. 100 ares equal 1 hectare.

Area Divider. In roof construction, a raised double wood member attached to a properly flashed wood base plate that is anchored to the roof deck; used to relieve the stresses of thermal expansion and contraction in a roof system where no expansion joints have been provided.

Area Drain. Any drain installed for the purpose of collecting rain water from an open area and channeling it to the storm drain.

Area Wall. 1. The masonry surrounding or partly surrounding an area. 2. The retaining wall around basement windows below grade.

Area. 1. The extent or measure of a surface, expressed in square units. 2. The space allocated to a particular function.

Areaway. A sunken space providing access, air, and light to a subterranean building area.

Ark. A repository in a synagogue for the scrolls of the Torah.

Armature. Part of an electric motor, generator, or other device moved by magnetism.

Armor Plate. A kick-plate made of metal installed on the bottom of a door to protect it from denting and scratching.

Armored Cable. Metal sheathed flexible electrical cable; BX cable.

Armored Faceplate. A metal plate which is fastened into the strike at the door jamb to provide a protection for the lock and keeper mechanisms.

Aromatic Hydrocarbons. Strong solvents such as benzene, toluene, xylene.

Aromatic. Derived from or belonging to a major class of organic compounds, many of which are useful as solvents.

Arrester, Lightning. A device connected to an electrical system to protect from lightning and/or voltage surges.

Arris. The sharp edge or salient angle formed by the meeting of two surfaces, as in a molding.

Art. Work exhibiting human creative skill or its application; creative activity; human skill or workmanship; an occupation requiring knowledge or skill.

Arterial Streets. Primary surface roads connecting to expressways with on and off ramps, also connecting sectors of cities, with surface crossings (controlled by traffic lights), normally over continuous long stretches of the cityscape, restricted for parking and to direct access often to adjacent commercial developments; pedestrian crossing controlled.

Artesian Well. 1. A bored well from which water flows from internal pressure. 2. A deep-bored well.

Artificer. 1. A skilled or artistic craft worker. 2. An inventor.

Artificial Intelligence. The ability of a computer to think and work like a human being; at present no computer has full artificial intelligence.

Artificial or Accelerated Drying. The process of drying block or brick with relatively warm, dry air, or other means.

Artificial Pozzolan. Fly ash and other similar substances such as rice hull ash and microsilica.

Artisan. A skilled worker in a trade such as carpentry, plumbing, or painting.

ASAHC. American Society of Architectural Hardware Consultants.

Asbestine. Natural fibrous magnesium silica, which is pure white in color; used as an extender pigment in paints.

Asbestos Abatement. Procedures to control fiber release from asbestos-containing materials in a building or to remove it entirely; may involve removal, encapsulation, repair, enclosure, encasement, and operations and maintenance programs.

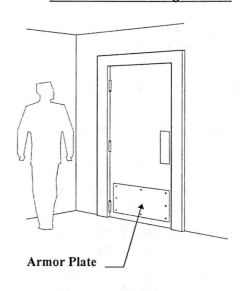

Armor Plate

Asbestos Cement. Cement asbestos.

Asbestos Program Manager. A building owner or designated representative who supervises all aspects of the facility asbestos management and control program.

Asbestos Removal. A special trade or occupation for the removal and disposal of hazardous asbestos.

Asbestos Vacuum. A filtered device using air suction to collect wetted down asbestos for easy removal into specially made bags.

Asbestos. A group of natural, fibrous, impure silicate materials, formerly used for its noncombustible, non-conducting, or chemically resistant properties; no longer used as it causes Asbestosis.

Asbestos-Containing Material (ACM). Any material containing more than one percent asbestos.

Asbestos-Containing Material. Any material containing more than one percent asbestos.

Asbestos-Free Compound. Joint treatment products that have no asbestos fiber.

Asbestosis. A lung disease caused by inhaling asbestos particles.

As-Built Drawings. See Record Drawings.

As-Builts. See Record Drawings.

ASCE. American Society of Civil Engineers.

Aseismic Region. A region that is relatively free of earthquakes.

Ash Dump. A metal access door in the floor of a fireplace firebox to dump the ashes into the ash pit below.

Ash Pit. The compartment below a fireplace in which ashes may be dumped from the firebox and stored for later removal.

Ashlar Veneer. An ornamental or protective facing of masonry composed of squared stones.

Ashlar. A pattern of masonry consisting of squared stones.

ASHRAE. American Society of Heating, Refrigeration, and Air-Conditioning Engineers, Inc.

Askarael. A synthetic nonflammable insulating liquid which, when decomposed by the electric arc, evolves only nonflammable gaseous mixtures.

ASLA. American Society of Landscape Architects.

ASME. American Society of Mechanical Engineers.

ASME Boiler Code. Standard specifications issued by the American Society of Mechanical Engineers for the construction of boilers.

ASPE. American Society of Plumbing Engineers.

Aspect Ratio. Ratio of length to width of a rectangular air grille or duct.

Asphalt Cement. A fluxed or unfluxed asphalt specially prepared as to quality and consistency for direct use in the manufacture of asphalt pavements.

Asphalt Curb. An extruded berm made from asphaltic concrete.

Asphalt Cut Back. Asphalt plus thinner; asphalt solution; asphalt coating formed by dissolving asphalt.

Asphalt Dampproofing. The application of asphalt to act as a water-resisting treatment to the surface of a concrete or masonry wall.

Asphalt Demolition. The destruction of roadways constructed of asphalt using large wheeled machinery that tears up the old asphalt into manageable pieces for easy removal.

Asphalt Emulsion. Asphalt dispersion; not a solution; a water emulsion of asphalt.

Asphalt Expansion Joint. Felt or fiberboard premolded and impregnated with asphalt and used as an expansion joint for cast-in-place concrete.

Asphalt Institute (AI). 6917 Arlington Road #210, Bethesda, Maryland 20814, (301) 656-5824.

Asphalt Leveling Course. A course asphalt aggregate mixture of variable thickness used to eliminate irregularities in the contour of an existing surface prior to a superimposed treatment or construction.

Asphalt Mastic. A mixture of asphaltic material, mineral aggregates, and fine mineral aggregates that can be poured when heated but needs mechanical manipulation to apply when cool.

Asphalt Membrane. A layer of asphalt used on a flat roof.

Asphalt Felt. An asphalt-saturated felt or an asphalt-coated felt.

Asphalt Prime Coat. A low viscosity liquid asphalt applied to prepare an untreated surface to penetrate into the voids, harden the top and help bind it to the overlying asphalt surface.

Asphalt Repair. The act or process of patching worn road surfaces or leaking roofs.

Asphalt Rock. Porous rock such as sandstone or limestone that has become impregnated with natural asphalt through geologic process; also called Rock Asphalt.

Asphalt Roof. A roof system which uses asphalt materials as a covering.

Asphalt Saturated Felt. A building and roofing felt sheet impregnated with a bituminous waterproofing material.

Asphalt Shingle. Saturated roofing felt either in large rolls or cut into composition shingles, impregnated with aggregate particles applied to a roof surfaces.

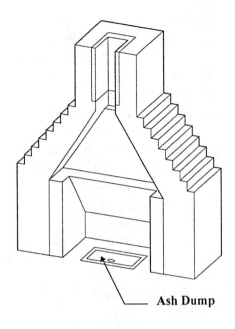

Ash Dump

Asphalt Subgrade, Improved. Subgrade, improved as a working platform (1) by the incorporation of granular materials or stabilizers such as asphalt, lime, or portland cement, prepared to support a structure or a pavement system, or (2) any course or courses of select or improved material placed on the subgrade soil below the pavement structure; subgrade improvement does not affect the design thickness of the pavement structure.

Asphalt Surface. A flat layer of asphalt.

Asphalt. A brown to black bituminous pitch occurring naturally or made from a residue of petroleum refining; also called asphaltum; used for pavements, roofing, and as a waterproofing cement; asphalt comes in a wide range of viscosities and softening points, from about 135° F. (used on dead level roofs) to 210° F. or more (to be used on special steep roofs); asphalt is a constituent in varying proportions of most crude petroleums; see Bitumen.

Asphaltenes. The high molecular weight hydrocarbon fraction precipitated from asphalt by a designated paraffinic naphtha solvent at a specified solvent-asphalt ratio.

Asphlatic Concrete. High quality, thoroughly controlled hot mixture of asphalt cement and well-graded, high quality aggregate, thoroughly compacted into a uniform dense mass.

Asphaltic Paint. 1. A liquid form of asphalt which can contain other materials such as aluminum flakes and mineral pigments, generally used for creating a water-resistant coating. 2. A liquid product used for weatherproofing.

Asphalt-Saturated Felt. A moisture-resistant sheet material, available in several different thicknesses, usually consisting of a heavy paper that has been impregnated with asphalt.

Aspirating Psychrometer. Device which draws sample of air through it to measure humidity.

ASSE. American Society of Sanitary Engineers.

Assemblage. 1. Something put together from parts. 2. A random, unordered assemblage is called a gathering. 3. An ordered assemblage is a system.

Assembly Area. An outdoor place for the gathering of a group of people exiting from a building or for an educational, sporting, or entertainment event.

Assembly Room. A room for the gathering of a group of people, such as an auditorium, gymnasium, restaurant, or meeting room.

Assessed Value. The value of a parcel of real property, usually a fraction of the market value, used for the purpose of determining the amount of property tax.

Asset. A possession or resource having value.

Assignment. Transfer of the rights and duties under a contract from one party to another.

Associated General Contractors of America (AGC). 1957 E Street, NW, Washington, DC 20006, (202) 393-2040.

Association of Iron and Steel Engineers (AISE). Three Gateway Center, #2350, Pittsburgh, Pennsylvania 15222, (412) 281-6323.

ASTM. American Society for Testing and Materials.

ASTM Standards. Materials specifications issued by the American Society of Testing Materials.

Astragal Weatherstripping. Fabric, rubber or plastic strips attached to the molding that is attached to one of a pair of doors or casement windows to cover the joint between the two stiles.

Astragal. An interior molding attached to one of a pair of doors or window sash in order to prevent swinging through; also used with sliding doors to insure tighter fitting where doors meet.

Asymmetrical. Not symmetrical; asymmetry.

ATH. Aluminum tri-hydrate, a mineral filler used in making Solid Surfacing.

Athletic Equipment. Various devices used for exercising or the playing of sporting events.

Atmospheric Corrosion. Galvanic corrosion that occurs between two adjoining dissimilar metals in a humid atmospheric condition.

Atmospheric Pressure. The pressure that atmospheric gases in air exert upon the earth; measured in pounds per square inch or grams per square centimeter; one standard atmosphere equals 14.69 psi of pressure and measures 760 mm (29.92 inches) in a barometer of mercury.

Atmospheric Vacuum Breaker. A simple mechanical device consisting essentially of a check valve in the supply line and a valve member (on the discharge side of the check valve) opening to the atmosphere when the pressure in the line drops to atmospheric; also called a siphon breaker.

Atom. The smallest particle of a chemical element; considered as a source of vast potential energy.

Atomic-Hydrogen Welding. An arc-welding process wherein coalescence is produced by heating with an electric arc maintained between two metal electrodes in an atmosphere of hydrogen; shielding is obtained from the hydrogen; pressure may or may not be used and filler metal may or may not be applied.

Atomize. Process of changing a liquid to minute particles or a fine spray; break steam into small particles.

Atomizer. Device by which air is introduced into material at the nozzle to regulate the texture of machine-applied plaster.

Atrium. A central courtyard with surrounding rooms opening off it.

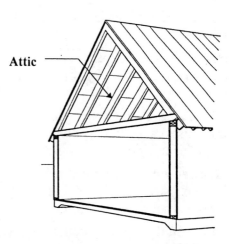

Attic

Attachment Bug. A device which, by insertion in a receptacle, establishes connection between the conductors of the attached flexible cord and the conductors connected permanently to the receptacle.

Attenuation Blanket. Material utilized to help in the reduction of the energy or intensity of sound.

Attenuation. The reduction of the energy or intensity of sound.

Attic Draft Stop. A partition in an attic, from roof to ceiling level, dividing the attic into discrete areas to prevent the spread of smoke and drafts.

Attic Insulation. Treated shreds of cellulose material that are blown into attic spaces or fiberglass rolls that are rolled out between ceiling joists to aid in weatherproofing a building or facility.

Attic Ventilators. 1. Openings in the roof or in gables for the purpose of allowing air to circulate. 2. Mechanical devices with power-driven fans to force the circulation of air.

Attic. A room or space immediately below the roof of a building.

Attorney. A qualified lawyer licensed to practice.

Attorney-in-Fact. A person acting for another under the authority of a power of attorney.

Audio Cable. A cable over which the transmission, reception, or reproduction of sound is carried.

Audio Sensor. A device that responds to the physical stimulus of sound and transmits a resulting impulse.

Audio Visual. Involving both sight and sound.

Audio. Acoustical, mechanical or electrical frequencies used in the transmission, reception or reproduction of sound.

Audit. A formal examination and report of a person's or entity's financial condition, usually by a qualified certified public accountant.

Auditorium. A building or room where an audience sits.

Auger, Fence. A rotating drill with a screw thread used to drill deep, straight, and narrow holes for the installation of fence posts.

Auger. 1. An instrument or device used for boring or forcing through materials or soil. 2. A woodworking tool.

Authorized Person. One who is properly authorized to perform specific duties under the conditions existing, usually in relation to a contract.

Auto Lift. 1. An apparatus for lifting automobiles in order to have access to the carriage underneath. 2. An apparatus used to move automobiles up and down in a parking structure.

Auto Transfer Switch. An electrical transfer switch that operates automatically.

Autoclave Curing. Steam curing of concrete products, sand-lime brick, asbestos-cement products, hydrous calcium silicate insulation products, or cement in an autoclave at maximum ambient temperatures generally between 340-420° F (170-215° C); also called High Pressure Steam Curing.

Autoclave. A pressure vessel in which an environment of superheated steam at high pressure may be produced; used for sterilizing, assisting a chemical reaction, in the curing of concrete products, and in the testing of hydraulic cement.

Autoclaved Aerated Concrete. A factory-produced building stone; pre-cast, lightweight concrete, in the form of large building blocks, panels, or planks; made of closed cell concrete steam cured in a pressurized autoclave.

Automatic Control. Valve action reached through self-operated or self-actuated means, not requiring manual adjustment.

Automatic Cyclic Control. A control system in which the opening and closing of the weigh hopper discharge gate, the bituminous discharge valve, and the pugmill discharge gate are actuated by means of self-acting mechanical or electrical machinery without any intermediate manual control. The system includes preset timing devices to control the desired periods of dry and wet mixing cycles.

Automatic Defrost. System of removing ice and frost from evaporators automatically.

Automatic Door. A door equipped with a power-operated mechanism and controls that open and close the door automatically upon receipt of a momentary actuating signal; the switch that begins the automatic cycle may be a photoelectric device, floor mat, or manual switch.

Automatic Dryer Control. In a concrete batching plant, a system that automatically maintains the temperature of aggregates discharged from the dryer within a preset range.

Automatic Expansion Valve (AEV). Pressure-controlled valve which reduces high-pressure liquid refrigerant to low-pressure liquid refrigerant.

Automatic Frost Control. Control which automatically cycles refrigerating system to remove frost formation on evaporator.

Automatic Ice Cube Maker. Refrigerating mechanism designed to automatically produce ice cubes in quantity.

Automatic Proportioning Control. In an asphaltic concrete plant, a system in which proportions of the aggregate and asphalt fractions are controlled by means of gates or valves which are opened and closed by means of self-acting mechanical or electronic machinery without any intermediate manual control.

Automatic Skimmer. See Surface Skimmer.

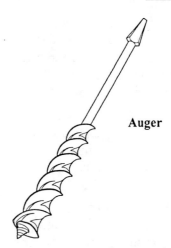

Auger

Automatic. 1. A device that works by itself without human intervention, usually actuated by some impersonal influence, as for example, a change in current strength, pressure, temperature, or mechanical configuration. 2. Done spontaneously without conscious intention, see Non-Automatic.

Automation. The use of automatic equipment to save mental and manual labor.

Automaton. A piece of mechanism with concealed motive power; a Robot.

Automotive Equipment. Implements or apparatus that pertain to the repair and care of the automobile.

Automotive Hoist. 1. An apparatus for hoisting automobiles in order to have access to the carriage underneath. 2. An elevator used to move automobiles up and down in a parking structure.

Autopsy Table. Surface upon which a body is placed for examination after death; a specified item installed in hospitals, laboratories, and police crime labs.

Autotransformer. Transformer in which both primary and secondary coils have turns in common; step-up or step-down of voltage is accomplished by taps on common winding.

Auxiliary Heat. The extra heat provided by a conventional heating system for periods of cloudiness or intense cold, when a solar heating system is insufficient.

Auxiliary Members. In a concrete shell structure, ribs or edge beams which serve to strengthen, stiffen or support the shell; usually, auxiliary members act jointly with the shell.

Auxiliary Switch. A standby device for switching.

Available Chlorine. Free or combined chlorine used to disinfect pool water.

Avenue. An access road or street.

Average. 1. An arithmetic mean. 2. An amount obtained by adding the elements of a set and dividing by the number in the set. 3. The ordinary standard.

Aviary. An enclosure or building for keeping birds confined.

Avoirdupois Weight. The weight system based on the pound of 16 ounces and the ounce of 16 drams.

Avoirdupois. 1. Weight. 2. Avoirdupois Weight.

Award. 1. Arbitration award. 2. Contract award.

Awarding Authority. The owner or the agent of the owner who awards an original building or construction contract, also known as the prime contract; this term is generally used with public works.

AWG. American Wire Gauge.

AWI. Architectural Woodwork Institute.

Awl. A small sharp pointed hand tool for marking lines or piercing small holes.

Awning Window. A window hinged at the top.

Awning. A sheet of canvas, plastic, or metal stretched on a frame to give shelter over a window, doorway, or other area.

AWPA. American Wood Preservers Association.

AWPI. American Wood Preservers Institute.

AWS. American Welding Society.

AWWA. American Water Works Association.

Axial Load. Force directly coincident with the primary axis of a structural member such as a beam.

Axial Stresses. Compressive or tensile stresses that are coincidental with the central axis of a structural member.

Axial. In a direction parallel to the long axis of a structural member; in line with the axis; an axial force causes tension or compression stress over the cross section of a member.

Axis of a Weld. An imaginary line through the length of a weld perpendicular to the cross section at its center of gravity.

Axis of Symmetry. A line dividing an area into two similar but opposite handed figures.

Axis, Neutral. Centroidal axis, transverse to longitudinal axis of a structural member, which is neither stretched nor shortened by bending of the member.

Axis. A straight line of reference; in three dimensions, the three axes are referred to as x, y, and z.

Axle. A fixed or rotating rod or spindle on which a wheel or group of wheels rotate.

Axminster Carpet. One of the four basic weaves used in making carpets; woven on an Axminster loom, the pile tufts in this weave are mechanically inserted and bound to the back in a manner similar to the hand knotting of Oriental rugs, making possible almost unlimited combinations of colors and patterns; see Woven Carpet.

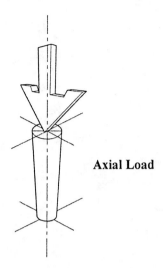

Axial Load

Axonometric Projection. A drafting projection in which objects on the drawing appear inclined with three sides showing and with horizontal and vertical distances drawn to scale but diagonal and curved lines distorted.

Azeotrope. Having constant maximum and minimum boiling points.

Azimuth. The horizontal angle or direction of a compass bearing.

B

Back Bar. A counter, shelf, or shelving for bottles and glasses extending along the wall of a barroom behind the bar.

Back Blocking. A short piece of gypsum wallboard adhesively laminated behind the joints between each framing member to reinforce the joint.

Back Clip. Specially designed clips attached to the back of gypsum board that fit into slots or other formations in the framing to hold the gypsum board in place; often used in demountable partition designs.

Back Filling. 1. Rough masonry built behind a facing or between two faces. 2. Filling over the extrados of an arch. 3. Masonry in spaces between structural timbers, sometimes called Nogging. 4. The filling with earth of the excavations after the concrete and masonry work below finish grade is completed.

Back Gouging. The forming of a bevel or groove on the other side of a partially welded joint to assure complete penetration upon subsequent welding from that side.

Back Mounted Tile. Mounted tile with perforated paper, fiber mesh, or other suitable bonding material applied to the backs or edges of the tile so that a relatively large portion of the tile area is exposed to the setting bed.

Back Nailing. In roofing, the nailing of felts under the overlap to prevent slippage of the felts.

Back Pressure. 1. In plumbing, a higher air pressure at the sewer side of the trap than on the fixture side of the trap; water can, under certain conditions, be forced out of the trap in the wrong direction towards the fixture. 2. In a refrigerating system, the pressure on the low side of the system, also called suction pressure or low-side pressure.

Back Priming. A coat of paint applied to the back of woodwork and exterior siding to prevent moisture from entering the wood and causing the grain to swell.

Back Saw. Small toothed, joint-cutting saw with a metal back strip to keep the saw rigidly in line.

Back Seam. While all carpet seams are located on the back or underside of the carpet, those made when the carpet is turned over or face-down are called back seams, while those made with the carpet face-up are called face seams.

Back Siphonage. The flow of water or other liquids, mixtures, or substances into the distributing pipes of a potable supply of water from any sources other than its intended course, due to a negative pressure in such pipe.

Back Splash. A protective panel installed on the wall behind a counter, sink or lavatory.

Back Vent. A branch vent installed for the purpose of protecting fixture traps from siphoning; back vents include most of the vents not installed specially to permit circulation between vent stacks and soil or waste stacks.

Back Wall. The wall facing an observer who is standing at the entrance to a room, shower, or tub shower.

Back. 1. The part or surface behind the front; the rear. 2. See Steel Square. 3. The back surface of gypsum board; the opposite side from the face; usually the side that would be concealed or the side in contact with the framing; also referred to as the Code Side.

Backband. A simple molding sometimes used around the outer edge of plain rectangular casing as a decorative feature.

Backboard. A vertical structure used as a guard against losing thrown objects, usually specified in the design of athletic playing fields and surfaces.

Backcharge. An offsetting charge against a bill, often asserted by an owner against a prime contractor or a by prime contractor against a subcontractor based on supposedly defective construction work.

Backcheck. A valve that allows flow of a liquid in only one direction; a backflow preventer; a backcheck valve.

Backer Board. See Backing Board.

Backfill, Hand. The act or process of placing excavated earth in a trench or back against the foundation of a structure, by a man and a hand shovel.

Backfill. Earth or earthen material used to fill the excavation around a foundation; the act of filling around a foundation.

Backfire. In welding, a short pop of the flame from the torch tip followed by immediate reappearance or complete extinguishment of the flame.

Backflow Connection. Any arrangement of pipe or fixtures which can cause backflow to occur.

Backflow Preventer. Backcheck valve; a device which prevents sewage from flowing backward into a building's plumbing system.

Backflow. The unintentional flow of water into the supply pipes of a plumbing system from a non-supply source; back-siphonage is one type of backflow, generally due to a temporary occurrence of negative pressure (suction) in the pipes.

Backfurrow. The first cut of a plow from which the slice is laid on undisturbed soil.

Background Noise. The sound level present in a room or space at any given time above which speech, music, desired signal, or sound must be raised in order to be heard or made intelligible.

Backhand Welding. A gas-welding technique wherein the flame is directed opposite to the progress of welding.

Backhoe. An excavating machine with a bucket rigidly attached to a hinged stick on a boom that is drawn toward the machine in operation.

Backhoe/Loader. An excavation machine combining a bucket on a hinged stick on a boom on one end, and a bucket or scoop at the other.

Backing Board. A gypsum board product designed for use as the first or base layer in a multilayer system; also called Backer Board.

Backing Filler Material. Filler metal in the form of a ring, strip or consumable insert, fused in a single-welded joint.

Backing Off. See Featheredging tile.

Backing Ring. Backing in the form of a ring, generally used in the welding of piping.

Backing Rod. A foam plastic rod inserted in a joint to be sealed, to regulate the depth of sealant.

Backing Up. Constructing the inside section of a brick wall after the facing brick of the same wall has been laid header high.

Backfill

Backing. 1. Something forming a back. 2. The wood blocking behind plastering for supporting the load of lighting fixtures, cabinets, and hardware attached to a wall or ceiling. 3. Any material used as a base over which a finished material is to be installed. 4. See Plaster Base. 5. Material (metal, weld metal, asbestos, carbon, granular flux, etc.) backing up the joint during welding to facilitate obtaining a sound weld at the root. 6. The carpet foundation of jute, kraftcord, cotton, rayon, or polypropylene yarn that secures the pile yarns and provides stiffness, strength and dimensional stability. 7. The part of the wall behind the face brick.

Backlog. Reserve of uncompleted work.

Back-Mopping. Mopping the back or underside of roofing.

Backnailing. In roofing, the practice of blind nailing (in addition to hot-mopping) all the plies to a substrate to prevent slippage on slopes of 1-1/2 inch or more for steep asphalt, 1/2 inch or more for coal-tar pitch and dead-level asphalt.

B

Back-Plastering. Plaster applied to one face of a lath system following application and subsequent hardening of plaster applied to the opposite face; used primarily in construction of solid plaster partitions and certain exterior wall systems.

Backset of a Hinge. The distance from the edge of the door to the hinge.

Backset of a Lock. The distance from the centerline of a tubular door lock or cylinder to the edge of the door, measured on the high side of a beveled door.

Backstep Sequence. A longitudinal sequence wherein the weld bead increments are deposited in the direction opposite to the progress of welding the joint.

Backstop. To support or bolster.

Backup Bar. A small rectangular strip of steel applied beneath a joint to provide a solid base for beginning a weld between two steel structural members.

Backup Block. Concrete block which is used as a non-exposed structural wall and backs a finished surface to provide a complete wall system.

Backup. That part of a masonry wall behind the exterior facing.

Backwash Pipe. See Filter Waste Discharge Piping.

Backwash. The process of cleaning the swimming pool filter by reversing the water flow.

Bacteria. Microorganisms present in all water supplies including swimming pools; chlorine and other chemicals are used to keep these microorganisms under control.

Bacteriacide. Any of a number of chemicals used to kill bacteria.

Bad Debt. An uncollectable debt.

Baffle. Plate or vane used to direct or control movement of fluid or air within a confined area.

Bain-Marie. A kitchen utensil that has a hot water compartment under an upper compartment that keeps food warm; also called steam table.

Bake Oven. A cooking device containing a chamber wherein food is baked by dry heat; a commercial or institutional oven.

Bakelite. Trademark; any of various thermosetting resins or plastics made from formaldehyde and phenol.

Baking Finishes. Baking at elevated temperatures improves certain types of coatings used on metal articles, such as automobiles and refrigerators; baking may be done in an oven, under infrared lamps or by induction heating according to the demands of shape, space and other requirements; the article that is coated must, of course, be able to withstand the temperature required for the proper baking of the finish.

Baking Japan. An enamel to which the artificial heat of an oven is usually applied in order to attain the maximum hardness or toughness of film.

Balance Box. A loaded box at the far side of a crane from the jib and the load, to counterbalance them.

Balance Sheet. A statement of financial condition of a specific date.

Balance. An instrument that measures mass, consisting of a central pivot, a horizontal beam, and two scales.

Balanced Construction. A method of constructing manufactured wood products so that moisture content changes will be uniformly distributed and therefore will not cause warping. An example would be symmetrical construction of plywood in which the grain direction of each ply is perpendicular to that of adjacent plies.

Balanced Cuts. 1. Cuts of tile at the perimeter of an area that will not take full tiles; the cuts on opposite sides of such an area shall be the same size. 2. The same sized cuts on each side of a miter.

Balanced Door. A door swung on an arm and pivot arrangement that is spaced out from the jamb so that some of the wind impinging on the door will assist in opening the door.

Balancing Valve. A pipe valve that is used to control the flow of a liquid rather than to shut it off.

Balcony. A platform that projects from the wall of a building.

Baldachin. An ornamental canopy over an altar; also called baldachino.

Ball Bearing Hinge. A hinge having ball bearings positioned between the protruding, cylindrical parts of the hinge, to prevent friction.

Ball Bearing. A bearing in which the two halves are separated by a ring of small metal balls which reduce friction.

Ball Check Valve. Valve assembly (ball) which permits flow of fluid in one direction only.

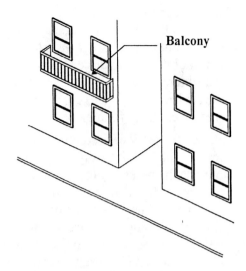

Balcony

Ball Clay. A secondary clay, commonly characterized by the presence of organic matter, high plasticity, high dry strength, long verification range, and a light color when fired.

Ball Cock. An automatic valve whose opening and closing are controlled by a spherical float at the end of a lever, such as on a flush toilet.

Ball Milling. A method of grinding and mixing material, with or without liquid, in a rotating cylinder or conical mill partially filled with grinding media such as balls or pebbles.

Ball Peen Hammer. A hammer which has a rounded head at its opposite end, used extensively in machine shops.

Ball Valve. A valve in which a ball regulates the opening by its rise and fall due to fluid pressure, a spring, or its own weight.

Ball, Tree. See Root Ball.

Ball. An ornament of rounded form, common as the termination of a cupola or lantern.

Ballast Replacement. 1. The replacement of broken stone, slag, and sand in railroads and highways to keep railroad ties in place and to provide drainage. 2. The replacement of the device that regulates current in a fluorescent lamp fixture.

Ballast. 1. A device used with a fluorescent and high intensity lamp, to provide the necessary circuit condition for starting and operation. 2. Any material used as non-structural fill or dead weight. 3. Heavy material, such as water, sand, or iron which has no function in a machine except increase of weight. 4. Crushed rock or gravel which is spread on a roof surface to form its final surface.

Balloon Frame. A wooden building frame composed of closely spaced members (studs) which are continuous from the sill to the top plate of the roof line; also called Eastern Frame; compare Platform Frame.

Balloon Payment. A large lump sum final payment to pay off an installment note, where the previous payments were much smaller.

Ballrace. Round or straight retainer to keep ball bearings in their proper operating position.

Ballroom. A large room or hall designed for dancing, balls, and similar festivities.

Baluster. Each of a series of often ornamental short posts or pillars supporting a rail or coping; also called banisters.

Balustrade. A row of balusters topped by a railing on a stair, porch, or balcony.

Bamboo. Giant tropical woody grasses with hollow stems that are used for making furniture.

Banc. The bench on which court judges sit.

Band Joist. A wooden joist perpendicular to the direction of the joists in a floor framing system, closing off the floor platform at the outside face of the building.

Band Stage. A raised platform where musicians perform.

Band. 1. A low, flat molding. 2. A group of bars distributed in a slab or wall or footing.

Band-Aid Approach. See Quick Fix.

Banding. Metal or plastic strapping to secure bundles of building products, such as gypsum wallboard, together in a shipping unit.

Bandsaw. A machine saw with a narrow endless blade that runs over pulleys.

Banister. A light baluster supporting a stair handrail.

Banjo Taper. A mechanical device which dispenses tape and taping compound simultaneously.

Bank Gravel. Gravel found in natural deposits, usually more or less intermixed with fine material, such as sand or clay, or combinations thereof; gravelly clay, gravelly sand, clayey gravel, and sand gravel indicate the varying proportions of the materials in the mixture.

Bank Measure. The volume of earth in its natural site.

Bank Run Gravel. Excavated material that is generally 1/4 inch minimum to 6 inches maximum.

Bank, Duct. See Duct Bank.

Bank. A building accommodating a financial establishment that receives and pays out money and deals generally in money and finance.

Bankruptcy. A legal proceeding by which a debtor may avoid legal and financial obligations.

Baptistry. A part of a church or formerly a separate building for baptism; also spelled baptistery.

Bar Chart. A simple construction scheduling technique which graphically shows the starting and finishing times for the various tasks which make up a job.

Bar Hanger. A metal bar, either straight or offset, to allow for the mounting of a ceiling outlet box between ceiling joists, or an outlet box, or switch box between wall studs; an adjustable hanger is one that can be made to fit the varying distances between floor and ceiling joists or rafters to securely hold electrical outlet boxes and devices.

Bar Hook. A semi-circular (180 degree) or a 90 degree turn at the free end of a steel reinforcing bar to provide anchorage in concrete.

Bar Joist. A truss-like floor joist or rafter fabricated from steel bars.

Bar Lap. The amount steel reinforcing bars must lap in order to develop sufficient bond to transfer their full load capacity.

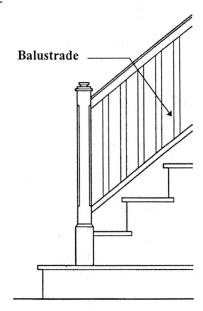

Balustrade

Bar Number. A number designating the size of a steel reinforcing bar, approximately the bar diameter in eighths of inches; for example, a #5 bar is approximately 5/8 inch in diameter; a #9 bar is approximately 1-1/8 inch in diameter (9/8); bar numbers are rolled onto the bar for easy identification.

Bar Spacing. Distance between parallel reinforcing bars measured from center to center of the bars perpendicular to their longitudinal axis.

Bar Support. A rigid device used to support or hold reinforcing bars in proper position to prevent displacement before or during concrete placement.

Bar Supports. Devices of formed wire, plastic or concrete, to support, hold and space reinforcing bars.

Bar, Parallel. A pair of bars on a support, adjustable in height and spacing, that are parallel to each other, and are used for gymnastic exercises.

Bar, Reinforcing. See Reinforcing Bar.

Bar. 1. A straight wood or metal piece used as a lever, support, barrier, or fastening. 2. A barrier of any shape. 3. A submerged or partly submerged sand bank along a shore or in a river often obstructing navigation. 4. The railing in a courtroom separating the judge, jury, lawyers, and witnesses from the spectators. 5. A particular system of courts. 6. The whole body of lawyers qualified to practice in a jurisdiction. 7. A building, room, or counter at which food or beverages are served. 8. A small shop or stall serving refreshments, such as a snack bar. 9. A specialized department in a large store, as a watch repair bar. 10. A steel member used to reinforce concrete. 11. Unit of pressure; one bar equals 0.9869 atmospheres (one million dynes per square centimeter).

Barbed Nail. A nail with a barbed shank to resist withdrawal.

Barbed Wire. Wire that is twisted with barbs or sharp points.

Barbican. The outer defense of a city or castle, often a part of a system of barrier walls and consisting of a double tower above a gate or drawbridge.

Bare Electrode. A filler-metal electrode used in arc welding consisting of a metal wire with no coating other than that incidental to the drawing of the wire.

Bare Solid Wire. Uninsulated single wire used as an electric conductor.

Bare Stranded Wire. Uninsulated group of fine wires used as a single electric conductor.

Barge. A floating platform or vessel from which construction activities may be performed; often used in rivers to install bridge piers and also used extensively in waterfront construction.

Bargeboard. A board, sometimes ornamental, fixed to the end of a gable, running from ridge to eave, to hide the ends of the roof timbers.

Barium Sulphate. Heavy, white, extender pigment made from the mineral, barite; unaffected by acids, alkalis.

Bark pocket. An opening between a tree's annual growth rings that contains bark.

Bark. The tough outer layer of a tree.

Barn. 1. A large farm building for storing hay and grain and housing farm animals and equipment. 2. An unusually large and bare building, pejoratively. 3. A large building for storing a fleet of trucks or trolleys.

Barometer. An instrument for measuring atmospheric air pressure, it may be calibrated in pounds per square inch, or in inches or millimeters of mercury in a column.

Baroque. Highly ornate, grotesque, flamboyant, and extravagant in style, especially of the architecture, art, and music of the 17th and 18th centuries.

Barracks. 1. A building or building complex used to house soldiers. 2. A large plain building.

Barratry. The persistent incitement of litigation.

Barrel Ceiling. A rounded or semi-circular ceiling.

Barrel Shell. A scalloped roof structure of reinforced concrete that spans in one direction as a barrel vault and in the other as a folded plate.

Barrel Tile. A type of ceramic or cast roofing tile.

Barrel Vault. A segment of a cylinder that spans as an arch; used as a structural technique to support a ceiling or roof by having all of the components act in compression as an arched ceiling; used extensively in ancient buildings and into the 19ᵗʰ century, because no structural steel or timber is needed.

Barrel. A unit of weight for cement. 376 lbs. net, equivalent to 4 U. S. bags of portland cement.

Barricade. A barrier; an obstruction to prevent passage or to prevent access.

Barrier Coating. 1. Shielding or blocking coating or film. 2. See Transition Coating.

Barrier, Vapor. A type of plastic sheeting that both eliminates drafts and keeps moisture from damaging a building or structure.

Barrier. A fence or other obstacle that bars advance or access.

Barroom. A room or establishment in which the principal feature is a bar for the serving of liquor.

Basalt Ware. A black unglazed vitreous ceramic ware having the appearance of basalt rock.

Basalt. A dark basic volcanic rock whose strata sometimes form columns.

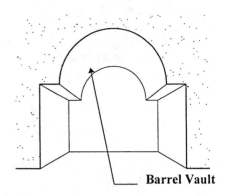

Barrel Vault

Base Bead. See Base Screed.

Base Bid. A stipulated construction sum based on the contract documents.

Base Board. See Base.

Base Cabinet. 1. Case, box, or piece of furniture which rests on the floor with sets of drawers or shelves with doors, primarily used for storage. 2. Floor-mounted cabinets, usually with a counter, sink or appliance installed.

Base Cap. A finish fitting for tile.

Base Coat Floating. The finishing act of spreading, compacting, and smoothing of the base coat plaster to a reasonably true plane.

Base Course. 1. The lowest row of masonry in a wall, pier, foundation, or footing. 2. In asphaltic concrete paving, the layer of material immediately beneath the surface or intermediate course; it may be composed of crushed stone, crushed slag, crushed or uncrushed gravel and sand, or combinations of these materials and may also may be bound with asphalt.

Base Flashing. 1. A waterproof membrane that is attached to a roof and bent up the side of a chimney. 2. A waterproof membrane that is attached at a joint between any vertical member and a roof.

Base Grouting. The injection of grout to fill voids in subfloor slabs, foundations or concrete slabs.

Base Isolation. A design of foundation footings that allows controlled movement of the superstructure, independent of the earth in the event of seismic movement.

Base Layer. The first or interior layer of gypsum board applied in a multilayer system.

Base Line. The main traverse or surveyed line running through the site of proposed construction, from which property lines, street lines, and buildings are located and plotted on the plan.

Base Lines. Part of a governmental land surveying grid system covering the country, with base lines running east and west and meridians running north and south; townships are located with reference to a specific base line and meridian; for example, T3N R12W, MDBM is read Township 3 North, Range 12 West, Mount Diablo Base Line and Meridian.

Base Metal. The metal to be welded, cut or brazed.

Base Molding. A strip of wood used to trim the upper edge of a baseboard.

Base of Structure. The level at which earthquake motions are assumed to be imparted to a building; this level does not necessarily coincide with the ground level.

Base Ply. The first layer in a built-up roof.

Base Screed. A preformed metal screed with perforated or expanded flanges; provides a ground for plaster and separates areas of dissimilar materials.

Base Shear. Total shear force acting at the base of a structure.

Base Sheet. This is the first ply in some multi-ply built-up roofing membranes; it is usually a saturated felt or a coated felt.

Base Shoe Molding. See Base Shoe.

Base Shoe. Wood strip, with a curved or projecting surface, used to finish intersection of base and floor; also called Base Shoe Molding, Shoe Mold, or Carpet Strip.

Base, Column. The plate beneath a column that distributes the load.

Base, Manhole. The cast iron frame into which a manhole cover fits.

Base, Mogul. A screw-in style base for an incandescent lamp of generally 300 watts or more.

Base, Stone. The beginning or starter stone of a fieldstone wall.

Base, Terrazzo. A subfloor slab or foundation using a flooring material made from marble or other stone chips set in portland cement and polished when dry.

Base. 1. A molding or board placed along the bottom of a wall next to the floor; also called Baseboard. 2. A vinyl, wood or metal trim applied at the floor line to protect the vertical wall from damage. 3. The bottom of a column. 4. One or more rows of tile installed above the floor. 5. See Plaster Base. 6. The lowest part, or the lowest main division, of a building, column, pier, or wall. 7. The lowest point of any vertical pipe.

Baseball Backstop. A construction, often of wood or chain-link fencing, that encloses the home plate area on three sides and from the top to prevent balls from leaving the field and hitting spectators.

Baseboard Heater. Heating strips that are installed at the juncture of the wall and floor and may be either recessed or surface-mounted; generally along the outside walls of rooms.

Baseboard Radiator. A heating unit installed along a baseboard in a building or structure, is usually hydronic (hot water or steam).

Baseboard. See Base.

Basecoat. Any plaster coat or coats applied prior to application of the finish coat.

Basement Soil. See Subgrade.

Basement. The part of a building that is wholly or partially below ground level; cellar.

Baseplate. A steel plate inserted between a column and the foundation, used to level the column and to spread the load of the column to a larger area of the foundation.

Basic Lead Carbonate. A type of white lead pigment.

Basic Lead Sulphate. A type of white lead.

Basilica. 1. An ancient Roman public hall with an apse and colonnades, used as a law court and place of assembly. 2. A similar building used as a Christian church.

Basin Wrench. A plumber's tool used for installing hard-to-get-at fittings.

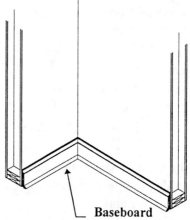

Baseboard

Basin. 1. A wide shallow open container. 2. A lavatory. 3. A hollow rounded depression. 4. Any sheltered area of water where boats can moor safely. 5. A round valley. 6. An area drained by rivers and tributaries.

Basis for Acceptance. The method of determining whether a lot of material is acceptable under given or accepted specifications.

Basket. Wire assembly to support and space dowel bars and expansion joints in concrete slabs on the ground.

Basketweave Bound. Module groups of brick laid at right angles to those adjacent.

Bas-Relief. Sculpture or carving in which the figures project only slightly from the background surface with no undercutting.

Bast. Fibrous plant material, such as jute or hemp, used in making rope, matting, and rough fabrics.

Bastard File. A course file for rough shaping of metal or wood.

Bastard Granite. A quarrier's term for nearly any stone which may not be considered a true granite, particularly applied to gneiss.

Bastard Sawn. Lumber, primarily hardwoods, in which the annual rings make angles of 30 to 60 degrees with the surface of the piece.

Bat. A broken brick; an end portion of a brick; approximately a half.

Batavia. Batavia dammar; see Dammar.

Batch Disposal. The orderly placement or distribution of freshly mixed concrete or masonry.

Batch Mixer. A machine which mixes batches of concrete or mortar in contrast to a continuous mixer.

Batch Plant. 1. A manufacturing facility for producing asphalt paving mixtures that proportions the aggregate constituents into the mix by weighed batches and adds asphalt material by either weight or volume. 2. An operating installation of equipment including batchers and mixers as required for batching or for batching and mixing concrete materials; also called mixing plant when equipment is included.

Bath Accessory. Bathroom equipment such as towel bars, grab bars, grab rails, soap dishes, toilet paper holders, and medicine cabinets.

Bath Chair. A wheelchair.

Bath Faucet. A valve used to draw hot or cold water into a bathtub.

Bath Tub Door. A folding or sliding door mounted on a bath tub rim to keep water spray within the tub area when there is a shower over the tub.

Bath Tub. The tub in a bath room.

Bath Vanity. A bathroom cabinet with a lavatory mounted in the counter top.

Bath. 1. A tub used for bathing. 2. The room that contains the tub. 3. Liquid solution used for cleaning, plating, or maintaining a specified temperature.

Bathhouse. A building with baths and dressing rooms for public use.

Bathroom Lock. A non-keyed privacy lock that can be locked from the inside by push-button and opened from the outside by a small tool; a privacy lock.

Bathroom. 1. A room containing a bath tub or shower and usually a toilet and lavatory. 2. A polite term for any toilet room.

Batt Insulation. Fiber or wool insulation in sheet form, usually with a paper lining; a preformed section of inorganic insulation sized to fit snugly in a framed cavity.

Batten Seam. A seam in a sheet metal roof.

Batten Siding. Vertical siding which has narrow strips of metal or wood covering the joints.

Batten. 1. A thin narrow strip of wood or metal to cover or reinforce a joint; a predecorated strip or joint covering designed to conceal the junction between adjacent boards; frequently used in demountable systems. 2. A cleat; a narrow strip of board used to fasten several pieces together.

Batter Board. A temporary framework used to assist in locating the corners when laying a foundation; also used to maintain proper elevations of structures, excavations and trenches in any kind of below ground construction.

Batter Pile. Pile driven at an angle to brace a structure against lateral or horizontal thrust.

Batter. 1. A slight receding upward slope to the outward face of a building wall or a retaining wall. 2. Stepping or sloping masonry back in successive courses; the opposite of corbel.

Battery of Fixtures. Any group of two or more similar adjacent plumbing fixtures which discharge into a common horizontal waste or soil branch.

Battery. One or more cells connected together to furnish electric current by interaction of metals and chemicals.

Battlement. A parapet with recesses along the top of a wall, originally as a fortification, now as ornamentation.

Baudelot Cooler. Heat exchanger in which water flows by gravity over the outside of the tubes or plates.

Bauhaus. An architectural school in Germany, founded by Walter Gropius in 1923, emphasizing technology, craftsmanship, and design aesthetics.

Bauxite. A claylike mineral containing varying proportions of alumina, the principal source of aluminum.

Bay Window. A rectangular, curved, or polygonal window projecting out from the face of a wall.

Bay. One of the intervals or spaces into which a building plan is divided by columns, piers, or division walls.

Bazaar. 1. A market in an oriental country consisting of rows of shops or stalls offering miscellaneous goods. 2. A large shop selling fancy goods. 3. A department store.

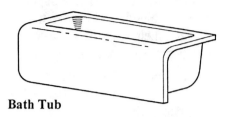

Bath Tub

Bead Weld. A type of weld composed of one or more string or weave beads deposited on an unbroken surface; the metal added in welding.

Bead, Molding. 1. A half-round narrow molding, attached or milled on a larger piece. 2. A square or rectangular trim less than 1 inch in width.

Bead, Parting. A narrow strip between the upper and lower sashes in a double-hung window frame.

Bead, Plaster. Built-in edging usually metal, to strengthen a plaster angle.

Bead, Sash. A strip with one edge molded, against which a sash slides.

Bead. 1. A preformed strip installed to reinforce the corners or ends of wall-board panels, or the raised metal section at a corner of this strip. 2. A narrow, convex molding profile. 3. A metal edge or corner accessory for plaster or gypsum board.

Beaded Moulding. A cast plaster string of beads planted in a moulding or cornice.

Beading. A ridge or raised linear deformation that may appear at finished gypsum board joints.

Beam Bolster. A fabricated wire device used to temporarily support reinforcing steel in structural formwork such as beams and slabs.

Beam Chair. A wire seat or support for reinforcing bars designed to maintain their location while concrete for a beam is poured around them.

Beam Clamp. A device which holds a horizontal structural member to a vertical member.

Beam Compass. A drafting tool consisting of a wood or metal bar fitted with a point and a movable pen or pencil holder to use in drawing large radius arcs and circles; a Trammel.

Beam Fireproofing. Fire-resistant materials that cover a horizontal structural member, to insure structural integrity in the event of a fire.

Beam Formwork. The system of support for freshly placed concrete for a horizontal structural member.

Beam Furring. Strips of wood or metal fastened to a horizontal structural member to form an airspace, to give the appearance of greater thickness, or for the application of an interior finish such as plaster.

Beam Hanger. 1. A strap wire or other device used to hang beam forms from another structural member. 2. A steel strap used for securing the end of a beam.

Beam Plaster. A combination of cementitious material and aggregate mixed with water and applied to a horizontal structural member, preserving in a rigid state the form or texture imposed during the period of elasticity.

Beam Rail. A solid wood band attached to a horizontal structural member.

Beam, Bond. A reinforced horizontal concrete masonry or concrete beam installed in place to strengthen a masonry wall and tie a masonry wall together.

Beam, Concrete. A horizontal structural member which transversely supports a load and transfers the load to vertical members, made of a composite material consisting of sand, coarse aggregate, cement and water.

Beam, Grade. An end-supported horizontal load-bearing foundation member that supports an exterior wall or other building load.

Beam, Precast. A concrete horizontal structural member that is cast and cured in other than its final position, on- or off-site.

Beam, Reinforcing. A horizontal member installed to strengthen and support the load of a structure.

Beam, Sheetrock. A horizontal member constructed of plasterboard.

Beam. A straight structural member that acts primarily to resist transverse loads; a structural element which sustains transverse loading and develops internal forces of bending and shear in resisting the loads; an inclusive term for joists, girders, rafters, and purlins.

Bearding. In carpeting, long fiber fuzz occurring on some loop pile fabrics, caused by fibers snagging and loosening due to inadequate anchorage.

Bearing Block. A piece of wood fastened to a column to provide support for a beam or girder.

Bearing Capacity. Allowable bearing capacity; the maximum allowable load on a structural element.

B

Bearing Pad. A block of metal, plastic, or synthetic material used to cushion the point at which one structural element rests upon another.

Bearing Partition. A partition that supports any vertical load in addition to its own weight.

Bearing Plate. A plate placed under a truss, beam, girder, or column to distribute the load.

Bearing Wall. A wall which supports any vertical loads in addition to its own weight.

Bearing. 1. Part of a machine that supports and aligns a rotating or other moving part. 2. That part of a lintel, beam, girder or truss, which rests upon a column, pier or wall.

Beating Block. A wooden block used to embed tiles in a flat plane; the method used is called beating in.

Becquerel. A unit for measuring radiation from radium and radon.

Bed Coat. See Bedding Coat.

Bed Joint. The horizontal layer of mortar in which a masonry unit is set.

Bed Mold. A flat area in a cornice, designed to have enrichment planted later; also called Bed.

Bed Molding. A molding in an angle, as between the overhanging cornice or eaves of a building and the sidewalls.

Bed. 1. The horizontal surface on which the bricks of a wall lie in courses, also the mortar on which the brick rests. 2. See Bed Mold.

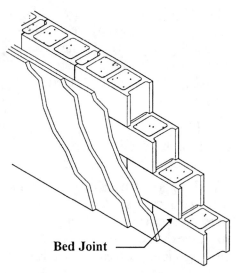

Bed Joint

Bedding Plane. A separation or weakness between two layers of rock, caused by changes during the building up of the rock- forming material.

Bedding. 1. A filling of mortar, putty, or other substance in order to secure a firm bearing. 2. Ground or supports in which pipe is laid.

Bedding Coat. 1. In gypsum board, the first coat of joint compound over tape, bead, and fastener heads; also called Bed Coat. 2. That coat of plaster to receive aggregate or other decorative material of any size, impinged or embedded into its surface before it sets.

Bedrock. A solid layer or stratum of rock beneath ground level; solid rock as distinguished from boulders.

Bedroom. A room for sleeping.

Beeswax. Wax produced by honey bee.

Behavioral Aspects of Management. These include human relations, cooperation, responsibility, communication and executive ability.

Belfry. A bell tower.

Bell and Spigot Joint. A type of joint used in cast iron pipe where a male pipe end (the spigot) slips into a female end (the bell) and is then caulked with oakum and sealed with lead.

Bell End. See Bell.

Bell Trap. A floor drain trap consisting of an inverted bell with a water seal.

Bell. That portion of a pipe which, for a short distance, is sufficiently enlarged to receive the end of another pipe of the same diameter for the purpose of making a joint; also called Bell End or Hub.

Belleek China. A highly translucent whiteware composed of a body containing a significant amount of grit and normally having a luster glaze.

Bellows Copper. A flexible joint in copper piping that can expand or contract to allow for thermal fluctuations.

Bellows Seal. Method of sealing the valve stem; the ends of the sealing material are fastened to the bonnet and to the stem; the seal expands and contracts with the stem level.

Bellows. An instrument or machine that draws in air through a valve or orifice by expansion and contraction and expels it through a tube.

Belly-Up. The condition of a contracting firm when it becomes insolvent and goes bankrupt.

Belt Course. 1. A course of bricks or other material projecting slightly from the face of the wall. 2. A horizontal board across or around a building, usually made of a flat member and a molding. 3. A horizontal course on the face of a building; when continuous with a row of window sills or lintels, is referred to as a Sill Course or a Lintel course; also called a String Course.

Belt Marks. A surface defect in gypsum board made by the machine's forming belt during manufacture.

Belt Sander. An electric sanding tool where the sand paper abrasive is a continuous belt.

Bema. The part of an eastern church containing the altar.

Bench Mark. A point of known or assumed elevation used as a reference in determining and recording other elevations.

Bench Saw. A power saw held securely on a stationary bench.

Bench. 1. A long seat for seating several people. 2. A judge's seat in a lawcourt. 3. A working table or counter for a carpenter, artisan, or scientist. 4. A low scaffold board that allows the gypsum board hanging crew to easily reach the ceiling area; also called Hanger's Bench.

Bend, Soil. A piece of short, curved pipe, like an elbow, used to connect two straight links of pipe in a sewage system.

Bending Moment. The sum of moments for all forces that occur above the neutral axis; the moment that causes a beam or other structural member to bend; see Moment.

Bending Spring. Coil spring which is placed on inside or outside of tubing to keep it from collapsing while bending it.

Bending Stress. A compressive or tensile stress resulting from the application of a nonaxial force to a structural member.

Bending. The result of a force which tends to cause curvature in a linear element; internal stresses of tension and compression are a result of this action.

Bends, The. Caisson disease

Beneficial Occupancy. The use of a premises or a portion for the uses intended, although the work of the project may not be completed.

Beneficiary. 1. The person designated to receive the proceeds of an insurance policy. 2. In a trust deed, the lender.

Benny. Slang for benzine.

Bent Bar. A steel reinforcing bar bent to a prescribed shape such as a truss bar, straight bar with hook, stirrup, or column tie.

Bent. A plane of framing consisting of beams and columns joined together, often with rigid joints; a single vertical framework consisting of horizontal and vertical members supporting the deck of a bridge or pier.

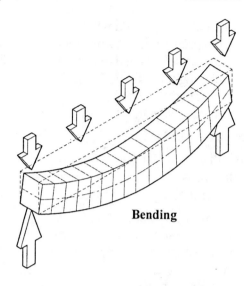

Bending

Bentonite Membrane. A thin, pliable sheet or layer manufactured from bentonite; used for waterproofing retaining walls.

Bentonite. A clay composed principally of minerals of the montmorillonoid group, characterized by high absorption and very large volume change with wetting or drying, commonly swelling to several times its dry volume when saturated with liquid.

Benzene. A material derived from coal tar and widely used a solvent; also used for cleaning after painting and other finishing operations; a very powerful aromatic solvent for many materials, its use is restricted because of its toxicity and because it is a fire hazard; often confused with Benzine due to similarity in pronunciation; also called Benzol.

Benzine. A highly flammable petroleum product used by painters as thinning solvent and diluent; also used in dyeing and as a cleansing agent; also called Varnish Makers' and Painters' Naphtha (VM&P).

Benzol. See Benzene.

Berm. A small artificial ridge of soil at the top of an earth bank to prevent water drainage onto the bank.

Bernoulli's Theorem. In a stream of liquid the sum of elevation head, pressure head, and velocity remains constant along any line of flow provided no work is done by or upon the liquid in the course of its flow and decreases in proportion to energy lost in flow.

Beryllium Oxide. An inorganic material of exceptionally high thermal conductivity which is toxic in the powder form; BeO; also called Berylla).

Bespoke. Custom-made.

Bevel Board. A board used in framing roof or stairway to lay out bevels; also called Pitch Board.

Bevel Gears. Meshing gears on intersecting axes.

Bevel of a Door. The angle of the lock edge in relation to the face of the lock stile; standard bevel is 1/8 inch in 2 inches (3.2 mm in 50.8 mm).

Bevel Siding. A type of finish siding used on the exterior of a house; usually manufactured by resawing a dry, squared, surfaced board diagonally to produce two wedge- shaped pieces; also called Lap Siding.

Bevel. 1. A slanted surface; an end or edge cut at an angle other than a right angle. 2. A type of edge preparation on metal where welding is to take place.

Beveled Concrete. An angle in concrete or inclination of any line in concrete or concrete surface that joins another.

Beveled Edge. The factory edge of gypsum board that has been angled to form a vee grooved joint when two pieces are placed together.

Beveling. A type of chamfering.

Bezel. 1. A sloping surface on a cutting tool, as on a chisel. 2. A frame around something like a glazed opening or a built-in appliance.

BHMA Hinge. A hinge that has the seal of approval of the Builders Hardware Manufacturers' Association.

BHMA. Builders Hardware Manufacturers Association.

BIA. Brick Institute of America.

BIA. Brick Institute of America; formerly Structural Clay Products Institute.

Bib. See Hose Bibb.

Bibb, Hose. See Hose Bibb.

Bicycle, Exercise. A stationary bicycle used for health and fitness purposes in gyms or health clubs.

Bid Bond. A bond, secured by a bidder, which guarantees that the bidder selected by the owner will accept the project, or the owner will have the project for the bid price as noted in the accepted bid.

Bid Date. See Bid Time.

Bid Package. The group of documents issued to general contractors who are bidding a construction project, consisting of the bidding requirements and the contract documents; also called the Bidding Documents.

Bid Rigging. Collusion; a fraudulent practice involving secret agreements between or among competing contractors or owners' agents to control the outcome of a bidding procedure.

Bid Shopping. A procedure wherein a general contractor contacts subcontractors in an attempt to receive lower subcontractor prices after having been awarded the contract for the project; considered unethical by the building industry.

Bid Time. The date and time designated by the architect and owner for the receiving of bid proposals; also called Bid Date.

Bid. An offer to perform; an offer to enter into a contract usually for a stipulated sum of money; a tender; a proposal.

Bidder. A prime contractor who submits a bid directly to an owner.

Bidding Documents. The bidding requirements and the proposed contract documents, including any addenda issued prior to receipt of bids.

Bidding or Negotiation Phase. One of the standard phases of architectural service (Schematic Design Phase, Design Development Phase, Construction Documents Phase, Bidding or Negotiation Phase, and Construction Phase-Administration of the Construction Contract).

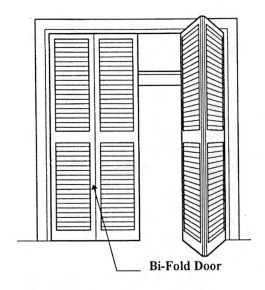

Bi-Fold Door

Bidding Period. The time period during which contractors can prepare their bid proposals.

Bidding Requirements. The group of documents issued to general contractors who are bidding a construction project, other than the contract documents; bidding requirements include the Advertisement or Invitation to Bid, Instructions to Bidders, Supplementary Instructions to Bidders, the bid form, and other sample bidding and contract forms.

Bidet. A plumbing fixture about the size and shape of a toilet, equipped with cold and hot running water, used for personal hygiene.

Bi-Fold Door. A door with two leaves, hinged together to close on itself; one edge of each leaf is hinged at the jamb and the other edge is connected and guided by an overhead track.

Bifunctional. Having two functions.

Bike Rack. A framework installed on a site to store bicycles for the building's inhabitants, commonly specified in educational complexes.

Billhook. A tool with a hooked blade used for cutting and pruning.

Bimetallic Strip. A strip made of two metals that bends as the temperature changes; a bimetallic strip of iron and brass, when heated, will bend, as brass expands more than iron, thus bending the strip, and unbending upon cooling; this action can be used to open or close a switch.

Bin Method. A method of computing cooling energy use requirements for commercial and industrial building with unusual operating needs and for residences utilizing passive heating/cooling design with high mass thermal storage.

Bin, Ice Storage. A box, frame or enclosed place used for the storage of ice and the maintaining of its solid properties.

Bin. A box, container, or enclosed place for storage.

Binder. 1. A substance that acts cohesively, see Cementitious Material. 2. Fines which hold gravel together when it is dry. 3. The nonvolatile portion of a paint which serves to bind or cement the pigment particles together; oils, varnishes and proteins are examples of binders; see vehicle. 4. A chemical additive to gypsum board core formulation, frequently starch, to improve the bond between the surfacing papers and the core.

Bindery. A building, workshop or factory where books are bound.

Binding. A strip sewed over a carpet edge for protection against unraveling.

Bioaerosols. Bacteriological debris in air, such as fungus, bacteria, pollen, or mites.

Biocide. Preservative used in making paint to keep bacteria from growing in the paint.

Biodegradability. The ability of a material to rot away after use, by the action of microorganisms.

Biological Contaminants. Agents derived from or that are living organisms, for example, viruses, bacteria, fungi, and mammal and bird antigens, that can be inhaled and can cause many types of health effects including allergic reactions, respiratory disorders, hypersensitivity diseases, and infectious diseases; also referred to as microbiologicals or microbials.

BIPV. Building Integrated Photovoltaic Cell.

Birch Door. A door constructed or faced with a strong fine-grained birch hardwood.

Birch Paneling. Rectangular sheets of paneling that have been constructed of a strong fine-grained birch hardwood.

Birch Veneer. Thin sheets of strong fine-grained hardwood used in furniture, flooring, and building paneling.

Birch. Hardwood, species Betula.

Bird peck. A small hole or patch of distorted grain in wood resulting from birds pecking through the growing cells in the tree.

Birdbath. A hollow rounded depression in asphaltic concrete paving or in portland cement concrete flatwork; see Ponding, 1.

Birdcage. Colloquial name for the end of a stair rail where the banisters are curved in a spiral to form a newel post.

Birdscreen. Wire screening attached to louvers, ventilators and openings in a building or structure to prevent birds and small animals from entering.

Bird's-Eye. Small, localized area in wood where the fibers are indented and otherwise contorted to form small circular or elliptical figures which look somewhat like birds' eyes; common in sugar maple and used for decorative purposes; rare in other hardwood species.

Bird's-Mouth. A notch cut on the underside of a rafter to fit it to the top plate; not a full notch if there is no rafter overhang.

Biscuit Chips. Glazed-over chips on the edge or corner of the body of a tile.

Biscuit Cracks. Any fractures in the body of a tile visible both on face and back.

Bisect. 1. To divide into two parts. 2. To divide into two equal parts.

Bisque Fire. The process of kiln-firing ceramic ware prior to glazing.

Bit. 1. A unit of computer information expressed as a choice between two possibilities. a 0 or 1 in binary notation; short for binary digit. 2. Of a key, the projecting blade, cut in a manner that actuates the lock tumblers. 3. The removable cutting edge of a tool, as a drill bit. 4. Of a screw gun, the replaceable tip or portion that seats in the slotted screw head.

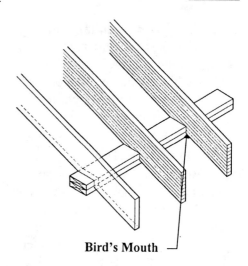

Bird's Mouth

Bite. The amount of the edge of a pane of glass that is covered by the stop.

Biting. Solvent in topcoat of a lacquer surface dissolves or bites into coat below; if lacquer solvent is too biting, the dried lacquer surface may be rough or produce an orange peel effect.

Bitting. The configuration of notches and hollows on the blade of a door lock key.

Bitumen. A class of black or dark-colored solid, semisolid, or viscous cementitious substances, natural or manufactured, composed principally of high molecular weight hydrocarbons, of which asphalts, tars, pitches and asphaltites are typical; a generic term for an amorphous, semi-solid mixture of hydrocarbons derived from petroleum or coal; in the roofing industry there are two basic bitumens. coal-tar pitch and asphalt; before use, the bitumens are heated to a liquid state, dissolved with a solvent, or emulsified.

B

Bituminous Dampproofing. Impregnated mixtures of hydrocarbons, like tar or asphalt, together with their nonmetallic derivatives used on a surface to prevent moisture from passing through; dampproofing will not ordinarily resist water under hydrostatic pressure; see Dampproofing.

Bituminous Emulsion. 1. A suspension of minute globules of bituminous material in water or in an aqueous solution. 2. Invert emulsion; a suspension of minute globules of water or an aqueous solution in a liquid bituminous material.

Bituminous Grout. A mixture of bituminous material and fine sand that will flow into place without mechanical manipulation when heated.

Bituminous Membrane. A thin layer or sheet of an impregnated mixture of hydrocarbons together with their nonmetallic derivatives; used as dampproofing.

Bituminous Sidewalk. A walkway constructed with an impregnated mixture of hydrocarbons together with aggregate such as sand or stone; commonly called blacktop.

Bituminous. Resembling, containing, or impregnated with various mixtures of hydrocarbons, like tar or asphalt, together with their nonmetallic derivatives.

Black Iron Cap. A steel fitting, with female threads, which seals the end of a pipe.

Blackboard. A panel or board with a smooth, usually dark, surface for writing on with chalk.

Blacktop. A term used in the trade for asphaltic concrete paving.

Blade Grader. A grader.

Blade. 1. The flattened part of a paddle or oar. 2. The projecting parts of a fan, turbine, or propeller. 3. The sharp cutting part of a knife or saw. 4. The flat or concave part of a bulldozer. 5. See Steel Square.

Blaine Fineness. The fineness of powdered materials such as cement and pozzolans, expressed as surface area usually in square centimeters per gram, determined by the Blaine air-permeability apparatus and procedure.

Blanc Fixe. Artificially prepared barium sulphate; an extender pigment.

Blanket Insulation. Thermal insulating material made of fibrous glass or mineral wool, sometimes with paper or foil surfacing, formed in batts or rolls.

Blanket Mortgage. A single trust deed or mortgage that covers two or more parcels of real property.

Blanket. Soil or broken rock left or placed over a blast to confine or direct throw of fragments.

Blast Angle. Angle of sand blasting nozzle with reference to surface.

Blast Cleaning. Cleaning with propelled abrasives.

Blast Freezer. Low-temperature evaporator which uses a fan to force air rapidly over the evaporator surface.

Blast Furnace Slag. A non-metallic waste product developed in the manufacture of pig iron, consisting basically of a mixture of lime, silica and alumina, the same oxides that make up portland cement, but not in the same proportions or forms; it is used both in the manufacture of portland blast furnace slag cement and as an aggregate for lightweight concrete.

Blast Furnace. A smelting furnace into which compressed hot air is driven.

Bleach. To remove color or stains from.

Bleacher, Gym. An uncovered series of tiered planks used to seat spectators in a school or public gymnasium.

Bleacher. An outdoor tiered stand of benches to provide seating at a sports ground.

Bleaching. Restoring discolored or stained wood to its normal color or making it lighter by using bleaching agents.

Bleb. A small blister or bubble.

Bleeder Gun. A spray gun with no air valve; trigger controls fluid flow only.

Bleeding Stain. Stain which works up or bleeds through succeeding coats of finishing materials.

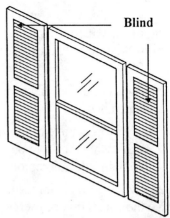

Blind

Bleeding. 1. The upward movement of asphalt in an asphalt pavement resulting in the formation of a film of asphalt on the surface; usually caused by too much asphalt in one or more of the pavement courses, resulting from too rich a plant mix, an improperly constructed seal coat, too heavy a prime or tack coat, or solvent carrying asphalt to the surface; usually occurs in hot weather; also called Flushing. 2. The autogenous flow of mixing water within, or its emergence from newly placed concrete or mortar; caused by the settlement of the solid materials within the mass; also called Water Gain. 3. In gypsum board, a discoloration, usually at a joint, which may occur on a finished wall or ceiling. 4. Penetration of color from the underlying surface; seeping of a stain or lower coat through the top coat, spoiling the appearance of the top coat. 5. When coloring material from the wood or undercoat works into succeeding coats and imparts to them a certain amount of color; see Extractive Bleeding. 6. Slowly reducing the pressure of liquid or gas from a system or cylinder by slightly opening a valve. 7. The draining or loosening of saturants from the roofing material.

Bleed-Valve. Valve with a small opening inside which permits a minimum fluid flow when valve is closed.

Blemish. A physical imperfection that affects appearance.

Blend. 1. To mix or make homogeneous. 2. A mixture, as of two pigments, to obtain a desired color.

Blending. Mixing one color with another so the colors mix or merge gradually.

Blind Header. A concealed header in the interior of a wall, not showing on the faces.

Blind Nailing. 1. Attaching boards to framing or sheathing with nails driven through the edge of each piece so as to be concealed by the succeeding board. 2. In installing tongue-and-groove flooring, the nails are placed at the root of the tongue where they will be hidden by the groove of the next piece; the nails pierce the subfloor at a 45 degree angle. 3. The practice of nailing the back portion of a roofing ply so that the nails will be concealed by the next ply of roofing.

Blind Stop. A rectangular wood molding, usually 3/4 by 1-3/8 inches, in a window assembly that receives the window screen frame.

Blind Story. A building story that has no windows.

Blind Vent. An ineffective, sometimes illegal, vent which stops in a wall thus giving the appearance of a vent but not actually functioning as a vent.

Blind. 1. A panel, shade, or screen used on a window to block out light, give protection, add insulation, or as decoration. 2. A shutter.

Blinding. Compacting soil immediately over a tile drain to reduce its tendency to move into the tile.

Blister. 1. A raised spot in a built-up roof caused by expansion of entrapped moisture, water vapor, or other gases, between any of the layers of roofing or mopping. 2. A loose raised spot on the gypsum board face usually due to an air space or void in the core.

Blistering. 1. In ceramic tile, the development, during firing, of enclosed or broken macroscopic vesicles or bubbles in a body or in a glaze or other coating. 2. A bulging of the finish plaster coat as it separates and draws away from the basecoat; the resulting protuberances are often termed turtle backs. 3. Formation of bubbles on surface of paint or varnish film, generally caused by moisture behind the film or excessive heat.

Blisters. 1. Cloudy or milky-looking raised spots on finished surfaces. 2. Protuberances on the finish coat of plaster caused by application over a too damp base coat or troweling too soon.

Block Dampproofing. The act or process of applying a water-resistant material to the surface of a concrete or masonry block to prevent passage or absorption of water or moisture; see Dampproofing.

Block Demolition. The act or process of tearing down an old block wall structure.

Block Grout. Mortar mixes used in block walls to fill voids and joints.

Block Plane. Woodworking hand tool, for final smoothing.

Block Sequence. A combined longitudinal and buildup sequence for a continuous multiple-pass weld wherein separated lengths are completely or partially built up in cross-section before intervening lengths are deposited; see Backstep Sequence.

Block Vent. An opening serving as an outlet or inlet for air in a block structure.

Block, Angle. A square of tile specially made for changing direction of the trim.

Block, Concrete. A hollow concrete masonry unit constructed of a composite material consisting of sand, coarse aggregate, cement, and water.

Block, Glass. A hollow masonry unit made of glass.

Block, Granite. A masonry unit consisting of a very hard natural igneous rock used for its firmness and endurance.

Block, Splash. A small masonry block placed in the ground beneath a downspout to receive roof drainage and prevent standing water or soil erosion.

Block, Terminal. A decorative element forming the end of a block structure.

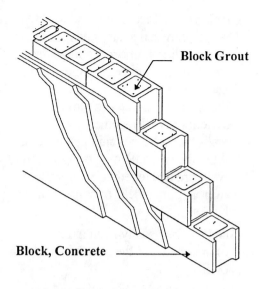

Block Grout

Block, Concrete

Block. 1. A compact, solid piece of substantial material that is worked or altered from its natural state to serve a particular purpose. 2. A wooden or metal case enclosing pulleys and having a hook, eye, or strap by which it may be attached. 3. A building divided into separate functional parts. 4. A part of a building or structure distinctive in some respect.

Blocked Diaphragm. A diaphragm in which all sheathing edges not occurring on framing members are supported on and connected to wood blocking.

Blocking, Wood. Wood blocks used as filler pieces or stabilization between framing members.

Blocking. 1. Pieces of wood inserted tightly between joists, studs, or rafters in a building frame to stabilize the structure, inhibit the passage of fire, provide a nailing surface for finish materials, or retain insulation. 2. A system of tying together two brick walls that were not built at the same time; the two adjoining or intersecting walls are tied together by offset and overhanging blocks of courses of bricks.

Blocks. The usually rectangular areas of land between the streets in a town or city.

Blood Analyzer. A mechanical apparatus which examines blood for medical or scientific purposes.

Blood Refrigerator. A mechanical cooling device or room used to keep stored blood at the proper temperature.

Bloom. 1. A visible exudation or efflorescence on the surface. 2. Whitening; blushing; clouded appearance on varnished surface. 3. Condition of clouding or fogging of paint film, usually caused by reactive materials in paint film coming into contact with dust, oil, deposits from gases in the air or soluble matter in rain.

Blots. Marks or stains on the face of a tile.

Blow Back. Rebound of atomized sprayed material.

Blow. In gypsum board, a large area of paper separated from the core during the manufacturing process; it may appear as a large puffy blister or a full loose sheet of paper.

Blown Oil. A vegetable or fish oil which has been thickened by air blown through it.

Blown-In Insulation. Loose cellulose insulation that is blown into an attic, crawl space, or walls by a blowing machine.

Blowout. A concrete form giving way.

Blowtorch. A gasoline torch used in burning off paint film; should be used only by experienced painters; it is a dangerous fire hazard when used by amateurs.

Blue Lead. A basic sulphate of lead containing small amounts of lead sulphide and carbon that impart a bluish-gray color; used primarily for its rust-preventive value.

Blue Stain. A bluish or greyish discoloration of the sapwood caused by the growth of certain dark-colored fungi on the surface and in the interior of wood.

Blueprint. 1. An obsolete method of copying construction drawings and maps; a wet process that produces a print with white lines on a blue background; a cyanotype or sun print. 2. Loosely, any construction drawing.

Bluestone. A sandstone of a dark-greenish to bluish-gray color that splits into thin slabs, commonly used to pave surfaces for pedestrian traffic.

Blue-Tops. Surveyor's stakes, marked with a blue lumber marking crayon, that should not be disturbed.

Blunder. A mistake or error made through clumsiness, stupidity, ignorance, or carelessness.

Blunging. The wet process of blending or suspending ceramic material in liquid by agitation.

Blushing. 1. A finish is said to blush when it takes on white or grayish cast during drying period; usually caused by the precipitation or separating of a portion of the solid content of the material, causing an opaque appearance. 2. Applied to lacquers when they become flat or opaque and white on drying; usually occurs when applied in a humid atmosphere.

BM. Bench mark.

Board and Batten. Linear vertical wood boards with wood strips covering vertical joints used as an exterior cladding for a framed wall.

Board Cement. An adhesive used to bond pieces of wood together.

Board Foot. A unit of lumber volume, a rectangular solid nominally 12" x 12" x 1".

Board Insulation. Rigid form of plastic foam used as a rigid application insulating surface, commonly of polyurethane.

Board Knife. A hand tool holding a replaceable blade to sharply score or trim gypsumboard products; a popular brand is a Stanley knife.

Board Measure. The standard system for the measurement of lumber; see Board Foot.

Board Rafter. The wooden structural member that makes up the sloping parallel beams used to support a roof.

Board Saw. A short hand saw with very coarse teeth for cutting gypsumboard for door and window frame openings.

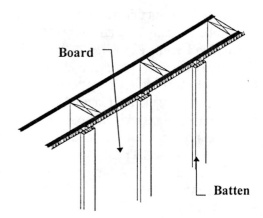

Board

Batten

Board Siding. A type of lumber installed on the exterior walls of a building or structure to act as the finish sheathing.

Board Sub-Flooring. A wooden member that is installed on floor joists to which the finished floor is fastened.

Board, Composite. A board that is made of several compressed materials; used for sheathing, wallboard, or as an insulation or acoustical barrier.

Board, Concrete Finish. Wooden boards placed in the concrete formwork as for liners to provide a wood-pattern finish to the completed reinforced concrete.

Board, Diving. A narrow platform, often equipped with a spring and raised off the ground by a ladder, attached to the edge of a swimming pool and extending over the water, used to give divers altitude and lift.

Board, Dock. Heavy timber used in the construction of the raised platform used for the loading and unloading of trucks.

Board, Ridge. The board against which the tips of rafters are fastened; the top line of a roof; the ridge.

Board. 1. A flat thin piece of sawn lumber, usually long and narrow; lumber less than 2 inches thick. 2. A group of persons having supervisory, investigatory, or advisory powers. 3. A sheet of insulating material carrying circuit elements and terminals so that it can be inserted in an electronic apparatus. 4. A flat panel of compressed fibers. 5. A flat panel used for posting bulletins or as a chalkboard.

Boarding In. The process of nailing boards on the outside studding of a house.

Boatswain's Chair. A trapeze-like seat that is slung from rigging to support a worker; a Bosun's Chair.

BOCA. Building Officials and Code Administrators International, an organization that publishes model building codes.

Bodega. 1. A wine vault or cellar; a wine shop where wine is drawn from barrels. 2. A storeroom or warehouse.

Bodied Linseed Oil. Linseed oil that has been thickened by suitable processing with heat or chemicals; bodied oils vary greatly in viscosity; some are little thicker than raw linseed oil; others are almost jelly like.

Body Coat. Intermediate coat of paint between priming and finishing.

Body Feed. The continuous addition of small amounts of filter aid during the operation of a diatomaceous earth filter.

Body. 1. The structural portion of ceramic articles; also refers to the material or mixture from which the article is made. 2. The structural portion of an article covered with ceramic tile. 3. Thickness, consistency, or viscosity of a fluid. 4. See Steel Square.

Boiled Linseed Oil. Linseed oil to which enough lead, manganese, or cobalt salts have been added to make the oil harden more rapidly when spread in thin coatings; drying properties accentuated by heating oil to 130 to 200° C.

Boiled Oil. Drying oil treated with driers to shorten the drying time.

Boiler Bow Off. An outlet on a boiler to permit emptying or discharging of water or sediment in the boiler.

Boiler Horsepower. Term now seldom used, meaning equivalent to a heating capacity of 33,475 Btu/hr. (983 watts).

Boiler Plate. Standardized, formulaic, or hackneyed language in a contract.

Boiler Room. The space provided for a hot water or steam boiler, circulating pumps, and other mechanical and electrical equipment; engine room.

Boiler, High-Pressure. Boiler furnishing steam at pressures of 15 pounds per square inch gauge or higher (1.05 kg/cm^2).

Boiler, Hot-Water and Low-Pressure Steam. A boiler furnishing hot water at pressures not more than 30 pounds per square inch gauge (2.12 kg/cm^2) or steam at pressures not more than 15 pounds per square inch gauge (1.06 kg/cm^2).

Boiler. 1. A fuel-burning apparatus for producing hot water or steam. 2. A heating system in which water is used as the distribution medium.

Boiling Point. The temperature at which a liquid boils; the point at which a liquid starts to change to gas; boiling temperature of a liquid under a pressure of 14.7 psia (760 mm); water boils at 100° C or 212° F.

Boiling Temperature. Boiling point; temperature at which a fluid changes from a liquid to a gas.

Boiling. Heated to the boiling point; the change of a liquid to a gas.

Bole. 1. The main stem of a tree of substantial diameter; roughly, capable of yielding saw timber, veneer logs, or large pole; seedlings, saplings, and small diameter trees have stems, not boles. 2. A fine soft clay, yellow or dark, colored by iron oxide, formerly used as a pigment.

Bollard Pipe. Short pipe length, placed vertically in the ground and filled with concrete to prevent vehicular access or to protect property from damage by vehicular encroachment.

Bollard. 1. Short steel post, usually filled with concrete, set to prevent vehicular access to or to protect property from damage by vehicular encroachment. 2. Steel or cast iron post to which ships are tied.

Bolster, Slab. Continuous, individual support used to hold steel reinforcing bars in the proper position.

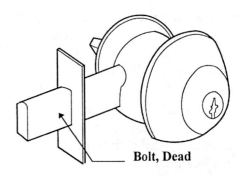

Bolt, Dead

Bolster. 1. A short piece of timber set horizontally across the top of a post, either to afford a greater bearing surface for a girder or girders, or to allow a post above to set between the ends of the girders, or to shorten the span of girders. 2. A long wire type chair used to support steel reinforcing bars in a concrete slab while the concrete is being placed.

Bolt Cutter. A hand tool, utilizing effective leverage, that can shear bolts and steel reinforcing rods.

Bolt, Anchor. Steel bolt used to secure wood construction to concrete or masonry construction.

Bolt, Carriage. A threaded bolt with a round smooth head and a square neck directly under the head to prevent rotation.

Bolt, Dead. A lock bolt having no spring action nor bevel, and which is operated by a key or a turn piece.

Bolt, Door. The tongue of a lock installed to prevent a door opening.

Bolt, Wood. Bolts specifically used in wood applications, that have an unslotted oval head and square shoulders that sink into the wood to prevent turning.

Bolt. 1. A threaded metal rod or pin for joining parts, having a head and usually used with a threaded nut. 2. a sliding bar for locking a door or gate. 3. A bar in a lock, moved by a key.

Bolted Steel. Steel structural system where the members are assembled / and connected with bolts, as opposed to welding.

Bolted Truss. Beams, frames, trusses, or other supports connected to support a roof, bridge, or floor system, that are fastened together with bolts.

Bolting Pattern. The arrangement, spacings, and dimensions of bolts used to attach two or more structural members together.

Bond Beam Block. A concrete masonry unit with the upper part of the ends and webs removed to make room for horizontal reinforcing bars and grout; . U-blocks are sometimes used to form bond beams, especially as over openings.

Bond Beam. A horizontal grouted element within a masonry wall in which steel reinforcement is embedded; a horizontal reinforced masonry beam, serving as an integral part of the wall.

Bond Breaker. A material used to prevent adhesion of newly placed concrete to other surfaces.

Bond Coat. A material used between the back of the tile and the prepared surface; suitable bond coats include pure portland cement. dry-set portland cement mortar, latex-type portland cement mortar, organic adhesive, and the like.

Bond Course. The course consisting of units which overlap more than one wythe of masonry.

Bond Plaster. A specially formulated gypsum plaster designed as a first coat application over monolithic concrete.

Bond. 1. The solid connection of one material to another; a substance which causes such a joining to take place. 2. The adhesion between masonry units and mortar or grout. 3. The patterns and methods in which brick and block are installed, for example, American bond, basket weave, Dutch cross bond, Flemish bond, running bond, and stack bond. 4. The adhesion between the surface of a reinforcing bar and the adjacent concrete, mortar, or grout. 5. The adhesion of cement paste to aggregate. 6. The degree of firmness with which the paper adheres to the gypsum board core. 7. The junction of the weld metal and the base metal. 8. The adherence of the bitumens between two layers of roofing felts. 9. See Chemical Bond. 10. See Mechanical Bond. 11. See Surety Bond. 12. See Completion Bond. 13. See Performance Bond. 14. See Payment Bond. 15. See Roof Bond. 16. See Bid Bond.

Bondability. 1. Indicating ease or difficulty in bonding a material with adhesive. 2. Ease or difficulty of a contractor in obtaining a surety bond.

Bonded Rubber Cushioning. Rubber or latex cushioning adhered to the carpet at the mill.

Bonded Stop Notice. A stop notice that is accompanied by a surety bond which guarantees any costs assessed against the claimant should the claimant lose its lawsuit; see Stop Notice.

Bonded Tendon. A prestressed tendon that is bonded to concrete either directly or through grouting.

Bonded Wall. A masonry wall in which two or more wythes are bonded to act as a structural unit; see Bonder.

Bonder. A bonding unit; see Header, 7.

Bonderized Flagpole. A metal flagpole that has been coated with a patented phosphate solution for protection against the elements.

Bonderizing. A five-step proprietary custom process for phosphatizing.

Bonding Agent. A substance applied to a suitable substrate to create a bond between it and a succeeding layer as between a subsurface and a terrazzo topping or a succeeding plaster application.

Bonding Jumper. 1. A reliable conductor to assure the required electrical conductivity between metal parts required to be electrically connected. 2. At the main service, the connection between the grounded circuit conductor and the equipment grounding conductor. 3. At equipment, the connection between two or more portions of the equipment grounding conductor. 4. In a circuit, the connection between portions of a conductor in a circuit to maintain required ampacity of the circuit.

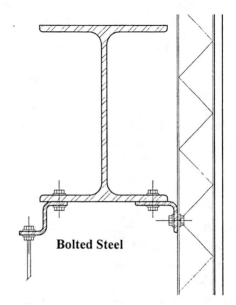

Bolted Steel

Bonding. Adhesion.

Bone Ash. Calcined bone consisting essentially of calcium phosphate.

Bone Black. Pigment made from calcined animal bones; dark in color, but does not have a strong tinting strength like lampblack.

Bone China. A translucent china made from a ceramic whiteware body composition containing a minimum of 25 percent bone ash.

Bonnet. The upper portion of the gate valve body into which the disc of a gate valve rises when it is opened.

Bonus. An extra payment to a contractor for achieving some specified goal, such as early completion.

Book Matched. Wood veneer where adjoining sheets are sliced from the same log so that the grain matches like an open book.

Bookkeeper. A person who records the accounts or transactions of a business.

Boom. 1. A long beam extending out from an upright to lift or carry something and guide it as needed; a derrick boom. 2. A barrier across a river or around an area of water to prevent floating logs from dispersing or to contain an oil spill. 3. To undergo swift, vigorous growth or development; flourish. 4. A period of business prosperity, or industrial expansion. 5. A sudden favorable turn in business prospects.

Booster. Common term applied to the use of a compressor when used as the first stage in the cascade refrigerating system.

Booth, Spray. An area in a building or structure used for spray painting; blocked off by walls to prevent dust and dirt from work surface.

Booth. 1. A small temporary roofed or unroofed structure used as a market stall. 2. A small enclosure used for various purposes, as for telephoning or voting. 3. A set of a table and benches in a restaurant or bar.

Border. The edge or boundary of anything; verge.

Bore. 1. To drill a hole. 2. The diameter of a tube. 3. Inside diameter of a motor cylinder. 4. A high tidal wave rushing up a narrow estuary.

Bored Lock. A door lock manufactured for installation in a circular hole.

Boring. 1. Making holes in wood or metal to aid in the insertion of bolts, nails or other fasteners. 2. Drilling into the ground to bring up samples of earth for testing. 3. Rotary drilling.

Borrow Pit. An area designated as the source of earth removal to be used elsewhere for fill.

Borrow. Excavated material that has been taken from one area to be used as fill at another location.

Borrowed Light Opening. A glazed window unit in an interior partition.

Boss. 1. A round knob, stud, or other protuberance on the center of a shield or in ornamental work. 2. A piece of ornamental carving covering the point where the ribs in a vault or ceiling cross. 3. The employer or contractor on a construction job.

Boston Ridge. Applying asphalt or wood shingles at the ridge or at the hips of a roof as a finish.

Bosun's Chair. See Boatswain's Chair.

Bottle Cooler. A container used for cooling or maintaining the coolness of bottled liquids.

Bottom Bars. The reinforcing bars that lie close to the bottom of a reinforced concrete beam or slab.

Bottom Beam. The lowest horizontal member supporting a building or structure.

Bottom Chord. The bottom member in a truss.

Bottom Dip. The lowest water or waste point in a trap.

Bottom Line. 1. The last line on a financial report which shows the profit or loss. 2. The most important factor or statement.

Bottom Pivot Hinge. A flexible pair of plates joined by a pin to allow swinging of a door or gate installed at the bottom.

Bottom Plate. In wood stud framing construction, the bottom continuous horizontal member that supports the studs.

Boulder. A large stone worn smooth by erosion; a rock which is too heavy to be lifted readily by hand.

Bounce-Back. Spray rebound similar to blow-back.

Boundaries. A separating line that indicates or fixes a limit or extent.

Boundary Elements. Portions along wall and diaphragm edges strengthened by longitudinal and transverse reinforcement; boundary elements do not necessarily require an increase in the thickness of the wall or diaphragm.

Boundary Survey. A survey of the property lines of a piece of land.

Bourdon Tube. Thin-walled tube of elastic metal flattened and bent into circular shape which tends to straighten as pressure inside is increased; used in pressure gauges.

Boutique. A small fashionable store or a small specialty shop within a larger store.

Bow Trowel. A finishing trowel with a slight curve for crowning the final application of gypsum board joint treatment.

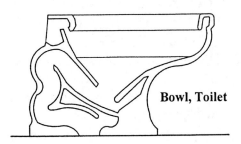

Bowl, Toilet

Bow Window. A bay window with a curved front.

Bow. The distortion of lumber in which there is a deviation, in a direction perpendicular to the flat face, from a straight line from end-to-end of the piece.

Bowl, Toilet. The oval part of a toilet which receives the waste and fills with water after flushing the toilet tank.

Box Beam. A beam of metal, concrete, or plywood which, in cross section, resembles a closed rectangular box.

Box Culvert Formwork. The temporary wooden structure which holds wet concrete in place for the final curing of a rectangular-shaped, reinforced concrete drainage system.

Box Culvert Reinforcing. Iron or steel rods that are embedded in the wet concrete of a rectangular-shaped drainage system to give additional strength.

Box Culvert. A concrete drainage structure rectangular shaped, reinforced and cast in place or made of precast sections.

Box Cutter. A specially designed hand tool for shear-cutting electrical outlet holes in gypsum board.

Box System. A framing system that is made up of bearing walls with lateral forces being resisted by shear walls and diaphragms.

Box Wrench. A type of end wrench in which the gripping end surrounds the nut or bolt head.

Box, Distribution. A box which contains the circuit breakers, connects to the service wires, and delivers current to the various outlets throughout a building or structure.

Box, Floor. A metal electrical rough-in box fed by conduits in or under the floor to provide for a floor outlet.

Box, Gang. Electrical rough-in box constructed of metal or hard plastic, to provide for two or more outlets or switches.

Box, Junction. A metal or hard plastic electrical rough-in box usually square or octagonal housing only wire or cable connections.

Box, Plastic. An electrical box for the joining of electrical wires, constructed of hard plastic and nailed in place.

Box, Pull. An electrical rough-in box placed in a length of conduit, through which cables can be pulled.

Box, Screw Cover. A removable ornamental or protective plate that is mounted to an electrical rough-in box.

Box, Tap. The electrical box where the public service electrical supply line is connected with a branch to serve a particular building or structure.

Box, Terminal. A metal electrical box, usually with a removable cover, that contains leads from electrical equipment ready for connection to a power source.

Box, Weatherproof. An electrical box, designed for exterior installation, that is impervious to the outside elements, such as water.

Boxing. Mixing by pouring back and forth from one container to another.

Boyle's Law. A physical law governing the behavior of gases, stating that the volume of a gas is directly proportional to its temperature and inversely proportional to its pressure.

Brace. 1. A cranked hand tool to hold a bit for drilling holes. 2. A diagonal member, either in tension or compression, to strengthen a structure.

Braced Frame. A truss system or its equivalent which resists lateral forces.

Braced Frame. One which is dependent upon diagonal braces for stability and capability to resist lateral forces.

Braced Wall Line. A series of braced wall panels in a single story.

Braces. Pieces fitted and firmly fastened to two others at any angle in order to strengthen the structure.

Bracing. 1. Diagonal members, either temporary or permanent, installed to stabilize a structure against lateral loads. 2. Structural member used to prevent buckling or rotation of wood studs.

Bracket Hanger. Hanger supporting a wall-hung sink.

Bracket, Wall. 1. A wall-mounted support for shelving or other object. 2. A wall-mounted lighting fixture.

Bracket. 1. A projecting support for a shelf or other structure. 2. In furring and lathing, a superficial structure usually in angles forming a frame to support lath; used to save material and weight in ornaments or cornices.

Brad. A small slender wire nail with a thickened top for a head.

Bradawl. A small handtool used by woodworkers for making small holes for brads or screws.

Brake Metal. Sheet metal that has been bent into a specified configuration, such as gravel stop, flashing, L-shapes, and Z-shapes; formed on a sheet metal brake.

Brake. A machine for flanging, bending, or folding sheet metal.

Branch Breaker. A switch which stops the flow of current by opening the circuit automatically when more electricity flows through the circuit than the circuit is capable of carrying; resetting may be either automatic or manual.

Branch Circuit Appliance. Circuits supplying energy either to permanently wired appliances or to attachment plug receptacle, that is appliance or convenience outlets, or to a combination of permanently wired appliances and additional attachment plug outlets on the same circuit; such circuits to have no permanently connected lighting fixtures not a part of an appliance.

Branch Circuit (General Purpose). A branch circuit that supplies a number of outlets for both lighting and appliances.

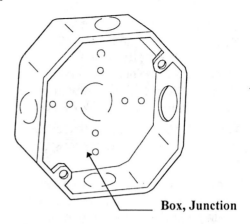
Box, Junction

Branch Circuit (Lighting). Circuits supplying energy to lighting outlets only.

Branch Circuit (Motor). Circuits from the motor branch circuit protective device to the motor, including the controller and overload protective device.

Branch Circuit (Multiwire). A multiwire branch circuit is a circuit consisting of two or more ungrounded conductors having a potential difference between them, and an identified grounded conductor having equal potential difference between it and each ungrounded conductor of the circuit and which is connected to the neutral conductor of the system.

Branch Circuit. 1. A circuit supplying several localized electrical outlets from a single breaker. 2. That portion of a wiring system extending beyond the final automatic overcurrent protective device, excluding any thermal cutout or motor running overload protective device that is not approved for short circuit protection.

B

Branch Intervel. This is the vertical distance, generally a floor or story in height, but never less than 8 feet, within which the horizontal branches from one floor of the building are connected to the main DWV stack.

Branch Vent. A vent pipe that connects a branch of the drainage system to the main stack.

Branch. A member or part of a system or structure which diverges from the main portion, as in heating, ventilation, or electrical installations; a smaller or subordinate duct, pipe, or circuit extending from the main line.

Brass Fitting. Threaded pipe connector made of brass, used to join two pieces of pipe together.

Brass. A metal alloy consisting essentially of copper and zinc in variable proportions.

Braze Welding. Sometimes known as Bronze Welding. Often carried out as in fusion welding except that the base metal is not melted. The base metal is simply brought up to what is known as a tinning temperature (dull red color) and a bead deposited over the seam with a bronze filler rod.

Braze. Solder with an alloy of brass and zinc at a high temperature.

Brazed Connection. Parts that are hardened and connected by soldering with an alloy.

Brazing. A welding process wherein coalescence is produced by heating to suitable temperatures above 800° F. and by using a nonferrous filler metal having a melting point below that of the base metals; the filler metal is distributed between the closely fitting surfaces of the joint by capillary attraction.

Breach of Contract. A material failure to perform an act required by contract.

Breadboard. A draw board in a kitchen cabinet used for kneading and slicing bread.

Break Joints. To arrange joints so that they do not come directly under or over the joints of adjoining pieces, as in shingling, siding, and brick laying.

Break. 1. To damage by separating into pieces under a blow or strain; shatter. 2. An interruption as in a circuit. 3. An interruption in the continuity of an element, such as a plastered wall or cornice.

Breakdown. An itemized list of building costs.

Breaker Strip. Strip of wood or plastic used to cover joint between outside case and inside liner of refrigerator.

Breaker, Circuit. A switch which stops the flow of current by opening the circuit automatically when more electricity flows through the circuit than the circuit is capable of carrying; resetting may be either automatic or manual.

Breaker, Main. A switch in a main electrical service panel where the service wires attach.

Breaker, Vacuum. An electrical breaker with a space that contains reduced air pressure.

Break-Even Point. The financial position wherein total revenue received equals the sum of the costs and expenses for a particular project showing neither profit nor loss.

Breaking Joints. The laying of bricks so that no two vertical joints come directly over one another; this is done for strength.

Breakpoint. The point at which a rising concentration of chlorine in swimming pools kills germs and bacteria by oxidizing organic matter; once all matter is oxidized, the amount of chlorine remaining is free or uncombined.

Breast Drill. A portable drill equipped with a plate that the operator leans against to provide pressure.

Breast. 1. The front part of a fireplace above and around the firebox opening. 2. The projecting portion of a chimney, especially when projecting into a room.

Breastplate. The pressure plate of a breast drill.

Breastsummer. A beam to carry the load above a fireplace opening, also called bressumer.

Breathing Zone. Area of a room in which occupants breathe as they stand, sit, or lie down.

Breeching. Space in hot water or steam boilers between the end of the tubing and the jacket.

Bressumer. See Breastsummer.

BRI. Building-Related Illness.

Brick Anchor. Fasteners that are designed to attach and secure a brick veneer to a concrete or brick wall.

Brick and Brick. The laying of bricks by which the bricks are touching each other and the mortar used is just enough to fill the irregularities of the bricks.

Brick Bat. Part of a brick, usually half a brick or less.

Brick Bond. The pattern or arrangement of bricks in a wall.

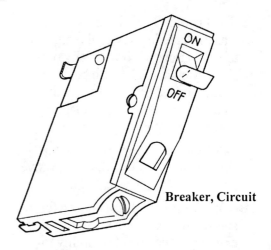

Breaker, Circuit

Brick Firewall. A masonry wall constructed to prevent or retard the spread of fire.

Brick Institute of America (BIA). 11490 Commerce Park Drive #300, Reston, Virginia 22901, (703) 620-0010.

Brick Manhole. A vertical access shaft from the surface to an underground area, constructed of bricks.

Brick Mason. A mason who builds in brick; also called a brick layer; see Mason.

Brick Molding. Milled trim piece designed to fill the gap between finished milled parts like door and window frames and irregular surfaces like masonry.

Brick Removal. The act or process of the demolition of a brick structure.

Brick Tongs. An iron grasping device consisting of two bars pivoted in the middle like a scissors and allowing two workers to lift and move a small pile of bricks.

Brick Trowel. The brick trowel is larger than the buttering trowel; the primary tool of masons; used when any preparatory brick work has to be done; its greater surface and weight are advantageous in the buttering and tapping in of block, brick, or larger tiles.

Brick Veneer. A one brick thick outside facing of brickwork used to cover a wall of some other material.

Brick, Chimney. Brick, chosen for the specific use in the construction of chimneys, because of its ability to withstand high temperatures without cracking.

Brick, Fire. Brick that has been tested for fire-resistance and then graded for specific construction uses; brick whose composition and characteristics make it suitable to use for masonry fireplace fire-boxes and fire chambers.

Brick, Masonry. Bricks that are shaped and molded in different sizes and shapes.

Brick, Paver. Brick units that are used in foot traffic areas; usually four inches wide, eight inches long, and 1-5/8 to 2-1/4 inches thick.

Brick. A solid masonry unit having the shape of a rectangular prism; usually made from clay, shale, fire clay, or a mixture of these.

Bricklayer. A brick mason.

Brickwork. Masonry of bricks and mortar.

Bridge Crane. A hoisting device spanning two overhead rails; the hoisting device moves laterally along the bridge with the bridge moving longitudinally along the rails.

Bridge Deck. The slab or other structure forming the travel surface of a bridge.

Bridge Glaze. Colorless or colored ceramic glaze having high gloss.

Bridge. A straightedge used as a starting line for the laying of tile; it can be blocked up to support tile over an opening.

Bridging Architect. An owner's architect who designs the project and then is replaced by the design/build entity's architect who prepares the construction documents.

Bridging. 1. Diagonal or longitudinal members used to keep horizontal members properly spaced, in lateral position, vertically plumb, and to distribute load; pieces fitted in pairs from the bottom of one floor joist to the top of adjacent joists, and crossed to distribute the floor load; sometimes pieces of width equal to the joists and fitted neatly between them. 2. In painting, forming a skin over a depression.

Briding. A section sized to fit inside the flanges of studs and channels to stiffen construction.

Bright Blast. White blast.

Bright Glaze. A high-gloss coating with or without color.

Brilliant Color. Very bright.

Brine. Water saturated or strongly impregnated with salt.

Brinell Hardness Test. A laboratory test for measuring the hardness of a material by hydraulically pressing a steel ball into the surface.

British Thermal Unit (BTU). Quantity of heat required to raise temperature of one pound of water one degree Fahrenheit.

Brittle Failure. Failure in material which generally has a very limited plastic range; material subject to sudden failure without warning.

Brittle. Easily broken; not tough.

Broach. 1. Any of various pointed or tapered tools, implements, or parts. 2. A drill bit. 3. A pointed tool for roughly dressing stone.

Broad Knife. A wide flexible finishing knife for applying joint finishing compound.

Broadcast. To sprinkle solid particles on a surface.

Broad-Leaved Trees. See Hardwoods.

Broadloom. Carpet woven on a broad loom in widths of 6 feet or more. usually 6, 9, 12, 15 and 18-ft. widths, and up to 30 ft. in Chenille; broadloom is not a type of weave of carpet, nor a pattern or color; it is simply a designation of width.

Broiler, Kitchen. A cooking device in a kitchen that cooks food by direct exposure to radiant heat.

Broker. An agent who, for a fee or commission, brings parties together for the leasing or sale of real property.

Bromide. A chemical compound containing bromine, a halogen; sodium or potassium bromide in solution will produce free bromine if chlorine is added to the pool.

Bronze Tools. Non-sparking tools; used when fire hazards are particularly acute.

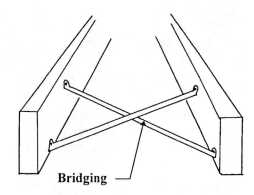

Bridging

Bronze Welding. See Braze Welding.

Bronze. An alloy of copper and tin and sometimes other elements.

Bronzing Liquid. A vehicle especially formulated for use as a binder for aluminum, gold, or bronze powder.

Bronzing. Formation of metallic sheen on a paint film.

Broom Finish. A finish applied to an uncured concrete surface, to provide skid or slip resistance, made by dragging a broom across the freshly placed concrete surface.

Broom. 1. A long handled brush of bristles for sweeping and brushing; also used for concrete finishing. 2. To spread out in a broom shape through separation of the fibers, as when a pile is partly crushed at its head under the blows of the pile driver.

Brooming. In roofing, embedding a ply by using a broom to smooth out the ply and ensure contact with the adhesive under the ply.

Brown Coat. The coat of plaster directly beneath the finish coat; in two-coat work, brown coat refers to the basecoat plaster applied over the lath; in three coat work, the brown coat refers to the second coat applied over a scratch coat; brown coats are applied with a fairly rough surface to receive the finish coat.

Brown Out. To complete application of basecoat plastering.

Brown Stain. Brown discoloration of the sapwood of some pines caused by a fungus.

Brownfield Site. A building site that has been previously built upon and is possibly polluted with toxic chemicals; compare with Greenfield Site.

Brownstone. 1. A building sandstone of prevailing brown color. 2. A dwelling faced with brownstone.

Brush Cutting. The act of removing unwanted plants to clear an area.

Brush Hand. A painter whose ability lies in his skill in applying material.

Brush Hook. A long handled tool for clearing brush.

Brush. 1. A painter's implement composed of bristles set into a handle, used for applying paint and other coatings to a surface. 2. An electrical conductor that makes contact with a moving part.

Brushability. 1. Adaptability of paint to application with a brush. 2. The ability or ease with which a paint can be brushed.

Brushcoating. The application of paint, stucco color, or other materials with a broad brush.

Brushed Surface. A sandy texture obtained by brushing the surface of freshly placed or slightly hardened concrete with a stiff brush for architectural effect or, in pavements, to increase skid resistance.

Brush-Off Blast. Lowest blast cleaning standard.

Brussels. A term formerly, but now rarely, used to describe a loop pile or round-wire carpet woven on the Wilton loom.

Brutalism. A heavy plain style of architecture.

BSSC. Building Seismic Safety Council.

BTU. British thermal unit; measurement of the heat energy required to raise one pound of water one degree Fahrenheit.

Bubble. A large void in the core of gypsum board caused by the entrapment of air while the core is in a fluid state during the manufacturing process.

Bubbling. Describes the appearance of bubbles on the surface while a coating is being applied.

Buck. Assembly of the framing that constitutes a rough door or window opening.

Bucket Trap. A mechanical steam trap operating on buoyancy that prevents the passage of steam through the mechanical system it protects.

Bucket. Large metal container into which concrete is discharged; the bucket is raised by crane to the placement area.

Buckle. To bend under compression; with very thin members, the bucking may be elastic, and the member will spring back if the load is removed; if the load is continued or if the buckling occurs with the stresses above the yield point, the member will fail by collapsing completely.

Buckles. In old plastering, raised or ruptured spots which eventually crack, exposing the lath beneath; most common cause for buckling is application of plaster over dry, broken or incorrectly applied wood lath.

Buckling. 1. Collapse, in the form of a sudden sideways deflection, of a slender element subjected to compression; structural failure by gross lateral deflection of a slender element under compressive stress, such as the sideward buckling of a long, slender column or the upper edge of a long, thin floor joist. 2. Wrinkling or ridging of the carpet after installation, caused by insufficient stretching, dimensional instability, or manufacturing defects.

Budget. A managerial plan of proposed operations to accomplish a financial objective.

Buffer. A pile of blasted rock left against or near a face to improve fragmentation and reduce scattering from the next blast.

Buffing Compound. Soft abrasive in stick form, bonded with wax.

Buffing. A router trimming of the shear cut end of gypsum board to smooth cut and adjust for length tolerances prior to the bundling tapes being applied.

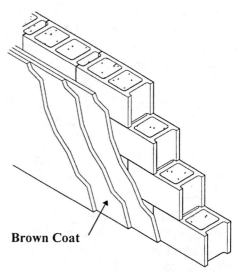

Brown Coat

Buglehead Screw. A special design screw that will seat beneath the plane of the gypsumboard surface without tearing the paper.

Build. Construct by putting materials or parts together.

Builder. One who constructs; a building contractor.

Builders Hardware Manufacturers Association (BHMA). 355 Lexington Ave., 17th Floor, New York, New York 10017, (212) 661-4261.

Builders' Hardware. Hardware used in construction, such as bolts, nuts, screws, nails, and other fastenings, metal and plastic parts, hinges, butts, catches, and similar parts.

Builder's Level. A simple form of transit-level for measuring and setting levels on a construction site; also called a contractor's level.

Building Brick. Brick for building purposes not especially treated for texture or color; also called Common Brick.

Building Code. A set of rules governing the quality of construction in a community; the purpose of these rules is to protect the public health and safety.

Building Coverage. The percentage of a specific parcel of land covered by buildings.

Building Demolition. The destruction, by means of explosives or otherwise, of a standing construction.

Building Drain. The lowest part of a drainage system which receives the discharge from soil, waste and other drainage pipes inside the walls of the building and conveys it to the building sewer beginning 3 feet outside the building wall.

Building Drainage System Complete system of piping used for carrying away waste water and sewage; also called House Drainage System.

Building Envelope. The outer limits of a building as encompassed by the exterior walls, roof, and foundation.

Building Excavation. The act or process of removing an area of earth and rock to make room for a foundation.

Building Inspector. A public official that examines the work in the field to determine compliance with the applicable building laws.

Building Integrated Photovoltaic Cell. Photovoltaic cells that substitute for building elements or materials, such as spandrel panels or roofing panels.

Building Lime. A lime whose chemical and physical characteristics and method of processing make it suitable for ordinary or special construction uses; also called construction lime.

Building Main. Water supply pipe that carries the water from the source of supply to the first branch of the water distributing system in the building.

Building Official. The official charged with administration and enforcement of the applicable building code, or his duly authorized representative.

Building Paper. Any of various usually water repellent sheet materials such as kraft paper assemblies and asphalt saturated felt, used for moisture proofing, draft stopping, and exterior plaster backing; a general term for papers, felts, and similar sheet materials used in buildings, without reference to their properties or uses.

Building Related Illness. Diagnosable illness whose symptoms can be identified and whose cause can be directly attributed to airborne building pollutants, for example, Legionnaire's disease, and hypersensitivity pneumonitis.

Building Residual. A real estate appraisal technique where a reasonable return on the land is first deducted from the income, the balance being attributable to the improvements.

Building Seismic Safety Council (BSSC). 1015 15th Street, NW #700, Washington, DC 20005, (202) 289-7800.

Building Sewer. The piping that takes the soil and waste water from the building drain and conveys it to the public sewer or private sewage disposal system; also called House Sewer.

Building Sub-Drain. That part of the drainage system of a house that receives the discharge from the fixtures and cannot drain into the building sewer because it is located below the building sewer or building drain; this piping usually drains into a sump and is pumped up into the house sewer or building drain.

Building Trap. A trap placed in the building drain to prevent entry of sewer gases from the sewer main.

Building, Metal. A building or structure constructed of a structural steel frame covered by metal roof and wall panels; commonly prefabricated in a factory and assembled at the site.

Building. A fixed structure forming an enclosure and providing shelter.

Build-Up Sequence. The order in which the weld beads of a multiple-pass weld are deposited with respect to the cross-section of the joint.

Built-In Oven. An oven installed directly into a wall or cabinet.

Built-In Shelving. Shelving that is permanently installed in a cabinet frame.

Built-Up Member. A single structural component made from several pieces fastened together.

Built-Up Plaque. Layers of a localized abnormal patch on a surface.

Built-Up Roof Removal. The act or process of removing many layers of old roofing material.

Built-Up Roof. A roof covering made of continuous rolls or sheets of saturated or coated felt, cemented together with bitumen, and may have a final coating of gravel or slag.

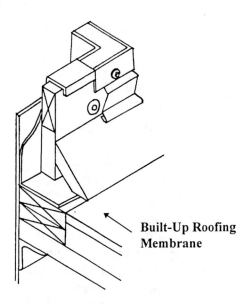

Built-Up Roofing Membrane

Built-Up Roofing Membrane (BUR). Roofing material applied in sealed waterproof layers or plies, or saturated or coated felts alternated with layers of bitumen, surfaced with mineral aggregate or asphaltic materials.

Built-Up Steel Lintel. Lintel fabricated of two or more pieces of structural steel secured together to act as one member.

Built-Up Timber. A timber made of several pieces fastened together and forming one of larger dimension.

Bulb Tee. Rolled steel in the form of a T with a formed bulb on the edge of the web.

Bulb, Sensitive. Part of sealed fluid device which reacts to temperature; used to measure temperature or to control a mechanism.

Bulge. An irregular swelling; a lump.

Bulk Density. The weight of a material per unit of volume.

Bulk Excavation. The digging out of large amounts of dirt and debris.

Bulk Mail Slot. An opening in a wall or door big enough to receive large pieces of mail.

Bulk. 1. Size; magnitude; large mass or quantity. 2. Large quantities as in bulk cement. 3. To cause to swell or bulge. 4. A small structure projecting from a building, as a booth or stall.

Bulked Continuous Filament (BCF). Continuous strands of synthetic fiber made into yarn without spinning; often extruded in modified cross section such as multi-lobal, mushroom or bean shape and/or texturized to increase bulk and covering power.

Bulkhead Formwork. The temporary formwork that blocks fresh concrete from a section of forms or closes the end of a form at a construction joint.

Bulkhead. 1. A vertical partition separating compartments, as on a ship or aircraft. 2. A structure or partition to resist pressure or shut off water, fire, or gas. 3. A retaining wall along a waterfront. 4. A projecting structure with a sloping door giving access to a cellar stairway or shaft.

Bulking Curve. Graph of change in volume of a quantity of sand due to change in moisture content.

Bulking Factor. Ratio of the volume of moist sand to the volume of the sand when dry.

Bulking Value. Of a pigment, its ability to add volume to a paint.

Bulking. Increase in the bulk volume of a quantity of sand in a moist condition over the volume of the same quantity dry.

Bull Float. A tool comprising a large, flat, rectangular piece of wood, aluminum, or magnesium usually 8 in. (20 cm) wide and 42 to 60 in. (100 to 150 cm) long, and a handle 4 to 16 ft. (1 to 5 m) in length used to smooth unformed surfaces of freshly placed concrete.

Bull Nose Brick. 1. A brick having one rounded corner. 2. Bricks that have their ends or corners rounded off.

Bull Nose. 1. In plastering, an external angle that is rounded in order to eliminate a sharp corner; can be made by running with plaster or using a bull nose corner bead with the proper radius. 2. Any material with a rounded edge such as a concrete block, ceramic tile, or brick. 3. See Step Return.

Bulldozer. A tractor driven machine with a horizontal blade for clearing land, road building, or similar work.

Bulletin Board. A wall hanging unit upon which information or messages are attached; a thin board, often of cork, hung on a wall for the attachment of public announcements.

Bulletproof Glass. A protective laminated glass sheet, heat and pressure-bonded with resin sheets to resist the passage of bullets.

Bulletproof Partition. A dividing wall which has been rendered bulletproof by using specific materials in its construction.

Bulletproof. Manufactured to prevent breakage or penetration from a strong external force.

Bullnose Block. A concrete masonry unit which has one or more rounded external corners.

Bullnose, Corner. A type of ceramic tile bullnose trim with a convex radius on two adjacent edges.

Bullnose, Glazed. A ceramic trim tile with a convex radius on one edge that has been given a glassy or glossy surface.

Bull's Eye. 1. A small circular window or opening; a bull's eye window. 2. A small thick disc of glass inserted, as in a deck, to let in light.

Bumper, Dock. Thick rubber units placed under loading dock openings to absorb the shock and prevent damage when trucks back in for loading or unloading.

Bumper, Door. Rubber tip devices mounted on walls or baseboards that prevent door knobs from marring walls.

Bund Wall. A wall built around a tank to contain its contents should the tank or its piping spring a leak.

Bundle. Two pieces of gypsumboard packaged face to face.

Bundled Bars. A group of not more than four parallel steel reinforcing bars in contact with each other, usually tied together.

Bundling Tape. End tape to secure two pieces of gypsum board into a bundle.

Bunker. Space where ice or cooling element is placed in commercial installations.

Buoyant Uplift. The force of water or liquefied soil that tends to raise a building foundation out of the ground.

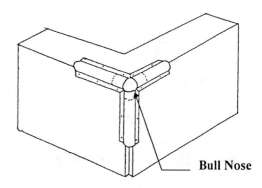

Bull Nose

BUR. Built-Up Roofing.

Burden. The amount of money that has to be added to cover overhead.

Burglar Alarm. A security system that signals when any of the contacts have been interrupted.

Burglar Bars. A grille of steel bars to protect a window or skylight opening from intrusion.

Burl. A hard woody outgrowth on a tree, good for highly figured veneers.

Burlap Rub. A finish obtained by rubbing burlap to remove surface irregularities from concrete.

Burlap, Concrete. A curing concrete surface that has had a coarse fabric of jute, hemp, or less commonly, flax applied, for use as a water-retaining covering.

Burlap. A coarse fabric of jute, hemp, or less commonly, flax, for use as a water-retaining covering in curing concrete surfaces; also called Hessian.

Burling. In carpet manufacturing, a hand-tailoring operation after weaving, to remove any knots and loose ends, to insert missing tufts of surface yarn and otherwise check the condition of the fabric; also, a repair operation on worn or damaged carpet.

Burn. To cut metal with a gas flame.

Burned Finish. Wood finish in which hard portion of grain stands out in relief; produced by using blowtorch and stiff bristled brush.

Burned. Over-dried, partially calcined gypsum board.

Burner. Device in which burning of fuel takes place.

Burning In. Repairing a finish by melting stick shellac into the damaged places by using a heated knife blade or iron.

Burnish. Polish by rubbing.

Burnoff Rate. See Melting Rate.

Burnt Sienna. An earthy substance containing oxides of iron and usually of manganese; orange red or reddish brown pigment, used in paint; sienna that has been roasted.

Burnt Umber. A brown earthy substance containing oxides of iron and manganese; a pigment, darker than ochre and sienna, used in paint; umber that has been roasted.

Burn-Thru. A term erroneously used to denote excessive Melt-Thru or a hole.

Burr. A sharp, roughened, in-turned edge on a piece of pipe which has been cut but not reamed.

Burst Strength. The internal pressure required to break a pipe or fitting; this pressure will vary with the rate of build-up of the pressure and the time during which the pressure is held.

Bus Bar. 1. A large, flat conductor, usually solid copper, used for carrying very high electrical currents. 2. An uninsulated bar or tube used as an electrical conductor at a circuit junction.

Bus Duct Connection. A metal bar serving as a common connection for two or more circuits in a prefabricated unit.

Bus Duct. A prefabricated unit containing one or more electric conductors, often a metal bar, that serves as a common connection for two or more circuits.

Bus Ground. In the main electrical service panel, where the neutral service wire, generally white, attaches and is linked to the earth by the ground wire.

Bush Hammer. In stone dressing, a steel hammer used in finishing the harder stones; it has a square-ended prismatic head divided into a number of pyramidal points.

Bushed Nipple. A pipe threaded at both ends to connect two pipes of different dimensions.

Bushing, Conduit. A threaded metal or plastic pipe connector used to connect conduit to a box or other housing where the hole is not threaded.

Bushing. 1. A removable cylindrical lining for an opening used to limit the size of an opening, resist abrasion, or serve as a guide. 2. An electrically insulating lining for a hole to protect a through conductor. 3. A pipe fitting with both male and female threads used in a fitting to reduce the size; used to connect pipes of different sizes.

Business Entity Concept. The assumption that a business is separate and distinct from its owner's financial operations and holdings.

Busway. A rigid assembly consisting of one or more busbars.

Butane. A gaseous hydrocarbon (C_4H_{10}) of the alkane series used in liquefied form as fuel; also used as a low temperature application refrigerant.

Butler's Pantry. A service room between kitchen and dining room

Butt Hinge. A type of hinge designed for mortising into the edge of the door and into the rabbet of a door frame, consisting of two plates with a removable connecting pin; also called a Butt.

Butt Joint. 1. A plain square joint between two members. 2. In wallpaper, a joint made by trimming both selvedges and butting the edges together; this is used in highest type of work. 3. The cut ends of gypsum board placed adjacent to one another. 4. The joint between two bricks placed end to end in the same course; also called Cross Joint, Head Joint, or Vertical Joint.

Butt Weld. A weld in a butt joint between two members lying approximately in the same plane.

Butt Welded Pipe. Pipe that is joined by welding.

Butt, Pile. See Pile Butt.

Butt. 1. The larger of the two ends of a log. 2. See Butt Hinge. 3. See Pile Butt.

Buttering Trowel

Butterflies. Color imperfections on a lime putty finish wall which smear out under pressure of the trowel; caused by lime lumps not put through a screen and insufficient mixing of the gauging.

Butterfly Reinforcement. Strips of metal reinforcement placed diagonally over the plaster base at the corners of openings before plastering.

Butterfly Roof. A roof shape that is like an inverted gable, the rain gutter being in the middle instead of the ridge.

Butterfly Tie. See Ties.

Butterfly Valve. A valve constructed with a disc that rotates 90 degrees within the valve body.

Buttering Trowel. The blade of the buttering trowel is approximately 4-1/2" wide and 7" long; used in buttering rich mortar to masonry, a method commonly used in the eastern states.

Buttering. 1. Spreading mortar on a masonry unit before it is laid. 2. The spreading of a bond coat to the backs of ceramic tile just before the tile is placed. 3. See Surfacing.

Buttonback Tile. Tile that have round or square projections on the bondable side.

Buttress. 1. A projecting structure of masonry or wood for supporting or giving stability to a wall or building wall and to react against horizontal outward forces. 2. The broadened base of a tree trunk or a thickened vertical part of it.

Butt-Welded Space. A reinforcing bar splice made by welding the butted ends.

Butyl Acetate. A lacquer solvent made from butyl alcohol by reaction with acetic acid.

Butyl Alcohol. An alcohol of higher boiling range than wood alcohol or grain alcohol; obtained from corn by fermentation.

Butyl Caulk. Caulking that is made from various synthetic rubbers derived from butanes.

Butyl Membrane. Pliable thin sheets or layers made from synthetic rubber.

Butylene Plastics. Plastics based on resins made by the polymerization of butene or copolymerization of butene with one or more unsaturated compounds, the butene being in greatest amount by weight.

Buzzer. An electric signaling device that produces a buzzing sound.

BX Cable. A type of indoor wiring consisting of two or more insulated wires protected by a wound, galvanized steel strip cover, the metal winding forming a flexible tube offering protection similar to rigid conduit.

BX Clamp. A clamping device to hold BX cable firmly in place against a wooden or metal member.

Bypass. 1. Passage at one side of, or around, a regular passage. 2. A secondary pipe or bridging of any plumbing fixture allowing this fixture to be disconnected and circulation maintained.

By-Product. A substance obtained during the manufacture of another substance; a secondary result, sometimes unexpected or unintended, of some process.

Byte. In computing, a group of eight binary digits.

C Channel. A C shaped steel or aluminum section shaped like a rectangular box with one side removed.

C Stud. A roll-formed metal channel stud.

C to C. Center to center.

C. 1. Celsius. 2. Centigrade. 3. In heat transfer, conduction of a material to the passage of heat; the reciprocal of resistance (C=1/R). 4. In seismic design, the numerical coefficient used which represents building acceleration. 5. Capacitance.

C Switch (Isolating). A switch intended for isolating an electric circuit from the source of power. It has no interrupting rating and is intended to be operated only after the circuit has been opened by some other means.

C/B Ratio. The ratio of the weight of water absorbed by a masonry unit during immersion in cold water for 24 hours to weight absorbed during immersion in boiling water for 5 hours; an indication of the probable resistance of brick to freezing and thawing; also called saturation coefficient.

Cab, Elevator. The enclosure in an elevator which carries passengers and freight up or down.

Cabana. A shelter near a swimming pool or beach.

Cabin. 1. A small one story dwelling of simple construction. 2. A compartment on a ship or aircraft.

Cabinet Drawer Guide. A wood strip used to guide the drawer as it slides in and out of its opening.

Cabinet Drawer Kicker. Wood cabinet member placed immediately above and generally at the center of a drawer to prevent tilting down when pulled out.

Cabinet Finish. Protective coatings to provide protection and decorative appearance for exposed portions of wood cabinets.

Cabinet Hardware. Metal and plastic fasteners and connectors used to facilitate the operation and movement of doors, drawers, and shelves in cabinets.

Cabinet Heater. A heating element enclosed in a metal housing, with openings for airflow, usually with a fan for controlling air flow.

Cabinet Knob. A handle, pull, or rounded protuberance for opening a cabinet door or drawer.

Cabinet Wall. The wall to which cabinets are attached or mounted.

Cabinet, Base. Floor-mounted cabinet, usually with a counter, sink or appliance installed.

Cabinet, Extinguisher. A case or cupboard having doors which contains a fire extinguishing device.

Cabinet, Kitchen. A case, box, or piece of furniture with sets of drawers or shelves, with doors, primarily used for storage, mounted on walls or floors in a kitchen area.

Cabinet, Laboratory. A case, box, or piece of furniture with sets of drawers or shelves, with doors, primarily used for storage, used in a place or area for scientific studies or commercial and institutional laboratories and testing facilities.

Cabinet, Monitor. A cabinet whose doors have louvered panels to allow for ventilation, light or finish design.

Cabinet, Transparent Finish. Protective clear or tinted coating to provide protection and decorative appearance for exposed portions of wood cabinets which allows grain of wood to be seen through coating.

Cabinet. 1. Case, box, or piece of furniture with sets of drawers or shelves, with doors, primarily used for storage. 2. An electrical switch enclosure designed either for surface or flush mounting, and provided with a frame, mat or trim in which swinging doors are hung; see Panelboard or Switchboard.

Cable Bus. An assembly of insulated cables.

Cable Camera. A camera that is controlled and activated with coaxial cable.

Cable Connector. A device used to connect lengths of cable together into one longer length.

Cable Coupler. A device for connecting two lengths of cable into one longer length.

Cable Fitting. Couplings, elbows, tees or unions used to form a junction or connect cable lines together.

Cable Lug. A connector for fastening the ends of cable to a terminal.

Cable Man-Hole. A vertical access shaft from the surface to the underground, leading to an area for the repair or installation of cable wires.

Cable Receptacle. An interrupting outlet box device installed in an electric supply circuit for the connection of electric cables.

Cable Roof. A circular structure in which the internal stresses of the members are subjected primarily to tensile stresses.

Cable Support. A structure which holds cable lines in place or at a safe overhead height.

Cable Tap Box. A box where public cable service supply lines are connected with a branch to serve a building or structure.

Cable Tray. Open track for support of insulated cables.

Cable TV. A system of distributing television signals to individual subscribers by use of subterranean cables or overhead wires, rather than by aerials.

Cable, Audio. A cable which carries and transmits audio signals.

Cable, BX. A type of indoor wiring consisting of two or more insulated wires protected by a wound, galvanized steel strip cover; the metal winding forms a flexible tube, offering protection similar to rigid conduit.

Cable, Coaxial. A cable consisting of two concentric conductors separated by an insulator; used to transmit telephone, television and computer signals.

Cable, Communication. A cable for transmission of telephone, television, and computer signals.

Cable, Computer. Coaxial cable which transmits computer signals.

Cable, Copper. Insulated, sheathed copper wires conducting power from a source to an electric appliance.

Cable, Fire Alarm. A specific electrical system cable which carries electric current to a warning horn or bell for use in the event of a fire or other catastrophe.

Cable, Guy. A wire used to secure a tall exterior mast, antenna, or other structure in place.

Cable, Thermostat. A specific electrical system cable which operates an automatic device for regulating the temperature in a room, space, or area.

Cable. 1. A thin, flexible line which carries only tensile forces. 2. A bundle of two or more electrical conductors.

CABO. Council of American Building Officials.

Cabriole Leg. A curved furniture leg rounded and swollen at the top and tapered down ending in an ornamental foot.

CAD. Computer Aided Design.

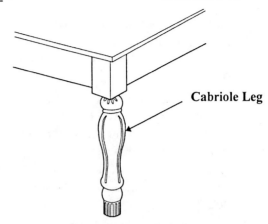

Cabriole Leg

CADD. Computer Aided Design and Drafting.

Cadastral. Pertaining to a register of ownership of lands, boundaries, and buildings for taxation purposes.

Cadmium Lithopone. A series of yellows and reds that are permanent to light and resistant to alkalis.

Cadmium Red. Non-fading red pigment made from cadmium and selenium metals; heat and alkali resistant.

Cadmium Yellow. Pigment prepared by precipitation from acid solution of soluble cadmium salt with hydrogen sulphide gas; fast to alkalis but not to acids.

Cadmium. A bluish white malleable ductile toxic metallic element used in protective platings and in bearing metals.

Cafe. A small coffee shop or simple restaurant.

Cafeteria. A restaurant where the diners collect their food on a tray and usually pay before eating.

Cage Ladder. A, usually wall-mounted, ladder that has, for safety, a surrounding structure to prevent the climber from falling off.

Caging. Metal furring used to enclose pipes, columns, beams or other configurations to be concealed by gypsum board.

Caisson Disease. A sometimes fatal disorder that afflicts workers in a compressed air atmosphere who return to normal air pressure too quickly; decompression sickness; also called The Bends, Air Embolism, or Aerembolism.

Caisson. 1. A type of drilled or augured piling. 2. A cylindrical, site-cast concrete foundation that penetrates through unsatisfactory soil to rest upon an underlying stratum of rock or satisfactory soil. 3. A foundation pier, either circular or rectilinear in plan, usually sunk to rock either by means of gravity, compressed air or by the open-well method. 4. A panel sunk below the normal surface in flat or vaulted ceilings.

Caking. Hard settling of pigment from paint.

Cal/OSHA. California Occupational Safety and Health Act.

Calcareous. Containing calcium or calcium carbonate; chalky.

Calcification. Buildup of calcium carbonate on swimming pool walls and equipment; caused by precipitation of calcium from hard water.

Calcimine. A white or tinted wash consisting of glue, whiting or zinc white, and water, used primarily on plastered surfaces; a type of tempera; sometimes written Kalsomine.

Calcine. 1. To alter composition or physical state by heating. 2. To drive off or lose chemically combined water by action of heat thereby altering the chemical and physical characteristics of a material. 3. To release part or all of the water of crystallization from gypsum by the application of heat. 4. A ceramic mineral or mixture fired to less than fusion for use as a constituent in a ceramic composition.

Calcined Board. Gypsum board which has been subjected to excessive heat.

Calcined Gypsum. Gypsum that has been partially dehydrated by heat.

Calcined. Heated to high temperature in absence of air.

Calcium Aluminate Cement. The product obtained by pulverizing clinker consisting essentially of hydraulic calcium aluminates resulting from fusing or sintering a suitably proportioned mixture of aluminous and calcareous materials.

Calcium Carbonate. Earth product obtained from deposits of chalk or dolomite; also called Whiting; used as extender pigment.

Calcium Driers. Used widely in combination with other metal driers to convert paint to hard films.

Calcium Gypsum. A dry powder, primarily calcium sulfate hemihydrate, resulting from calcination of gypsum; cementitious base for production of most gypsum plasters; also called Plaster of Paris; sometimes called Stucco.

Calcium Hypochlorite. A chemical compound of chlorine and calcium used as a bacteriacide in swimming pools; available in white granular or tablet form and releases 70 percent of its weight as available chlorine.

Calcium Insulation. A type of insulation, made of hydrated calcium silicate, that can withstand 1200° Fahrenheit and is not affected by moisture.

Calcium Silicate. A sand and lime solution.

Calcium Sulfate. 1. The chemical compound $CaSO_4$. 2. White inert pigment which provides very little color or opacity. 3. A drying agent or desiccant in liquid line driers.

Calculus. A method of computation or calculation in a special notation; the use of algebra to calculate changing quantities.

Calendar Day. Each and every day on the calendar without deduction for weekends or holidays.

Calendered Papers. Wallpapers with hard finish.

Caliber. 1. The internal diameter of a tube. 2. The character and capacity of a firm.

Calibrate. To verify the graduations of an instrument and adjust them if necessary.

Calidarium. The room containing the warm bath in Roman baths, also called Thermae.

California Redwood Association (CRA). 405 Enfrente Drive #200, Novato, California 94949, (415) 382-0662, FAX (415) 382-8531.

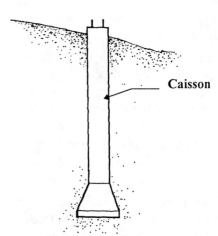

Caisson

Caliper. 1. A measuring instrument with two legs resembling a divider, for measuring thicknesses, diameters, and distances between surfaces; with in-turned points to measure convex surfaces and out-turned points for measuring internal dimensions. 2. The precise measured thickness of gypsumboard. 3. The diameter of a tree trunk.

Call System. A communicating device to connect one place with another.

Calorie. 1. Small calorie or gram calorie, used by medical science; the amount of heat needed to raise the temperature of one gram of water one degree Celsius. 2. Large calorie or great calorie; used by engineering science; the amount of heat needed to raise the temperature of one kilogram of water one degree Celsius; a kilocalorie equals 1,000 calories.

Calorimeter. Device used to measure quantities of heat or determine specific heats.

Cam. A projection on a rotating part in machinery, shaped to impart reciprocal or variable motion to the part in contact with it.

C

Camber Rod. A tension rod installed under a trussed beam.

Camber. 1. A deflection that is intentionally built into a structural element or form, usually a beam, to improve appearance or to nullify and offset the deflection of the beam under the effects of loads, shrinkage, and creep. 2. A slight rise at the center of a flat arch.

Cambium. A thin layer of tissue in a tree between the bark and wood that repeatedly subdivides to form new wood and bark cells.

Came. Slender grooved lead rod used in stained glass windows.

Camera Cable. A device for the transmission or recording of visual images.

Campanile. A free-standing bell tower.

Campus. The grounds and buildings of a university, college, or school.

Canadian General Standards Board (CGSB). 222 Queen Street, 14th Floor, Ottawa, Canada K1A 1G6, (613) 941-8648.

Canadian Standards Association (CSA). 178 Rexdale Boulevard, Rexdale, Toronto, Canada, M9W 1R3, (416) 747-4000.

Candela. An international unit of luminous intensity; also called a Candle.

Candelabrum. (Pl. candelabra) A branched candle stick or lamp with several lights.

Candelilla Wax. Wax obtained from small shrub grown in Texas and Mexico; softer than Carnuba wax.

Candle. See Candela.

Candlepower. The unit of luminous intensity of a light source, expressed in Candelas.

Cane Bolt. An L-shaped rod, mounted on a swinging or sliding door or gate, that drops into a pipe sleeve below the floor surface to secure the door or gate.

Cannibalization. The process of operating an income property by borrowing parts, fixtures, and equipment from vacant spaces.

Cannular. Tubular.

Canopy. An overhanging shelter; a marquee.

Cant Strip. 1. A strip of material, usually treated wood or fiber, with a sloping face used to ease the transition from a horizontal to a vertical surface at the edge of a flat roof; prevents the roofing material from abruptly stopping at the parapet wall and also helps prevent leakage at that juncture. 2. A triangular shaped strip of wood used under shingles at gable ends or under the edges of roofing on flat decks.

Cantilever Wall. A retaining wall in which the wall and footing resists earth pressure by cantilever effect.

Cantilever. A structural shape, beam, truss, or slab, that extends beyond its last point of support.

Cantilevered Beam. A beam that is supported at one end only.

Cap Base. Wood strip applied to the base of a wall to protect wall surface and finish the intersection of wall and baseboard.

Cap Block. A solid flat slab usually 2-1/4 inches thick used as capping units for parapet and garden walls; also used as a Paving Unit.

Cap Flashing. See Counterflashing.

Cap Rate. See Capitalization Rate.

Cap Sheet. The top sheet of roofing in a built up roof, usually made of organic or inorganic fibers, saturated and coated on both sides with a bituminous compound and factory-coated with mineral granules, mica, talc, iliminite, asbestos, or other inorganic fibers, or similar material.

Cap, Black Iron. A steel fitting, with female threads, which seals the end of a pipe.

Cap, Pile. A structural member usually fastened to, and placed on the top of a slender timber, concrete, steel pile; used to transmit loads into the pile or group of piles and to interconnect them.

Cap, Post. A fitting which joins the end of a wooden post to a joist or girder connected to the post.

Cap, Welded. A fitting that is fastened by welding to seal the end of a pipe.

Cap. 1. A caplike part or thing; cover, or top. 2. An upper limit set on a budget or cost; a ceiling. 3. A trim tile with a convex radius on one edge; used for finishing the top of a wainscot or for turning an outside corner, a bullnose. 4. The upper member of a column, pilaster, door, cornice molding, and the like. 5. A female pipe fitting, solid at one end; used to close off the end of a piece of pipe. 6. Masonry units laid on top of finished masonry wall or pier. 7. Flashing for tops of parapet walls.

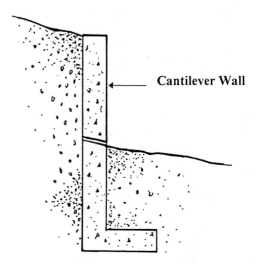

Cantilever Wall

Capacitance Sensor. A sensor which shows a device's ability or hold or store electrical energy.

Capacitance. 1. Property of a nonconductor, condenser or capacitor, that permits storage of electrical energy in an electrostatic field; a measure of this is the farad. 2. Property of an electric circuit which tends to oppose a change in voltage.

Capacitive Reactance. The opposition or resistance to an alternating current as a result of capacitance; expressed in ohms.

Capacitor Motor. Single-phase induction motor with an auxiliary starting winding connected in series with a condenser (capacitor) for better starting characteristics.

Capacitor. 1. A device which introduces capacitance into an electric circuit. 2. Type of electrical storage device used in starting and/or running circuits on many electric motors.

Capacitor-Start Motor. Motor which has a capacitor in the starting circuit.

Capacity. 1. The ability to contain or receive. 2. The maximum amount that can be accommodated or contained. 3. Refrigeration rating system; usually measured in Btu per hour or watts (metric). 4. The requirements of a lender that indicate borrowers' physical and financial ability to proceed with a project and be able to withstand any reasonable loss.

Capillarity. The action by which the surface of a liquid where it is in contact with a solid, as in a capillary tube, is elevated or depressed depending on the relative attraction of the molecules of the liquid for each other and for those of the solid.

Capillary Attraction. 1. The force of adhesion between a solid and a liquid in capillarity. 2. In brazing, the phenomenon by which adhesion between the molten filler metal and the base metals, together with surface tension of the molten filler metal, distributes the filler metal between the properly fitted surfaces of the joint to be brazed.

Capillary Break. A slot or groove intended to create an opening too large to be bridged by a drop of water, and thereby to eliminate the passage of water by capillary action.

Capillary Movement. Movement of underground water in response to capillary attraction.

Capillary Space. In cement paste, any space not occupied by anhydrous cement or cement gel; air bubbles, whether entrained or entrapped, are not considered as part of the cement paste.

Capillary Tube. See Capillary and Choke Tube.

Capillary Water. Underground water held above the water table by capillary attraction.

Capillary. A tube with a small bore.

Capital Gain or Loss. The gain or loss in the sale or disposition of a capital asset.

Capital Stock. The total amount invested in the business by the owner in exchange for shares of common stock at par value.

Capital, Column. The uppermost member of a column crowning the shaft and taking the weight of the slab, beam, or girder.

Capital. 1. The money or other assets with which a company starts in business and uses in operation. 2. The head or cornice of a pillar or column; also called Column Cap. 3. A city serving as the seat of government.

Capitalization Rate. The rate of interest to be used in the capitalization process, reflecting risk and rates of return on alternative investments; also called Cap Rate.

Capitalization. 1. The process of determining the capital value of a real property investment by relating its annual income to an assumed capitalization rate; for example, a property with an annual income of $12,000 and a capitalization rate of 8 percent would have a capitalized value of $150,000 (Income/Capitalization Rate = Capitalized Value). 2. Total of financial resources available to the business.

C

Capitol. 1. A building in which a state legislative body meets. 2. The building in which the US Congress meets in Washington, DC.

Car Wash. A building housing a mechanized method for washing cars.

Carbon Black. Jet black, non-bleeding pigment, made by burning natural gas in insufficient supply of air.

Carbon Dioxide. Compound of carbon and oxygen (CO_2) which is sometimes used as a refrigerant; refrigerant number is R-744.

Carbon Filter. Air filter using activated carbon as air cleansing agent.

Carbon Monoxide. A colorless, odorless toxic gas, CO, a product of incomplete burning of carbon.

Carbon Steel. Low carbon or mild steel.

Carbon Tetracholride. Colorless non-flammable and very toxic liquid used as a solvent; it should never be allowed to touch skin and fumes must not be inhaled.

Carbon. A nonmetallic element found native, as in the diamond and graphite, or as a constituent of coal, petroleum, and asphalt, of limestone and other carbonates, and of organic compounds; when combined with iron, forms various kinds of steel; in solid form, it is used as an electrode for arc welding; as a mold, it will hold weld metal; motor brushes are made from carbon.

Carbon-Arc Cutting. An arc-cutting process wherein the severing of metals is effected by melting with the heat of an arc between a carbon electrode and the base metal.

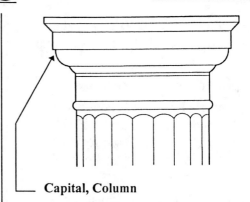

Capital, Column

Carbon-Arc Welding. An arc-welding process wherein coalescence is produced by heating with an electric arc between a carbon electrode and the work and no shielding is used; pressure may or may not be used and filler metal may or may not be applied.

Carbonation. Reaction between the products of portland cement and carbon dioxide to produce calcium carbonate.

Carbonizing Flame. See Reducing Flame.

Carborundum. Trademark. A compound of carbon and silicon used as an abrasive.

Carcinogen. Any substance that produces cancer.

Card Access Control. An entry device that is operated by a small magnetized plastic card, similar to a credit card.

Carillon. A set of bells sounded either from a keyboard or manually.

Carnuba Wax. A hard wax obtained from species of palm grown mostly in Brazil.

Carpenter Pencil. A sturdy pencil with a thick rectangular lead for marking lumber.

Carpenter. A craft worker skilled in woodwork, especially of the structural kind.

Carpenter's Saw. See Handsaw.

Carpenter's Level. A hand tool consisting of a wood or metal bar with spirit levels attached, used for establishing plumbness and levelness of construction members.

Carpentry, Finish. The finish woodwork installation such as base, casing, doors, stairs, paneling; all woodwork installed after plastering or drywall.

Carpentry, Rough. The preliminary framing, boxing, and sheeting of a wood frame building.

Carpet & Rug Institute (CRI). PO Box 2048, Dalton, Georgia 30722, (706) 278-3176.

Carpet Base. 1. Vinyl or rubber base attached to the wall with adhesive and installed as a finish for carpeting. 2. A base made of carpeting serving the same purpose.

Carpet Lining. See Carpet Padding.

Carpet Padding. A jute, felt, hair, foam, or plastic rubber underlayment installed under carpeting to increase underfoot comfort, to absorb pile-crushing forces and to reduce impact sound transmission; also called Cushioning, Lining, or Underlay.

Carpet Plate. A thin piece of ornamental metal that the rough edges of carpet is connected to at doorways or at the junction of carpet to another flooring material to form a clean ending point.

Carpet strip. See Base Shoe.

Carpet Tack. A small tack used for attaching carpeting to wood floors.

Carpet Tile. Carpet that comes in sheets or small squares and is installed with the use of adhesives.

Carpet. General designation of fabric constructions which serve as soft floor coverings, especially those which cover the entire floor and are fastened to it, as opposed to rugs; see also Woven, Tufted, Knitted, Punched, and Flocked carpets.

Carport. An open-sided shelter with a roof for a car, usually beside a house.

Carrel. A small cubicle for an individual reader in a library.

Carrene. Refrigerant in Group One (R-11); chemical combination of carbon, chlorine and fluorine.

Carriage Bolt. A bolt with a round smooth head that is threaded; a bolt with a square neck directly under the head to prevent its rotation.

Carriage. The framing support of the steps and risers of a flight of stairs; also called Rough Horse and Stair Horse.

Carrier Channel. The main supporting metal members used in the construction of suspended ceilings.

Carrier, Fixture. A mounting frame built into a wall to support a plumbing fixture.

Carrying Channels. See Channels, Carrying.

Cart, Laundry. A push or pull institutional vehicle with wheels, for the transport of clean or soiled laundry.

Cartouche. An ornamental frame.

Cartridge Filter. A swimming pool filter which operates through a disposable cartridge; these are of two general types: the surface or area type where the suspended matter is removed at the surface and the depth type in which the interstices vary from large to small in depth.

Cartridge, Extinguisher. The cylindrical container of a fire prevention apparatus which contains the chemicals used in the suppression of fire.

Cartridge. 1. Any of various small containers, holding a supply of material for a larger device into which it is inserted. 2. Disposable element containing filtering media and used in some pool filters.

Carved Carpet. See Sculptured Carpet.

Carved Door. A door that has been finished with either factory or hand cutting onto its surface to improve its appearance.

Cascade Sequence. In welding, a combined longitudinal and buildup sequence wherein weld beads are deposited in overlapping layers; in manual shielded metal arc welding a backstep sequence is normally used. See Block Sequence and Buildup Sequence.

Cascade System. Arrangement in which two or more refrigerating systems are used in series; uses the evaporator of one machine to cool the condenser of another machine; produces ultra-low temperatures.

Case Joint. A type of joint used in cabinetmaking in which the two pieces are butted together at an angle and fastened by dowels.

Carriage Bolt

Case Mould. Plaster shell used to hold various parts of a plaster mould in correct position; also used with gelatin and wax moulds to prevent distortions during pouring operation.

Case, Refrigerated. A storage case kept cold by a mechanical device. For use to store perishable items.

Casein Glue. An adhesive substance composed of casein (the curd of milk), lime, and sodium salt; it comes as dry powder to which water is added.

Casein Paints. Paint in which casein solution has replaced the binder.

Casein. Protein obtained from milk, soluble in alkaline water solution; casein is used extensively in the manufacture of water paints.

Casement. A window in which the sash opens with hinges and pivots on an axis along the vertical line of the frame; casement window.

Casework. Assembled cabinetry or millwork.

Cash Accounting. A method of keeping accounting records in which income is recorded when actually received and expenses are recorded when cash is paid out; also called Cash Basis.

Cash Basis. See Cash Accounting.

Cash Flow. The actual cash income after all cash outlays and reserves have been deducted from the gross income.

Casing Bead. Metal or wood molding used to separate different materials, used as an edge or used around openings to provide a stop.

Casing Trim. Metal or wood material that is attached around windows and doors to act as the decorative finish.

Casing, Ranch. An architectural style of exposed millwork enclosure of cased beams, posts, pipes, and the exposed molding or lining around doors and windows.

Casing. 1. The wood finish pieces surrounding the frame of a window or door, or the finished lumber around a post or beam. 2. A cylindrical steel tube used to line a drilled or driven hole such as a well or caisson.

Casino. 1. A building or room used for gambling or other amusements. 2. A small summerhouse.

CASSB. Cedar Shake & Shingle Bureau.

Cassiterite. An inorganic mineral of the tetragonal form used as a source of tin and tin oxide; SnO_2.

Cast Iron Soil Pipe Institute (CISPI). 5959 Shallowford Road, #419, Chattanooga, Tennessee 37421, (615) 892-0137.

Cast Iron Wheel Guard. Lineal component placed at intersection of wall and horizontal surface to restrain wheels of vehicles from coming close to wall surface, protecting it from vehicular damage.

Cast Iron. Iron with a high carbon content, which cannot, because of the percentage of carbon, be classified as steel.

Cast Stone. Concrete cast in molds for ornamental use in construction.

Cast. Inclination of one color to look like another; for example, sulphur is yellow with a greenish cast.

Castellated Beam. A steel, wide-flange section whose web has been cut along a zigzag path and reassembled by welding in such a way as to create a deeper section.

Casting Bed. A form in which precast concrete units are constructed.

Casting Mold. Use of gelatin, wax, or plaster molds to make plaster ornamentation

Casting Plaster. A fast-setting gypsum plaster that is used to anchor marble to walls; see Gypsum Molding Plaster.

Casting, Solid. Forming castings by introducing a body slip into a porous mold which usually consists of two major sections, one section forming the contour of the inside of the ware and allowing a solid cast to form between the two mold faces.

Casting. Something cast in a mold, like cast iron or cast aluminum.

Cast-In-Place Concrete. Concrete that is poured in its intended location at a site.

Cast-In-Place. Mortar or concrete which is deposited in the place where it is required to harden as part of the structure, as opposed to precast concrete.

Castor Oil. Nondrying oil obtained from the castor bean; may be converted to a drying oil by chemical treatment.

Casts. Finished plaster products from a mold, sometimes referred to as staff; used generally as enrichments and stuck in place.

Catalyst. A substance that initiates a chemical reaction and enables it to proceed under different conditions (as at a lower temperature) than otherwise possible.

Catch Basin. A receptacle for catching water runoff from a designated area; usually a shallow concrete box with a grating and a discharge pipe leading to a plumbing or stormwater system.

Catch. A piece of hardware for fastening a door, window, or cabinet door.

Catenary. The curve assumed by a completely flexible string or cable loaded only by its own weight.

Catface. Blemish or rough depression in the finish coat of plaster caused by variations in base coat thickness.

Cathedral. A church that is the official seat of a diocesan bishop.

Cathode. 1. The negative electrode in an electrolytic cell. 2. The positive terminal of a primary cell such as a battery.

Cathode-Ray Tube. A high-vacuum tube in which cathode rays produce a luminous image on a fluorescent screen.

Catwalk. A narrow walkway, such as used in an attic for access.

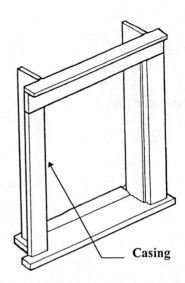

Casing

Caulk, Masonry. A resilient material applied where masonry work butts against other materials to seal cracks or openings.

Caulk. 1. To seal and waterproof cracks and joints, especially around window and exterior door frames. 2. To seal small openings in wall or ceiling systems to prevent leakage of sound or to effect a finished appearance and seal between dissimilar materials.

Caulking Compound. A soft, plastic material used for sealing joints in buildings and other structures where normal structural movement may occur; retains its plasticity for an extended period after application; available in forms suitable for application by gun and knife and in extruded preformed shapes.

Caulking Gun. A device, usually hand-powered, which dispenses liquid caulking into joints and seams.

Caulking. 1. A composition of vehicle and pigment, used at ambient temperatures for filling joints; remains plastic for an extended time after application. 2. A method of making a bell and spigot pipe joint watertight by packing it with oakum and lead or other materials.

Caustic Lime. Lime.

Caveat Emptor. Let the buyer beware. A warning that the buyer purchases at its own risk

Caveat. A warning or proviso.

Cavetto. A quarter hollow molding, the converse of a quarter round

Cavitation. Localized gaseous condition or partial vacuum that is found within a liquid stream; caused by mechanical force, as in a pump impeller, or in fluids at high velocities.

Cavity Wall Anchor. A metal device mounted on a masonry wall used to secure other attachments or masonry to an existing or back-up wall.

Cavity Wall. A masonry wall that includes a continuous airspace between its outermost wythe and the remainder of the wall.

Cavity. A hollow or void space within a mass.

CBD. Certified Bath Designer.

CBD. Commerce Business Daily; a daily newspaper that carries listings of governmental contracts available for bidding.

CC&Rs. Covenants, conditions, and restrictions.

C-Clamp. A clamp in the shape of a "C" with jaw capacities usually ranging from 1 to 8 inches used for the securing of wood or metal pieces in a fixed position and for temporary assemblies.

CCMCA. California Conference of Mason Contractor Associations, Inc.

CCTV Cable. Cable that is used for the transmission of closed-circuit television.

CCTV. Closed Circuit Television.

CDA. Copper Development Association.

CDX Plywood. Plywood used in exterior applications that is graded C and D, for sheathing.

CDX. A grading system mark for plywood which means. grade C and D, exterior glue.

CE. Civil Engineer.

Cedar Closet. A closet that is lined with thin pieces of cedar wood; used for its fragrance and its ability to repel insects.

Cedar Deck. A flat-floored roofless area adjoining a structure constructed of cedar wood; a platform serving as a structural element constructed of cedar wood; planks for flooring, from cedar, usually 2" nominal thickness.

Cedar Shake & Shingle Bureau (CASSB). 515 116th Avenue, NE, #275, Bellevue, Washington 98004-5294, (206) 453-1323.

C

Cedar Shake. A shingle made by splitting a block of cedar along its grain and thereby creating a shingle which may be used for roofing or siding; also called Handsplit Shingles.

Cedar Shingle. A thin piece of cedar wood with one end thicker than the other for laying in overlapping rows as a covering for a roof or the sides of a building or structure.

Cedar Siding. Boards milled from cedar wood, used for the finish covering on the exterior walls of a building or structure; used for its appearance and resistance to moisture and aging.

Cedar. An aromatic durable softwood, genus Cedars, of the pine family.

Ceiling Access Door. A hinged door or loose fitting panel that allows for admittance to an attic.

Ceiling Blocking. Wood pieces installed between ceiling joists and rafters to provide nailing surfaces for finishing ceiling materials.

Ceiling Diffuser. A mechanical device through which warm or cold air is blown into an enclosure, for the purpose of to distributing conditioned air.

Ceiling Framing. Wood or metal pieces which form the rough framing of ceilings.

Ceiling Furring. Wood or metal strips applied to a ceiling or rafter to make the ceiling or rafter level, provide a nailing surface, or create an air space.

Ceiling Grille. A grating, screen, or louvered panel that allows air into a ventilating duct.

Ceiling Heater. An electric heater installed in a ceiling, often in a bathroom.

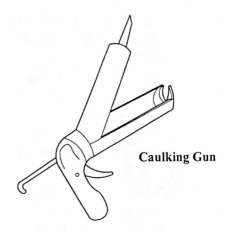

Caulking Gun

Ceiling Insulation. Loose, blown-in material or fiberglass rolls that are in installed at the ceiling plane.

Ceiling Joist. The horizontal members in a building or structure to which the ceiling material is fastened.

Ceiling Lath. Sheets of expanded metal, gypsum or in older structures, wood lath, which are attached to a ceiling to provide a plaster base.

Ceiling Molding. Molding that is used to form a projection at the top of a wall.

Ceiling Mortar. Extra-rich wall mortar.

Ceiling Painting. The actual physical process of applying paint, either by brush, roller, or spray gun to the ceiling section of a structure.

Ceiling Plenum. Space below the flooring and above the suspended ceiling that accommodates the mechanical and electrical equipment and that is used as part of the air distribution system.

Ceiling Price. The maximum price that an informed buyer would pay to purchase or lease property.

Ceiling Removal. The demolition and removal of ceiling materials in order to replace or remodel.

Ceiling Sound Transmission Class. A measure of reduction in sound transmission via plenum path between two rooms.

Ceiling. 1. The overhead inside lining of a room; classified by structural type, contact, furred, or suspended.

Cell. 1. Any void space. 2. A single room in a prison or jail. 3. The anatomical units of plant tissue, including wood fibers, vessel members, and other elements of diverse structure and function. 4. One of the hollow openings in building tile or cement blocks. 5. In electrical raceways, a single, enclosed tubular space in a cellular metal floor member, the axis of the cell being parallel to the axis of the metal floor member.

Cellar. See Basement.

Cellular Concrete. A lightweight product consisting of cement, cement-pozzolan, sand, lime-pozzolan or lime-sand pastes, or pastes having a homogenous void or cell structure, attained with gas- forming chemicals or foaming agents; for cellular concretes containing binder ingredients other than or in addition to portland cement, autoclave curing is usually employed; also called Foam Concrete or Gas Concrete.

Cellular Decking. Metal floor or roof deck panels made of steel sheets corrugated and welded together in such a way that hollow longitudinal cells are created within the panels. Deck which during construction supports wet concrete and construction loads, but after concrete cures does not perform structural function in completed construction; deck also is fabricated of two sheets to form linear voids.

Cellular Raceway. The hollow spaces of cellular metal floors, together with suitable fittings, that are used as enclosures for electrical and telephone conductors.

Cellulose Acetate. A binder made by chemical reaction of acetic acid on cellulose (cotton linters).

Cellulose Nitrate. A binder made by chemical reaction of nitric acid on cellulose (cotton linters); also called Nitrocellulose or Pyroxylin.

Cellulose. 1. The carbohydrate that is the principal constituent of wood and forms the framework of the wood cells. 2. An organic substance obtained from the cotton plant and used as raw material in the manufacture of paints and other materials.

Celsius Temperature Scale. The temperature scale used in metric system in which the freezing point of water is 0° and the boiling point is 100°; see Celsius.

Celsius. International thermometric scale where 0.01 degrees represents the triple point of water and 100 degrees the boiling point; similar to Centigrade.

Cement Asbestos. A material composed of portland cement, fine aggregate, and asbestos fibers; it is formed into flat and corrugated building boards used for roofing and siding, pipes and fittings, and water tanks.

Cement Base Paint. A paint composed of portland cement, lime, pigment, and other modifying ingredients; sold as dry powder to be mixed with water for application.

Cement Body Tiles. Tiles with the body made from a mixture of sand and portland cement; the surface may be finished with portland cement, spheroids of marble, or other materials.

Cement Color. Colored powdered or liquid pigments added to a mix to integrally color concrete.

Cement Content. The quantity of cement contained in a unit volume of concrete or mortar, ordinarily expressed as pounds, barrels, or bags per cubic yard.

Cement Factor. The number of bags or cubic feet of cement per cubic yard of concrete; see Cement Content.

Cement Fiber Board. A prefabricated concrete building sheet that is compressed and bonded.

Cement Fiber. A threadlike structure added to cement to stiffen and strengthen it.

Cement Gel. The colloidal, glue like, material that makes up the major portion of the porous mass of which hydrated cement paste is composed.

Cement Grout. 1. A cementitious mixture of portland cement, sand or other ingredients, and water which produces a uniform paste used to fill joints and cavities between masonry units. 2. A thin mortar used for pointing-up and finishing joints between tile units.

Ceiling Joist

Cement Mortar. A mixture of cement, lime, sand, or other aggregates, and water, used for plastering over masonry or to lay block, brick or tile.

Cement Paste. The mixture of portland cement, water, pozzolans and other admixtures, if any, and air which surround the aggregates in concrete; also called the matrix.

Cement Plaster. 1. Plaster having portland cement as its binder; used on exterior surfaces or in damp areas. 2. Gypsum plaster made to be used with the addition of sand for basecoat plaster; also called Neat or Hardwall plaster.

Cement, Keene's. See Keene's Cement.

Cement, Masonry. A hydraulic cement for use in mortars for masonry construction, containing one or more of the following materials: portland cement, portland blast-furnace, slag cement, portland-pozzolan cement, natural cement, slag cement or hydraulic lime; and in addition usually containing one or more materials such as hydrated lime, limestone, chalk, calcereous shell, talc, slag, or clay, as prepared for this purpose.

Cementing. In roofing, a solidly mopped application of hot asphalt, cold liquid asphalt compound, hot coal-tar pitch, or other cementing material.

Cementitious Material. A component material of plaster, mortar, or concrete which when mixed with water provides plasticity necessary for placement; upon subsequent setting or hardening it serves to bind aggregate particles together into a rigid heterogeneous mass.

Cementitious Topping. A compound that is capable of setting like concrete when applied on a concrete base to form a floor surface.

Cementitious. Having cementing properties; usually used with reference to inorganic substances, such as portland cement and lime.

Cenotaph. A tomb or monument erected in honor of a person or group of persons whose remains are elsewhere.

Center Matched. See Tongue and Groove.

Center of Gravity. The point at which the weight of a body may be considered to act; center of mass.

Center Pole. Column in center of spiral stair which supports stair treads.

Center Punch. A hand punch consisting of a short steel bar with a hardened conical point at one end used for marking the centers of holes to be drilled.

Center to Center (C to C). The dimension from the centerline of one member to the centerline of the next member.

Center. 1. The middle point of a line, circle, or sphere, equidistant from the ends or from any point on the circumference or surface; a pivot or axis of rotation. 2. A place or group of buildings forming a central point in a district or city. 3. A temporary structure to support the arch while it is being built; see Centering.

Center-Hung Sash. A sash hung on its centers so that it swings on a horizontal axis.

Centering Shims. Small blocks of synthetic rubber or plastic used to hold a sheet of glass in the center of its frame.

Centering. Temporary formwork for an arch, dome, vault, or other overhead surface.

Centerline. A real or imaginary line that is equidistant from the sides of some object; it is usually represented on drawings as a line of alternate dots and dashes.

Centesimal Measure. 1. Division into hundredths. 2. Division of the circle into 400 grads.

Centigrade. Thermometric scale where 0 degrees represents the freezing point of water and 100 degrees the boiling point, similar to Celsius.

Centimeter. A metric unit that equals one-hundredth of a meter or 10 millimeters and is equivalent to 2.54 inches.

Centipose. A metric unit of viscosity.

Central Inverter. A device for converting direct current into alternating current by mechanical or electronic means.

Central System. A system of conditioning air supplied to various areas or space, serviced by the same source of heat or cooling; all equipment in central systems is indoors except air-cooled condensers, evaporative condensers, and cooling towers.

Centrifugal Compressor. Pump which compresses gaseous refrigerants by centrifugal force.

Centrifugal Force. An apparent force that acts outwards on a body moving about a center.

Centrifugal Pump. A pump which draws water into the center of a high speed impeller and forces the fluid outward with velocity and pressure.

Centrifugal. Away from the center; opposite from centripetal.

Centrifuge, Laboratory. A laboratory apparatus using centrifugal force for separating substances of different densities, for removing moisture, or for simulating gravitational effects.

Centripetal Force. A force that keeps a body moving about a center from flying outwards.

Centripetal. Towards the center; opposite from centrifugal.

Centroid. Center of mass.

Ceramic Adhesive. Used for bonding tile to a surface; rubber solvents; rubber- and resin-based emulsions used as adhesives.

Ceramic Color Gaze. An opaque colored glaze of satin or gloss finish obtained by spraying the clay body with a compound of metallic oxides, chemicals and clays; it is burned at high temperatures, fusing glaze to body, making them inseparable.

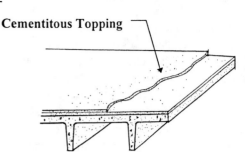

Cementitous Topping

Ceramic Insulator. A device made of ceramic non-conductive material which is used in electrical installations.

Ceramic Mosaic Tile. An unglazed tile formed by either the dust-pressed or plastic method, usually 1/4 to 3/8 in. (6.4 to 9.5 mm) thick, and having a facial area of less than 6 in, usually mounted on sheets approximately 2 by 1 ft. (0.3 by 0.6 m) to facilitate setting; ceramic mosaic tile may be of either porcelain or natural clay composition and may be either plain or with an abrasive mixture throughout.

Ceramic Process. The production of articles or coatings from essentially inorganic, nonmetallic materials, the article or coating being made permanent and suitable for utilitarian and decorative purposes by the action of heat at temperatures sufficient to cause sintering, solid-state reactions, bonding, or conversion partially or wholly to the glassy state.

Ceramic Tile Institute (CTI). 700 North Virgil Avenue, Los Angeles, California 90029, (213) 660-1911.

Ceramic Tile. A thin surfacing unit made from clay and/or a mixture of clay and other ceramic materials; the tile has either a glazed or unglazed face; it is fired above a red heat in the course of manufacture to a temperature sufficiently high to produce specific physical properties and characteristics.

Ceramic Veneer. A type of architectural terra cotta, characterized by larger face dimensions and thinner sections ranging from 1-1/8 in. to 2-1/2 in. in thickness.

Ceramic. Made of clay and permanently hardened by heat.

Ceramics. A general term applied to the art or technique of producing articles by a ceramic process, or to the articles so produced.

Ceresin. A hydrocarbon wax which possesses considerable flexibility.

Certificate for Payment. A written document forwarded to the general contractor by the architect, engineer, or owner approving payment for work completed.

Certificate of Insurance. A certificate provided by the general contractor verifying that he has obtained the required insurance for the project; the certificate is issued by the insurance company or its agent and confirms the existence of the insurance, the coverage, and its expiration date.

Certificate of Substantial Completion. A written document forwarded to the general contractor by the architect, engineer, or owner indicating that the project is substantially complete; this document initiates the time period for the final payment to the contractor.

Certified Bath Designer (CBD). A designation issued by the National Kitchen and Bath Association (NKBA) to persons who have taken specified courses and have relevant design experience.

Certified Check. A depositor's check guaranteed for payment by the bank.

Cessation of Work. The ending of work on a construction project without completion.

Cesspool. A subterranean container for temporary storage of septic tank effluent while it soaks into the adjoining soil.

CFC. Chlorofluorocarbon.

CFM. Cubic feet per minute.

CFS. Cubic feet per second.

CGL Insurance. Comprehensive General Liability Insurance.

CGSB. Canadian General Standards Board.

Chain Binders. In carpet making, yarns running warpwise (lengthwise) in the back of the carpet, binding all construction yarns together; the chain binder runs alternately over and under the weft binding and filling yarns, thereby pulling the pile yarn down and the stuffer yarns up for a tightly woven construction.

Chain Hoist, Door. A chain in a grooved pulley or sheave with a chain hook used to hoist a large door.

Chain Link Fence Manufacturer's Institute (CLFMI). 1101 Connecticut Avenue, NW, Washington, DC 20036, (202) 857-1140.

Chain Link Fence. A fence made of a wire mesh fabric.

Chain Trencher. A self-propelled machine with blades attached to a continuous chain, used to excavate trenches.

Chain Warp. In carpet making, zigzag warp yarn that works over and under the shot yarns of the carpet, binding the backing yarns together; see Warp.

Chain. A flexible series of connected metal links, to support a load.

Chair Rail. A wood molding separating the dado or wainscot from the upper wall; usually at a convenient height to prevent chair backs from abrading the wall.

Chair, Hydrotherapy. A device for the immersion of a medical patient in water.

Chair, Lifeguard. A raised chair, equipped with a ladder, that affords a lifeguard an elevated view of a large area.

Chair, Reinforcing. Metal supports made of fabricated wire, made to hold reinforcing steel in place until concrete is poured.

Chair. 1. A separate seat for one person, usually with four legs and a back. 2. A device used to support reinforcing bars while concrete is being poured.

Chalet. 1. A Swiss alpine dwelling with exposed structural elements and wide roof overhangs on front and sides. 2. A small suburban house or bungalow, in the chalet style, particularly with a broad roof overhang.

Chalk Line. A straight working line made by snapping a chalked cord between two points; also called a Snap Line; see Chalk Reel.

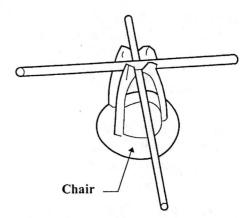

Chair

Chalk Rail. A trough mounted under a chalkboard to store chalk and erasers and to catch the chalk dust.

Chalk Reel. A carpenter's implement consisting of a string reel and chalk in a container, used as a method of chalking a snap line and storing the string.

Chalk. 1. A form of natural calcium carbonate; see whiting. 2. A lump of soft limestone used by carpenters for impregnating a snap line. 3. A crayon for marking materials on a construction site or in a workshop.

Chalkboard. Panel for writing on with chalk or liquid chalk.

Chalking. 1. The decomposition of a paint film into a loose powder on the surface; mild chalking, accompanied by satisfactory color retention in tinted paint, is considered a desirable characteristic; heavy chalking which washes off to leave an unprotected surface is highly undesirable; before recoating a heavily chalked surface, all of the chalk should be removed by vigorous brushing. 2. The dusty powdering on an asphalt roof surface that is subject to ultraviolet degradation.

Chamber. A room or space; a bedroom.

Chamfer Strip. An insert that is triangular or curved, placed in an inside corner to produce a rounded or flat beveled edge at the right angle corner of a construction member; also called Chamfering Strip.

Chamfer. A beveled surface cut on the corner of a piece of wood.

Chamfering Strip. See Chamfer Strip.

Chamfering Strip. Piece of stock placed in an inside corner of a form to produce a beveled edge.

Chamfering. The preparation of a contour other than for a square groove weld on the edge of a member for welding.

Chandelier. An ornamental branched hanging fixture for several candles or electric light bulbs.

Change Of State. Condition in which a substance changes from a solid to a liquid or a liquid to a gas caused by the addition of heat, or the reverse, in which a substance changes from a gas to a liquid, or a liquid to a solid, caused by the removal of heat.

Change Order. An order to change the work to be performed under a construction contract, usually given by an owner to a prime contractor or a by prime contractor to a subcontractor; a revision in the contract documents after the execution of the owner-contractor contract.

Change Trailer. A temporary vehicle that is used by personnel on a jobsite.

Changes. Key changes.

Channel Block. 1. A concrete masonry unit with a solid bottom and sides but no ends or webs, for use in a lintel. 2. A hollow unit with web portions depressed less than 1-1/4 inches to form a continuous channel for reinforcing steel and grout.

Channel Closure Strip. A U-shaped resilient strip used to close openings in metal panels and flashing.

Channel Door Frame. A U-shaped or L-shaped channel used as a door frame.

Channel Furring. A formed sheet metal furring strip.

Channel Slab. A manmade watercourse of molded, layered, plain or reinforced concrete.

Channel Strap. A U-shaped or L-shaped iron plate used to connect two or more timbers.

Channel Valve. A valve which controls the flow of water from a natural or artificial water course.

Channel, Manhole. The bottom of a sanitary or stormwater manhole which creates a channel between the incoming and outgoing pipes.

Channel. A U-shaped rolled steel or extruded aluminum section shaped like a rectangular box with one side removed.

Channels, Carrying. The heaviest integral supporting member in a suspended ceiling; carrying channels, or main runners, are supported by hangers attached to the building structure, and in turn, support various grid systems and furring channels or rods to which lath is fastened.

Channels, Furring. The smaller horizontal member of a suspended ceiling, applied at right angles to the underside of carrying channels and to which lath is attached; the smaller horizontal member in a furred ceiling; in general; the separate members used to space lath from any surface member over which it is applied.

Channels. 1. In asphalt paving, ruts or grooves that may develop in the wheel tracks of a pavement; may result from consolidation or lateral movement under traffic in one or more of the underlying courses, or by displacement in the asphalt surface layer itself; they may develop under traffic in new asphalt pavements that had too little compaction during construction or from plastic movement in a mix that does not have enough stability to support the traffic. 2. Hot or cold-rolled steel, of various sizes, used for furring, studs and in suspended ceilings.

Chapel. 1. A place for private worship in a large church or attached to a house or institution. 2. A room for services in a funeral home.

Character. A requirement by a loan officer to evaluate experience with similar jobs and locations, business reputation with lenders, suppliers, and subcontractors and reasonableness of bid.

Charcoal. An amorphous form of carbon consisting of a porous black residue from partially burnt wood, bones, or other substances; made by charring in a kiln from which air is excluded.

Charge. Amount of refrigerant placed in a refrigerating unit.

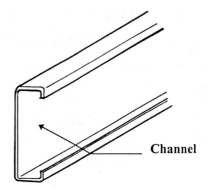

Channel

Charging Board. Specially designed panel or cabinet fitted with gauges, valves, and refrigerant cylinders used for charging refrigerant and oil into refrigerating mechanisms.

Charles' Law. For a constant pressure, the volume varies directly as the absolute temperature, and for constant volume, pressure varies directly as absolute temperature.

Charette. The final intense efforts of architectural students in producing presentations of their design solutions.

Chase. A groove or indentation cut into masonry to accommodate electric or plumbing lines.

Chattel. Moveable personal property.

Check Cracks. See Craze Cracks.

Check Valve. A device which allows fluid or air to pass through in only one direction; a valve which prevents the back-flow of water or other liquid by automatically closing.

Check. A lengthwise separation of the wood that usually extends across the rings of annual growth and commonly results from stresses set up in wood during seasoning.

Checking. 1. The pattern of irregular surface cracks on the top pour of an asphalt roof, a preliminary stage of alligatoring. 2. Cracks or fissures that appear with age in many exterior paint coatings; superficial at first, they may in time penetrate the coating entirely; these cracks may assume many patterns, but the usual ones resemble the print of a bird's foot or small squares.

Checkrail. A meeting rail sufficiently thicker than a window sash to fill the opening between the top and bottom sash made by the parting stop in the frame of double-hung windows; checkrails are usually beveled.

Checkroom Shelf. A horizontal mounted surface upon which objects can be stored and kept track of in a checkroom.

Chemical Bond. 1. The bond produced by cohesion between separate laminates of similar crystalline materials; based on formation and subsequent interlocking of crystals. 2. The adherence of one plaster coat to another or to the base which implies formation of interlocking crystals or fusion between the coats or to the base.

Chemical Brown Stain. A chemical discoloration of wood which sometimes occurs during the air drying or kiln drying of several species apparently caused by the concentration and modification of extractives.

Chemical Extinguisher. A wheeled device or hand-held cylinder that contains chemicals to extinguish certain types of fires.

Chemical Feed. Injection of chemicals into pool water circulation for pollution control.

Chemical Piping. Piping which conveys concentrated chemical solutions from a feeding apparatus to the circulation piping.

Chemical Porcelain. Vitreous ceramic whitewares used for containing, transporting, or reacting of chemicals.

Chemical Refrigeration System. System of cooling using a disposable refrigerant; also called an expendable refrigerant system.

Chemical Resistance. The effect of specific chemicals on the properties of various materials with respect to concentration, temperature and time of exposure.

Chemical Toilet. A self-contained portable toilet, not attached to a sewer line, for temporary use on a construction site.

Chemically Active. 1. Of paint pigments, those which react with oil of vehicle to form soaps which influence toughness of film, and increase durability. 2. Pigments such as red lead which react with acids formed at metal surface to prevent rust.

Chemically Pure (CP). Of the highest grade but not necessarily 100 percent pure.

Chenille. A pile fabric woven by the insertion of a prepared weft row of surface yarn tufts in a fur or caterpillar form through very fine but strong cotton catcher warp yarns, and over a heavy woolen backing yarn.

Cherry Veneer. A thin layer of cherry wood used as a finished surface material.

Chestnut Veneer. A thin layer of chestnut wood used as a finished surface material.

Chevron Bracing. That form of bracing wherein a pair of braces located either above or below a beam terminates at a single point within the clear beam span.

Chicken Ladder. A lightweight ladder that can be hung over the ridge for work on a steep roof.

Chicken Wire. Thin, galvanized, hexagonal, woven wire mesh mounted on an exterior wall as a base for stucco plaster.

Chill Factor. See Windchill.

Chilled Water System. A cooling system in which the entire refrigeration cycle occurs within a single piece of equipment; water is used to bring the heat from the space to the evaporator section of the chiller, and water is also used to carry the heat from the condenser to the outside.

Chiller, Absorption. A water cooling system similar to a vapor compression chiller with the exception that it does not use a compressor, but uses thermal energy - low pressure steam, hot water, or other hot liquids to produce the cooling effect.

Chiller. A piece of equipment that produces chilled water for circulation through a building and contains a compressor, condenser and evaporator tank.

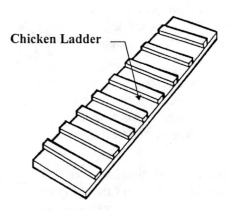

Chicken Ladder

Chime. 1. An apparatus for chiming a bell or a set of bells. 2. An electrical device used as a door bell, actuated by a push button.

Chimney Breast. The projecting portion of a chimney, especially when projecting from the exterior wall of a building or into a room.

Chimney Brick. Brick used for the construction of chimneys because of its ability to withstand high temperatures without cracking.

Chimney Connector. Pipe connecting a heating appliance, such as a furnace, with the vertical flue.

Chimney Effect. Tendency of air or gas to rise when heated.

Chimney Flue. A channel or shaft in a chimney for conveying smoke and exhaust gases to the exterior atmosphere.

Chimney Lining. Fire clay or terra cotta material, or refractory cement, made to be built inside of a chimney.

Chimney Pot. An earthenware or metal pipe at the top of a chimney, narrowing the aperture and increasing the updraft.

Chimney Throat. The narrowest part of a fireplace chimney, adjacent to the smoke shelf; the location of the damper.

Chimney, Masonry. A vertical noncombustible structure with a flue or flues to remove smoke and other gases, constructed of shaped or molded masonry units.

Chimney. A vertical, noncombustible structure with one or more flues to carry smoke and other gases of combustion into the atmosphere.

China Clay. Kaolin.

China Process. The method of producing glazed ware by which the ceramic body is fired to maturity, following which the glaze is applied and matured by firing at a lower temperature.

China. A glazed or unglazed vitreous ceramic whiteware used for nontechnical purposes; designation of such products as dinnerware, sanitary ware, and art ware when they are vitreous.

China Wood Oil. Tung oil.

Chinese Blue. A form of iron blue.

Chinese Red. Chrome orange, deep.

Chip Cracks. See Eggshelling.

Chipboard. A paperboard used for many purposes that may or may not have specifications for strength, color, or other characteristics; normally made from paper stock with a relatively low density in the thickness of 0.006 inch and up.

Chipped. In tile work, caused by rough handling and confined to the corners and edges of the tile; the scaling or breaking off at the edges of fragments from the surface of a tile.

Chipping Hammer. A mason's hand tool, often capped with tungsten carbide for durability; used to chip excess material from the backs and edges of block, brick, stone, or tile.

Chipping. 1. Removing welding defects and surface slag by use of a chipping chisel. 2. Cleaning steel using special hammers. 3. Type of paint failure.

Chisel Edge. A slanted factory edge on gypsumboard.

Chisel, Cold. See Cold Chisel.

Chisel, Wood. See Wood Chisel.

Chloramine. See Ammonia.

Chlordane. A chlorinated substance used as a pesticide.

Chlorinated Isocyanurate. Chlorine and cyanuric acid compound used to maintain chlorine level in pool water and prevent chlorine from dissipating in sunlight. See Conditioned Water.

Chlorinated Polyvinyl Chloride. A type of plastic used to make pipe to carry hot water and chemicals.

Chlorinated Rubber. A particular film former used as a binder, made by chlorinating natural rubber.

Chlorine Demand. The amount of chlorine necessary to oxidize all organic material present in pool water at a given moment or over a period of time.

Chlorine Residual. The amount of chlorine remaining in pool water after the chlorine demand has been satisfied at a given moment; this chlorine is available to oxidize other bacteria in water.

Chlorine. A poisonous greenish-yellow gaseous element of the halogen group, used for purifying water, bleaching, and the manufacture of many organic chemicals.

Chlorofluorocarbon. A compound that contains carbon, chlorine, and fluorine; known commonly as CFCs, they were widely used in refrigerators and aerosol sprays, but it is now known that they have harmful effects on the earth's atmosphere.

Chlorosulfonated Polyethylene. Hypalon single-ply roofing.

Chock. Heavy timber or wooden block, fitted under tires or wheels to prevent movement.

Choir Loft. A gallery or balcony in a church to be occupied by a choir.

Choir. The part of a cathedral or large church between the altar and the nave.

Choke Tube. Throttling device used to maintain correct pressure difference between high-side and low-side in refrigerating mechanism. Capillary tubes are sometimes called choke tubes.

Chopper, Food. A device that chops food and blends it into smaller pieces.

Chord. 1. One of the main members of a truss, braced by web members of the truss. 2. Perimeter member of a building or structure which resists lateral forces.

Chroma. Saturation, purity, or intensity of color.

Chrome Green. Mixture of chrome yellow and Prussian blue, one of industry's most important green pigments.

Chrome Orange. An orange pigment composed principally of basic lead chromate.

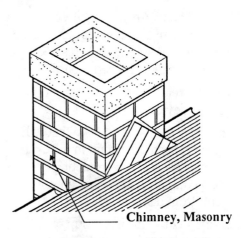
Chimney, Masonry

Chrome Yellow. Important inorganic yellow pigment made by mixing solutions of lead acetate and potassium bichromate; highly corrosion-inhibiting.

Chrome. Chromium.

Chromium Oxide Green. Green pigment which is extremely permanent in color and has good resistance to both alkali and heat.

Chromium Oxide. See Chromium Oxide Green.

Chromium. A blue-white metallic element, used as a shiny decorative electroplated coating.

Chromometer. An instrument used to indicate the color of light liquids and oils; also called a Colorimeter.

Chronometer. A timepiece; a clock.

Chuck. An attachment for holding a tool in a machine, as a bit on a drill.

Chute, Mail. See Mail Chute.

Chute, Trash. See Trash Chute.

Chute. An inclined plane, sloping channel, or passage down or through which materials may pass.

CI. Cast iron.

Circle Cutter. An adjustable scribe tool for cutting circular patterns or openings for lighting fixtures and other devices in gypsum board.

Circuit Breaker. An overcurrent protection device.

Circuit Protector. A device that will open an electrical circuit in the event of an overload, thus protecting operating equipment and other components from damage.

Circuit Vent. This is a group plumbing vent that extends from the front of the last fixture of the horizontal group to the vent stack; this type of vent may be used when the circuit carries the drainage of from two to eight urinals, water closets, stall showers, or sinks.

Circuit, Parallel. See Parallel Circuit.

Circuit, Pilot. See Pilot Circuit.

Circuit, Series. See Series Circuit.

Circuit. The path of an electric current from the source.

Circular Mil. A unit of length used in wire sizes.

Circular Saw. A power saw with a circular cutting blade, either in the form of a portable hand tool or a stationary table saw.

Circulating Fireplace. A fireplace that has cold air intakes and hot air outlets into and out of a heat exchanger that is built into the firebox to enhance the effectiveness of a fireplace as a heating system.

Circulation Path. An exterior or interior way of passage from one place to another for pedestrians, including, but not limited to, aisles, walks, hallways, courtyards, stairways, and stair landings.

Circulation Piping System. The piping between a pool, spa or hot tub structure and the mechanical equipment; usually includes suction piping, face piping and return piping.

Circulation Pump. A pump that moves fluids in a piping system such as in a domestic hot water system or in a hot water heating system.

Circulation System. Entire flow arrangement of fittings, pipework, and equipment.

Circumference. The enclosing boundary of a circle or other figure enclosed by a curve.

CISPI. Cast Iron Soil Pipe Institute.

Cistern. A tank for storing water.

Citronella Oil. An oil with a peculiar odor, obtained from a species of grass grown in Asia.

City Hall. A building housing city government administrative offices.

Civil Engineer. An engineer who designs public works such as roads, harbors, piping, earthwork, and waterworks.

Cladding Panel. A panel applied to a structure to provide durability, weathering, corrosion and impact resistance.

Cladding. 1. Metal exterior building surfacing panels. 2. In welding, see Surfacing.

Clamp, Beam. A device which holds a horizontal structural member to a vertical member.

Clamp, Ground Rod. The device that attaches the main ground wire to a cold water pipe in an electrical system.

Clamp. A mechanical device used to hold two or more pieces of material together.

Clapboard. A type of wood siding consisting of narrow boards thicker on one side than the other.

Clarified Sewage. Sewage from which part or all of the suspended matter has been removed.

Clarity. The transparency of pool water.

Class A, B, C, Roofing. Roof covering materials classified according to their resistance to fire when tested in accordance with ASTM E108; Class A being the highest and Class C the lowest.

Classified Product. A product labeled and listed by an approved laboratory having a factory follow-up and inspection service.

Classroom Lock. A doorknob in a classroom that has the ability to be locked from the inside.

Classroom Lockset. An assembly mounted in a door in a classroom that contains both a lockable doorknob and deadbolt.

Classroom. A room or building housing a class of students; part of a school.

Claw Hammer. A hammer with one end of the head forked for extracting nails.

Clay Brick Floor. Hard, baked or fired brick used in flooring applications.

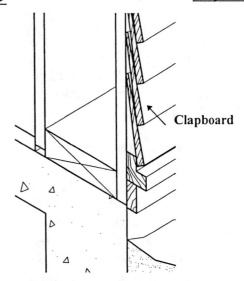
Clapboard

Clay Brick. A type of brick manufactured from fine-grained materials mainly from hydrated silicates of aluminum; soft and cohesive when moist, but becomes hard when baked or fired.

Clay Coping. The use of clay to form a cap or finish on top of a wall, pier, pilaster, or chimney.

Clay Court. A type of surface used for athletic competition, usually tennis or track.

Clay Floor Tile. Quarry tile that is fired and used for flooring.

Clay Pipe. Pipe used for drainage systems and sanitary sewers made of earthenware and glazed to eliminate porosity.

Clay Tile Partition. An assembly of hollow clay units for constructing interior partitions; the surface is often grooved to receive plaster.

Clay Tile. 1. Earthenware tile that is fired; formed for use on roofs and floors. 2. Quarry Tile.

Clay. 1. A natural mineral aggregate, consisting essentially of hydrous aluminum silicates; it is plastic when sufficiently wetted, rigid when dried, and vitrified when fired to a sufficiently high temperature; used in the manufacture of brick. 2. A heavy soil composed of particles less than 1/256 mm in diameter.

Clean Out Holes. Openings in first course of one wythe of a brick wall, or cutouts in face shell of first course of block wall, to enable cleaning out of mortar protrusions and droppings.

Clean out. Removable drainage fitting which permits access to the inside of drainage piping for the purpose of removing obstructions.

Clean Room. A dust-free environment that is required for some types of manufacturing, assembly, or fabrication.

Clean Surface. One free of contamination.

Clean. 1. Free of foreign material. 2. In sand or gravel, lack of binder.

Cleaner. 1. Detergent, alkali, acid, or other cleaning material; usually water or steam borne. 2. Solvent for cleaning paint equipment.

Cleaning Masonry. The final removal of excess grout, excess concrete, and construction soil from an exterior masonry structure.

Cleanout Door. Cast iron door located at base of chimney to allow access for cleaning the ash pit.

Cleanout Plug. A removable plug in a pipe fitting to enable cleaning a drainage pipe from blockages.

Cleanout, Storm Drain. An opening to a storm drain that allows for the removal of debris.

Cleanout. An opening to the bottom of a space of sufficient size and spacing to allow the removal of debris; in plumbing, a fitting in a pipeline which can be easily accessed to remove foreign objects or provide an opening to insert cleaning type devices.

Clear Ceramic Glaze. A colorless or colored transparent ceramic glaze; same as Ceramic Color Glaze except that it is translucent or slightly tinted, with a gloss finish.

Clear Dimension. The dimension between opposing inside faces or walls of an opening or a room; also called Clear Opening.

Clear Glaze. A colorless or colored transparent ceramic glaze; see Clear Ceramic Glaze.

Clear Opening. See Clear Dimension.

Clear Span. The horizontal distance between supports of any structural member.

Clearance Pocket. In a compressor, a small space in a cylinder from which compressed gas is not completely expelled; also called Clearance Space; for effective operation compressors are designed to have as small a clearance space as possible.

Clearance Space. See Clearance Pocket.

Clearing. An area that has all vegetation and objects removed.

Clearstory. Clerestory.

Cleat. A small strip or block of material, usually wood, which is fastened to a secure surface and used for attachment or as a toe hold or stopping device for another supporting member.

Cleavage Membrane. A layer of 15 lb. roofing felt, or an equivalent type of construction paper or polyethylene sheeting, used to isolate a wire reinforced mortar bed for tile from the concrete substrate.

Cleavage Membrane. In tile setting, a membrane such as saturated roofing felt, building paper, or 4 mil polyethylene film installed between the backing and mortar bed to permit independent movement of the tile finish.

Cleavage. The natural tendency of certain materials, especially of stones and crystals, to fracture or split in certain definite directions determined by the molecular or physical structure of the material.

Cleft, Natural. See Natural Cleft.

Clerestory. An upper row of windows in a high ceiling room, above the level of a lower adjoining roof; also called clearstory.

Clerk of the Works. Architect's representative on the jobsite; an obsolete term; now called Architect's Project Representative.

Clevis Hanger. A U-shaped metal hanger with the ends drilled to receive a pin or bolt used for attaching or suspending parts or piping.

Clevis Pin. The bolt or pin used to complete the connection in a Clevis Hanger.

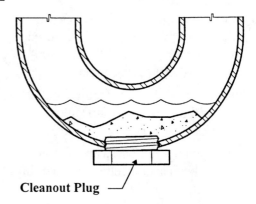

Cleanout Plug

CLFMI. Chain Link Fence Manufacturer's Institute.

Clinch. To turn over or flatten the protruding point of a nail to prevent withdrawal.

Clinker. Generally a fused or partly fused by-product of the combustion of coal, but also including lava and portland-cement clinker, and partly vitrified slag and brick.

Clinker Brick. A very hard-burned brick whose shape is distorted or bloated due to nearly complete vitrification.

Clinometer. A device for measuring angles of slope or inclination.

Clip Course. The course of bricks that is resting on a clip joint.

Clip Joint. A brick masonry joint that is thicker than usual to bring that particular course up to a necessary height; the joint should not be any thicker than 1/2 inch.

Clipped Header. In brick masonry, a bat placed to look like a header for purposes of establishing a pattern; also called a false header.

Clips. A classification of devices usually made of wire or sheet metal used to attach various types of lath to supports or to secure adjacent lath sheets.

Clock Receptacle. An electrical outlet box for the connection and support of a clock.

Clock System. A clock which controls the mechanical or electrical devices in a building or facility.

Clock-Timer, Hospital. A clock system in a hospital which can be set at different time gradients to announce, with sound, the passage of a specific amount of time.

Cloister. A covered passage on the side of a court.

Close Nipple. The shortest length of a given size pipe which can be threaded externally from both ends; used to closely connect two internally threaded pipe fittings.

Close-Coupled Toilet. A toilet directly connected to its water tank.

Closed Circuit Telephone. See Closed Circuit, 2.

Closed Circuit Television. See Closed Circuit, 2.

Closed Circuit. 1. Electrical circuit in which electrons are flowing. 2. A television or telephone system where the signal is transmitted by wire to a restricted number of receivers.

Closed Container. Container sealed by means of a lid or other device so that neither liquid nor vapor will escape from it at ordinary temperatures.

Closed Joint. In welding, a zero root opening.

Close-Grained Wood. Wood with narrow, inconspicuous annual rings; wood having fine fibers and small closely spaced pores.

Closer. 1. A Door Closer. 2. The last masonry unit laid in a course. 3. A partial masonry unit used at the corner of a course to adjust the joint spacing, sometimes spelled Closure.

Closet Bend. An elbow drainage fitting connecting a water closet to a branch drain.

Closet Flange. The fitting attached to a subfloor onto which the toilet bowl is attached.

Closet Pole. A horizontal, round member or rod installed in a closet to provide a place to hang clothes.

Closet, Cedar. See Cedar Closet.

Closet. A storage cabinet or room.

Closing Entries. Entries made at the end of the financial period to close all temporary accounts (income and expense) and to transfer the net profit (or loss) to the owner's equity or retained earnings account.

Closing the Ledger. The operation of closing all income and expense accounts at the end of the period and transferring the net profit (or loss) to the capital or surplus accounts.

Closure Strip. A resilient strip, used to close openings created by joining metal panels and flashing.

Closure. In brick masonry, supplementary or short length units used at corners or jambs to maintain bond pattern; see Closer, 3.

Clothes Hanger Rod. See Closet Pole.

Clothing Locker. A cabinet with a lockable door used for storing clothes.

Clothing Presser. A two-part mechanical ironing board that removes wrinkles from clothing with heated, compressive force.

Cloud. 1. A defect in the title to real estate or property; when a property has a cloud on it, it is difficult to sell or complete escrow. 2. A marking on a drawing consisting of a billowing line surrounding portions of the drawing involved in a change, usually identified with a numbered delta symbol.

Cluster Development. A system of close grouping of residential units leaving larger portions of open land around the group.

Clutch, Magnetic. See Magnetic Clutch.

CM. 1. Construction manager. 2. Construction management. 3. Center-Matched.

CMAA. Crane Manufacturers Association of America.

CMACN. Concrete Masonry Association of California and Nevada.

CMU Grout. Concrete masonry unit grout.

CMU. Concrete Masonry Unit.

CO. 1. Carbon monoxide. 2. Cleanout.

CO_2 Extinguisher. A portable fire-fighting device which dispenses carbon dioxide to extinguish small fires.

CO_2 Indicator. Instrument used to indicate the percentage of carbon dioxide in stack gases.

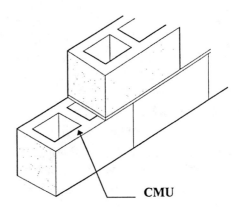

CMU

CO_2 Sprinkler System. An overhead sprinkler system containing Carbon Dioxide, installed and set to turn on when excess heat or smoke activates built-in sensors.

CO_2. Carbon dioxide.

Coagulant. A chemical compound, usually alum, used in swimming pools for the purpose of gathering and precipitating suspended matter.

Coagulate. To change from a liquid into a dense mass; solidify; curdle.

Coagulation. Precipitation of colloids into a single mass; usually caused by excessive heat or catalytic agents.

Coal Hopper Door. Steel door to allow for addition of coal to burning chamber.

Coal Tar. A dark brown to black cementitious material produced by the destructive distillation of bituminous coal.

Coal Tar Bitumen. Coal Tar Pitch.

Coal Tar Epoxy Paint. Paint in which the binder or vehicle is combination of coal tar with epoxy resin.

Coal Tar Felt. See Tarred Felt.

C

Coal Tar Pitch. A roofing pitch made from the distillation of bituminous coal; used mainly in dead-level or low-slope roofs; coal-tar pitch comes in a narrow range of softening points from approximately 140° F to 155° F.

Coal Tar Solvent. Derived from the distillation of coal tar.

Coal Tar Urethane Paint. Paint in which the binder or vehicle is a combination of coal tar with polyurethane resin.

Coalesence. In welding, the growing together, or growth into one body, of the base metal parts.

Coarse Aggregate. Concrete aggregate over 1/4-inch diameter; crushed stone, gravel, slag, or other inert materials.

Coarse Strainer. Basket within a pipeline to trap large debris before the pump.

Coarse-Graded Aggregate. One having a continuous grading in sizes of particles from coarse through fine with a predominance of coarse sizes.

Coarse-Grained Wood. Wood with wide conspicuous annual rings in which here is considerable difference between springwood and summer-wood; wood with large pores such as oak, ash, chestnut, and walnut; also called Coarse-Textured and Open-Grained.

Coarse-Textured Wood. See Coarse-Grained Wood.

Coat of Paint. One layer of dry paint, resulting from a single wet application; single layer of paint spread at one time and allowed to harden.

Coat Rack. A piece of furniture that is used to hang coats.

Coat. 1. A thickness, covering or layer of plaster applied in a single operation. 2. See Coat of Paint.

Coated Base Sheet. A felt that has been impregnated and saturated with asphalt and then coated on both sides with harder, more viscous asphalt to increase its impermeability to moisture; a parting agent is incorporated to prevent the material from sticking in the roll.

Coated Roof. A roof, usually flat, that has an asphaltic material applied to it to seal against the elements.

Coating In. Applying a coat of paint.

Coating. A layer or covering applied to a surface.

Coatings. Surface coverings; paints; barriers.

Coaxial Cable. A cable composed of two concentric conductors separated by an insulating layer; used for transmitting low voltage electronic signals.

Cobalt Blue. Blue pigment, stable in color; made by heating a mixture of cobalt oxide and aluminum hydrate.

Cobalt Drier. Powerful drier which is soluble in all drying oils; known as surface dryer.

Cobble. 1. To make or put together hastily. 2. A small naturally rounded stone of a size used for paving; a Cobblestone.

Cobblestone. A Cobble.

Cobwebbing. Premature drying of a liquid surface causing a spider web effect.

Cock. A device, as a faucet or valve, to regulate the flow of a liquid.

Cockle. A crease-like wrinkle or small depression in gypsum board face paper usually running in the long or machine direction; also called a wrinkle.

Code Blue System. An electronic warning device installed in hospital rooms or health care facility rooms used to notify caregivers of health emergencies in the building.

Code Installation. Refrigeration or air conditioning installation which conforms to the local code and/or the national code for safe and efficient installations.

Code Side. See Back, 3.

Code. 1. A set of regulations which has been adopted by a governmental unit for the purpose of protecting the public health and safety. 2. The identification marking on the back of sheets of gypsum board; denotes manufacturing plant, date, time, and other details.

Coefficient of Static Friction. The ratio of the limiting friction to the normal pressure (the weight of the moving body).

Coefficient of Conductivity. Measure of the relative rate at which different materials conduct heat; copper is a good conductor of heat and, therefore, has a high coefficient of conductivity.

Coefficient of Cubical Expansion. See Coefficient of Volumetric Expansion.

Coefficient of Expansion. Change in unit length, area or, volume for one degree rise in temperature.

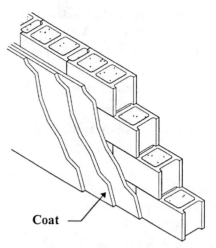

Coat

Coefficient of Friction. The mathematical relationship between the weight of an object and the force required to slide it, considering the characteristics of the two materials, the angle of the surfaces, and the angle of the force.

Coefficient of Heat Expansion. The rate of heat loss in BTU per hour through 1 square foot of a wall or other building surface when the difference between the indoor and outdoor air temperatures is 1° F; U-Value.

Coefficient of Linear Expansion. The change in unit length for a rise in temperature of 1° F.

Coefficient of Performance (COP). Ratio of work performed or accomplished as compared to the energy used.

Coefficient of Thermal Expansion. Change in unit length, area, or volume per degree change of temperature.

Coefficient of Volumetric Expansion. The change in unit volume for a rise in temperature of 1° F; also called Coefficient of Cubical Expansion.

Coefficient. 1. A multiplier in any algebraic expression. 2. A multiplier that measures a property of a material or operation.

Coffee House. A place serving coffee and other refreshments.

Coffee Room. A room where workers can make and drink coffee, tea, and other refreshments.

Coffee Urn. A device used for the brewing, storage, and serving of coffee, to keep it fresh and hot.

Coffer. A recessed panel in a ceiling or dome.

Cofferdam. A watertight enclosure from which water is pumped to expose the bottom of a body of water and permit construction.

Coffered Ceilings. Ornamental ceilings made up of sunken or recessed panels.

Cog. A gear tooth.

Cohesion. The act or condition of sticking together tightly; tendency to cohere; attractive force between polymers of similar nature which tends to hold them together; property of holding a film together; the soil quality of sticking together.

Cohesive Failure. Rupture of an adhesive joint, such that the separation appears to be within the adhesive.

Cohesive Soil. A soil, such as clay, the particles of which will adhere to one another by means of cohesive and adhesive forces.

Coil Deck. Insulated horizontal partition between refrigerated space and evaporator space.

Coil. A winding arrangement of a conductor around a core to convert low voltage to high voltage as in a transformer or to create a magnetic field as in a solenoid.

Coiled Pipe. Tubing in rows, layers, or windings in steam heating, water heating, refrigeration condensers, and evaporators.

Coin Dryer. A coin-operated public appliance for the drying of clothes.

Coin Washer. A coin-operated public appliance for the washing of clothes.

Cold Applied. Roofing products that do not have to be heated before application, unlike tar or asphalt.

Cold Chisel. A hand tool made from a steel bar with a sharpened tip for chipping concrete, stone, and similar materials.

Cold Color. See Cool Color.

Cold Formed Steel. Process of shaping steel without using heat.

Cold Joint Lines. Visible lines on the surfaces of formed concrete indicating the presence of joints where one layer of concrete had hardened before subsequent concrete was placed.

Cold Joint. 1. A visible lineation which forms when the placement of concrete is delayed; the concrete in place hardens prior to the next placement of concrete against it. 2. Any point in a tile installation when tile and setting bed have terminated and the surface has lost its plasticity before work is continued. 3. In road construction, a paving joint in which one strip of asphalt is installed at a different time from the other and bonding is not enhanced.

Cold Junction. That part of a thermoelectric system which absorbs heat as the system operates.

Cold Patch. A roof repair done with cold applied material.

Cold Rolled Steel. Steel rolled to its final form at a temperature at which it is no longer plastic.

Cold Wall. Refrigerator construction which has the inner lining of refrigerator serving as the cooling surface.

Cold Water Paint. The paint in which the binder or vehicle portion is composed of casein, glue or a similar material dissolved in water; usually employed on concrete, masonry or plaster surfaces.

Cold Worked Steel. Steel formed at a temperature at which it is no longer plastic, as by rolling or forging.

Cold. The absence of heat; a temperature considerably below normal.

Cold-Checking. Checking caused by low temperature.

Cold-Pressed Plywood. Interior-type plywood manufactured in a press without external applications of heat.

Cold-Process Roofing. A built-up roof consisting of layers of coated felts bonded with cold-applied asphalt roof cement and surfaced with a cutback or emulsified asphalt roof coating.

Cold-Setting Resin Glue. A resin-base glue that comes in powder form and is mixed with water.

Coliseum. A large sports stadium or building designed like the Roman Coliseum for public entertainments.

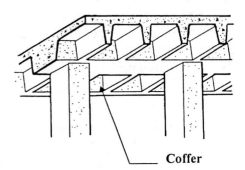

Coffer

Collage. An artistic composition in which various materials such as photographs, pieces of fabric, metals, and wood are arranged and glued to a backing.

Collapse. To cave or fall in or give way; failure of a structure.

Collar Beam. A horizontal beam near the top of a trussed rafter system attached to opposing rafters to resist their spreading; also called a Collar Tie or a Collar Beam.

Collar Joint. The vertical mortar joint between wythes of masonry.

Collar Tie. See Collar Beam.

Collar. A compression ring around a small circular opening.

Collateral. Property pledged or in the possession of a creditor to guarantee payment of an obligation by a debtor.

Collector Efficiency. The ratio of heat energy extracted from a collector to the quantity of solar energy striking the cover expressed in percent.

Collector Elements. Elements that serve to transmit the inertial forces with the diaphragms to members of the lateral-force- resisting systems.

Collector Streets. Connecting roads between arterial streets, not necessarily continuous, to discourage through traffic; intersections often controlled by traffic lights; parking permitted under various conditions; pedestrian crossing controlled.

Collector. 1. In structural analysis, a force transfer element that collects loads from a diaphragm (horizontal element) and transfers them to the shear walls (vertical element). 2. Any of a wide variety of devices (flatplate, concentrating, etc.) used to collect solar energy and convert it to heat; a Solar Collector.

Collet. A metal band, flange, or ferrule.

Colloid. A mixture containing ultramicroscopic particles of one substance scattered evenly throughout another; also known as a dispersion.

Colloidal Suspension. A substance divided into fine particles which remains in permanent suspension in a liquid.

Collusion. A secret combination or conspiracy between two or more persons having a fraudulent or deceitful purpose; the illegal practice of contractors agreeing to submit higher bids so that a chosen contractor's bid will be low.

Colonial Base. An architectural style of a board or molding used against the bottom of walls to cover the joint with the floor and to protect the walls from dents and scuffs.

Colonial Molding. A style of molding from eighteenth century English Georgian, reproduced and incorporated into buildings in America.

Colonnade. A set of columns occurring at regular intervals.

Color Coding. The use of different colors to identify piping or wiring.

Color Man. In painting, the individual, either the journeyman or contractor, who is an expert in tinting and matching colors.

Color Pigments. Pigments such as blue, red, etc. which absorb a portion of the light which falls upon them and reflect or return to the eye certain groups of light bands which enable us to recognize various colors.

Color Retention. When a paint product exposed to the elements shows no signs of changing color it is said to have good color retention.

Color, Complementary. See Complementary Color.

Color, Primary. See Primary Color.

Color. The visual appearance of objects and light sources in terms of hue, lightness, and saturation for objects and hue, brightness, and saturation for light sources; a hue as contrasted with black, white, and gray.

Colored Finishes. Plaster finish coats containing integrally mixed color pigments or colored aggregates.

Colored Grout. Commercially prepared grout consisting of carefully graded aggregate, portland cement, water dispersing agents, plasticizers, and color fast pigments.

Color-Fast. Non fading.

Colorimeter. See Chromometer.

Coloring Strength. The relative strength or ability of pigments to color base material which is white or light in color.

Color-In-Japan. A paste formed by mixing a color pigment with Japan drier; used principally for tinting.

Color-In-Oil. A paste formed by mixing a color pigment in linseed or other vegetable oil; used principally for tinting.

Colossus. A statue of colossal size or proportions.

Columbarium. A structure containing vaults for cinerary urns.

Column Base Plate. The part of a structure on which the column base is set.

Column Base. 1. The part which forms the bottom of a vertical supporting pillar. 2. The plate beneath a column that distributes the load.

Column Cage. As assembly of vertical reinforcing bars and ties for a concrete column.

Column Cap. See Capital, 2.

Column Capital. The uppermost member of a column crowning the shaft and taking the weight of the beam or girder.

Column Cover. A structure that forms the uppermost part of a column.

Column Fireproofing. The act or process of applying fire-retardant materials to a column.

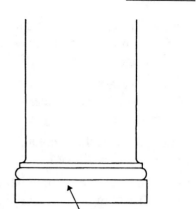

Column Base

Column Footing. Concrete support for a column; commonly known as individual footing, generally square or rectangular in shape.

Column Formwork. The mold or sheathing that forms the outline of a column into which the concrete is placed or poured.

Column Furring. Strips of wood or metal applied to a column to provide a fastening surface for a finish covering.

Column Pier. A foundation member of plain or reinforced concrete to support a column.

Column Reinforcing. The embedded steel bars to strengthen a concrete column.

Column Spiral. A continuous coil of steel reinforcing used in a concrete column.

Column Strip. The zone of a two-way concrete floor or roof structure that is centered on a line of columns.

Column Tie. A single loop of steel bar, usually bent into a rectangular configuration, used to tie the reinforcing cage in a concrete column.

Column, Concrete. A long, relatively slender, supporting pillar made from concrete and reinforcing steel.

Column, Precast. A column that has been cast and cured in other than its final position.

Column, Sheetrock. A column made from gypsum wallboard.

Column, Timber. Structural lumber, five inches or more in its least dimension used as a vertical compression member.

Column, Wood. Vertical wood structural member, usually carrying a beam.

Column. 1. A structural member used primarily to support axial compression loads and with a height of at least three times its least lateral dimension; an upright structural member acting primarily in compression. 2. A square, rectangular, or cylindrical support for roofs, ceilings, and so forth, composed of base, shaft, and capital.

Columniation. The arrangement of columns in a building.

Comb Board. See Saddle Board.

Combination Column. A column in which a structural steel member, designed to carry the principal part of the load, is encased in concrete which carries the remainder of the load.

Combination Doors or Windows. Doors or windows with self- storing or removable glass and screen inserts; the need for handling a different unit each season is thus eliminated.

Combination Frame. In light wood framing, a combination of the principal features of the full and balloon frames.

Combination Receptacle. An electrical fixture with an outlet for plugging in an electrical appliance, along with an electric switch for operating another circuit.

Combination Switch. A switch that includes a switch and a receptacle.

Combined Chlorine. Chlorine combined with other substances; though available to disinfect pool water; chlorine in this form is less effective than free chlorine.

Combined Footing. A concrete footing which supports two or more columns.

Combined Stress. The combination of axial and bending stresses in a structural member.

Combined Water. The water, chemically held as water of crystallization, by the calcium sulphate dihydrate, or hemihydrate crystal.

Combustible. Capable of being burned.

Combustion. Burning; consumption by fire; the development of light and heat from the chemical combination of a substance with oxygen.

Comfort Chart. Chart used in air conditioning to show the dry bulb temperature, humidity and air movement for human comfort conditions.

Comfort Cooler. System used to reduce the temperature in the living space in homes; these systems are not complete air conditioners as they do not provide complete control of heating, humidifying, dehumidification, and air circulation.

Comfort Zone. Area on psychrometric chart which shows conditions of temperature, humidity, and sometimes air movement in which most people are comfortable.

Commerce. 1. The exchange or buying and selling of commodities, especially on a large scale involving transport within a country or between countries. 2. The principles and techniques of business and office systems.

Commercial Carpet. Carpet that is highly resistant to heavy traffic.

Commercial Door. A type of door used for strength and durability in commercial building applications.

Commercial Facilities. Facilities that are intended for nonresidential use and whose operations will affect commerce, including factories, warehouses, office buildings, and other buildings in which employment may occur.

Commercial Matching. Matching of colors within acceptable tolerances, or with a color variation that is barely detectable to the naked eye.

Commercial Standard (CS). U.S. Dept. of Commerce, Govt. Printing Office, Washington, DC 20402.

Commercially Acceptable Standards. Of workmanship, the standard of work produced by the average competent craftworker, allowing a reasonable amount of imperfections; not perfect.

Commingling of Funds. To combine funds belonging to different accounts into a single account

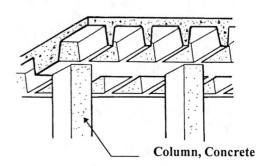

Column, Concrete

Commission. 1. The professional fee paid to a broker or agent for services, usually calculated as an agreed percentage of the sales or lease consideration. 2. A fee paid to an agent or employee for transacting a piece of business or performing a service.

Commissioning. Start-up of a building that includes testing and adjusting HVAC, electrical, plumbing, and other systems to assure proper functioning and adherence to design criteria; also includes the instruction of building representatives in the use of the building systems.

Commitment. 1. A legally enforceable agreement to do something in the future. 2. A pledge by a lending institution to make a real estate loan under certain stated conditions.

Commode. 1. A chest of drawers. 2. A chamber pot concealed in a chair with a hinged cover.

Common Bolt. An ordinary carbon steel bolt.

Common Bond. Brickwork laid with each five courses of alternating stretchers followed by one course of headers.

Common Brick. Inexpensive brick, not selected for appearance, that is used as filler or backing. See Building Brick.

Common Enemy Doctrine. The legal doctrine that flood waters are a common enemy and that property owners may fight to protect their property regardless of the damage to neighboring property.

Common Rafter. A rafter that is square with the plate and extends to the ridge.

Common Seal-P Trap. A P-trap with a water seal depth of 2 to 4 inches.

Common Vent. A vent that connects at the junction of two fixtures, acting as a vent for both fixtures.

Common Waster Pipe. Used when there are two sets of bathroom fixtures on opposite sides of a wall with their drain connections opposite each other; the fixtures may drain into the same waste and also have dual vents; fixtures which are directly across from each other may have common wastes and dual vents.

Communication Cable. A cable for transmission of telephone, television, and computer signals.

Communication Circuit. A circuit which is part of a central station system; such circuits include telephone, telegraph, district messenger, intercommunications, public address systems, fire and burglar alarms, watchmen and sprinkler supervisory circuits with their associated operating and signaling power supply equipment.

Communication Manhole. Any of various types of manholes used in the installation of communication conduits or cables.

Communication. An act or instance of transmitting information.

Community Kitchen. A room in a hotel or lodging house used or intended to be used by the occupants of two or more guest rooms for cooking or preparing food.

Community Shopping Center. An intermediate sized shopping center.

Commutator. Part of rotor in electric motor which conveys electric current to rotor windings.

Compact Borrow. Fill acquired from excavation that has been compacted.

Compacted Concrete. Freshly poured concrete that has been packed tighter by vibration, tamping, or a combination of both to remove voids.

Compacted Yards. Cubic measurement of soil or rock after it has been placed and compacted in a fill.

Compaction Tile. A hard tile surfacing unit made from a mixture of chemicals; the finished surface can be the mixture of chemicals or can be marble chips to create a terrazzo finish; the unit is made hard by the set of the chemicals and the product is not fired as in the manufacture of ceramic tile.

Compaction. The process whereby the volume of freshly placed material is reduced or flattened by vibration or tamping, or some combination of these; reduction in bulk of fill by rolling, tamping or soaking; insufficient compaction of asphalt pavement courses may result in channeling on the pavement surface.

Compactor. 1. A mechanical device that compresses objects into smaller units. 2. A machine in an industrial setting that compacts or compresses materials. 3. A machine in a kitchen that compresses or compacts materials by using hydraulic weight, force or vibration.

Companion Flange. A pipe connection device that is machined and drilled to match another flange on a pipe or fitting.

Comparative Negligence. The legal doctrine that wrongdoers should pay damages proportional to their fault.

Compartment Kiln. A kiln in which the total charge of lumber is dried as a single unit; it is designed so that, at any given time, the temperature and relative humidity are essentially uniform throughout the kiln; the temperature is increased as drying progresses, and the relative humidity is adjusted to the needs of the lumber.

Compartment, Shower. See Shower Compartment.

Compass. 1. A drafting tool; an adjustable device for drawing circles; similar to a dividers but with a point on one leg and a pen or pencil on the other; often called compasses or a pair of compasses. 2. A Magnetic Compass.

Common Bond

Compatibility. Ability to mix with or adhere properly to other components or substances.

Compensating Errors. Errors that are small in magnitude and in which the pluses and minuses (overs and unders) tend to offset each other; compare with Cumulative Errors.

Compensatory Damages. An amount calculated to compensate a party for economic loss caused by the wrongful act of another.

Competitive Bidding. The process of two or more contractors submitting proposals for the same work at the same time.

Complementary Color. A color that combined with its complement yields gray; complementary pairs include red-green, yellow-violet, and blue-orange.

Complete Fusion. In welding, fusion which has occurred over the entire base- metal surfaces exposed for welding.

Completion Bond. A bond guaranteeing to the lender that the project will be completed free of liens.

Completion Date. The date stipulated in the construction contract for substantial completion.

Completion of Construction When the work of the construction contract is complete.

Component Depreciation. A method of computing depreciation of property by its individual parts rather than as a whole.

Component. 1. A constituent part or ingredient. 2. One of two or more forces which, acting together, have the same effect as a single force, called the resultant.

Composite Beam. A beam that is composed of two different materials; for example, a wood and steel beam, or a steel beam and concrete slab, in which the two act as one.

Composite Board. A board that is made of several compressed materials; used for sheathing, wallboard, or as an insulation or acoustical barrier.

Composite Column. A column in which a steel or cast-iron structural member is completely encased in concrete containing spiral and longitudinal reinforcement.

Composite Concrete Flexural Composition. A precast concrete member and cast-in-place reinforced concrete so interconnected that the component elements act together as a flexural member.

Composite Decking. A type of decking construction made up of different materials, such as concrete, wood, and steel.

Composite Metal Decking. Corrugated steel decking manufactured in such a way that it bonds securely to the concrete floor fill to form a reinforced concrete deck.

Composite Steel Deck Shoring. Shoring used on composite steel decks during the placement of concrete.

Composite Wall. A masonry wall that incorporates two or more different types of masonry units, such as clay bricks and concrete blocks.

Composite. An order of Greek architecture consisting of elements of the Ionic and the Corinthian orders.

Composition Roofing. Any asphaltic roofing.

Composition Shingles. Shingles made or formed from composition roofing material.

Composition. 1. The factors or parts of any substance or material. 2. The quantitative or qualitative makeup of any chemical.

Compound Gauge. Instrument for measuring pressures both above and below atmospheric pressure.

Compound Refrigerating Systems. System which has several compressors or compressor cylinders in series; used to pump low-pressure vapors to condensing pressures.

Compreg. Wood in which the cell walls have been impregnated with synthetic resin and compressed; this process reduces swelling and shrinking and increases density and strength; compare with Impreg.

Comprehensive General Liability. Insurance policy covering comprehensive general liability.

Compression Bars. Steel used to resist compression forces.

Compression Block. An extremely short wooden column.

Compression Connector. A connecting device which when attaching uses a force that pushes together and squeezes.

Compression Coupling. A connecting device which creates a force that pushes together and crushes.

Compression Elbow. A connecting device which joins two pipes at 90 degrees which uses a force that pushes together and crushes.

Compression Failure. Deformation of the wood fibers resulting from excessive compression along the grain in direct end compression or in bending; it may develop in standing trees due to bending by wind or snow or to internal longitudinal stresses developed in growth, or it may result from stresses imposed after the tree is cut.

Compression Faucet. One designed to stop the flow of water by the action of a flat washer closing against a seat.

Compression Fitting. Bends, couplings, crosses, elbows, tees, and unions which use a force when connecting that pushes together and squeezes a metal or rubber gasket.

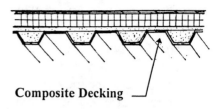
Composite Decking

Compression Gasket. A synthetic or rubber strip or washer that seals by being squeezed tightly.

Compression Gauge. An instrument used to measure positive pressures, those above atmospheric pressures, only; these gauges are usually calibrated from 0 to 300 psig (0-21.1 kg./cm^2).

Compression Lug. A connector for fastening the end of a wire to a terminal that uses a force that pushes together and crushes.

Compression Ratio. In a refrigeration compressor, the ratio of the volume of the clearance space to the total volume of the cylinder, in refrigeration it is also used as the ratio of the absolute low-side pressure to the absolute high-side pressure.

Compression Reinforcement. Steel reinforcing bars in a reinforced beam that are placed near the top of the beam to assist in resisting compressive forces.

Compression Ring. A structural element, circular in plan, which is in compression because of the action of the rest of the structure; for a dome, it would occur as a collar around a circular opening, or oculus, at the top; for a suspended roof, it would be around the outside edge.

Compression Splice. A connection of two similar materials by a force which tends to shorten a member.

Compression Strength. The ability of a structural material to withstand compression forces; the measured maximum resistance of a concrete or mortar specimen to axial loading; expressed as force per unit cross-sectional area; or the specified resistance used in design calculations, in the U.S. customary units of measure expressed in pounds per square inch.

Compression Valve. One designed to stop the flow of water by the action of a flat washer closing against a seat.

Compression Wood. Abnormal wood formed on the lower side of branches and inclined trunks of softwood trees; compression wood shrinks excessively lengthwise, as compared with normal wood.

Compression. 1. Force which tends to crush adjacent particles of a material together and cause overall shortening in the direction of its action; stress which tends to shorten a member. 2. The increase of pressure on a fluid by using mechanical energy.

Compressive Strength. The measured resistance of a concrete or mortar specimen to axial loading expressed as pounds per square inch of cross-sectional area; the maximum compressive stress which material, such as portland cement, concrete, or grout is capable of sustaining.

Compressor Displacement. Volume, in cubic inches, represented by the area of the compressor piston head or heads multiplied by the length of the stroke.

Compressor Seal. Leakproof seal between crankshaft and compressor body in open type compressors.

Compressor, External Type. See Compressor, Open Type.

Compressor, Hermetic. See Hermetic Compressor.

Compressor, Multiple Stage. Compressor having two or more compressive steps. Discharge from each step is the intake pressure of the next in series.

Compressor, Open Type. Compressor in which the crankshaft extends through the crankcase and is driven by an outside motor; also called External Drive Compressor.

Compressor, Reciprocating. See Reciprocating Compressor.

Compressor, Rotary. Compressor which uses vanes, eccentric mechanisms or other rotating devices to provide pumping action.

Compressor, Single Stage. Compressor having only one compressive step between low-side pressure and high-side pressure.

Compressor. A machine that compresses gases or air; pump of a refrigerating mechanism which draws a low pressure on the cooling side of the refrigerant cycle and squeezes or compresses the gas into the high-pressure or condensing side of the cycle.

Computer Cable. Coaxial cable which transmits computer signals.

Computer Floor. Special flooring designed to prevent electrostatic buildup and sparking in a computer room; usually elevated above the existing floor, to facilitate the running of wires between the components in the room.

Computer Hardware. All of the physical equipment, including the input units (keyboard and mouse), processing unit, and output units (screen, and printer).

Computer Language. A code used to write a computer program.

Computer Software. The programs that make a computer function.

Computerese. Arcane jargon used by computer technologists.

Computer-Room AC. An air-conditioning unit used in a computer room.

Concave Fillet Weld. A fillet weld having a concave face.

Concave Joint. In masonry, a mortar joint formed with a special tool or a bent iron rod; this joint is weather resistive and inexpensive.

Concave. Curved or rounded inward like the inside of a bowl; the opposite of convex.

Concavity. The maximum distance from the face of a concave fillet weld perpendicular to a line joining the toes.

Concealed Grid. A suspended ceiling framework that is completely hidden by the tiles or panels it supports.

Concealed Picture Mold. A recessed horizontal slot in a plaster wall, formed by a sheet metal screed, used to hang pictures and other objects.

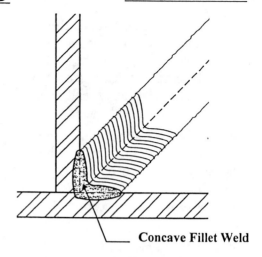

Concave Fillet Weld

Concealed Z Bar. A hidden z-shaped bar that is used as a wall tie.

Concealed. Rendered inaccessible by the structure or finish of the building.

Concentrated Load. A load which acts at one point or small area of a structure or member.

Concentrating Collector. A device that uses reflective surfaces to concentrate the sun's rays onto a smaller area, where they are absorbed and converted to heat energy.

Concentration. 1. The strength of a solution. 2. A concentrated mass or thing.

Concentric. Having a common center or axis.

Concrete Accessory. An implement or device used in the formwork, pouring, spreading, or finishing of concrete surfaces.

Concrete Admixture. A substance added to concrete to aid in imparting color, control workability, help in waterproofing, control setting, and to entrain air.

Concrete Beam. A horizontal structural member which transversely supports a load and transfers the load to vertical members, made of a composite material consisting of sand, coarse aggregate, cement and water.

Concrete Block Bar Supports. Precast concrete blocks, with or without tie wires used to support bars above the sub-grade or to space bars off vertical forms and above horizontal forms.

Concrete Block Removal. The act or process of demolition of a concrete block structure.

Concrete Block. A hollow concrete masonry unit made from portland cement and suitable aggregates such as sand, gravely crushed stone, bituminous or anthracite cinders, burned clay or shale, pumice, volcanic scoria, air-cooled or expanded blast furnace slags with or without the inclusion of other materials.

Concrete Brick. A solid concrete masonry unit the same size and proportion as a clay brick.

Concrete Burlap. A curing concrete surface that has had a coarse fabric of jute, hemp, or less commonly, flax applied, for use as a water-retaining covering.

Concrete Column. A long, relatively slender, supporting pillar made from concrete and reinforcing steel.

Concrete Cover. The distance from a reinforcing bar to the outside of a concrete member; also referred to as fireproofing, clearance, or concrete protection.

Concrete Cutting. Scoring or cutting of concrete or masonry with a saw equipped with a carborundum blade; commonly done by specialty subcontractors with customized equipment.

Concrete Dowel. A pin of reinforcing steel embedded in concrete to strengthen two pieces where they join or to create a place where other pieces can be fastened to it.

Concrete Filled Pile. A long slender construction element filled with concrete, driven in the ground for the purpose of supporting a load.

Concrete Finish, Board. Wooden boards placed in the concrete formwork as for liners to provide a wood-pattern finish to the completed reinforced concrete.

Concrete Finish. The act or process of the final compaction and finishing operations of curing concrete.

Concrete Headwall. The end of a culvert or drain constructed of concrete.

Concrete Manhole. A vertical access shaft from the surface to the underground, constructed of concrete.

Concrete Masonry Association of California and Nevada (CMACN).

Concrete Masonry Unit (CMU). See Concrete Block.

Concrete Mat. A grid of metal reinforcement for concrete foundations, slabs, or mats.

Concrete Mix. The amount of each material specified, portland cement, fine aggregate, coarse aggregate, admixture, and water.

Concrete Nail. A hardened steel nail that may be driven into concrete or masonry.

Concrete Parking Barrier. A concrete structure placed to act as a barrier against vehicular encroachment.

Concrete Paving. The use of concrete to make a hard surface in areas such as walks, roadways, ramps, and parking areas.

Concrete Pile. A precast slender reinforced concrete member that is embedded in the soil, by driving or inserting into a predrilled hole.

Concrete Pipe. Pipe manufactured from concrete; the manufacturing is done in a plant under controlled conditions; usually used for drainage but may also be used for sanitary sewers.

Concrete Placement. The placing and finishing of concrete during a continuous operation; also known as pouring.

Concrete Plank. A solid or hollow-core, flat-beam used for floor or roof decking; usually precast and pre-stressed.

Concrete Planter. A concrete reservoir to hold soil for plantings.

Concrete Pole. A vertical member made of concrete.

Concrete Post. A vertical structure made of concrete which carries stresses in compression.

Concrete Pump. An apparatus which forces concrete to the placing position through a pipeline or hose.

Concrete Receptor. A precast concrete structure which forms the drain area of a shower.

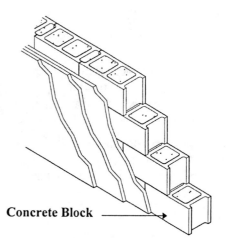

Concrete Block

Concrete Reinforcement. Steel rods that are embedded in wet concrete to give additional strength.

Concrete Reinforcing Steel Institute (CRSI). 933 North Plum Grove Road, Schaumburg, Illinois 60173-4758, (708) 517-1200.

Concrete Removal. The act or process of demolition of old concrete into manageable pieces.

Concrete Repair. To restore concrete by replacing a section or repairing what is broken.

Concrete Restoration. The rebuilding of the surface of concrete to approach as nearly as possible the original form.

Concrete Saw Cut. A cut in hardened concrete utilizing diamond or silicone-carbide blades or discs.

Concrete Sheet Piling. A row of concrete piles driven in close contact to provide a tight wall to resist the lateral pressure of water, adjacent earth or other materials.

Concrete Sleeper. Strips of wood placed on a rough concrete floor to which the finished wood floor is nailed.

Concrete Testing. Testing to determine the plasticity or strength of concrete.

Concrete Topping. A rich mixture of fine aggregate concrete used to top concrete floor surfaces for durability, safety and appearance.

Concrete Waterproofing. An act or process of adding a material to concrete to make it impervious to water or dampness.

Concrete, Cellular. See Cellular Concrete.

Concrete, Fibrous. See Fibrous Concrete.

Concrete, Field. See Field Concrete.

Concrete, Foamed. See Foamed Concrete

Concrete, Green. See Green Concrete

Concrete, Lightweight. Concrete that has substantially lower weight than that made from gravel or crushed stone.

Concrete, Normal Weight. See Normal Weight Concrete.

Concrete, Plain. See Plain Concrete.

Concrete, Precast. Concrete parts that are cast on- or off-site and, after curing and hardening, are installed in their final position of use.

Concrete, Prestressed. See Prestressed Concrete.

Concrete, Pumped. See Pumped Concrete.

Concrete, Reinforced. See Reinforced Concrete.

Concrete, Structural Lightweight. Concrete containing lightweight aggregate, not exceeding 115 pounds per cubic foot.

Concrete, Terrazzo. See Terrazzo.

Concrete. A mixture of portland cement, fine aggregate, coarse aggregate, admixtures, air, and water.

Concurrent Heating. The application of supplemental heat to a structure during a welding or cutting operation.

Concurrent. The point at which the line of action of several forces meet.

Condemnation. A procedure by which private property is acquired for public use by the legal process of eminent domain.

Condensate Pump. A pump that removes water or condensation from an air-cooling unit.

Condensate. A product obtained by cooling vapors of a substance being distilled.

Condensation Drain. A drain pipe to carry off water condensed from the atmosphere.

Condensation. The process of changing from a gaseous to a liquid state, especially as applied to water; liquid drops which form when a vapor is chilled below its boiling point; water droplets that deposit on surfaces whose temperature is below the dew point.

Condense. Action of changing a gas or vapor to a liquid.

Condenser Comb. Comb-like device, metal or plastic, used to straighten the metal fins on condensers or evaporators.

Condenser Fan. Forced air device used to move air through air- cooled condenser.

Condenser, Air-Cooled. Heat exchanger which transfers heat to surrounding air.

Condenser, Water Cooled. Heat exchanger designed to transfer heat from hot gaseous refrigerant to water.

Condenser. A heat exchanger in a refrigeration cycle used to discharge heat to the outside; commonly used types are water cooled, air cooled, and evaporative; condenser water is normally circulated through a cooling tower through which heat is distributed to the atmosphere.

Condensing Pressure. Pressure inside a condenser at which refrigerant vapor gives up its latent heat of vaporization and becomes a liquid; this varies with the temperature.

Condensing Temperature. Temperature inside a condenser at which refrigerant vapor gives up its latent heat of vaporization and becomes a liquid; this varies with the pressure.

Condensing Unit Service Valves. Shutoff valves mounted on condensing unit to enable service technicians to install or service unit.

Condensing Unit. Part of a refrigerating mechanism which pumps vaporized refrigerant from the evaporator, compresses it, liquefies it in the condenser and returns it to the refrigerant control.

Conditional Sales Contract. A contract for sale of land, usually with installment payments and where title does not pass until a certain amount, or the full amount, has been paid; also called Agreement of Sale, Contract for Deed, Installment Land Sale, or Land Sale Contract.

Conditioned Air. Air that has been heated, cooled, humidified, or dehumidified to maintain an interior space within the comfort zone; also called Tempered Air.

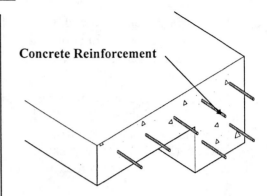

Concrete Reinforcement

Conditioned Water. In swimming pools, water treated with cyanuric acid or chlorinated isocyanurate to prevent chlorine from dissipating in sunlight.

Condominium. Individual ownership of a unit in a multi-unit development, as dwellings, offices, storage, or manufacturing spaces; also includes ownership of an interest, in common with other owners, of common areas and facilities that serve the structure.

Conductance. 1. The property of a material to conduct electric current. 2. A property of a material to conduct heat; a property of a slab of material equal to the quantity of heat, in BTU. per hour, that flows through 1 square foot of the slab, when a 1 °F. temperature difference is maintained between the two sides.

Conduction. Heat transfer from one particle to another in direct contact with it.

Conductive Floor. Flooring material specifically designed to prevent electrostatic buildup and sparking.

Conductive Mortar. A tile mortar to which specific electrical conductivity is imparted through the use of conductive additives.

Conductive Terrazzo. A type of tile that prevents electrostatic buildup and sparking, that is made from specific conductive materials.

Conductive Tile. Tile made from special body compositions or by methods that result in specific properties of electrical conductivity while retaining other normal physical properties of ceramic tile.

Conductive. Having the quality or power of conducting or transmitting heat, electricity, or static electricity.

Conductivity (k). Ability of a substance to conduct or transmit heat or electricity; the reciprocal of resistivity, $1/r$.

Conductor, Aluminum. An aluminum wire or cable for transmitting electrical current.

Conductor, Copper. A copper wire or cable for transmitting electrical current.

Conductor, Stranded. A number of fine wires twisted around a center wire or core, used as a single electric conductor.

Conductor. A substance or body capable of transmitting electricity, heat or sound.

Conductors. 1. Pipes for conducting water from a roof to the ground or to a receptacle or drain; downspout. 2. Any electrical wire used to convey electricity.

Conduit Bushing. A threaded metal or plastic pipe connector used to connect conduit to a box or other housing where the hole is not threaded.

Conduit Cap. A cap placed on the end of a length of conduit to protect the threads or terminate the conduit run.

Conduit Locknut. A threaded connector used where conduit enters an electrical box or other housing where the hole in not threaded; the locknut is screwed onto the threaded end of the conduit on the outside of the box and connected to a bushing on the inside of the box until secure.

Conduit Locknut. A threaded connector used where conduit enters an electrical box or other housing where the hole is not threaded; the locknut is screwed onto the threaded end of the conduit on the outside of the box and connected to a bushing on the inside of the box until secure.

Conduit Plug. A fitting that is screwed into the ends of conduit or conduit fittings.

Conduit, Aluminum. A pipe constructed of a light alloy material used to enclose electric wires to protect them from damage.

Conduit, EMT. See EMT Conduit.

Conduit, Flexible. See Flexible Conduit.

Conduit, Galvanized. See Galvanized Conduit.

Conduit, Plastic Coated. See Plastic Coated Conduit.

Conduit, PVC. See PVC Conduit.

Conduit, Steel. See Steel Conduit.

Conduit. A protective sleeve or pipe commonly used for individual electrical conductors.

Cone, Slump. See Slump Cone.

C

Cone. 1. A solid figure with a circular (or other curved) plane base, tapering up to a point. 2. The dry fruit of a conifer; a pine cone. 3. The conical part of a gas flame next to the orifice of the tip.

Conference Room. A room for meetings.

Configuration. An arrangement of parts or elements in a particular form or figure.

Congo Gum. A gum resin obtained from the Congo region of Africa.

Conifer. Any evergreen of a group usually bearing cones; conifers produce softwood lumber.

Connection, Bus Duct. See Bus Duct Connection.

Connection. The union, or joint, of two or more distinct elements.

Connector Set Screw. A screw on a connector fitting that when tightened connects two components together.

Connector, Compression. See Compression Connector.

Connector, Die Cast. A connecting device that has been cut, formed and threaded by die tools.

Consent of Surety. A written consent of the surety to pay the final payment to the contractor, reduce or eliminate the retainage, or make any change in the contract conditions.

Conservation. The careful preservation and protection of something, such as a building or the environment.

Conservatism. A principle of accounting, the object of which is to place each asset item on the balance sheet at a conservative figure or low end of its price range.

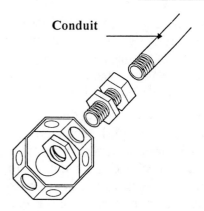

Conduit

Conservatory. A greenhouse for tending and displaying plants.

Consideration. A benefit (or money) coming from a promisee to a promisor in exchange for the promisor's agreement to perform an act.

Consistency, Normal. Of gypsum plaster or gypsum concrete, see Normal Consistency.

Consistency. 1. The fluidity or viscosity of a liquid or paste; resistance of a product to flow. 2. The degree of plasticity of fresh concrete or mortar; the normal measure of consistency is slump for concrete and flow for mortar. 3. The degree of fluidity or plasticity of asphalt cement at any particular temperature; the consistency of asphalt cement varies with temperature; therefore, it is necessary to use a common or standard temperature when comparing the consistency of one asphalt cement with another, the standard test temperature being 140° F (60° C).

Console. 1. An ornamental bracket supporting a shelf. 2. A cabinet or panel containing controls and switches for operating lighting, sound, television, or radio equipment.

Consolidation Compaction. This is usually accomplished by vibration of newly placed concrete to its minimum practical volume, to mold it within form shapes and around embedded parts and reinforcement and to eliminate voids other than entrained air.

Consolidation. Process of working fresh concrete so that a closer arrangement of particles is created and the number of voids is decreased or eliminated.

Consortium. A combination of companies organized to invest in an enterprise usually too large for them individually.

Constant Air Volume System. Air handling system that provides a constant air flow while varying the temperature to meet heating and cooling needs.

Constant. A component of a relationship between variables that does not change its value.

Constrictor. Tube or orifice used to restrict flow of a gas or a liquid.

Construct. Make by putting materials or parts together; build; erect.

Constructibility. The capacity of a certain design of being developed into construction.

Construction Change Directive. A document that directs a change in the work of the project, prepared by the architect and signed by the owner and architect. This change can adjust the contract sum and/or contract time. This document may be used in the absence of the contractor's agreement on the terms of a change order.

Construction Documents Phase. One of the standard phases of architectural service (Schematic Design Phase, Design Development Phase, Construction Documents Phase, Bidding or Negotiation Phase, and Construction Phase-Administration of the Construction Contract).

Construction Documents. The drawings and specifications that describe the construction requirements.

Construction Industry Arbitration Rules. Arbitration rules administrated by the AAA and referred to in AIA's standard agreements for construction and architectural services.

Construction Joint. The contact between the placed concrete and concrete surfaces, against or upon which concrete is to be placed and to which new concrete is to adhere, that has become so rigid that the new concrete cannot be incorporated integrally by vibration with that previously placed.

Construction Lender. An institution or individual that lends money to a borrower for the purpose of constructing improvements on property.

Construction Lime. An industrial form of calcium oxide that is added to give it cling or adhesion.

Construction Loan. A short term interim loan to pay for the construction of a building.

Construction Management. Activities over and above normal architectural and engineering services, conducted during the predesign, design, and construction phases, that contribute to the control of time and cost.

Construction Manager. A person or entity who provides construction management services, either as an advisor or as a contractor.

Construction Phase-Administration of the Construction Contract. One of the standard phases of architectural service (Schematic Design Phase, Design Development Phase, Construction Documents Phase, Bidding or Negotiation Phase, and Construction Phase-Administration of the Construction Contract).

Construction Schedule. A time table setting out the times for starting and completing each of the operations required for the construction of a building or other project.

Construction. 1. The act or process of building. 2. Buildings and structures that have been built. 3. The method by which the carpet is made (loom or machine type) and other identifying characteristics, including pile rows per inch, pitch, wire height, number of shots, yarn count and plies, pile yarn weight, and density.

Constructive Notice. The giving of notice by recording it in the office of the county recorder or placing it in a newspaper.

Consultant. One who provides services as an advisor.

Consumable Insert. See Backing Filler Metal.

Consumer Product Safety Commission (CPSC). Washington, DC 20207, (301) 492-6800.

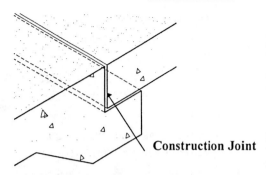

Construction Joint

Contact Ceiling. A ceiling which is secured in direct contact with the construction above without use of furring.

Contact Cement. Neoprene rubber-based adhesive which bonds instantly upon contact of parts being fastened.

Contact Fireproofing. See Fireproofing.

Contact Splice. A means of connecting reinforcing bars by lapping in direct contact.

Contract for Deed. See Conditional Sales Contract.

Contractor. An electrical relay to provide power to electrical equipment; an electrically operated switch, usually by a coil or solenoid, containing one or more sets of contacts which control one or more circuits.

Contaminant. 1. A substance that taints, infects, or pollutes; an impurity. 2. In a refrigeration system, substances such as dirt, moisture, or other matter foreign to refrigerant or refrigerant oil in system.

Contaminated. 1. No longer as originally manufactured or no longer pure because of contact with some foreign substance. 2. In tile work, stained tile as a result of carton and tile being saturated by moisture, oils, solvents or other materials.

Contingency Allowance. A sum in a construction budget to cover unforeseen expenses.

Continuity. 1. Uninterrupted electrical path in a circuit. 2. Degree of being intact or pore free.

Continuous Beam. A beam that is supported by more than two supports.

Continuous Cycle Absorption System. System which has a continuous flow of energy input.

Continuous Duty. In an electrical system, a requirement of service that demands operation at a substantially constant load for an indefinitely long time.

Continuous Footing. A concrete footing that supports a wall or two or more columns; the footing can vary in width and depth, also called a Strip Footing.

Continuous High Chairs. Welded wire bar supports consisting of a top longitudinal supporting wire with evenly spaced legs welded thereto and used to support bars near the top of slabs.

Continuous Hinge. A hinge designed to be the same length as the moving part (cabinet door or lid) to which it is applied; also called Piano Hinge.

Continuous Load. An electrical load where the maximum current is expected to continue for three hours or more during any one period of time.

Continuous Mix Plant. A manufacturing facility for producing asphalt paving mixtures that proportions those aggregate and asphalt constituents into the mix by a continuous volumetric proportioning system without definite batch intervals.

Continuous Ridge Vent. A screened, water-shielded ventilation opening that runs continuously along the ridge of a gable roof.

Continuous Vent. A vertical vent pipe that is a continuation of the waste drain.

Continuous Waste and Vent. A vertical vent pipe that is in a continuous line with the waste line; below the point where the fixture drains into the pipe is the waste line; above this point is the vent.

Continuous Waste. Two or more fixtures that use a single trap.

Contour Interval. The difference in vertical measurement between adjacent contour lines on a topographic map.

Contour Lines. The lines connecting points of equal elevation on a topographic map.

Contour Map. See Topographic Map.

Contract Award. An owner's notice to a contractor that a bid has been accepted and a contract will be entered into.

Contract Bonds. Performance Bond and Payment Bond, required by some construction contracts.

Contract Documents. The documents that comprise a construction contract, consisting of the agreement between owner and contractor, the conditions of the contract (general, supplementary, and other), drawings, specifications, and addenda issued before the contract is signed; also included are modifications issued after execution of the contract.

Contract of Adhesion. A contract offered to a person of inferior bargaining position who has no effective means of negotiating more favorable terms.

Contract of Sale. See Land Contract.

Contract Sum. The amount stated in the contract for the owner to pay to the contractor for doing the work of the contract.

Contract Time. The amount of time stated in the contract for substantial completion of the work of the contract.

Contract. An agreement or covenant between two or more persons in which each is bound to do or refrain from doing some act, and each acquires a right to what the other promises.

Contraction Joint. Formed, sawed, or tooled groove in a structure to create a weakened plane and regulate the location of cracking resulting from the dimensional change of different parts of the structure.

Contraction. 1. Becoming smaller from any cause. 2. A shrinking in size when something gets colder. 3. Reduction in the size of a business.

Contractor CM. A construction manager who acts as a contractor and who may guarantee the construction cost.

Contractor Manager. The role of the contractor as a business manager.

Contractor. One who enters into a contract for construction of a building or part of a building; a builder.

Contractor's Level. Builder's level.

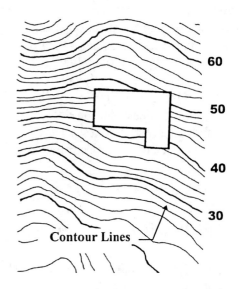

Contour Lines

Control Circuit. In a control apparatus or system, the circuit which carries the electric signals directing the performance of the controller, but does not carry the main power.

Control Joint. An intentional linear discontinuity in a structure or component, designed to form a plane of weakness where cracking or movement can occur in response to various forces so as to minimize or eliminate cracking elsewhere in the structure; see Control Joint and Relief Joint.

Control Panel. A panel, cabinet, or enclosure containing two or more controllers, contractors, relays, or other control devices for the control of electrical circuits, equipment, apparatus or system.

Control System. All of the components required for the automatic control of a process variable.

Control Valve. A mechanical device used to regulate an operation or function; a valve which regulates the flow or pressure of a medium which affects a controlled process; such valves are operated by remote signals from independent devices using any of a number of control media such as pneumatic, electric, or electrohydraulic.

Control, Compressor. See Motor Control.

Control, Defrosting. See Defrosting Control.

Control, Low Pressure. Low pressure Control.

Control, Motor. See Motor Control.

Control, Pressure Motor. High or low-pressure control connected into the electrical circuit and used to start and stop motor. It is activated by demand for refrigeration or for safety.

Control, Refrigerant. See Refrigerant Control.

Control, Temperature. See Temperature Control.

Controller. A device, or group of devices, which serves to govern in some predetermined manner the electric power delivered to the apparatus to which it is conducted.

Controls. 1. A mechanism used to regulate or guide the operation of a machine, apparatus, or system. 2. A functional area of management established to inform the manager when the actual experience of the business is different from that which was planned and to help maintain original budget and plan.

Conurbation. The growing together of previously separate towns creating a large community.

Convection Oven. An oven operating by the principal of heat transfer through automatic heated air circulation.

Convection, Forced. See Forced Convection.

Convection, Natural. See Natural Convection.

Convection. Heat transferred by fluid motion.

Convector. A heat exchange device that uses the heat in steam, hot water, or an electric resistance element to warm the air in a room; also called a radiator.

Conventional Installation. The method of installing ceramic tile with portland cement mortar.

Conventional Light-Frame Construction. A type of construction whose primary structural elements are formed by a system of repetitive wood-framing members.

Conventional Loan. A mortgage not insured by the FHA or guaranteed by the VA.

Converter. That which causes change to different state; catalyst; curing agent; promoter.

Convex Fillet Weld. A fillet weld having a convex face.

Convex. A protruding rounded surface, curved or rounded like the exterior of a sphere or circle.

Convexity. The maximum distance from the face of a convex fillet weld perpendicular to a line joining the toes.

Conveyance. The transfer of property from one owner to another, usually by deed.

Conveying System. A device used to move material from one place to another, including elevators, escalators, moving walks, dumbwaiters, and conveyors.

Cooking Unit, Counter-Mounted. An assembly of one or more domestic surface heating elements for cooking purposes designed for flush mounting in, or supported by, a counter, and which assembly is complete with inherent or separately mountable, controls and internal wiring; see Oven, Wall-Mounted.

Cool Colors. Hues or colors in which blue-green predominates; so termed because of the association with ice, water, and sky; also called Cold Colors.

Cooled Loop Run-Around. A method of heat recovery from exhaust air which uses finned-tube coils installed in the incoming and exhaust air supply.

Cooler, Nail. A nail with special size and head configuration for use in gypsum board applications.

Cooler, Bottle. See Bottle Cooler.

Cooler, Water. See Water Cooler.

Cooler. Heat exchanger which removes heat from a substance.

Cooling Coil. A heat exchange between a cooling element and the air stream; the cooling medium may be chilled water or refrigerant gas.

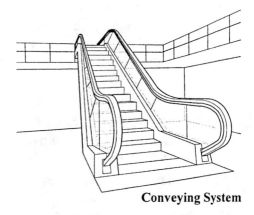

Conveying System

Cooling Tower. A structure in which warm water is circulated for cooling by evaporation by exposure to the air; the water is cooled to the wet bulb temperature of the air.

Cooper. A barrel maker.

Cooperage. 1. Barrels and kegs. 2. A place where barrels are made.

Cooperation. 1. The act of working together for the common good. 2. A behavioral aspect of management; this is achieved by leadership and should be encouraged by management.

Coordinates. A system of magnitudes to fix the position of a point, line, or plane.

COP. Coefficient of Performance.

Copals. Group of resinous substances exuding from various tropical trees; collected from living trees and also dug from the ground as a fossil; includes resins such as amber, congo, kauri, manila, Pontianak, West India gum, and zanzibar.

Cope. To cut or shape the end of a molded wood member so it will cover and fit the contour of an adjoining piece of molding; see Scribe and Scribing.

Coped Joint. See Scribing, 2.

Coping Saw. A handsaw with a very narrow blade held under tension in a U-shaped frame; used for cutting curves in wood.

Coping, Clay. See Clay Coping.

Coping, Limestone. The use of limestone to form a cap or finish on top of a wall, pier, pilaster, or chimney.

Coping, Precast. A precast concrete member used to form a cap or finish on top of a wall, pier, pilaster, or chimney.

Coping. 1. The material or units used to form a cap or finish on top of a wall, pier, pilaster, or chimney; a protective cap at the top of a masonry wall; it should be waterproof and weather resistant and sloped to shed water. 2. Perimeter edging around a swimming pool.

Copolymer. A product of copolymerization; substance obtained when two or more types of monomers polymerize.

Copolymerization. Simultaneous polymerization of two compounds which have properties different from polymer obtained with either monomer separately; see Polymerization.

Copper Bellows. A flexible joint in copper piping that can expand or contract to allow for thermal fluctuations.

Copper Braid. Three or more strands of copper intertwined.

Copper Development Association (CDA). 260 Madison Avenue, New York, New York 10016, (212) 251-7200.

Copper Piping. Pipe and tubing manufactured of copper, classified as Type K, L or M. Type K being the thickest walled, Type M, the thinnest walled.

Copper Plating. Abnormal condition developing in some units in which copper is electrolytically deposited on some compressor surfaces.

Copper Sulfate Test. In test for Mill Scale, copper color indicates absence of mill scale when steel is swabbed with 5 to 10 percent solution.

Copper Sulfate. 1. A chemical compound, $CuSO_4$, used in fungicides, electroplating solutions, textile dyeing, and as a timber preservative. 2. An algaecide declining in popularity because of its toxicity and incompatibility with some other compounds found in swimming pools.

Copper. a common reddish metallic element that is ductile and malleable and one of the best conductors of heat and electricity.

Copperas. See Ferrous Sulphate.

Copyright. The exclusive legal right to publish or sell artistic compositions.

Corbel Out. To construct a corbel of brick or stone.

Corbel. 1. A masonry unit such as brick or stone which projects beyond the unit below; a spanning device in which masonry units in successive courses are cantilevered slightly over one another; a projecting bracket of masonry or concrete. 2. A projection from the face of a beam, girder, column, or wall used as a beam seat or a decoration.

Cord Pendant. A cord hanging freely in the air in a vertical position, which has a fixed connection to a permanent wiring enclosure at the upper end and having a suitable cord connection body or lampholder attached to the lower end.

Cord, Electric. See Electric Cord.

Cord. 1. A small, very flexible insulated electrical cable. 2. A flexible thick string or thin rope consisting of woven thin strands. 3. A volume measure of cut firewood of 128 cubic feet (3.63 cubic meters).

Cordierite Porcelain. A vitreous ceramic whiteware for technical application in which cordierite, 2MgO 2A$_{12}$O3 - 5SiO$_2$ is the essential crystalline phase.

Cordless Tool. A handtool, such as a drill or screwdriver, containing a rechargeable battery.

Cordwood. Wood cut in 4 foot or shorter lengths to be used as fuel; see Cord.

Core Barrel. A hollow cylinder containing a socket and choker springs for holding a section of drilled rock.

Core Bracing. Vertical elements of a lateral bracing system such as the walls for stairs, elevators, or duct shafts.

Core Cock. A type of valve through which the flow of water is controlled by a circular core or plug that fits closely in a machined seat; the core has a part bored through it to serve as a water passageway; also called Plug Valve.

Core Drill. A rotary drill, usually a diamond drill, equipped with a hollow bit and a core lifter.

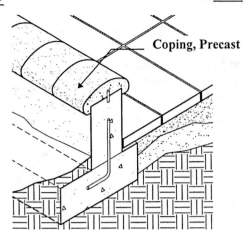

Coping, Precast

Core Drilling. The process of drilling which extracts a cylindrical sample of concrete, rock or soil; sometimes used to install pipe or conduit through an existing concrete or masonry wall.

Core Hardness. The resistance of the core to penetration by a steel punch as measured by ASTM C 473 hardness test.

Core Separation. A split in the gypsumboard core often accompanying an over calcined condition.

Core Stock. A solid or discontinuous center ply used in panel-type glued structures such as solid or hollowcore doors.

Core, Air. See Air Core.

Core, Magenetic. See Magnetic Core.

Core. 1. In plywood, the center of the panel; it may be either veneer or lumber. 2. A cylindrical piece of an underground formation cut and raised by a rotary drill with a hollow bit. 3. A hollow space within a concrete masonry unit formed by the face shells and webs. 4. The holes in clay masonry units. 5. The gypsum structure between the face and back papers of gypsum board.

C

Coreboard. A gypsum board product used primarily in shaftwall systems, normally 1 inch thick or less, either laminated or homogenous, usually manufactured in 24 widths and lengths as per job requirements; may have either square, rounded, or tongue and groove edges.

Cored Block. See Gypsum Block.

Cored Brick. A brick in which the holes consist of less than 25 percent of the section.

Cored Plug. A wooden plug inserted into a drilled hole where a nail, screw or other fastener has been driven.

Cored Slab. A concrete slab that has been drilled with holes for the installation of plumbing pipe or conduit.

Cored Tile. See Gypsum Tile.

Coring Concrete. To drill concrete to obtain samples for testing or to drill a hole in concrete masonry for conduits or pipe.

Corinthian. The order of Greek architecture characterized by ornate decoration and flared capitals with acanthus leaves.

Cork Running Track. Compressed cork particles used as the base in a running track.

Cork Tile. A flooring material made of a thin sheet of cork cut into a tile shape and attached by adhesive.

Cork Wall Covering. Cork particles that are bound and pressed into sheets and used to deaden sound and add insulation value.

Corner Bead. 1. A metal or plastic strip used to form a neat, durable edge at an outside corner of two walls of plaster or gypsum board. 2. A small, usually curved, wood mold for covering an inside corner.

Corner Block. A large triangular piece of wood or metal used for added strength at the corners of frames or where legs and rails join, see Glue Block.

Corner Boards. Boards used as trim for the corners of a house or other frame structure and against which the ends of the siding are finished.

Corner Bullnose. A type of ceramic tile bullnose trim with a convex radius on two adjacent edges.

Corner Cabinet. A cabinet wall unit that extends down two walls from the inside corner point.

Corner Guard. Type of molding that is mounted on outside corners in a room or space for finishing and for the protection of the corner from damage.

Corner Joint. A joint between two members located approximately at right angles to each other in the form of an L or 90 degrees.

Corner Lath. See Corner Reinforcement.

Corner Post. 1. A vertical post located in the corner of a timber structure. 2. A glazing mullion in the corner of a structure that retains glazing in both walls.

Corner Protection. The act or process of attaching molding to the outside corner of two walls for protection from bumping.

C

Corner Reinforcement. Plaster reinforcement used at re-entrant or internal angles to provide continuity between two intersecting plaster planes, usually a strip of diamond mesh metal lath bent to form a right angle; also called Cornerite or Corner Lath.

Corner Studs. The arrangement of studs on a corner of a wood frame building that provides nail backing for the lathing or finishes both inside and out.

Corner Tool. In gypsum wallboard finishing, an angular finishing knife to allow the simultaneous application of joint treatment to both sides of a 90 degree interior angle.

Corner Weld. Weld in a joint between two members located approximately at right angles to each.

Cornerite. See Corner Reinforcement.

Cornerstone. 1. A stone in a corner of a wall. 2. A ceremonial stone with names of the sponsors, designers, and constructors and the date of construction.

Cornice Molding. Molding that is used to form a projection at the top of a wall; a crowning member at the top course of a wall.

Cornice. 1. The exterior detail at the meeting of a wall and a roof overhang. 2. A decorative molding at the intersection of a wall and a ceiling. 3. The molded projection which finishes the top of the wall of a building.

Corporation. A group of people authorized by law to act as an individual; an artificial being, created by the State through a charter to engage in a particular kind of business.

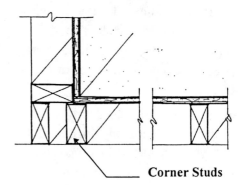

Corner Studs

Corridor. A passage or hallway from which doors lead into rooms.

Corrosion. Chemical reaction that causes deterioration of metal.

Corrosion-Resistant. Ability of a metal to withstand the effects of corrosion; corrosion resistant metals include any nonferrous metal or any metal having an unbroken surfacing; also includes nonferrous metal or steel with not less than 10 percent chromium or with not less than twenty-hundredths percent copper.

Corrosive Flux. A flux with a residue that chemically attacks the base metal; it may be composed of inorganic salts and acids, organic salts, and acids of activated rosins or resins.

Corrugated Fastener. A type of nail used to hold butt joints together; the slight taper of the corrugations tend to pull the joint together.

Corrugated Glass. Glass that is in a corrugated shape and is translucent but not transparent.

Corrugated Panel. Metal or fiberglass sheeting formed into alternating ridges and valleys in parallel and applied as siding on a building or structure.

Corrugated Roof. Metal or fiberglass formed into alternating ridges and valleys in parallel and mounted on rafters to serve as sheet roofing.

Corrugations. In asphalt paving, a type of pavement distortionion; a form of plastic movement typified by ripples across the pavement surface; these distortions usually occur at points where traffic starts and stops, on hills where vehicles brake on the downgrade, on sharp curves, or where vehicles hit a bump and bounce up and down; they occur in asphalt layers that lack stability, usually caused by a mixture that is too rich in asphalt, has too high a proportion of fine aggregate, has coarse or fine aggregate that is too round or too smooth, or has asphalt cement that is too soft; it may also be due to excessive moisture, contamination due to oil spillage, or lack of aeration when placing mixes using liquid asphalts; see Shoving.

Corundum. Extremely hard crystallized alumina, used as an abrasive.

Cost Approach. A real estate appraisal method by which the replacement cost is estimated and then reduced by the accumulated depreciation.

Cost Estimate. A preliminary statement of approximate cost, determined by one of the following methods: (1) Area and volume method; cost per square foot or cubic foot of the building. (2) Unit cost method; cost of one unit multiplied by the number of units in the project; for example, in a hospital, the cost of one patient unit multiplied by the number of patient units in the project. (3) In-place unit method; cost in-place of a unit, such as doors, cubic yards of concrete, and squares of roofing.

Cost Plus Contract. A type of construction contract where the contract price is the sum of the costs of labor, materials, and subcontracts plus a fixed or percentage fee.

Cost Plus Fee Agreement. A construction contract where the contractor is paid stipulated direct and indirect costs plus a fee. Also called cost plus.

Cost. The amount paid or charged for something; price.

Cottage. A small, simple house, especially in the country.

Cotter Pin. A longitudinally split pin which is inserted into a pre-drilled hole at the end of a rod-type fastener; the two ends of the cotter pin are then spread apart to resist removal of the pin.

Coulomb. An SI unit of electric charge, equal to the quantity of electricity conveyed in one second by a current of one ampere.

Coumarone-Indene Resins. Resins derived as by-products in making coke from coal.

Council of American Building Officials (CABO). 5303 Leesburg Pike, #798, Falls Church, Virginia 22401, (703) 931-4533.

Count. A number identifying yarn size or weight per unit of length, or length per unit of weight, depending on the spinning system used, such as denier, woolen, worsted, cotton, or jute systems.

Counter EMF. Tendency for reverse electrical flow as magnetic field changes in an induction coil.

Counter Griddle. A cooking device that has been installed in a kitchen counter.

Counter Lavatory. A sink that is installed in a counter top of a base cabinet.

Counter Top Range. A burner assembly mounted on the top of a kitchen counter allowing for additional cabinet space below.

Counter, Laboratory. See Laboratory Counter.

Counter. A table or case over which transactions are made or work is done.

Counterbalance. A weight or force that balances another

Counterboring. Enlarging a hole so that the head of a screw or bolt inserted in it can be completely covered.

Counterbrace. Bracing installed in opposite directions.

Counterflashing. An inverted L-shaped metal strip built into a wall to overlap base flashing and make a roof or wall watertight; also called Cap Flashing.

Counterflow. Flow in opposite direction.

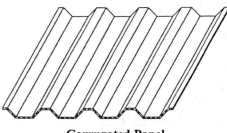

Corrugated Panel

Counterfort. A buttress or portion projecting from a wall and upward from the foundation to provide additional resistance to thrusts.

Countersink. A funnel-shaped enlargement of a drilled hole to allow a screw head to be flush with the surface of the drilled material.

Countersunk Plug. A wooden peg used to fill a drilled hole in a wooden surface.

Countertop. The work surface placed on base cabinets in a kitchen, lavatory, or laboratory.

Couple. Where a pair for forces of equal magnitude acting in parallel but opposite directions are capable of causing rotation.

Coupling Set Screw. A screw used to secure a fitting in place.

Coupling, Compression. A connecting device which when attaching uses a force that pushes together and squeezes.

Coupling, Split. A coupling that is split longitudinally and is assembled and secured with screws.

Coupling, Threaded. A fitting for joining two lengths of pipe that is threaded for connection to another fitting with connecting threads.

Coupling, Threadless. A fitting for joining two lengths of pipe that is slid over for connection by soldering, welding or cementing.

Coupling. A pipe fitting containing female threads on both ends; used to join two or more lengths of pipe in a straight run or to join a pipe and fixture.

Course Textured Wood. Wood with large pores, such as oak, ash, chestnut, and walnut.

Course. 1. In masonry, a continuous layer of bricks or block. 2. A continuous row of shingles in a roof. 3. Any layer in a waterproofing system.

Coursed. In masonry, laid in courses with straight bed joints.

Courtyard. A court or enclosure adjacent to a building.

Cove Base, Glazed. Cove base tile that has a ceramic coating that is hard, thin, and glossy.

Cove Base, Sanitary. A trim tile having a concave radius on one edge and a convex radius with a flat landing on the opposite edge; used as the only course of tile above the floor tile.

Cove Base, Tile. Tile that is placed in the bottom course of a tiled wall.

Cove Base. A flexible strip of plastic or synthetic rubber used to finish the junction between floor and wall.

Cove Molding. Molding that is concave-shaped; used to cover interior angles, such as that between the ceiling and a wall.

Cove. A trim tile unit having one edge with a concave radius; used to form a junction between the bottom wall course and the floor or to form an inside corner.

Coved Stair Riser. A concave-shaped surface on the vertical part of a stairway riser where the riser meets the horizontal or stair tread.

Covenant. A promise.

Covenants, Conditions, and Restrictions (CC&Rs). The basic rules establishing the rights and obligations of owners of real property within a subdivision in relation to other owners within the same tract and in relation to an association of owners organized for the operation and maintenance of property owned in common by the individual owners.

Cover, Catch Basin. A cast iron plate or grate on top of a receptacle or reservoir that catches water runoff or aids in drainage.

Cover, Manhole. A heavy, usually round, steel or iron cover used to gain access to underground work through a manhole.

Cover, Pool. A tarp used to cover a pool when not in use.

Cover. Concrete cover.

Coverage. 1. The scope of an insurance policy. 2. Amount of surface a given quantity of paint will cover. 2. The ability of paint to conceal the surface being painted. 3. The surface area to be continuously coated by a specific roofing material after allowance is made for a specified lap of the material.

Coverage. A measure of the amount of material required to cover a given surface.

Covered Electrode. In arc welding, a filler-metal electrode consisting of a metal core wire with a relatively thick covering which provides protection for the molten metal from the atmosphere and improves the properties of the weld metal and stabilizes the arc.

Covering Power. 1. Of paint, the ability to cover the surface to which it is being applied. 2. Of ceramic tile, the ability of a glaze to uniformly and completely cover the fired surface.

Covering Up. Burying, sealing, or otherwise covering work before it has been inspected.

Coverplate. A sheet of glass or transparent plastic that sits above the absorber in a flatplate solar collector.

CP. Chemically Pure.

CPA. Certified Public Accountant.

CPE. Chlorinated polyethylene, a type of single-ply roofing material.

CPM. Critical Path Method.

CPSC. Consumer Product Safety Commission.

CPSE. A type of single-ply roofing material, a self-curing non-vulcanized elastomer. Available as a liquid coating or a membrane sheet. May be reinforced with polyester scrim or laminated to felt backing.

CPVC. Chlorinated Polyvinyl Chloride.

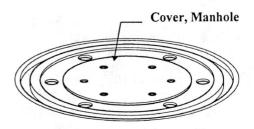

Cover, Manhole

Cr. Credit.

CRA. California Redwood Association.

Crab. A hand device used for stretching carpet in a small area where a power stretcher or knee kicker cannot be used.

Crack Control Reinforcement. Reinforcement in concrete construction designed to prevent cracks, often effective in limiting them to uniformly distributed small cracks.

Crack Length Method. A method of calculating the quantity of infiltration air into a building; this method requires specific information about dimensions and construction details of windows, doors, and other openings.

Crack Monitor. A two piece adhesive paper pattern that may be attached to a building wall over a crack to record differential structural movement, over time, of the wall on each side of the crack.

Crack Patch. To repair a crack in a surface by using plaster, concrete, asphalt, etc.

Crack Repair. To patch a crack in a surface by using such materials as plaster, concrete, or asphalt.

Crack Sealing. To close the opening between two materials to prevent moisture or air from passing through.

Crack. 1. To break, split, snap apart or develop fissures. 2. A fracture in the monolithic surface of gypsum board. 3. A separation or fracture occurring in a roof membrane or in a roof deck, generally caused by thermally induced stress or substrate movement; also called a Split. 4. A break in the surface of an asphalt pavement; see Alligatoring, 3.

Crackage. Joint in a structure which permits movement of a gas or vapor through it, even under a small pressure difference.

Cracking a Valve. Opening a valve a small amount.

Cracking. Form of paint failure in which breaks in film extend through all coats down to building material.

Crackle Finish. A finish in which alligatoring is produced, allowing the undercoat to show through the cracks. Cracking is produced by rapid drying of topcoat over slow drying undercoat.

Craft Union. A labor union whose membership is limited to workers in the same trade or craft.

Craft. An occupation or trade requiring artistic skill and manual dexterity.

Craftsman. A person skilled in a trade or handicraft; an artisan; sometimes called Craftsperson.

Craftsmanship. Quality of work done in a trade or handicraft.

Craftsperson. A Craftsman, regardless of gender.

Crane Jib. The projecting arm of a crane.

Crane Manufacturers Association of America (CMAA). 8720 Red Oak Boulevard., #201, Charlotte, North Carolina 28217, (704) 522-8644.

Crane Mobilization. To assemble and put a crane into movement to work at a construction site.

Crane, Travelling. See Travelling Crane.

Crane. A machine for raising, shifting and lowering heavy weights by means of a projecting swinging arm or with the hoisting apparatus supported on an overhead track.

Crank Throw. Distance between center line of main bearing journal and center line of the crankpin or eccentric.

Crankshaft Seal. Leakproof joint between crankshaft and compressor body.

Crater. A depression at the termination of an arc weld bead.

Cratering. Formation of holes or deep depressions in paint film.

Crawl Space Insulation. Insulation that is applied between the first floor joists just above the ground.

Crawler Crane. A crane mounted on a pair of tracks used to support and propel the crane.

Crawler. Any vehicle propelled by caterpillar tracks, such as a crane.

Crawling. 1. A parting and contraction of the glaze on the surface of a ceramic surface during drying or firing, resulting in unglazed areas bordered by coalesced glaze. 2. Varnish defect in which poor adhesion of varnish to surface in some spots causes it to crawl or gather up into globules instead of covering the surface.

Crawlspace. Space under a floor or in an attic that is accessible by crawling.

Craze Cracks. Fine, random fissures or cracks which may appear in a plaster surface caused by plaster shrinkage; also called check cracking; these cracks are generally associated with a lime finish coat that has not been properly gauged or troweled.

Crazing. 1. Minute, interlacing cracks on the surface of a finish. 2. Hairline cracks in paint film. 3. The cracking which occurs in fired glazes or other ceramic coatings due to critical tensile stress.

Cream Paper. A highly sized and calendered paper used as the face paper in the manufacture of gypsum board.

Credit. 1. A person's financial standing. 2. The sum of money at a person's disposal in a bank or other financial institution. 3. The power to obtain goods or services before payment, based on the trust that payment will be made. 4. A reputation for solvency and honesty in business. 5. The acknowledgement of being paid by an entry on the credit side of an account; the sum entered. 6. An acknowledgement of a contributor's services to the design or construction of a building project.

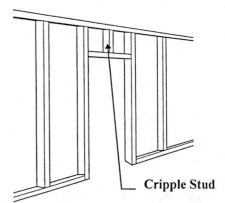

Cripple Stud

Creditor. Persons or companies to whom money is owed.

Creditor's Equity. The claims of outsiders against the business; the amounts that the business owes to persons or companies other than the owners.

Creek. A small river.

Creep, Seismic. See Seismic Creep.

Creep. 1. A permanent, inelastic, deformation in a material due to changes in the material caused by the prolonged application of structural stress; plastic deformation that proceeds with time when certain materials, such as concrete, are subjected to constant, long-duration stress. 2. The shrinking or stretching of a roofing membrane due to heat or moisture changes.

Crematorium. A building containing a furnace for cremating the dead; Crematory.

Crematory. See Crematorium.

Creosote Oil. Distillate, heavier than water, from coal tar, used largely as a wood preservative.

Creosote Stain. Creosote, made mostly from wood and coal tars, mixed with linseed oil and drier and thinned with benzine or kerosene.

Creosote. A brownish oily liquid obtained by distillation of coal tar, used as a wood preservative.

Creosoted Pole. A wooden pole that has been impregnated with creosote to help with its preservation.

Crescent Wrench. An adjustable wrench.

Crescent. 1. The curved sickle shape of the waning moon, with a convex and a concave edge. 2. Anything of this shape.

Crest. 1. Of a weir, the surface over which the liquid flows. 2. The ridge of a roof.

Crew Trailer. A trailer provided on a job site for use by workers.

CRI. Carpet & Rug Institute.

Cricket. A chimney flashing on the uphill side, resembling a small roof ridge, to divert the rainwater around the chimney.

Crimping Pool. A hand operated tool to apply metal corner beads or fasten steel studs to track by clinching part of metal.

Crimping. Method of texturizing staple and continuous filament yarn to produce irregular alignment of fibers and increase bulk and covering power; also facilitates interlocking of fibers, which is necessary for spinning staple fibers into yarn.

Cripple Rafter. The fillers in a roof framing system which connect the valley and hip rafters.

Cripple Stud. 1. A short wood stud occurring in less than full height walls. 2. A short wood stud occurring over door or window headers or under window dsills.

Crisper. Drawer or compartment in refrigerator designed to provide high humidity along with low temperature to keep vegetables, especially leafy vegetables, cold and crisp.

Criterion. (Pl. criteria) A standard on which a judgment or decision may be based.

Critical Path Method (CPM). A construction scheduling system.

Critical Pressure. Compressed condition of refrigerant which gives liquid and gas the same properties.

Critical Temperature. Temperature at which vapor and liquid have same properties.

Critical Vibration. Vibration which is noticeable and harmful to structure.

Crook. The distortion of lumber in which there is a deviation, in a direction perpendicular to the edge, from a straight line from end-to-end of the piece.

Crooked Edges. A curvature of the sides, either convex or concave, measured along the sides of flat surface. The degree of crook is the departure from the straight line between two corners, expressed in percentage of length.

Crossband. In a flush wood door, the thin layer of wood between the core and the outer veneer layer.

Cross Brace. Bracing with two intersecting diagonals; slender diagonal member within a framed wall or partition, to support the wall or partition and to withstand structural loads imposed by wind and suction loads, building loads, movement, and deflection of structure.

Cross Break. A crack or separation of wood cells across the grain of a board; may be caused by unequal shrinkage or by external forces.

Cross Connection. 1. The improper interconnection of potable and contaminated water piping. 2. The improper interconnection of electrical or communication wiring.

Cross Examination. Questions framed to undermine the testimony of a witness.

Cross Fitting. A plumbing fitting consisting of two short pipes meeting at right angles on a pipe run.

Cross Furring. Term used to denote furring members attached to other structural components to support lath in suspended ceilings; generally 3/4 inch steel channels or pencil rods.

Cross Grain. Wood incorporated into a structure in such a way that its direction of grain is perpendicular to the principal loads on the structure.

Cross Hair. The fine vertical and horizontal lines used for sighting of an optical instrument such as a transit or builder's level.

Cross Joint. See Butt Joint, 4.

Cross Peen Hammer. A hammer where the face opposite the flat face is shaped for cutting or grooving metal; the peen is flat and parallel to the handle.

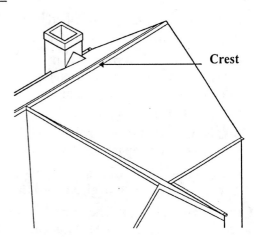

Crest

Cross Slope. A slope that is perpendicular to the direction of travel.

Cross. A pipe fitting with four female threaded openings at right angles to one another.

Crossarm Pole. A vertical pole which supports a railroad warning cross-arm on a pivot, which descends when an oncoming train trips an electrical signal to stop oncoming traffic on either side of the railroad tracks.

Crossarm. A horizontal member attached to a utility pole at a 90° angle, used to support cables.

Crossband. 1. To place the grain of layers of wood at right angles in order to minimize shrinking and swelling. 2. In plywood of three or more plies, a layer of veneer whose grain direction is at right angles to that of the face plies.

Cross-Cut Saw. A hand saw whose teeth are designed for cutting across the wood grain.

Cross-Grained Wood. Wood in which the fibers deviate from a line parallel to the sides of the piece; cross grain may be either diagonal or spiral grain or a combination of the two.

Crosslinking. A particular method by which chemicals unite to form films.

Crosslot Bracing. Horizontal compression members running from one side of an excavation to the other, used to support sheeting.

Cross-Seam. Seam made by joining the ends of carpet together.

Cross-Section. A drawing representing an orthogonal slice through an object.

Cross-Sectional Area. The area of a section of a member taken at right angles to its axis, such as a steel angle or beam, expressed in square units as square inches or square millimeters.

Cross-Spray. Spraying first in one direction and second at right angles.

Crow Hop. Tile joints that are out of alignment.

Crowbar. An iron or steel bar with a flattened wedge at one end, used for prying.

Crowding the Line. The laying of the brick so that the line is not free of the brick wall; this will cause the line to be inaccurate and the wall may bulge or overhang.

Crown Glass. Glass sheet formed by spinning an opened globe of heated glass.

Crown Molding. A decorative type of molding used to make ceiling to wall transitions.

Crown Venting. A vent connection no longer permitted on modern installations, consisting of a vent pipe connected at the top or crown of the curve which forms the trap.

Crown Weir. The lower part of the outlet of a plumbing fixture trap.

Crown. 1. In gypsum wallboard installation, the buildup of joint compound over a joint to conceal the tape; the higher the crown the wider the compound must be feathered to make the joint less visible; also called Crowned Joint or Hump Joint. 2. The increased elevation of the center of a roadway to facilitate drainage to the edges. 3. The highest point of an arch or vault.

Crowned Joint. See Crown, 1.

CRSI. Concrete Reinforcing Steel Institute.

CRT. Cathode Ray Tube.

Crucible. A heat-resistant vessel for melting or calcining materials that require high temperatures.

Cruciform. Forming or arranged in a cross; said of the floor plan of a church with nave and transepts.

Crumbs. Ragged chunks of gypsum on cut outs or cut ends of gypsum wallboard.

Crushed Gravel. The product resulting from artificial crushing of gravel with substantially all fragments having at least one fracture face.

Crushed Marble. Marble that has been crushed to smaller and more uniform sizes so substantially one face of each stone is fractured.

Crushed Stone. The product resulting from the artificial crushing of rocks, boulders, or large cobblestones substantially all faces of which have resulted from the crushing operation.

Crusher Run. Gravel, rock, boulders, or blasted rock that has been reduced in size by a machine, but has not been sorted for size; the total unscreened product of a stone crusher.

Crust. The outer part of the earth, the lithosphere, the outer 80 kilometers of earth's surface made up of rocks, sediment, and basalt; general composition is silicon-aluminum-iron.

Cryogenic Fluid. Substance which exists as a liquid or gas at ultra-low temperatures, minus 250° F. or lower.

Cryogenics. 1. The branch of physics dealing with the production and effects of extremely low temperatures. 2. Refrigeration which deals with producing temperatures of minus 250° F. and lower.

Crypt. A vault under the main floor of a church.

Cryptometer. An instrument used to measure the opacity of paint.

Crystalline Glaze. 1. Glazed tile with an extra heavy glaze produced for use on counter tops and light duty floor surfaces where abrasion or impacts is not excessive. 2. A glaze that contains microscopic crystals.

Crystallizing Lacquer. Novelty finish which crystallizes forming unusual crystal and floral patterns as it dries.

CS. Commercial Standard.

CSA. Canadian Standards Association.

CSPE. Chlorosulfonated Polyethylene.

CSTC. Ceiling Sound Transmission Class.

CTI. Ceramic Tile Institute.

Cubage. The cubic contents of a vessel, space, or building, expressed in cubic units, such as cubic feet or cubic meters; also called Cubature.

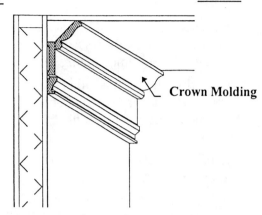

Crown Molding

Cubature. See Cubage.

Cubicle Adapter. A device for connecting partitions into different positions.

Cubicle, Hospital. See Hospital Cubicle.

Cubicle, Office. See Office Cubicle.

Cul De Sac. A dead end road no more than 400 feet long with a turn-around of no less than 80 feet in diameter.

Culling. Sorting of the brick for quality, size and color.

Culls. The rejected bricks in culling.

Cultured Marble. A plastic imitation marble for interior use.

Culvert Formwork. The temporary support for the pouring, placing and curing of a concrete drain pipe constructed under a road or embankment to provide a waterway.

Culvert Reinforcement. The placing of metal or steel bars in concrete forms for drainage systems or pipe; usually refers to a box culvert shape constructed under a road or embankment to provide a waterway.

Culvert. A drain pipe or small bridge for drainage under a road or structure.

C

Cumulative Errors. Errors that are always positive or always negative, as contrasted with compensating errors.

Cup. A distortion of a board in which there is a deviation flatwise from a straight line across the width of the board.

Cupola. A small domed structure adorning a roof.

Cupped Taper. In gypsum wallboard, a condition where the outer edge of the taper is in the same plane as the surface causing a ridge to appear in the tapered edge.

Cuprous. Of or like copper.

Curb and Gutter. Concrete or stone structure that forms the edging of a sidewalk, separating it from the paved street; the adjacent, usually integral, gutter provides a drainage channel.

Curb Cock. A valve placed on the water service usually near the curb line; also called the Curb Stop.

Curb Cut. An interruption of a curb at a driveway pedestrian way.

Curb Form. A retainer or mold with the necessary shape to make a concrete curb.

Curb Formwork. The temporary support for a freshly placed or poured curb system.

Curb Granite. Extremely hard rock used in a curb system; available in sections.

Curb Inlet Frame. The steel or metal framing surrounding an opening in a curb through which water flows and drains.

Curb Inlet. The opening in a curb through which water flows and drains.

Curb Ramp. A sloping pedestrian way, intended for pedestrian traffic, which provides access between a walk or sidewalk to a surface located above or below an adjacent curb face.

Curb Removal. The act or process of the demolition of a curb into manageable pieces.

Curb Roof. See Gambrel Roof.

Curb Stop. See Curb Cock.

Curb, Terrazzo. A curb made of decorative mosaic material made by embedding small pieces of marble or granite in mortar and polishing.

Curb. 1. A protective rim. 2. A concrete edging raised above a roadway and forming the edge of a drainage gutter.

Cure. 1. The process of concrete hardening, rubber vulcanizing, or adhesive achieving its maximum strength. 2. To change the properties of an adhesive by chemical reaction, which may be condensation, polymerization, or vulcanization, and thereby develop maximum strength; generally accomplished by the action of heat or a catalyst, with or without pressure. 3. Toughening or hardening of paint film. 4. To provide conditions conducive to completion of the hydration process in portland cement concrete or plaster.

Cured. Completely dry; moisture free.

Curing Agent. Hardener; promoter.

Curing Blanket. A built-up covering of sacks, matting, hessian, straw, waterproof paper, or other suitable material placed over freshly finished concrete to prevent premature dehydration.

Curing Compound. A liquid that can be applied as a sprayed coating to the surface of newly placed concrete to retard the loss of water or, in the case of pigmented compounds, also to reflect heat so as to provide an opportunity for the concrete to develop its properties in a favorable temperature and moisture environment; forms a water-resistant layer to prevent premature dehydration of the concrete.

Curing Concrete. To control the humidity and temperature of freshly finished concrete to assure the proper drying and hardening of the concrete.

Curing Paper. Waterproof paper placed over freshly finished concrete, to help control the humidity and temperature, aiding in the proper curing of concrete.

Curing Period. The amount of time that concrete should be kept damp after placing, usually about a week.

Curing Temperature. The temperature to which an adhesive or an assembly is subjected to cure the adhesive.

Curing, Electrical. See Electrical Curing.

Curing, Steam. See Steam Curing.

Curing. Maintenance of humidity and temperature of freshly placed concrete during some definite period following placing, casting, or finishing to assure satisfactory hydration of the cementitious materials and proper hardening of the concrete; the hardening of concrete or plaster.

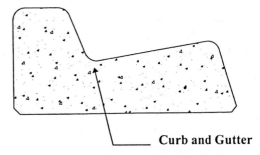

Curb and Gutter

Curling. The distortion of an originally essentially linear or planar member into a curved shape such as the warping of a slab due to creep or to differences in temperature or moisture content in the zones adjacent to its opposite faces.

Curly-Grained Wood. Wood in which the fibers are distorted so that they have a curled appearance, as in birdseye wood; the areas showing curly grain may vary up to several inches in diameter.

Current Assets. Cash and other assets that can be easily and quickly converted to cash, usually within one year.

Current Liabilities. Liabilities to be paid within a certain time, usually one year.

Current Ratio. The ratio of current assets to current liabilities.

Current Regulator. An automatic electrical control device for maintaining a constant current in the primary of the welding transformer.

Current Relay. Device which opens or closes a circuit; made to act by a change of current flow in that circuit.

Current. 1. Electrical flow through conductors, measured in amperes. 2. Belonging to the present time, as current week or month.

Current-Limiting Overcurrent Protective Device. A device which, when interrupting a specified circuit, will consistently limit the short-circuit current to a specified magnitude substantially less than that obtainable in the same circuit if the device was replaced with a solid conductor having comparable impedance.

Curtain Drain. See Intercepting Drain.

Curtain Rod. A horizontal bar that supports window coverings.

Curtain Wall. An exterior non-bearing wall between columns, sometimes containing windows or are all glass.

Curtain, Gym. See Gym Curtain.

Curtain. 1. A single layer of vertical and horizontal reinforcing bars in a wall. 2. Window drapery.

Curtilage. The ground or yard surrounding a house up to the boundary fence.

Curvature Factor. A factor applied to the allowable bending stress of glue-laminated beams where a curved portion occurs.

Curvature Friction. The friction resulting from bends or curves in the specified prestressing tendon profile.

Curved Curb Form. A curve shaped support for the placing, pouring and curing of a concrete curb.

Curved Stair Terrazzo. Terrazzo tile used in a curved stair system.

Curved Wall Form. A curved shaped support for the placing, pouring, and curing of concrete in a curved wall system.

Cushion-Edged Tile. Tile on which the facial edges have a distinct curvature that results in a slightly recessed joint.

Cushioning, Carpet. See Carpet Padding.

Cusp. A projecting point between small arcs in Gothic tracery.

Custom and Usage. The way the majority of those in the trades and professions commonly conduct themselves in doing their particular work.

Custom Door. A door that has been manufactured to fit a specific size opening or manufactured from specific materials.

Cut and Cover. A work method which involves excavation in the open, and placing of a temporary roof over it to carry traffic during further work.

Cut and Fill. Excavated material removed from one location and used as fill material in another location.

Cut End. The end of the gypsum board with the exposed core.

Cut in the Sash. Painting the window sash; ordinarily done with a brush, often called a sash tool, which permits the painter to get a clean edge.

Cut Joints. Masonry bed and head joints cut flush with trowel.

Cut Loop Pile. In carpet making, pile surface in which the tufts have been cut to reveal the fiber ends.

Cut Out. 1. A mechanical or electrical device used to break an electrical circuit because of an overload. 2. An opening in a wall or surface for access. 3. A piece stamped out of metal. 4. An opening in a countertop for access or installation of a piece of equipment.

Cut Stock. Softwood stock comparable to dimension stock in hardwoods; see Dimension Stock.

Cut Stone. Building stone cut to size and shape for specific applications and designated locations in a building or structure.

Cut. Dispersion of a certain number of pounds of shellac or resin per gallon of volatile liquid; for example, a 4 pound cut of shellac contains 4 pounds of dry shellac and 1 gallon of alcohol.

Cutback Asphalt. Asphalt cement which has been liquefied by blending with petroleum solvents, as for the RC and MC cutback asphalts; upon exposure to atmospheric conditions the solvents evaporate, leaving the asphalt cement to perform its function.

Cutback. Asphalt or tar that has been diluted with solvents and oils so the resulting material becomes fluid.

Cut-In. The temperature value or the pressure value at which the control circuit closes.

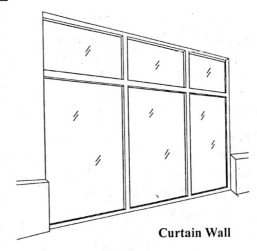

Curtain Wall

Cutoff. In roofing installation, the procedure to prevent lateral water or moisture entering the insulation where it terminated at the end of the work day; usually a felt strip hot-mopped to the deck and to the insulation edge and to the horizontal insulation surface; at the start of the next day's work the strip is taken off.

Cut-Out. Temperature value or pressure value at which the control circuit opens.

Cutting Attachment. A device which is attached to gas welding torch to convert it into an oxygen-cutting torch.

Cutting Tip. That part of an oxygen-cutting torch from which the gases issue.

Cutting Torch. A nozzle or device used in oxygen cutting for controlling and directing the gases used for preheating and the oxygen used for cutting the metal.

Cutting. 1. Excavating. 2. Sawing. 3. Slicing.

Cuttings. In hardwoods, a portion of a board or plank having the quality required by a specific grade or for a particular use; obtained from a board by crosscutting or ripping.

CW. Cold Water.

Cyanic. Containing blue or pertaining to blue color.

Cyanuric Acid. Acid used in pool water to prevent chlorine loss.

Cybernetics. The science of communications and automatic control systems in both machines and living things.

Cycle. 1. One complete performance of an electrical oscillation or current alternation. 2. A series of events or operations which have the tendency to repeat in the same order.

Cyclone. A storm or system of winds that rotate about a center of low atmospheric pressure.

Cyclopean Aggregate. Concrete aggregate where the individual pieces are over 100 pounds.

Cyclotron. An apparatus in which charged atomic and subatomic particles are accelerated by an alternating electric field while following an outward spiral or circular path in a magnetic field.

Cylinder Glass. Glass sheet produced by blowing a large, elongated glass cylinder, cutting off its ends, slitting it lengthwise, and opening it into a flat rectangle.

Cylinder Piling. Concrete filled steel pipes or tubes, driven through the ground to reach bedrock and used as a foundation for tall buildings and in underpinning.

Cylinder Plug. The round part of a door lock containing the keyway and rotated by the key to transmit motion to the bolt.

Cylinder Test. A test to determine the compressive strength of concrete.

Cylinder, Refrigerant. Refrigerant Cylinder.

Cylinder. 1. The surface traced by a straight line moving parallel to a fixed straight line and intersecting a fixed planar closed curve. 2. Of a lock, the cylindrical-shaped assembly containing the tumbler mechanism and the keyway. 3. A device which converts fluid power into linear mechanical force and motion; this usually consists of movable elements such as a piston and piston rod, plunger or ram, operating within a cylindrical bore. 4. A closed container for fluids. 5. A portable cylindrical container used for transportation and storage of a compressed gas.

Cylindrical Commutator. Commutator with contact surfaces parallel to the rotor shaft.

Cylindrical Lock. A door lock that can be installed in a cylindrical hole rather than in a mortise; also called a Tubular Lock.

Cyma Recta. An ogee molding combining a concave curve with a convex curve; when the concave curve is on top and the convex curve is on the bottom it is a cyma recta; see Cyma Reversa.

Cyma Reversa. An ogee molding combining a concave curve with a convex curve; when the convex curve is on top and the concave curve is on the bottom it is a cyma reversa; see Cyma Recta.

Cyma. A type of cove molding in which the surface of the face changes from concave to convex; a Cyma Recta or Cyma Reversa; commonly made of wood, plaster, or masonry.

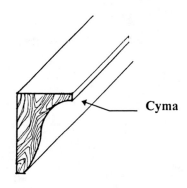

Cyma

D

D Cracking. The progressive formation on a concrete surface of a series of fine cracks at rather close intervals, often of random patterns, but in slabs on grade paralleling edges, joints, and cracks and usually curving across slab corners; also called D-Cracks and D-Line cracks.

D Cracks. See D Cracking.

D Line Cracks. See D Cracking.

D Load. A constant load that in structures is due to the mass of the members, the supported structure, and permanent attachments or accessories.

D&M. Dressed and Matched.

Dado. (pl. dadoes) 1. Part of a column base. 2. The lower part of an interior wall when differently surfaced. 3. A groove in a piece of wood made by dadoing.

Dado Joint. A joint in which one piece is grooved to receive the piece which forms the other part of the joint.

Dairy. A building or room for the storage, processing, or sales of milk and milk products.

Dalton's Law. Vapor pressure created in a container by a mixture of gases is equal to the sum of individual vapor pressures of the gases contained in mixture.

Damages. 1. Injuries. 2. An amount of money awarded to a plaintiff to compensate for loss caused by the wrongful conduct of a defendant.

Dammar. A natural resin used extensively in the preparation of varnishes and lacquers; usually classified according to the place from which it is shipped to market, for example, Singapore dammar or Batavia dammar.

Damp Check. See Damp Course.

Damp Course. A course or layer of impervious material which prevents capillary entrance of moisture from the ground or a lower course; also called Damp Check.

Damp Location. A location subject to a moderate degree of moisture, such as some basements, some barns, some cold storage warehouses, and the like.

Damper. A flap to control or obstruct the flow of air or other gasses; specifically, a metal control flap in the throat of a fireplace, or in an air duct; controls that vary airflow through an air outlet, inlet, or duct; a damper position may be immovable, manually adjustable, or part of an automated control system.

Damper, Seismic. See Seismic Damper.

Damping. 1. Dissipation of structure-borne noise by conversion to some other form of energy, usually heat; usually accomplished by using a material with a high internal energy-absorbing capacity. 2. A rate at which natural vibration decays as a result of absorption of energy.

Dampproofing. The treatment of concrete or mortar to help prevent the passage or absorption of water in the absence of hydrostatic pressure; not necessarily waterproof.

Dap. To cut and form a recess in timber to form a dapped joint; also called a let-in joint, as where 1" x 6" diagonal bracing is cut into 2" x 4" studs.

Darby. 1. A hand-manipulated straightedge, usually 3 to 8 feet. (1 to 2.5 m) long, used in the leveling operation of the early stage of concrete placement, preceding supplemental floating and finishing. 2. A stiff straightedge, about 4 inches wide and 42 inches long with handles, of wood or metal, used to level the surface of wet plaster.

Darkroom Dryer. A mechanical drying device used by a photographer in a darkroom to dry prints.

Darkroom Equipment. Equipment used for the developing of photographs.

Darkroom Sink. A basin in a photographic darkroom for the dispensing or holding of liquids.

Darkroom. A room for photographic work, with normal light excluded.

Dash Bond Coat. A thick slurry of portland cement, sand, and water flicked on surfaces with a paddle or brush to provide a base for subsequent portland cement plaster coats; sometimes used as a final finish on plaster.

Dash Brush. A long-bristled brush for flinging a plaster mixture on the wall as a dash coat.

Database Management System (DBMS). Computer software designed to manage usually large quantities of data.

Datum. 1. The fixed starting point for surveying or measuring. 2. A piece of information.

Daub. A glob of adhesive.

Daylight. A painting defect where there is a lack of coverage of a coat of paint.

Daylighting. Illuminating the interior of a building by natural means.

Days. Calendar days are consecutive days on the calendar; working days are calendar days less non-working days such as week ends and holidays.

db. Decibel.

DB. Dry Bulb.

DBA. Doing Business As.

DBIA. Design/Build Institute of America.

Dbl Conduit. Double thick conduit.

Dbl Duct. Double thick metal duct.

DC. Direct Current.

De Facto. In reality; actually; in fact, whether by right or not.

De Jure. By right or of right.

DE. Diatomaceous Earth.

Dead Burned. Removal of all water content during calcining of gypsum.

Dead End. A branch leading from a soil, waste or vent pipe, building drain sewer which is terminated at a developed length of 2 feet or more by means of a plug or other fitting. A dead end is not used to admit water or air into the piping system.

Dead Flat. In paint, having no gloss at all.

Dead Knot. Loose knot in wood, usually dark in color.

Dead Level Asphalt. A roofing asphalt that has a softening point of 140° F (60° C).

Dead Level. An absolutely horizontal roof. slope.

Dead Load. The load due to the vertical weight of all permanent structural and nonstructural elements of a building, such as walls, floors, and roofs.

Dead Man's Switch. One that stays in on or off position only as long as it is held there manually.

Dead Space. 1. A space with an abundance of absorptive surfaces that does not support sound. 2. A space between walls or a floor and ceiling.

Deadbolt. A lock bolt having no spring action nor bevel, and which is operated by a key or a turn piece.

Deadening. Construction intended to prevent the passage of sound.

Dead-Front. As applied to switches, circuit breakers, switchboards, control panels, and panelboards, so designed, constructed, and installed that no current-carrying parts are normally exposed on the front.

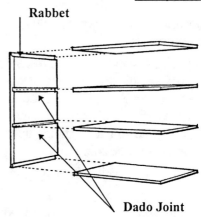

Rabbet

Dado Joint

Deadlock. A lock equipped with a deadbolt only.

Deadman. A large or heavy object buried in the ground as an anchor.

Deaeration. Act of separating air from substances.

Deal. Softwood timber.

Debris. Accumulated rubbish, trash and fragments of waste.

Debt, Long Term. See Long Term Debt.

Debt, Short Term. See Short Term Debt.

Debt. Money that is owed.

Debtors. Persons or companies that owe money to the business; accounts or notes receivable.

Decal. Decalcomania.

Decalcomania. Pictures, diagrams and designs, on specially prepared paper, that can be transferred to glass or other surfaces; also called a Decal.

Decay. The decomposition of wood or other substance by fungi or bacteria.

Decay Fungus. Requires high moisture content and will eat wood. See also Non-Decay Fungus.

Dechlorination. The removal of chlorine from water.

Decibel (db). A standard unit of measure for classifying the transmission of airborne sound levels; the ratio of sound pressure to a base level chosen at the threshold of hearing and ranging up to 130 for the average pain level.

Deciduous. Trees which annually lose their leaves.

Decimal Fraction. A fraction or mixed number in which the denominator is a power of 10, usually expressed by use of the decimal point, as $1/4 = 25/100 = 0.25$.

Decimal. Any real number expressed in base 10.

Deck Insulation. Thin sheets of insulation placed on a concrete surface before a flooring system is installed.

Deck Level Pool. Water surface level with the deck.

Deck Mud. See Floor Mud.

Deck Paint. An enamel with a high degree of resistance to mechanical wear; designed for use on such surfaces as porch floors.

Deck, Coil. See Coil Deck.

Deck. 1. The roof surface to be covered, the substrate or substrata. 2. A small platform used for walking. 3. The form on which concrete for a slab is placed. 4. A floor or roof slab.

Decking, Cellular. See Cellular Decking.

Decking, Metal. See Metal Decking.

Decking, Open. See Open Decking.

Decking, Wood. See Wood Decking.

Decking. 1. A material used to span across beams or joists to create a floor or roof surface. 2. Heavy plank floor of a pier or bridge. 3. Surfaced area surrounding a swimming pool.

Declaration of Homestead. See Homestead, 3.

Declining Balance. A form of accelerated depreciation where a larger amount is taken in the beginning years gradually declining to a smaller amount in the later years.

Decor. The furnishing, layout, style, and decoration of building interiors.

Decorated. Adorned, embellished, or made more attractive by means of color or surface detail.

Decoration, Inglaze. See Inglaze Decoration.

Decoration, Overglaze. See Overglaze Decoration.

Decoration, Underglaze. See Underglaze Decoration.

Decoration. 1. The act of decorating. 2. The finish used over the gypsumboard face.

Decorative Painting. Architectural painting; aesthetic painting.

Decorative Tile. Tile with a ceramic decoration on the surface.

Decoupling. Separating elements to retard the transmission of structuralborne sound, thermal conductance, or physical loads.

Dedication. The appropriation of land or easements by a private owner for public purposes or use.

Deductive Alternate. An alternate bid that, if accepted, reduces the contract price.

Deed of Reconveyance. See Reconveyance Deed.

Deed of Trust. See Trust Deed.

Deed Restrictions. Restrictive covenants often included in a deed which may limit future uses of the land, and other restrictions such as height, size, and aesthetics, as long as they are not against the public interest.

Deeds. A general term which refers to all documents conveying property, from one person to another.

Deep Color. Intense or strong color with no apparent presence of black.

Deep Seal-P Trap. A P-trap with a water seal depth of more than 4 inches.

Default. A material failure to perform the requirements of a contract.

Defect. An imperfection.

Defendant. A party against whom a legal claim is made in a civil law suit.

Deferred Maintenance. Repair and upkeep that has been neglected or postponed.

Deficit. An economic shortfall, where expenses exceed income; a debit balance in the retained earnings account.

Deflection. 1. A variation in position or shape of a structure or structural element due to effects of loads or volume change, usually measured as a linear deviation from an established plane rather than an angular variation. 2. Displacement or bending of a structural member due to application of external force.

Deformation, Eutectic. See Eutectic Deformation.

Deciduous Tree

Deformation. The change in shape of a body brought about by the application of a force. Deformation is proportional to the force within the elastic limits of the material.

Deformed Bar. A steel reinforcing bar manufactured with deformations (bumps, lugs, or ridges) to provide a locking anchorage with the surrounding concrete.

Deformed Reinforcement. Deformed reinforcing bars, bar and rod mats, and deformed wire fabric.

Defrost Device. Refrigerating cycle in which evaporator frost and ice accumulation is melted.

Defrost Timer. Device connected into electrical circuit which shuts unit off long enough to permit ice and frost accumulation on evaporator to melt.

Defrosting Control. Device to automatically defrost evaporator; it may operate by means of a clock, door cycling mechanism, or during off portion of refrigerating cycle.

Defrosting Type Evaporator. Evaporator operating at such temperatures that ice and frost on surface melts during off part of operating cycle.

Defrosting. Process of removing frost accumulation from evaporators.

Degradation. A decline in quality; degeneration.

Degreaser. Chemical solution for grease removal.

Degreasing. Solution or solvent used to remove oil or grease from refrigerator parts.

Degree Day Method. A method of computing fuel requirements for HVAC systems, see Degree Day.

Degree Day. Unit that represents one degree of difference from inside temperature and the average outdoor temperature for one day and is often used in estimating fuel requirements for a building.

Degree of Saturation. The amount of water present in the air relative to the maximum amount it can hold at a given temperature without causing condensation.

Degree. 1. A unit on one of the temperature scales. 2. A unit of measurement of angles, equal to 60 minutes; a right angle is 90 degrees.

Dehumidification. The removal of moisture from air.

Dehumidifier. A machine which removes moisture from the air.

Dehydrated Castor Oil. A drying oil prepared from castor oil.

Dehydrated Oil. Lubricant which has had most of its water content removed, also called Dry Oil.

Dehydration. The process of removing water from a substance.

Dehydrator. See Drier, 1.

Dehydrator Receiver. Small tank which serves as liquid refrigerant reservoir and which also contains a desiccant to remove moisture; used on most automobile air conditioning installations.

Deice Control. Device for operating a refrigerating system in such a way as to provide melting of the accumulated ice and frost.

Delamination. 1. Separation of one layer from another. 2. The separation of layers in laminated wood because of failure of the adhesive, either within the adhesive itself or at the interface between the adhesive and the adherend. 3. The failure of a built-up roofing membrane, resulting in the separation of the felt plies, resulting in cracking and wrinkling. 4. The separation of the paper plies or surface coverings on gypsum board.

Delay. Time lost in the construction schedule by any reason.

Delicatessen Case. A refrigerated cabinet used for the storing and displaying of perishable foods.

Delignification. Removal of part or all of the lignin from wood by chemical treatment.

Deliquescence. The natural absorption of atmospheric moisture by a solid substance.

Delivery Cart. A heavy or lightweight, two or four-wheeled vehicle used to transport goods.

Delivery Tolerances. In asphaltic concrete making, permissible variations from the exact desired proportions of aggregate and bituminous material as delivered into the pugmill.

Delta Transformer. Three-phase electrical transformer which has ends of each of three windings electrically connected to form a triangle.

Delta. 1. The 4th letter in the Greek alphabet, the capital delta being shaped like an equilateral triangle. 2. Anything so shaped. 3. The alluvial deposit at the mouth of a river. 4. A numbered marking on contract drawings to identify changes.

Delustered Nylon. Nylon on which the normally high sheen has been reduced by surface treatment.

Demand Factor. The demand of any system; the ratio of the maximum demand of the system to the total connected load of the system; the loads of a system are practically never on at the same time due to the many uses of the power; somewhere between the maximum connected load and the actual usage is a load that is considered the maximum demand.

Demand Meter. Instrument which measures the kilowatt-hour usage of a circuit or group of circuits.

Demise. Transfer property by will, grant, or lease.

Demised Premises. The property leased.

Demising Wall. See Party Wall, 3.

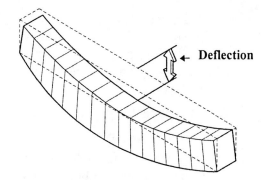

← Deflection

Demography. The study of the statistics of a human population in respect to trends in size, density, age, and economic characteristics, to identify markets.

Demolish. To raze or tear down a building or structure.

Demountable Partition. An assembly designed to be dismantled and reassembled with a minimal loss of components.

Demur. To take exception; object.

Demurrage. A charge for detaining a ship, freight car, or truck during the period of unloading.

Demurrer. A response in a court proceeding in which the defendant does not dispute the truth of the allegation but claims it is not sufficient grounds to justify legal action; an objection.

Den. A private comfortable room for pursuit of hobbies or recreation.

Denatured Alcohol. Grain or ethyl alcohol made unsuitable for beverage purposes by adding compounds of a poisonous nature.

Dendriform. Resembling a tree in structure.

Dendrochronology. The science of dating events and variations in environment in former periods by comparative study of growth rings in trees and aged wood.

Dendroid. Resembling a tree in form.

Dendrology. The study of trees.

Denier. System of yarn count used for synthetic fibers; the weight in grams per 9,000 meters of yarn length; one denier equals 4,464,528 yards per pound or 279,033 yards per ounce.

Dense-Graded Aggregate. An aggregate that has a particle size distribution such that when it is compacted, the resulting voids between the aggregate particles, expressed as a percentage of the total space occupied by the material, are relatively small.

Densification. The act of increasing the density of a mixture during the compaction process.

Density, Pile Yarn. See Pile Yarn Density.

Density. 1. The quantity per unit volume, unit area, or unit length; in wood it is expressed as pounds per cubic foot, kilograms per cubic meter, or grams per cubic centimeter at a specified moisture content. 2. Closeness of texture or consistency of particles within a given substance.

Dental Equipment. Equipment used by dentists for dental work on patients.

Dental Light. An adjustable lamp used by dentists, having an arm-like extension enabling it to be maneuvered in any direction, and which focuses light in one direction.

Dentil. One of a series of small projecting rectangular blocks forming a molding, part of a cornice.

Deodorizer. Device which absorbs or adsorbs various odors, usually by the principle of absorption; activated charcoal is commonly used.

Deposit. 1. A natural resource such as iron ore. 2. Settled material, such as sediment. 3. A returnable sum paid usually for safekeeping or to guarantee the return of something.

Deposited Metal. Filler metal that has been added during a welding operation.

Deposition. 1. The extra-judicial testimony of a witness given under oath. 2. The written record thereof.

Depository, Night. See Night Depository.

Depreciation. The decrease in value of a property from any cause.

Derrick. A hoisting frame with a long beam, ropes, gear, and pulleys that can lift a heavy load and swing it to the side.

Desiccant. 1. A drying agent. 2. Substance used to collect and hold moisture in refrigerating system; common desiccants are activated alumina and silica gel.

Design Concept. An architectural idea that is depicted on the drawings and in the specifications with the expectation that it will be faithfully executed by the contractor.

Design Development Drawings. Drawings prepared by architects for the Design Development Phase.

Design Development Phase. One of the standard phases of architectural service (Schematic Design Phase, Design Development Phase, Construction Documents Phase, Bidding or Negotiation Phase, and Construction Phase-Administration of the Construction Contract).

Design Displacement. The design-basis earthquake lateral displacement, excluding additional displacement due to actual and accidental torsion, required for design of the seismic isolation system.

Design Filter Rate. See Design Rate of Flow.

Design Head. The total head requirement of the circulation system at the design rate of flow.

Design Heat Load. The total heat loss from a building under the most severe winter conditions likely to occur.

Design Pressure. Highest or most severe pressure expected during operation; sometimes used as the calculated operating pressure plus an allowance for safety.

Design Professional. A person educated and skilled in the field of building design or the related structural and environmental systems.

Design Rate of Flow. In a pool, spa, or hot tub, the rate of flow in gallons divided by the number of minutes in the turnover time; also called the Design Filter Rate.

Design Temperature. A temperature close to the lowest expected for a location, used to determine design heat load.

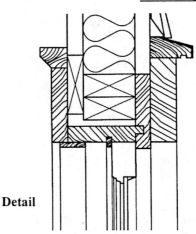

Detail

Design. To conceive or plan out in the mind.

Design/Build Institute of America. 1010 Massachusetts Avenue, NW, Suite 350, Washington, DC 20001, (202) 682-0110

Design/Build. A project delivery system in which the same entity is responsible for both design and construction.

Destructive Distillation. Distilling a product at a temperature so high that products obtained are of different chemical composition than existed in the original material.

Destructive Testing. Testing of building materials in a manner that destroys the material.

Detail. 1. Extended treatment or attention to particular items. 2. Drawing of a small portion of an assembly.

Detectable Warning. A standardized surface or feature built in or applied to walking surfaces or other elements to warn visually impaired persons of hazards in the path of travel.

Detector, Leak. See Leak Detector.

Detector, Smoke. See Smoke Detector.

Detergent. Cleaning agent.

Deterioration. Depreciation; loss of value of real property due to physical wear and tear.

Determinate. 1. Having defined limits; definite. 2. Statically Determinate; see Statically Determinate Structures.

Developed Length. The length as measured along the center line of the pipe and fittings.

Development Length. In reinforced concrete, the length of steel reinforcing required to develop the required bond between the reinforcing and the adjacent concrete.

Device Cover. A protective or ornamental removable cover over an electrical box or device.

Device. 1. A piece of equipment or a mechanism designed for a special purpose or function. 2. An ornamental design. 3. A unit of an electrical system which is intended to carry but not utilize electric energy.

Devil's Float. In plastering, a wooden float with two nails protruding from the toe, used to roughen the surface of the brown coat.

Dew Point. Temperature at which vapor, at 100 percent humidity, begins to condense and deposit as liquid on cold surfaces.

Dew. Condensed atmospheric moisture deposited in small drops on cool surfaces.

Dewatering. To remove water from the ground or excavations with pumps, wellpoints, or drainage systems.

Dew-Point Temperature. The temperature at which condensation occurs.

Dextral. Right handed.

DF. 1. Douglas Fir. 2. Drinking Fountain.

DFPA. Douglas Fir Plywood Association.

DG. Decomposed Granite.

DHI. Door & Hardware Institute.

Diagnostic Architecture. The science and practice of analyzing building failures to determine the causes and devise repair methods.

Diagonal Bond. This is a form of raking bond in which the bricks are laid in an oblique direction in the middle section of a thick wall; the bricks may also be laid in this fashion in paving.

Diagonal Bracing. That form of bracing that diagonally connects joints at different levels.

Diagonal Sheathing. Wood sheathing with the individual boards running at a 45 degree angle to the studs, joists, or rafters.

Diagonal. 1. Running in an oblique direction from a reference line. 2. Inclined member of a truss or bracing system used for stiffening or wind bracing.

Diagonal-Grained Wood. 1. Wood in which the annual rings are at an angle with the axis of a piece as a result of sawing at an angle with the bark of the tree or log. 2. A form of cross grain.

Diameter. A chord passing through the center of a figure or body, such as a circle.

Diamond Mesh Window Guard. Guard fabricated of diamond-shaped mesh to provide protection over the face of a window to prevent damage to glass and to prevent intrusion.

Diamond Mesh. One of the common types of metal lath having a characteristic geometrical pattern produced by the slitting and expansion of metal sheets.

Diaper. 1. An allover pattern consisting of one or more small repeated units of design, as geometric figures, connecting with one another or growing out of one another with continuously flowing or straight lines. 2. Any repeated and continuous pattern in brickwork, usually applied to diamond or other diagonal patterns.

Diaphragm Action. A bracing action that derives from the stiffness of a thin plane of material when it is loaded in a direction parallel to the plane; diaphragms in buildings are typically floor, or roof surfaces of plywood, reinforced masonry, steel decking, or reinforced concrete.

Diaphragm Pump. A water pump used to continuously remove water from excavations containing mud and small stones.

Diaphragm. 1. Flexible material usually made of thin metal, rubber or plastic. 2. A thin, usually rectangular or square element of a structure that is capable of resisting lateral forces in its plane, such as a floor or roof.

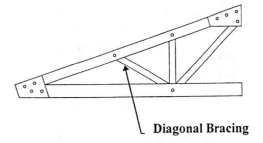

Diagonal Bracing

Diatomaceous Earth (DE). Sedimentary rock deposits consisting of or abounding in diatoms or their siliceous remains; diatomite; used in water filters.

Diatomaceous Earth Filter. Pool filter using DE as filtering medium; one designed to filter water through a thin layer of filter aid such as diatomaceous earth or volcanic ash; diatomite filters may be of the Pressure or Vacuum type.

Dichlorodifluoromethane. Popular refrigerant known as Freon 12.

Dichlorodifluromethane. Refrigerant commonly known as R-12.

Dichroic Lens. A lighting fixture lens that reflects light of one color and transmits light of other colors.

Dichromatic. Having two colors.

Die Casting. Process of molding low-melting temperature metals in accurately shaped metal molds.

Die. 1. To cease living. 2. To stop. 3. An engraved metal device for stamping coins or medals. 4. A perforated steel block against which a material pressed to form a shape by extrusion. 5. An internally threaded tool used to cut threads on a pipe or rod.

Diecast Set Screw. A set screw cast by forcing molten metal into a mold.

Dielectric Fluid. Fluid with high electrical resistance.

Diesel Generator. An electrical generator operated by diesel fuel.

Different Systems. Those which derive their supply from different sources, or from individual transformers or banks of transformers which do not have their secondary windings interconnected, or from individual service switches.

Differential Settlement. The uneven sinking of different parts or sections of a building's foundations.

Differential. The temperature or pressure difference between cut-in and cut-out temperature or pressure of a control.

Diffuse Radiation. Sunlight that is scattered from air molecules, dust and water vapor and comes from the sky vault.

Diffuse Transmission. Transmitted light that is scattered evenly in all directions as it passes through a material.

Diffuse-Porous Wood. Certain hardwoods in which the pores tend to be uniform in size and distribution throughout each annual ring or to decrease in size slightly and gradually toward the outer border of the ring.

Diffuser, Suction. See Suction Diffuser.

Diffuser. 1. A device for distributing air in a forced air heating/cooling system; often flush-mounted on a ceiling, it has slats to direct the conditioned air evenly into the room or space; components of the ventilation system that distribute and diffuse air to promote air circulation in the occupied space. 2. A lens in a light fixture to diffuse light.

Diffusion. The natural mingling of two or more substances to form a mixture or solution.

Digital. A digital instrument shows a measurement directly as a number, such as the LED display on a digital watch; compare with Analog.

Dike. A long wall or embankment built to prevent flooding.

Dilapidation. A building in a state of neglect, disrepair, decay, ruin, or deterioration through neglect or misuse.

Diluent. In lacquer, the volatile portion of vehicle not capable by itself of dissolving nitrocellulose.

Diluents. See Thinners.

Dimension Lumber. Lumber with a nominal thickness of from 2 inches up to but not including 5 inches and a nominal width of 2 inches or more; includes joists, planks, rafters, studs, and small timbers; see also Dimension Stock.

Dimension Ratio. The diameter of a pipe divided by the wall thickness; each pipe can have two dimension ratios depending on whether the outside or inside diameter is used; in practice, the outside diameter is used if the standards requirement and manufacturing control are based on this diameter; the inside diameter is used when this measurement is the controlling one.

D

Dimension Stock. Hardwood stock processed to a point where the maximum waste remains at the dimension mill and the maximum utility is delivered to the user; stock of specified thickness, width, and length; according to specifications, it may be solid or glued up, rough or surfaced, semifabricated or completely fabricated; commonly known as Hardwood Dimension Lumber.

Dimension. 1. A measurable extent of any kind, such as width, length, height, or volume.

Dimensional Stability. 1. The ability of a fabric to retain its dimensions in service and wet cleaning. 2. The ability of any material to maintain its essential dimensions in the conditions in which it is used.

Dimensional Stabilization. Measures taken to prevent or reduce warping, swelling, or shrinking.

Dimensions of Concrete Masonry Units. In practice, the first dimension of a CMU represents the thickness, the second, height; and the third, length.

Dimer. Formed by chemical combination of two similar molecules.

Diminish. To make or appear less or smaller.

Diminishing Returns. An increase in expenditure, investment, or taxation, beyond a certain point, ceases to produce a proportionate yield.

Diminution in Value. The amount by which the market value of property is reduced by any cause.

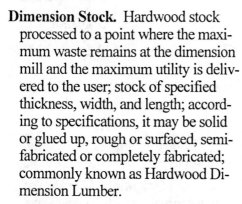
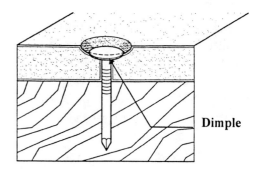

Dimple

Dimmer Switch. A switch which controls the brightness of a light.

Dimple. The depression in the surface of gypsum board caused by a hammer as the nail head is set slightly below the surface or plane of the gypsum board to permit concealment with joint compound.

Diode. A semiconductor allowing the flow of current in one direction only and having two terminals.

Diorite. A granular, crystalline igneous rock commonly of acid plagioclase and hornblende, pyroxene, or biotite.

Dip Coating. Process of finishing an article by immersing it in finishing material.

Dip of a Trap. The lowest portion of the inside top surface of the channel through a plumbing trap; compare with Weir of a Trap.

Dipentene. Solvent made by destructive distillation of pine stumps; is stronger than turpentine.

Direct Burial Cable. Electrical cable that is suitable for burial directly in the earth without being in a conduit.

Direct Burial Conduit. Electric conduit suitable for burial in exterior applications with an outer surface that resists moisture, fungus, and corrosion.

Direct Costs. Costs which may readily be attached or directly related to the unit of production.

Direct Current (DC). An electric current, of constant voltage, flowing in one direction only.

Direct Drive Fan. A fan that is positively connected in line to a motor to operate at the same speed.

Direct Examination. Questions directed to a friendly witness.

Direct Expansion (DX). Interior air cooled by directly passing over an evaporator in which refrigerant is expanding from a fixed reference point.

Direct Expansion Evaporator. Evaporator using either an automatic expansion valve (AEV) or a thermostatic expansion valve (TEV) refrigerant control.

Direct Job Overhead. A firm's overhead expense that is directly attributable to completing a work in process and that would not be incurred if no projects were under way.

Direct Labor. The actual cost of labor payroll for all jobs worked on during the period covered by the income statement.

Direct-Labor Burden. Includes all payroll taxes, insurance, and employees' benefits which are associated with the labor payroll.

Direct Methods. Techniques of solar heating in which sunlight enters a house through the windows and is absorbed inside.

Direct Nailing. See Face-Nailing.

Direct Process. See American Process Zinc Oxide.

Direct Radiation. Solar radiation that comes straight from the sun, casting shadows on a clear day.

Direct Transmission. Light passing through a clear, transparent material.

Direction of Irrigation. Direction of flow of irrigation water; usually at right angles to the supply ditch or pipe.

Directional Arrow. A symbol used to denote direction.

Directional Sign. A publicly displayed notice which indicates by use of words or symbols a recommended direction or route of travel.

Directory Board. A panel that is used to display information.

Disability. 1. A physical or mental impairment that substantially limits one or more of the major life activities of an individual. 2. A record of such an impairment. 3. Being regarded as having such an impairment.

Disappearing Stair. A folding, hinged stairway that is pulled down from the ceiling and commonly hidden in a trap door.

Discharge Valve. A special purpose valve.

Discharging Arch. An arch built over a lintel or header to help relieve the pressure from the wall above the arch; also known as a Relieving Arch or Safety Arch.

Discoloration. Change of color from that which is normal or desired.

Disconnect. An electrical switch.

Disconnecting Means. A device, or group of devices, or other means whereby the conductors of a circuit can be disconnected from their source of supply.

Discount. 1. A deduction in price or the amount of a bill. 2. To sell a note for less than its principal sum; this has the effect of increasing the yield to the buyer.

Discovery. A process in which parties to a lawsuit are required to divulge information.

Dishwasher. An electric appliance for washing dishes.

Disinfect Pipe. The act or process of transmitting an antimicrobial agent through a piping system.

Disinfectant. Chemical (also called bacteriacide) used to destroy germs and bacteria.

Disintegration. Decomposition into constituent elements or fragments.

Disk Sander. An electric sanding tool where the sand paper abrasive is mounted on a rotating disk.

Dismissal. The act of dropping a lawsuit.

Dispensary. A building or place where medicines and medical or dental aid are dispensed.

Dispenser, Tissue. See Tissue Dispenser.

Dispersed. Finely divided or colloidal in nature.

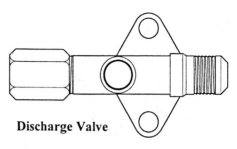
Discharge Valve

Dispersing Agent. An admixture capable of increasing the fluidity of pastes, mortars, or concretes by reduction of interparticle attraction; material used to aid in holding finely divided matter in dispersed state.

Dispersion. 1. Suspension of one substance in another. 2. A suspension of very fine particles in a liquid medium, as in paints.

Displacement, Piston. See Piston Displacement.

Displacement. Movement away from a fixed reference point.

Disposal Field. See Leaching Field.

Disposal Suit. A complete body suit worn by persons entering a hazardous materials area or handling hazardous materials.

Disposal. 1. The process of disposing of something. 2. Garbage Disposal Unit.

Distemper. 1. A water paint in which the pigment is mixed with white or yolk of egg, casein or size and which is used for mural decoration. 2. Broadly any of numerous water-based paints for general, especially for household use.

Distensibility. Ability to be stretched.

Distillate. A condensed product produced by cooling vapors of a material heated sufficiently to drive off part of the material in the form of vapor.

Distillation. The process of vaporizing a liquid with heat and condensing the vapor to a liquid for the purpose of purification, fractionation, and forming new substances.

Distillery. A building or place where alcoholic liquor is distilled.

Distilling Apparatus. Fluid-reclaiming device used to reclaim used refrigerants; reclaiming is usually done by vaporizing and then re-condensing the refrigerant.

Distortion. 1. The quality or state of being twisted out of a natural, normal, or original shape or condition. 2. A lack of proportion in an optical image resulting from defects in the optical system. 3. Faulty reproduction of radio sound or a television picture, caused by change in the wave form of the original signal. 4. Pavement distortion is any change of the pavement surface from its original shape.

Distributed Load. A load on a structural member that is evenly applied along its entire length.

Distributing Pipes. The piping in a building that distribute the water received from the Service Pipe.

Distribution Box. The main feed line terminus of an electrical service to which branch circuits are connected.

Distribution Panel. The main electrical control center, which contains switches or circuit breakers, is connected to the service wires, and delivers current to the various branch circuits.

Distribution Rib. A transverse thickening of a one-way concrete joist structure used to allow the joists to share concentrated loads.

Distribution. 1. The movement of fresh concrete to its placement point. 2. The circulation of collected heat to living areas from collectors or storage. 3. The movement of any fluid or current to its end point.

Distyle. A building or structure having two columns in front.

Ditch. A long narrow excavation dug in the earth for foundation walls, sewer systems, irrigation systems, electrical cable lines, and utilities; a trench.

Divider, Hospital. See Hospital Divider.

Divider, Terrazzo. See Terrazzo Divider.

Dividers. A drafting tool, like a compass, but with two adjustable sharp points; used for marking out and scribing edges.

Diving Board. A narrow platform, often equipped with a spring and raised off the pool deck by a ladder, attached to the edge of a pool and extending over the water, used to give divers altitude and lift.

Diving Platform. Rigid stand for diving.

DIY. Do it yourself.

DLH Series. A standard Steel Joist Institute designation for longspan steel joists.

DOC. U.S. Department of Commerce, c/o National Institute of Standards & Technology, Gaithersburg, Maryland 20899, (301) 975-2000.

Dock Board. Heavy timber used in the construction of the raised platform used for the loading and unloading of trucks.

Dock Bumper. Thick rubber units placed under loading dock openings to absorb the shock and prevent damage when trucks back in for loading or unloading.

Dock Leveler. A mechanical device at a loading dock that adjusts to accommodate the different heights of truck beds.

DOE. U.S. Department of Energy.

Dog Leg. Any device that is crooked or bent like a dog's hind leg.

Dog. Fits on a steel bar to form a tool for cramping or clamping.

Dogging Device. A mechanism used on exit devices that fastens the cross bar in the fully depressed position, and also retains the latch bolt or bolts in a retracted position, thus permitting free operation of the door from either side.

Dog's Tooth. Brick so laid that their corners project from the face of the wall.

DOI. U.S. Department of the Interior.

DOJ. U.S. Department of Justice.

Dolerite. Any of various coarse basalts or diabase.

Dolly Varden Siding. Beveled wood siding that is rabbeted on the bottom edge.

Dolomite. A mineral of the double carbonate of lime and magnesia having the general formula $CaCO_3 MgCA_3$.

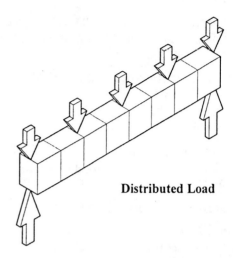

Distributed Load

Dolomitic. A type of lime or limestone containing calcium carbonate in combination with up to 50% magnesium carbonate.

Dome Light. An overhead light.

Dome. An arch rotated about a vertical axis passing through its crown, or highest point.

Dome-Hat. Sealed metal container for the motor compressor of a refrigerating unit.

Domestic Marble. Marble which comes from the country or area where the structure in which it used is built.

Domestic Well. A water well dug for household use.

Domicile. 1. A place of residence; a home. 2. A person's fixed, permanent, and principal home for legal purposes.

Dominant Color. Color that predominates or is outstanding.

Doodlebug. A device, as a divining rod, used in attempting to locate underground piping and conduits.

Doohickey. A small mechanical object whose name is not known or has been forgotten.

Door & Hardware Institute (DHI). 14170 Newbrook Drive, Chantilly, Virginia 22021-2223, (703) 222-2010.

Door and Hardware Institute. 14170 Newbrook Drive, Chantilly, Virginia 22021-2223.

Door Buck. A door frame.

Door Bumper. Rubber tip device mounted on walls or baseboards that prevent door knobs from marring walls.

Door Closer. A mechanical device attached to a door to make it close automatically.

Door Coordinator. A device used on a pair of doors to ensure that the inactive leaf is permitted to close before the active leaf; necessary when an overlapping astragal is present and exit devices and automatic or self-latching bolts are used with closers on both door leaves.

Door Frame Grout. Grout to fill in the voids where a masonry structure meets a metal door frame.

Door Frame. The members that completely surround a door, made of wood or metal, to which the hinges are attached.

Door Framing. The rough construction of the wall immediately surrounding the area of a door using wood or metal materials.

Door Guard. Guard fabricated of steel components to provide protection over interior face of door or prevent damage to glass and to prevent intrusion.

Door Head. The assembly of parts at the top of a door frame including the frame, stop, casing, shims, and flashing.

Door Header. Horizontal beam placed on vertical jack studs which form the uppermost portion of the framing of a door opening.

Door Holder. A device used to hold a door in an open position.

Door Jamb. The vertical members of a door frame.

Door Knob. A projecting knob for operating a lock.

Door Louver. Slanted fins, either fixed or movable, provided on a door for ventilation.

Door Motor. A mechanical device operated by electric current which opens or closes a door.

Door Opener. An electronic device which when activated opens a door by mechanically unlatching the throw bolt of a doorknob or which opens a door on a pivot point on the hinge side of the door.

Door Paint. A specific paint, usually a latex or alkyd enamel, used for its ability to be washed, to withstand cracking and scratch resistance.

Door Pull. A handle by which a door is pulled.

Door Rail Hanger. Structural steel shape used to support large rolling door.

Door Removal. The act or process of removing a door from its hinges.

Door Seal. Rubberized material attached to door head, jamb, and bottom of a door to aid in preventing air drafting.

Door Stop. A molding nailed to the faces of the door frame jambs to prevent the door from swinging through.

Door Switch. An electric switch installed in a door jamb which turns on a light in a closet when the door is opened and turns off a light when the door is closed.

Door Threshold. See Threshold, Door.

Door Trim. 1. Wood or metal finishing work, often ornamental, used for covering the joints between door jambs and plaster walls. 2. The locks, knobs and hinges on a door.

Door, Access. See Access Door.

Door, Aluminum. A glazed door constructed with aluminum stiles and rails.

Door, Chain Hoist. A door that is operated by a chain placed in a pulley or sheave.

Door, Commercial. See Commercial Door.

Door, Flush. See Flush Door.

Door, Folding. See Folding Door.

Door, Hollow Core. See Hollow Core Door.

Door, Overhead. See Overhead Door.

Door, Revolving. See Revolving Door.

Door, Roll-Up. See Roll-Up Door.

Door, Shower. See Shower Door.

Door, Sliding Glass. See Sliding Glass Door.

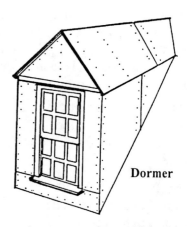

Dormer

Door, Solid Core. See Solid Core Door.

Door, Special. See Special Door.

Door, Vault. See Vault Door.

Door, Vertical Lift. See Vertical Lift Door.

Door. A hinged, sliding, or revolving barrier for closing and opening an entrance to a building, room, or cupboard.

Dope Coat. In tile setting, neat cement applied to the setting bed.

Dope. A term used by plasterers for mortar additives of any type, such as those used to retard or accelerate set.

Dormer Window. A window in a dormer.

Dormer. A projecting roofed housing in a sloping roof.

Dormitory Wardrobe. A closet in student housing where clothes are kept; commonly a free-standing cabinet made to match the cabinets of the dormitory room.

Dormitory. A room occupied by more than two guests.

Dot. A small lump of plaster placed on a surface between grounds to assist the plasterer in obtaining the proper plaster thickness and aid in aligning the surface; see Screed, 1.

DOT. U.S. Department of Transportation, 400 Seventh Street, SW, Washington, DC 20590-0001, (202) 366-4488.

Dote. Wood rot or decay.

Double Back. A webbed backing cemented to the backing of tufted, knitted and some woven carpets as additional reinforcement, to provide greater dimensional stability; also called Scrim Back.

Double Bullnose. A type of trim with the same convex radius on two opposite sides.

Double Duty Case. Commercial refrigerator in which a part of space is for refrigerated storage and part is equipped with glass windows for display purposes.

Double Glass. Double Glazing.

Double Glazing. Two parallel sheets of glass with an airspace between; also called Dual Glazing or Double Glass.

Double Headed Nail. A nail with two closely-spaced heads to permit easy removal; widely used in concrete formwork as a temporary fastener; also called a Duplex Nail.

Double Hung Window. The most common style of operable window, it has two sashes that slide vertically in parallel tracks; a window with two overlapping sashes that slide vertically in tracks.

Double Layer. Two layers of gypsum board; various thicknesses may be combined to improve the fire, sound, or structural characteristics.

Double Nailing. A method of applying gypsum board by using two nails spaced approximately 2 inches apart every 12 inches in the field along the framing member to insure firm contact with the framing.

Double Offset. In piping, two offsets installed in succession in the same line.

Double Oven. An oven with two baking compartments.

Double Plate. The two wood plates on the top of a wood framed wall.

Double Regulation. Regulation of both pot and gun air pressure.

Double Reinforcement. A concrete beam with steel on both sides of the neutral axis to resist tension and compression.

Double Reinforcing. To double-up on certain materials to increase strength and stability.

Double Spread. See Spread.

Double Stud. Two adjoining studs in a wood framed wall.

Double Tee. A precast concrete slab element that resembles the letters TT in cross section.

Double Thickness Flare. Copper, aluminum or steel tubing end which has been formed into two-wall thickness, 37 to 45 degrees bell mouth or flare.

Double Walled Tank. A tank constructed with two walls for leak protection and to increase strength and stability.

Double Wye. A fitting or branch pipe in plumbing, constructed of either cast or wrought iron, which has two side outlets at any angle except a right angle; usually a 45 degree angle.

Double-Back. See Double-Up.

Double-Bevel Groove Weld. See Groove Weld.

Doubled-Up. See Double-Up.

Double-J Groove Weld. See Groove Weld.

Double-U Groove Weld. See Groove Weld.

Double-Up. A method of plaster placement characterized by application in successive operations with no setting or drying time allowed between coats; also called Double-Back, Doubled-Up, Laid Off, Laid On, or Two Coat Work.

Double-Vee Groove Weld. See Groove Weld.

Douglas Fir Plywood Association (DFPA). c/o American Plywood Association), PO Box 11700 Tacoma, Washington 98411, (202) 272-2283.

Douglas Fir. A tall, evergreen tree that furnishes long straight framing lumber; also known as Red Fir, Douglas Spruce, and Yellow Fir.

Douglas Spruce. Douglas Fir.

Dovecote. A shelter for nesting doves.

Dovetail Anchor Slot. A matching interlocking strip or slot which is used with a dovetail fastener.

Dovetail Joint. A joint in which one piece has dovetail shaped pins or tenons which fit into corresponding openings on the other piece.

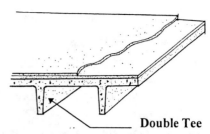

Double Tee

Dovetail Saw. Saw used for cutting very accurate joints; similar to a tenon saw but with a thinner blade and finer teeth.

Dowel Pin. Dowel.

Dowel Sleeve. Cap of light metal or cardboard on one end of a steel dowel bar to allow free movement of an expansion joint.

Dowel. 1. A steel bar, which extends into two adjoining portions of a concrete or masonry construction, as at a joint in a pavement slab, so as to connect the portions and transfer shear loads. 2. As used in the construction of column and wall sections, a deformed steel reinforcing bar placed so as to transmit tension or compression as well as shear loads. 3. A steel reinforcing bar that projects from a foundation to tie it to a column, wall, slab, porch, or steps. 4. A short cylindrical headless rod of wood or steel inserted into adjoining members to keep them in alignment and hold them together. 5. A long cylindrical strip of metal or wood to be cut into dowels.

Down Payment. The money paid by the buyer, being the difference between the full price and the borrowings.

Down Time. The time during which equipment is out of service for essential maintenance or emergency repairs.

189

Downlight. A lighting fixture that directs its beam downward.

Downspout Boot. Device at bottom of downspout to receive water from downspout and provide transition to drainage piping.

Downspout Bracket. A metal bracket for securing a downspout to a wall.

Downspout. Vertical pipe usually made from sheet metal or plastic which carries water from the roof gutters to the ground or a storm drain; also called a Leader.

Dozer Excavation. The digging out of large areas of earth with a bulldozer.

Dozer. A term used in the trade for a bulldozer.

Draft Gauge. Instrument used to measure air movement by measuring air pressure differences.

Draft Indicator. Instrument used to indicate or measure chimney draft or combustion gas movement; draft is measured in units of .10 in. of water column.

Draft Regulator. Device which maintains a desired draft in a combustion-heated appliance by automatically controlling the chimney draft to the desired value.

Drafter. Draftsman.

Drafting Board. A flat surface used by drafters on which to produce drawings.

Drafting Machine. A mechanical device with adjustable straight edges that may be clamped onto a drawing board for use by a drafter.

Draftsman. 1. One who prepares architectural, construction, or engineering drawings; draftsman. 2. A person who drafts documents.

Drag Strut. A structural member used to transfer lateral forces from one vertical element to another.

Drag. In welding, the distance between the point of exit of the cutting oxygen stream and the projection on the exit surface of the point of entrance.

Dragging. See Floating.

Dragline. A bucket attachment for a crane commonly used in a marsh or marine area, that digs soft materials that must be excavated at some distance from the crane, and draws the bucket towards itself using a cable.

Dragon's Blood. A red gum exuded from the fruit of a species of palm; used for coloring varnish; the color is not permanent.

Drain Board. A countertop adjoining a sink.

Drain Field. A system of trenches filled with sand, gravel or crushed stone, and a series of pipes to distribute septic tank effluent into the surrounding soil.

Drain Pipe. Any pipe that carries waste or water-borne wastes in a building drainage system.

Drain Tile. Short-length sections of burned clay, plastic, or concrete pipe, laid with open joints and surrounded with aggregate and covered with asphaltic paper or straw to drain the water from an area.

Drain, Intercepting (Curtain Drain). A drain that intercepts and diverts ground water before it reaches the area to be protected.

Drain, Intercepting. See Intercepting Drain.

Drain, Waste and Vent (DWV) System. The assemblage of pipes which facilitates the removal of liquid and solid wastes as well as the dissipation of sewer gases.

Drain. 1. A trench, ditch, or pipe designed to carry away waste water. 2. A device that allows for the flow of water from a roof area.

Drainage Head. The farthest or highest spots in a drainage area.

Drainage Pipe. Any pipe in a plumbing system which removes waste water, rainwater, or sewage.

Drainage System. All drain and waste pipes that carry water and waste away from the fixtures.

Drainage, Underslab. The process of continuous interception and removing of ground water from under a concrete slab with the installation of crushed stone and perforated pipe.

Drainage. 1. The process or means of draining a building or land area of atmospheric moisture or sewage. 2. A system of drains, artificial or natural.

Drape. A window covering; curtain.

Drapery Rod. A wall-mounted wood or metal shaft for supporting drapery.

Drapery Track. A U-shaped device mounted to a door or window header, ceiling or floor, used as a guide for draperies.

Drapery. Fabric hangings for use as a curtain.

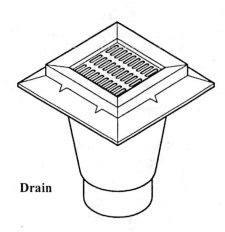

Drain

Draughtsman. Draftsman; drafter.

Draw Board. A board fitted in a kitchen cabinet that may be pulled out to create a supplementary work surface, like a breadboard.

Draw. A small valley or a gully.

Drawboard Joint. A mortise-and-tenon joint with holes so bored that when a pin is driven through, the joint becomes tighter.

Drawbridge. A bridge over water, hinged at one end so that it may be raised or lowered to allow or prevent passage of ships.

Drawer Knob. A cabinet knob.

Drawer Pull. A handle with which to withdraw a drawer.

Drawer Roller. A device used to ease the sliding of a drawer open or shut, usually with a metal or fiber wheel rotating in a metal frame.

Drawer Slides. A mechanism employing guides and rollers that guide and support the drawer, permitting easy operation.

Drawer. A sliding boxlike storage compartment, usually part of a cabinet.

Drawing Room. A formal reception room; a withdrawing room.

Drawings, Shop. Drawings of specific items of the project provided by subcontractors or fabricators.

Drawings. Graphic representations of buildings and their parts; plans or blueprints.

Drawn Glass. Glass sheet pulled directly from a container of molten glass.

Drayage. The charge for transportation or delivery of goods.

Dredge. Deepen a waterway by use of a dredging machine, usually mounted on a barge.

Dressed and Matched (D&M). Tongued and grooved; also called Center Matched; see Tongue and Groove.

Dressed Size. The dimensions of lumber after being surfaced with a planing machine; the dressed size is usually 1/2 to 3/4 inch less than the nominal or rough size; a 2- by 4-inch stud, for example, actually measures about 1-1/2 by 3-1/2 inches.

Dressing Room. 1. In a home, a room for storage of clothes and for dressing. 2. In a theater, a room for storing and changing costumes and for applying and removing make-up.

Drier. 1. Substance or device used to remove moisture from a refrigeration system. 2. A solution added to drying oils in paint to quicken the drying. 3. Composition of certain metals that accelerate drying action of oil when added to paint or varnish; some driers are in dry form, others in paste form; most are solutions of metallic soaps in oils and volatile solvents; they are known as driers, oil driers, Japan driers, liquid driers, and Japans; the metallic soaps most commonly used are those of lead, manganese and cobalt.

Drift Pin. A tapered steel shaft used to align bolt holes in steel connections during erection.

Drift. 1. Lateral deflection of a building caused by wind or earthquake loads; the horizontal displacement or movement of structure when subjected to lateral forces. 2. Spray loss, in using a spray gun; also called Overspray.

Drill. Common name for small boring bit and for the electric drill.

Drilled Well. The act or process of using a rotary drill to dig for water.

Drilling Rock. The act or process of boring holes into rock.

Drip Cap. A piece of metal or wood trim which is placed on the top of an exterior window header to shed off rain.

Drip Edge. A discontinuity or strip installed at roof eaves or over a window or wall component to force adhering drops of water to fall free of the face of the building rather than run toward the interior; the projection of a window head member or water table to allow the water to drip clear of the side of the building below it.

Drip Line. The line made on the earth or other surface when water drips off a roof or other building element.

Drip Loop. A low spot in an aerial wire to allow atmospheric moisture and rainwater to drip off before the wire enters the building.

Drip Pan. Pan-shaped panel or trough used to collect condensation from evaporator.

Drip. 1. A painting defect where the paint has started to run. 2. A projecting piece of material so shaped as to throw off water and prevent its running down the face of the wall or other surface of which it is a part.

Drippage. Leakage of roofing bitumen through cracks in the roof deck or over the edges of the roof.

Drive Screw. A screw which is driven in with a hammer but removed by a screwdriver.

Drive-Up Window. An opening in a wall through which transactions can be made with persons in motor vehicles.

Driveway Apron. A sloping transition from the public street to a private driveway.

Driveway. A passageway for automobiles on private property.

Driving Cap. A steel cap placed on the top of a pile for protection from damage during driving.

Drop Black. Bone Black.

Drop Cloth. A canvas, plastic, or paper protective sheet used by painters to cover floors and furniture.

Drop Match. See Set Match.

Drop Panel. A thickening of a two-way concrete floor structure at the head of a column.

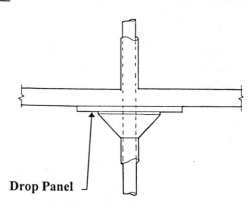

Drop Panel

Dropback. Softening Point Drift.

Drum Trap. A cylindrical plumbing trap, closed at the bottom, with a cover for cleaning; the pipe from the fixture enters the trap near the bottom; the space between where the pipe enters and the bottom is a settling area; the outlet to the waste pipe is higher up in the trap.

Dry Air. Air that contains no water vapor.

Dry Bulb (DB) Temperature. The temperature of air as registered by an ordinary thermometer; compare with Wet Bulb Temperature.

Dry Bulb Thermometer. An ordinary thermometer for measuring ambient air temperature; compare with Wet Bulb Thermometer.

Dry Capacitor Condenser. Electrical device made of dry metal and dry insulation; used to store electrons.

Dry Cell Battery. Electrical device used to provide DC electricity, having no liquid in the cells.

Dry Cleaner. An apparatus for the washing of fabrics which uses substantially non-aqueous organic solvents.

Dry Edging. Rough edges and corners of glazed ceramic ware due to insufficient glaze coating.

Dry Film Thickness. Depth of applied coating when dry, expressed in mils.

Dry Ice. Refrigerating substance made of solid carbon dioxide which changes directly from a solid to a gas (sublimates). Its subliming temperature is -109° F. (-78° C.).

Dry Kiln. A chamber having controlled air-flow, temperature, and relative humidity for drying lumber, veneer, and other wood products.

Dry Lap. The absence of bitumen at the overlap of two roofing felts.

Dry Location. A location not normally subjected to dampness or wetness; a location classified as dry may be temporarily subject to dampness or wetness, as in the case of a building under construction.

Dry Mix. A concrete or mortar mix in which there is little or no water.

Dry Oil. See Dehydrated Oil.

Dry Pack. A mixture of cement and fine aggregate with enough moisture for hydration but dry enough to be rammed into place.

Dry Packing. Placing of zero slump, or near zero slump, concrete, mortar, or grout by ramming into a confined space.

Dry Press Brick. Brick formed in molds under high pressures from relatively dry clay, 5 to 7 percent moisture content.

Dry Rodded Weight. The weight of dry aggregate rodded into a cylindrical container of diameter approximately equal to the height, each of 3 layers rodded 25 times, and the excess aggregate struck off level with the top of the container.

Dry Rot. A term loosely applied to any dry, crumbly rot but especially to that which, when in an advanced stage, permits the wood to be crushed easily to a dry powder; the term is actually a misnomer for any decay, since all fungi require considerable moisture for growth.

Dry Set Mortar. A water-retentive hydraulic cement mortar usable with or without sand; when this mortar is used, neither the tile nor walls have to be soaked during the installation process.

Dry Set Tile. Tile set into an adhesive that seems dry but that adheres on contact.

Dry Sheet. A ply of roofing felt mechanically attached to the deck to prevent asphalt or pitch from leaking into the building below; dry sheets are not part of built-up roofing assemblies.

Dry Spots. Small areas on the face of tile which have been insufficiently glazed.

Dry Spray. Overspray or bounce back; sand finish due to spray particle being partially dried before reaching the surface.

Dry Sprinkler System. A fire extinguishing sprinkler system whose pipes remain empty of water until the system is activated.

Dry Standpipe. A fire fighting pipeline in a building that is dry until filled by external connection to a fire hydrant.

Dry Stone Wall. A wall of stone that has been constructed without the use of mortar or concrete in its joints.

Dry System. Refrigeration system which has the evaporator liquid refrigerant mainly in the atomized or droplet condition.

Dry Tape. The application of tape over gypsum board joints with adhesives other than conventional joint compound.

Dry To Handle. 1. A film of paint is dry to handle when it is hardened sufficiently so that it may be handled without being damaged. 2. Time interval between application and ability to pick up without damage.

Dry To Recoat. In painting, time interval between application and ability to receive next coat satisfactorily.

Dry To Touch. 1. A film of paint is dry to touch when it is hardened sufficiently so that it may be touched lightly without any of it adhering to the fingers. 2. Time interval between application and tack-free time.

Dry Type Extinguisher. A fire extinguishing system that discharges a dry powder by means of compressed gas.

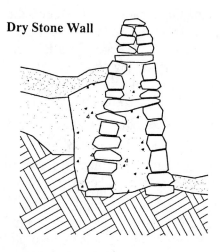

Dry Stone Wall

Dry Type Spray Booth. A spray booth where an inhibitor has been added to paint to prevent the surface from drying too rapidly, causing wrinkling or cracking.

Dry Type Transformer. A transformer whose coils and core are not immersed in an oil bath.

Dry Vent. A plumbing vent that does not carry any liquids; it acts as a vent only and carries only air.

Dry Well. A deep hole, covered and usually lined or filled with rocks, that holds drainage water until it soaks into the ground.

Dryer. An apparatus that will dry the aggregates and heat them to the specified temperatures.

Dryer Receptacle. An electrical outlet wired to match the specific electric current needs and configuration of a clothes dryer.

Dryer, Coin. See Coin Dryer.

Dryer, Darkroom. See Darkroom Dryer.

Drying Oil. An oil which, when a thin film of it is exposed to the air, takes on oxygen and becomes hard, tough, and elastic; drying oils are used in the manufacture of paints and varnishes; linseed oil is a common drying oil.

Drying Shrinkage. A decrease in the volume of concrete upon drying.

Drying Time. Time interval between application and final cure.

Drying. Act of changing from liquid to solid state by evaporation of uncombined water and volatile thinners and by oxidation of oils.

Dryness. 1. The condition of being free of liquid, especially water. 2. The degree of free moisture contained within a gypsum board product.

Dry-Out. A condition occasionally occurring in gypsum plaster work which by excessive evaporation or suction has lost some or all of the water necessary for crystallization; appears as a light colored, friable area.

Drywall Removal. The act or process of demolishing and carrying away old drywall systems in a building or structure.

Drywall Studs. Light gauge metal studs that fit into a top and bottom track to support sheets of gypsum wallboard for constructing interior non-bearing partitions.

Drywall Track. The horizontal light gauge metal top and bottom wall members in an interior gypsum wallboard non-bearing partition.

Drywall. 1. Wall materials that do not involve the use of plaster, such as gypsum wallboard or wood paneling. 2. A gypsum board product, usually gypsum wallboard.

Drywood Termites. See Termite.

DS Glass. Double strength window glass.

DS. Downspout.

Dual Duct Air Handling System. A system similar to a multi-zone system, except that instead of mixing the hot and cold at the air handling unit, separate hot and cold ducts are run to the space to be conditioned, and a terminal mixing box is provided at the space to be conditioned; a constant supply of air is supplied to various zones.

Dual Glazing. Double Glazing.

Dual Vents. A plumbing vent that connects at the point where two fixture drains come together and acts as a back vent for both fixtures' traps.

Duck Tape. Duct Tape, mistakenly or humorously.

Duckboard. Wood board or slats to provide dry footing over mud or a wet work area.

Duct Bank. A group, series, or tier of round or rectangular metal pipes used to distribute warm or conditioned air throughout a building or structure.

Duct Heater. A heating element in a duct of an air-handling system.

Duct Insulation. The material installed on the ductwork in an HVAC system, for the reduction of fire hazard, or as thermal insulation.

Duct Shaft. A lined vertical shaft in a building, usually of a fire-rated construction, through which air ducts and piping are run.

Duct Tape. A tough adhesive tape used to assemble and to repair light gauge air ducts.

Duct, Bus. See Bus Duct.

Duct, Feeder. See Feeder Duct.

Duct, Underfloor. See Underfloor Duct.

Duct. A pipe, tube, or channel carrying air, gas, liquid, or wires.

Ductile Iron Pipe. Iron pipe that is manufactured so as to render it able to be flexed; ductile iron pipe has the non-corrosive qualities of cast iron but is not brittle and has the handling characteristics of steel.

Ductile Moment Resisting Space Frame. A three dimensional structural system without bearing walls and composed of interconnecting members that resist lateral forces with or without the aid of horizontal diaphragms.

Ductile. 1. In structures, the ability of a material to deform under tensile load. 2. Of metals, capable of being drawn out into wires, rods, or thin shapes.

Ductility. 1. The quality of being able to deform considerably under load before fracturing; contrasted to the quality of brittleness. 2. The ability of a substance to be drawn out or stretched thin; while ductility is considered an important characteristic of asphalt cements in many applications, the presence or absence of ductility is usually considered more significant than the actual degree of ductility. 3. The property of a material which allows it to be formed or bent without breaking; copper is more ductile than steel.

Ductwork, Metal. The rigid sheet material out of which the ducts of an HVAC system are manufactured, commonly, galvanized sheet steel, aluminum, or stainless steel.

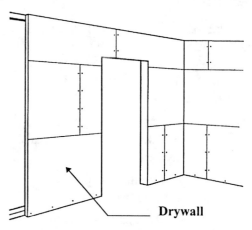

Drywall

Due Process. Notice, and an opportunity to be heard.

Dull Rubbing. Act of rubbing a dried film of finishing material to a dull finish, usually with abrasive paper, pumice stone, steel wool, and oil or water.

Dull. Term applied to colors that have a neutral or grayed quality.

Dulling. Loss of gloss or sheen.

Dumbwaiter. A small elevator used for conveying food, dishes, or materials from one floor to another in a building.

Dummy Joint. A joint placed strictly for design, in sidewalks and patios.

Dummy Trim. Hardware trim only, without a lock, usually used on the inactive door in a pair of doors.

Dump Fee. The amount charged for dumping jobsite debris in a landfill or dumpsite.

Dump Truck. A truck with a tiltable body used for transporting and dumping loose materials.

Dumpster. A large, heavy metal container used to hold and haul rubbish.

Dumpy Level. An Engineer's Level.

Dunnage. 1. Loose materials used around a cargo to prevent damage. 2. Scrap gypsum board used for protection of a shipping unit of gypsum wallboard.

Dunting. The cracking that occurs in fired ceramic bodies due to thermally induced stresses.

Duplex Apartment. An apartment having rooms on two floors.

Duplex Box. An electrical box for wiring switches or duplex receptacles.

Duplex Nail. A Double Headed Nail.

Duplex Paper. Wallpaper which consists of two separate papers pasted together; used to create a highly embossed effect.

Duplex Plate. A protective, finished, plate installed over duplex switches or receptacles.

Duplex Receptacle. A double electric outlet.

Duplex. 1. A two-family house. 2. A Duplex Apartment.

Durability. 1. Ability to exist for a long time without significant deterioration. 2. The ability of concrete to resist weathering action, chemical attack, and abrasion. 3. The property of an asphalt paving mixture that describes its ability to resist disintegration by weathering and traffic; under weathering are changes in the characteristics of the asphalt, such as oxidation and volatilization, and changes in the pavement and aggregate due to the action of water, including freezing and thawing.

Duress. Compulsion.

Durham. A type of pipe system composed of iron or steel pipe and all of the joints are threaded screw joints; the fittings used in this system are recessed drainage fittings.

Dust Cap. A protective cap placed over a device to protect it from the intrusion of foreign materials.

Dust Free. A film of paint is dust free when dust no longer adheres to it.

Dust Mask. A fabric mask worn to filter out dust and other foreign matter.

Dusting. 1. The application of dry portland cement to a wet floor or deck mortar surface; a pure coat is thus formed by suction of the dry cement. 2. The development of powdered material on the surface of hardened concrete caused by inadequate curing.

Dustproof. So constructed or protected that dust will not interfere with its successful operation.

Dust-Tight. So constructed that dust will not enter the enclosing case.

Dutch Bond. See English Bond.

Dutch Bond. In masonry, a bond having the courses made up alternately of headers and stretchers; same as English Bond.

Dutch Door Bolt. Device for locking together the upper and lower leaves of a Dutch door.

Dutch Door. A door with two separately hinged leaves, one above the other, enabling one to be open while the other stays shut.

Dutch Metal. Thin leaves of bright brass which are used for overlaying in the same manner in which gold leaf is applied.

Dutchman. 1. A small wood patch in woodwork. 2. A cut tile used as a filler in the run of a wall or floor area. 3. A narrow strip of carpet side-seamed to standard-width broadloom to compensate for unusual offsets and sloping walls but never used as a sub-stitute for good planning, accurate cutting, and proper stretching tech-niques.

Duty, Continuous. See Continuous Duty.

Duty. An obligation that is imposed by law or by contract.

Dwarf Wall. A short wall or partition that does not reach the ceiling.

Dwelling Unit. One or more habitable rooms which are intended or designed to be occupied by one family with fa-cilities for living, sleeping, cooking, and eating.

Dwelling. A place in which people live.

DWV. Drainage, waste, and vent sys-tem.

DX. Direct Expansion.

Dye. A material used for dyeing or staining.

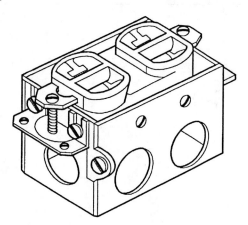

Duplex Receptacle

Dynamic Force. A moving force; F=ma, or force equals mass times ac-celeration.

Dynamic. Having to do with bodies in motion; implying motion or change of state; for example, an earthquake force is a dynamic force; the opposite of Static.

Dynamite. A type of explosive; a slang term used by tilesetters when referring to a mortar accelerator.

Dynamo. An electrical power genera-tor.

Dynamometer. Device for measuring power output or power input of a mechanism.

Dyne. A unit of force that, acting on a mass of one gram, increases its veloc-ity by one centimeter per second along the direction that it acts.

E&O Insurance. Errors and Omissions Insurance.

E. 1. Represents the modulus of elasticity. 2. Represents volts (electromotive force).

E. Switch (Snap, General Use). A form of general use switch so constructed that it can be installed in flush device boxes or on outlet box covers, or otherwise used in conjunction with wiring systems recognized by these regulations.

E.P.S. Expanded Polystyrene.

EA. Exhaust Air.

Eagle Beak. A 6-inch x 3/4 inch outside corner ceramin tile trim shape tile.

Earlywood. Springwood.

Earth Pigment. Pigments which occur as deposits in earth and are removed by mining; such pigments as a whole are permanent in color, non-bleeding, and are not readily changed by heat, light, moisture, and alkalis.

Earthquake. Movement in the earth's crust that produces horizontal and vertical movement on the ground surface.

Earthwork. An embankment or other construction made of earth; any work involving movement or use of soil and other earthen material.

Eased Edges. 1. The slight rounding of the corners of a piece of finish lumber. 2. A tapered, slightly rounded factory edge of gypsum board.

Easement. 1. A right to utilize real property owned by another. 2. An interest in land owned by another that entitles its holder to a specific limited use.

Eastern Frame (Balloon Frame). A type of wood framing where the wall studs in a two story structure are continuous from the foundation to the roof line. Rarely used in today's construction.

Eastern Method. See Pick and Dip.

Eave Trough. Gutter for catching rain water at the eaves of a sloping roof.

Eave Vent. A usually screened opening at the eave line to allow a free flow of ventilation air into the underroof area.

Eaves. The part of a roof that extends beyond the exterior walls below it.

Eccentric. Not having common centers; offcenter.

Eccentrically Braced Frame (EBF). A diagonal braced frame in which a least one end of each bracing member connects to a beam a short distance from a beam-to-column connection or from another beam-to-brace connection.

Eccentricity. The distance from the application of a structural load to the axis or centroid of the carrying member.

Eclectic. Architectural design based on picking and choosing from various styles; see Pastiche.

Ecology. A branch of science concerned with the interrelationship of organisms and their environments.

Economic Life. The period of time a property will produce sufficient income after expenses to justify its continued operation.

Economic Obsolescence. Depreciation; the reduction in value of real property caused by extrinsic factors such as a declining neighborhood, a factory closure, or unsuitable zoning.

Economic. Profitable; careful, efficient, and prudent use of resources.

Economizer Cycle. Not an individual system, it is a modification of the single zone system, terminal reheat, multi-zone, dual duct, and variable air volume (VAV); it is a modification that controls and adjusts motorized dampers to draw in outside air when it is advantageous for reduced cooling energy.

Economy Brick. A brick larger than standard.

Economy. Efficient and concise use of resources.

Ecosystem. A biological community of interacting organisms and their physical environment.

Eco-tech. Green design.

Edge Bead. A strip of metal or plastic used to make a neat, durable edge where plaster or gypsum board abuts another material.

Edge Distance. The distance from a rivet, bolt, screw, or nail to the edge of a structural member.

Edge Form. A forming member used to limit the horizontal spread of fresh concrete on flat surfaces.

Edge Joint Cracks. In concrete or asphalt road paving, the separation of the joint between the pavement and the shoulder, commonly caused by the alternate wetting and drying beneath the shoulder surface; other causes are shoulder settlement, mix shrinkage, and trucks straddling the joint.

Edge Joint. The place where two pieces of wood are joined together edge to edge, commonly by gluing; the joint may be made by gluing two squared edges as in a plain edge joint or by using machined joints of various configuration, such as tongued-and-grooved joints.

Edge Mounted Tile. A type of mounted tile wherein tile is assembled into units or sheets and are bonded to each other at the edges or corners of the back of the tiles by an elastomeric or resinous material which becomes an integral part of the tile installation.

Edge Preparation. The prepared shape on the edge of metal for welding.

Edge Repair. The repair of an edge of a construction member by the use of plaster or concrete.

Edge Sheets. Felt strips that are cut to widths narrower than the standard width of the full felt roll; used to start the felt-shingling pattern at a roof edge.

Edge Stripping. Application of felt strips cut to narrower widths than the normal 36 inch width of a felt roll, used to start the felt-shingling pattern at a roof edge.

Edge Venting. The practice of providing regular spaced protected openings around a roof's perimeter to relieve the water vapor.

Edge, Drip. See Drip Edge.

Edge. 1. The cutting side of a blade. 2. The degree of sharpness of a tool. 3. The line where an object or area begins or ends. 4. The extreme verge or brink of a cliff. 5. Of gypsum board, the paper bound edge as manufactured.

Edge-Grained Lumber. Lumber that has been sawed so that the wide surfaces extend approximately at right angles to the annual growth rings. Lumber is considered edge grained when the rings form an angle of 45 degrees to 90 degrees with the wide surface of the piece.

Edge-Matched. 1. See Tongue and Groove. 2. Lumber that has been rabbeted on both edges of each piece; in either case, the purpose is to provide a close joint when fitting two pieces together.

Edger. A finishing tool used on the edges of fresh concrete to provide a rounded corner.

Edging. Striping.

Edifice. A large building.

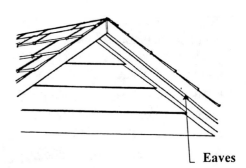

Eaves

EDM. Electronic Distance Measuring.

EE. Electrical Engineer.

Effective Area of Concrete. The area of a section which lies between the centroid of the tension reinforcement and the compression face of the flexural member.

Effective Area of Reinforcement. The area obtained by multiplying the cross sectional area of the reinforcement by the cosine of the angle between its direction and the direction for which the effectiveness is to be determined.

Effective Area. Actual flow area of an air inlet or outlet; gross area minus area of vanes or grille bars.

Effective Depth of Section. The distance from the extreme compression fiber to the centroid of tension reinforcement.

Effective Opening. The minimum cross-section area of the opening where water is discharged from a water supply pipe.

Effective Prestress. The stress remaining in prestressing tendons after all losses have occurred, excluding effects of dead load and superimposed load.

Effective Stiffness. The value of the lateral force in the isolation system, or an element thereof, divided by the corresponding lateral displacement.

Effective Temperature. Overall effect on a human of air temperature, humidity, and air movement.

Effective. 1. Capable of producing a decided, decisive, or desired effect. 2. Capable of performing the particular function specified with safety.

Efficiency Living Unit. Any room having access to bathroom facilities and having cooking facilities and intended or designed to be used for combined living, dining, and sleeping purposes.

Efficiency. The ratio of useful work performed to the total energy expended or heat taken in.

Efflorescence. A deposit of water soluble salts on the surface of masonry or plaster caused by the dissolving of salts present in the masonry; migration of the solution to the surface and deposition of the salts when the water evaporates; the surest preventative of efflorescence is to keep water out of masonry; the process of efflorescing is also referred to as Whiskering or Salt-petering.

Effluent. Fluid discharge from a sewage system.

Egg and Dart. A decorative carved molding consisting of alternating egg-shaped and arrow-shaped parts.

Egg Shell. Finish that closely resembles the luster of an egg shell.

Eggshelling. 1. The texture of a fired glaze similar in appearance to the surface of an eggshell. 2. Chip-cracked plaster, either base or finish coat; the form taken is concave to the surface and the bond is partially destroyed.

Egress. The way out; exit.

EIA. 1. Environmental Impact Assessment. 2. Electronic Industries Association.

EIFS. Exterior Insulation and Finish System.

Eight Bend. A pipe fitting which allows the run of pipe to make a 45 degree bend.

EIP. Membrane is generally reinforced with polyester fabric. Resists fire, chemicals, oils and tears.

EIR. Environmental Impact Report.

EIS. Environmental Impact Statement.

Ejector. Device which uses high fluid velocity, such as a venturi, to create low pressure or vacuum at its throat to draw in fluid from another source.

Elapsed Time Indicator. A mechanical or electronic device which measures the passage of time.

Elastic Analysis. An analysis of deformations and internal forces based on equilibrium, compatibility of strains, and assumed elastic behavior and representing to suitable approximation the three dimensional action of the structural shell together with its auxiliary members.

Elastic Limit. The largest stress which a material can withstand without being permanently deformed.

Elastic Sheet Roof. A thin, pliable roofing material that is able to expand and contract to its original size regardless of weather conditions.

Elastic Shortening. In prestressed concrete, the shortening of a member which occurs on the application of forces induced by prestressing.

Elastic. Able to return to its original size and shape after removal of stress.

Elastomer. Any of various elastic substances resembling rubber; a material which at room temperature can be stretched repeatedly to at least twice its original length and, upon release of the stress, will return with force to its approximate original length.

Elastomeric Flashing. A rubber-like material used as flashing on a roof system.

Elastomeric. Having the characteristics of an elastomer.

Elastomeric/Plastomeric Membrane. A rubber-like sheet material used as a roof covering.

Elbow Catch. A spring-loaded device consisting of a rocker arm and angle strike, for locking the inactive leaf of a pair of cabinet doors.

Elbow. A pipe fitting having two openings which allows a run of pipe to change directions.

Electric Boiler. A tank in which water is heated or hot water is stored, controlled by an electric current.

Electric Cord. A small flexible insulated electrical cable having a plug at one or both ends used to connect a power tool or mechanical device with a receptacle which supplies electric current.

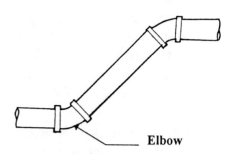

Elbow

Electric Defrosting. Use of electric resistance heating coils to melt ice and frost off evaporators during defrosting.

Electric Elevator. See Electric Traction Elevator.

Electric Eye. 1. Photoelectric cell; a miniature cathode ray tube.

Electric Field. A region of electrical influence.

Electric Furnace. An enclosed structure in which heat is produced, controlled by electric current.

Electric Grinder. A mechanical device powered by electric current that wears down, polishes, or sharpens by friction.

Electric Heater. A heat-producing unit powered by electricity.

Electric Heating. System in which heat from electrical resistance units is used to heat the building.

Electric Hoist. An apparatus for lifting people or materials powered by electric current.

Electric Insulation. Substance which has almost no free electrons.

Electric Lockset. A doorknob and deadbolt assembly in a door locked or unlocked by electric current.

Electric Manhole. An access hole for the service of underground electric lines; also used for pull stations when laying electric cable.

Electric Meter. An instrument for measuring consumption of electrical power.

Electric Pole. A vertical wooden pole used to carry electric utility wires. Opposite charged terminals, as in an electric cell or battery.

Electric Range. A cooking stove with an oven and a flat top with burners, powered by electric current.

Electric Sign. A fixed or portable, self-contained electrically- illuminated appliance with words or symbols designed to convey information or attract attention.

Electric Steamer. A device which produces steam to strip old wallpaper off of wall surfaces.

Electric Strike. An electric device that permits releasing of the door latch from a remote control.

Electric Traction Elevator. Elevator operated by electric motor and suspended from cables.

Electric Water Valve. Electrically operated solenoid valve controlling water flow.

Electrical Conductivity. The conducting power of a specified material; the reciprocal of electrical resistivity.

Electrical Curing. A system in which a favorable temperature is maintained in freshly-placed concrete by supplying heat generated by electrical resistance.

Electrical Engineer. An engineer who designs electrical systems.

Electrical Fee. The amount of money charged for the inspection or installation for the electrical wiring work in a building or structure.

Electrical Metallic Tubing. Unthreaded light weight piping for running electrical conductors; easier to handle than rigid conduit and installed more rapidly because of the type of non-threaded fittings used with it; also called Thin Wall Conduit.

Electrical Porcelain. Vitrified whiteware having an electrical insulating function.

Electrical Potential. Electrical force which moves, or attempts to move, electrons along a conductor or resistance; measured in volts.

Electrical Resistance (R). The difficulty electrons have moving through a conductor or substance.

Electrical Resistivity. The resisting quality of a specified material; the reciprocal of electrical conductivity.

Electrical Socket. A receptacle for connecting electrical appliances to the electrical supply.

Electrical. Relating to, or operated by electric current.

Electrically Welded Wire Fabric. Large dimension wire mesh used for reinforcing concrete slabs on grade. Same as Electrically Welded Wire Mesh.

Electrically Welded Wire Mesh. Electrically Welded Wire Fabric.

Electrician. A craft worker who installs, maintains, and repairs electrical systems in buildings.

Electrochemical Coating. A coating on metal by means of electron transfer by electrical current; electroplating; electrodeposition.

Electrode Force. In spot, seam, and projection welding, the force, in pounds, between the electrodes during the actual welding cycle.

Electrode Holder. A device used for mechanically holding the electrode and conducting current to it.

Electrode. Terminal point to which electricity is brought in the welding operation and from which the arc is produced to do the welding; in electric arc welding, the electrode is usually melted and becomes part of the weld.

Electrolier. Pole-mounted street light.

Electrolysis. Production of chemical changes, such as decomposition, by the passage of current through an electrolyte, an acidic liquid, or damp earth; corrodes metals.

Electrolytic Condenser-Capacitor. Plate or surface capable of storing small electrical charges.

Electromagnet. A soft metal core made into a magnet by the passage of electric current through a coil surrounding it.

Electromotive Force. The force that makes electrons move in an electric current; voltage.

Electron. Elementary particle or portion of an atom which carries a negative charge.

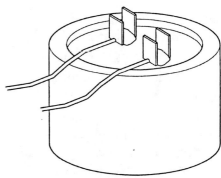

Electrolytic Condenser-Capacitor

Electronic Air Filter. A filter that attracts dust particles by static charge; also called Electrostatic Filter.

Electronic Distance Measuring. A high-precision surveyor's instrument that measures distance by use of radio-frequency or light-frequency electromagnetic waves that are reflected back to source, the elapsed time being precisely measured and converted to distance.

Electronic Industries Association (EIA). 2500 Wilson Boulevard, #300, Arlington, Virginia 22201, (703) 907-7500.

Electronic Leak Detector. Electronic instrument which measures electronic flow across gas gap; electronic flow changes indicate presence of refrigerant gas molecules.

Electronic. Pertaining to electrons.

Electronics. A branch of physics and technology concerned with the behavior and movement of electrons in a vacuum, gas, or semiconductor and with electronic devices.

Electroplate. Coat by electrolytic deposition with chromium, silver, copper, or other metal.

Electrostatic Coating. Painting with a spray that utilizes electrically charged particles to ensure complete coverage.

Electrostatic Filter. See Electronic Air Filter.

Element. 1. A component part. 2. Any of the hundred or so substances that cannot be resolved by chemical means into simpler substances.

Elevated Floor. A floor system not supported by a subgrade.

Elevated Slab Formwork. The system of support for a freshly poured or placed concrete elevated slab.

Elevated Slab Reinforcing. Metal or steel bars embedded in freshly poured concrete to strengthen an elevated slab.

Elevated Slab, Concrete. A concrete roof system or concrete flooring system supported by structural members.

Elevated Slab. A roof slab or floor supported by structural members.

Elevated Stairs. A stair system not supported by the subgrade.

Elevated Temperature Testing. Tests on plastic pipe above 23° C.

Elevation. 1. The height above a given level, such as sea level. 2. High place or position. 3. A drawing or diagram made by projection on a vertical plane; a flat drawing of the front, side, or back of a building.

Elevator Penthouse. A roof structure containing the machine room for an electric traction elevator.

Elevator Shaft. A lined vertical shaft in a building, usually of a fire-rated construction, in which the elevator cab ascends and descends.

Elevator. 1. A hoisting machine in a shaft; a cable or chain hoist conveying system used for raising material or passengers in a cab, cage, or platform. 2. A building for elevating, storing, discharging, and sometimes processing grain; a grain elevator.

Ellipse. 1. A closed plane curve generated by a point moving in such a way that the sums of its distances from two fixed points is a constant. 2. An oval.

Ellipsoid. A solid like a flattened sphere.

Elliptical. Shaped like an ellipse.

Elongation. The state of being lengthened.

Embankment. A fill whose top is higher than the adjoining surface.

Embed. In application of gypsum wallboard, to apply and cover joint tape with joint compound.

Embedment Length. The length of embedded steel reinforcement provided beyond a critical section.

Embedment. 1. The process of pressing a felt, aggregate, fabric, mat, or panel uniformly and completely into hot bitumen or adhesive. 2. The process of placing a material into another material so that it becomes an integral part of the whole material.

Embezzlement. Theft of property which has become a possession of the thief; to appropriate property, that is entrusted to one's care, fraudulently for one's own use.

Emboss. 1. Carve or mold in relief. 2. Form designs so that they stand out on a surface. 3. Make protuberant. 4. Adorn or embellish.

Embossed Paper. Wallpaper run through rollers with raised areas to provide a light relief effect.

Embossed. In carpet, the type of pattern formed when heavy twisted tufts are used in a ground of straight yarns to create an engraved appearance; both the straight and twisted yarns are often of the same color.

Emergency Generator. A gasoline powered motor and electrical generator provided for emergency lighting and power during interruption of the normal electrical supply.

Emery. Slow-cutting, short-lived abrasive.

EMC. Equilibrium Moisture Content.

EMF. Electromotive Force.

Eminent Domain. The power of the state to take private property for public use, upon payment of a fair price.

Emittance. A measure of the propensity of a material to give off thermal radiation.

Empirical Formula. A formula that is developed from experience rather than from scientific theory.

Empty-Cell Process. A method of obtaining deep penetration of a wood preservative with a relatively low net retention of the preservative; surplus preservative is removed from the wood cells by vacuum.

EMT Conduit. Electrical Metallic Tubing.

Emulsified Asphalt Mix (Cold Mix). A mixture of emulsified asphalt and aggregate; produced in a central plant (plant mix) or mixed at the road site (mixed-in-place).

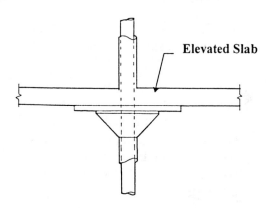

Elevated Slab

Emulsified Asphalt. An emulsion of asphalt cement and water that contains a small amount of an emulsifying agent, a heterogeneous system containing two normally immiscible phases (asphalt and water) in which the water forms the continuous phase of the emulsion, and minute globules of asphalt form the discontinuous phase. Emulsified asphalt may be of either the anionic, electronegatively charged asphalt globules, or cationic, the electropositively charged asphalt globule types, depending upon the emulsifying agent.

Emulsifier. Material which, when added to a mixture of dissimilar materials will produce a stable emulsion.

Emulsifying Agents. Substances of chemical nature that intimately mix and disperse dissimilar materials ordinarily immiscible, such as oil and water, to produce a stable emulsion; a substance which when added to a liquid permits suspension of fine particles or globules in the liquid.

Emulsion Paint. Water-thinned paint with an emulsified oil, resin, or latex vehicle.

Emulsion. 1. A fine dispersion of one liquid in another, as in paint; emulsion paint is a water-thinned paint containing a non-volatile substance as its binding medium. 2. In roofing, a coating consisting of asphalt and fillers suspended in water.

Enamel Paint. A paint which dries to a hard gloss or semi-gloss smooth finish.

Enamel. Type of paint made by grinding or mixing pigments with varnishes or lacquers.

Enameling. The process of painting with enamel.

Encased Burial Conduit. Metal or plastic conduit EB for outdoor wiring with type TW wires encased, or type UF cable.

Encased Knot. A knot whose rings of annual growth are not intergrown with those of the surrounding wood.

Enclosed Switch. A electric switch which is protected by thin metal shields on either side of the switch to prevent accidental tripping of the device.

Enclosed. Surrounded by a case which will prevent a person from accidentally contacting wiring, equipment or live parts contained therein.

Enclosure, Telephone. Partitions which provide privacy on two sides of a public telephone, or which completely surround the user; a phone booth.

Enclosure. The case or housing of electrical or mechanical apparatus, or the fence or walls surrounding an installation to prevent personnel from accidentally contacting energized parts, or to protect the equipment from physical damage.

Encroachment. Personal property of one person intruding upon real estate owned by another.

Encumbrance. A charge against real property.

End Bell. 1. Cast iron pipe with a wide opening at one end, to receive the small end of an adjoining pipe; see Bell and Spigot Joint. 2. End structure or plate of electric motor which usually holds motor bearings.

End Burn. Over-calcined gypsum board resulting in easily damaged, soft, fragile ends.

End Distance. The distance from a bolt, screw, or nail to the end of a wood structural member.

End Joint. A joint made by bonding two pieces of wood together end to end, commonly by finger or skarf joint or butt joint.

End Lap. The overlapping of roofing felts at the ends of sheets cut off the roll.

End Matched. A board with a tongue and groove joint on the ends as well as on the sides.

End Nail. To drive a nail through one piece of lumber and into the end grain of another.

End Play. Slight movement of a shaft along its center line.

End Support. Bearing point for a metal open-web joist located at the end of the joist, usually constructed of a steel plate attached to a supporting component to uniformly distribute the load to a supporting component.

End Truss. A factory made wooden truss which is used at the gable end of a building and to which the sheathing and siding are fastened.

End Wrench. A hand tool with one or both ends shaped to grip a nut or bolt head that is to be held or turned.

End. 1. The extreme point of an object. 2. The termination of an operation. 3. Of gypsumboard, the end perpendicular to the paper- bound edge; the gypsum core is always exposed.

End-Bearing Pile. A pile calculated to carry all of its load on its point, neglecting skin friction.

End-Grained Wood. The grain as seen on a cut made at a right angle to the direction of the fibers.

Endothermal. Chemical reaction in which heat is absorbed.

Endothermic Reaction. A process that requires and absorbs heat.

Energy Absorption. Energy is absorbed as a structure distorts.

Energy Conservation. Measures taken to reduce if not eliminate unnecessary use of all forms of energy and the consumption of non-renewable resources.

Energy Dissipation. Reduction in intensity of earthquake shock waves with time and distance.

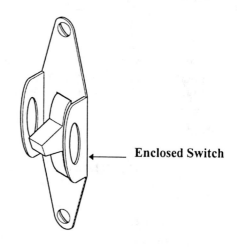

Enclosed Switch

Energy Efficient Standards. Building code, which sets standards for energy conservation in buildings and facilities to which the code applies.

Energy. Usable power such as heat or electricity; actual or potential ability to do work.

Enforcing Agency. The designated department or agency of any city, county or the state as specified in the statutes.

Engaged Column. A building column that is partially subsumed into the wall.

Engineer. A person who is professionally qualified in a branch of engineering.

Engineered Fill. Earth compacted into place in such a way that it has predictable physical properties, based on laboratory tests and specified, supervised, installation procedures.

Engineering Fee. The amount charged for engineering services.

Engineering. The application of science and mathematics by which the properties of matter and the sources of energy in nature are made useful to human beings in structures, machines, products, systems, and processes.

Engineer's Level. An accurate telescopic transit for measuring horizontal and vertical angles; used in ordinary surveying work; see Dumpy Level.

English Bond. Brickwork laid with alternating courses, each consisting entirely of headers or stretchers; also called Old English Bond or Dutch Bond.

English Chalk. Chalk obtained from the cliffs of England.

Engraved Plaque. A commemorative or identifying tablet where figures or letters are inscribed.

ENR. Engineering News Record; a construction industry publication.

Enrichments. Any cast ornament which cannot be executed by a running mould.

Entablature. In classical architecture, the horizontal parts just above the columns, consisting of the cornice, frieze, and architrave.

Entasis. A convex curvature added to the taper of the shaft of a column in the Greek and Roman styles as a design refinement to make the columns appear straight; also applies to similar adjustments to high walls and spires.

Enthalpy. A measure of the energy content of a system per unit mass; the sum of sensible and latent heat of a material.

Entrained Air. Microscopic air bubbles intentionally incorporated into mortar or concrete during mixing, usually by use of a surface-active agent, typically between 10 and 1,000 μm in diameter.

Entraining Agent. A substance added to concrete, mortar, or cement that produces air bubbles during mixing, making it easier to work with and increasing its resistance to frost and freezing.

Entrance Door. The door which provides access to a building or structure.

Entrance Mat. A woven fabric or ribbed rubberized mat placed on the outside of a door threshold for the wiping of shoes.

Entrance. The way in to a property, a building, or a room.

Entrapped Air. Air in a concrete mix that enters from the atmosphere during mixing; after the concrete hardens and the excess moisture has evaporated, irregular holes remain; entrapped air bubbles are normally much larger and more irregular than entrained air bubbles.

Entrepreneur. One who assumes the risk of starting and operating a business.

Entropy. 1. A measure of the disorder or randomness of a system. 2. Mathematical factor used in engineering calculations. 3. Energy in a system.

Entry Lock. A deadbolt assembly mounted in a door which when unlocked allows access to a building or structure.

Entry Lockset. A combination door-knob and deadbolt assembly mounted in a door that when unlocked allows access to a building or structure.

Entry. 1. An entrance room. 2. The recording of a business transaction in a record.

Envelope. In roofing, a continuous edge that is formed by folding an edge of the base felt over the plies above and securing the base felt to the top felt, or if above-deck insulation is used, to the top surface of the insulation; the envelope thus formed will prevent bitumen dripping through the exposed edge joints of the laminated roofing membrane and also prevent water seeping into the insulation.

Environment. Physical surroundings, circumstances, and conditions that affect peoples' lives.

Environmental Factors. Conditions other than indoor air contaminants that cause stress, comfort, or health problems, for example, humidity extremes, drafts, lack of air circulation, noise, and overcrowding.

Environmental Impact Assessment. See Environmental Impact Report.

Environmental Impact Report. A report that examines and considers all likely effects on the environment caused by a proposed development of land, and proposes measures avoiding, reducing, or offsetting any significant adverse effects; also called Environmental Impact Assessment or Environmental Impact Statement.

Environmental Impact Statement. See Environmental Impact Report.

English Bond

Environmental Protection Agency (EPA). 401 M Street, SW, Washington, DC 20460, (202) 260-4700.

Environmental Sustainability. Land development and construction utilizing materials that are replaceable, like wood, and minimizing use of non-replaceable materials and energy resources.

Enzyme. Complex organic substance, originating from living cells, that speeds up chemical changes in foods; enzyme action is slowed by cooling.

Eolian. Borne, eroded, or deposited by the wind.

EPA. Environmental Protection Agency.

EPDM. Ethylene Propylene Diene Monomer.

Epicenter. The point on the earth's surface vertically above the origin, focus, or hypocenter of an earthquake.

Epigraph. An engraved inscription on a building, statue, or monument.

Episcopal Throne. The bishop's ceremonial chair in a cathedral.

Epoxy Adduct. Epoxy resin having all of the required amine incorporated but requiring additional epoxy resin for curing.

Epoxy Adhesive. A two-part adhesive system employing epoxy resin and epoxy hardener used for bonding of ceramic tile to back-up materials.

Epoxy Amine. Amine-cured epoxy resin.

Epoxy Concrete. Concrete with added adhesive resin to aid in binding.

Epoxy Ester. Epoxy-modified oil; single package epoxy.

Epoxy Flooring. A resin spread on flooring to make a hard, tough surface.

Epoxy Grout. A two-part grout system consisting of epoxy resin and epoxy hardener, especially formulated to have impervious qualities, stain, and chemical resistance, used to fill joints between tile units.

Epoxy Mortar. A two-part mortar system consisting of epoxy resin and epoxy hardener used to bond tile to back-up material where chemical resistance and high bond strength is a consideration.

Epoxy Paint. Paint with added resin to give it adhesive and improved bonding characteristics in certain applications.

Epoxy Resin. An epoxy composition used as a chemical-resistant setting adhesive or chemical-resistant grout.

Epoxy Terrazzo. A two-part adhesive, employing epoxy resins, an epoxy hardener used for bonding marble or other stone chips set in portland cement to a backup material.

Equalizer. One that equalizes by distributing evenly or uniformly, as of force.

Equation. 1. A chemical equation shows what happens in a chemical reaction; the reactants equal the products. 2. A mathematical equation is a statement that two quantities or expressions are equal.

Equilateral Triangle. A triangle having three equal sides.

Equilibrium Moisture Content. The moisture content at which wood neither gains or loses moisture when surrounded by air at a given relative humidity and temperature.

Equilibrium. A state of rest due to balanced forces; state of being in balance; implies no tendency to change.

Equipment Mobilization. The assembly and movement of equipment to a jobsite.

Equipment Pad. A thick slab-type stone or precast concrete block placed under mechanical equipment to spread the weight and load of the machinery evenly and to prevent excessive vibration.

Equipment Rack. A device mounted on a wall to hold or store the implements used in an operation or activity.

Equipment, Architectural. See Architectural.

Equipment, Insulation. See Insulation Equipment.

Equipment. The implements used in an operation or activity.

Equitable Lien. A lien that is given to an improver of property to prevent the unjust enrichment of a property owner or a construction lender.

Equity. 1. The value of a firm's assets in excess of its liabilities. 2. The prevention of injustice that might result from strict enforcement of law.

Equivalent Static Force Analysis. A method by which a dynamic force is translated into an equivalent static force that produces a similar effect.

Equi-Viscous Temperature. The critical temperature at which asphalt reaches the optimum state of viscosity for adhesion to roofing felt.

Eraser. A rubber-like substance used by drafters to remove pencil or ink marks on drawings.

Erasing Machine. A small hand-held electric motor that rotates an eraser, used by drafters.

Erasing Shield. A thin metal stencil with various sizes and shapes of holes to enable a drafter to erase with precision.

Erect. 1. To build or construct by fitting parts together. 2. Vertical.

Erection. A building or structure.

Erg. A unit of work or energy, equal to the work done by a force of one dyne when its point of application moves one centimeter in the direction of action of the force.

Ergonomics. Applied science that investigates the impact of people's physical environment on their health and comfort; for example, determining the proper chair height for computer operators.

Erode. Gradual wearing away or destruction.

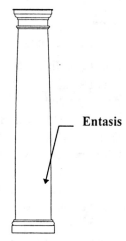

Entasis

Erosion. 1. The gradual wearing away of material as a result of abrasive action. 2. Wear caused by moving water or wind. 3. Wearing away of paint films; heavy chalking tends to accelerate erosion.

Errors and Omissions Insurance. Insurance carried by architects and engineers to indemnify their clients against losses caused by the professional negligence of the architect or engineer; also called E and O Insurance, Professional Indemnity Insurance, or Malpractice Insurance.

Escalation. Increase in building cost due to inflation or other factors.

Escalator. A power-driven set of stairs arranged like an endless belt that ascend or descend continuously.

Escheat. The reversion of property to the state when a person dies without heirs and without a will.

Escrow. A neutral facilitator who follows the instructions of parties to a transaction.

Escutcheon. The protective, sometimes ornamental, plate around a door lock or pipe.

Esquisse. A preliminary sketch of an architectural or sculptural design.

Essential Facilities. Structures or buildings which have been determined essential, and are intended to be safe and usable after an earthquake; such facilities include, but are not limited to. hospitals and other medical facilities having surgery or emergency treatment areas, fire and police stations, and government disaster and communication centers.

Essential Oils. Oils which have an odor, such as cedar oil or camphor oil.

Essential Services Act. An act that provides that essential services buildings shall be capable of providing essential services to the public after a disaster, shall be designed and constructed to minimize fire hazards, and to resist, insofar as possible, the forces generated by earthquake, gravity, and winds.

Estate. 1. The degree, quality, nature, or extent of one's ownership in property. 2. The assets of a deceased person. 3. Landed property, usually with a large house on it.

Ester Gum. Resin produced synthetically by rosin reacting with glycerine.

Ester. Organic compound formed from an alcohol and an organic acid by eliminating water.

Estimate. 1. A prediction of the cost of performing work; compute; calculate cost of a job. 2. A value judgement based on experience. 3. An approximation of construction costs.

Estoppel. The doctrine that a person may not contradict one's own positive representations.

ET. See Net Effective Temperature.

Etch Acid. A chemical agent used in etching.

Etch. The art of producing designs on metal or glass by the use of the corrosive action of an acid; the use of acid to cut lines into metal or remove the surface of concrete.

Etched Nails. Chemically treated nails to improve their holding power in wood framing.

Ethane. Low temperature application refrigerant.

Ethanol. Alcohol.

Ethyl Acetate. Rapid evaporating solvent made from ethyl alcohol and acetic acid.

Ethyl Alcohol. Alcohol produced by the distillation of fermented grain.

Ethyl Chloride. Toxic refrigerant now seldom used.

Ethyl Lactate. A solvent made by a reaction between ethyl alcohol and lactic acid.

Ethylene Plastic. Plastics based on resins made by the polymerization of ethylene or copolymerization of ethylene with one or more other unsaturated compounds.

Ethylene Propylene Diene Monomer (EPDM). A single-ply membrane of synthetic rubber, usually black or white, highly resistant to damage from ozone, ultraviolet radiation, weathering and abrasion; resists contamination from acids, alkalis, animal and vegetable oils, and oxygenated solvents such as ketones, esters, and alcohols; can be loosely laid, ballasted, or fully adhered.

ETS. Environmental Tobacco Smoke.

Eutectic Deformation. The composition within a system of two or more components which, on heating under specific conditions, develops sufficient liquid to cause deformation at minimum temperature.

Eutectic Point. Freezing temperature for eutectic solutions.

Eutectic Salts. 1. A group of materials that melt at low temperatures, absorbing large quantities of heat and then, as they recrystallize, release that heat. 2. One method used for storing solar energy.

Eutectic. That certain mixture of two substances providing lowest melting temperature of all the various mixes of the two substances.

Evacuation. Removal of air, gas, and moisture from a refrigeration or air conditioning system.

Evaporate. To pass off in vapor; to change a liquid into vapor or gas.

Evaporation Rate, Final. Time interval for complete evaporation of all solvents.

Evaporation Rate, Initial. Time interval during which low boiling solvent evaporates completely.

Evaporation Rate. Rate at which a solvent evaporates.

Evaporation. 1. The change of a liquid to a gas below the boiling point; heat is absorbed in this process. 2. Loss of water to the atmosphere; one concerning plumbers is the loss of drainage trap seal by the evaporation of the trap water.

Escalator

Evaporative Condenser. Device which uses open spray or spill water to cool a condenser; evaporation of some of the water cools the condenser water and reduces water consumption.

Evaporative Cooling. A process in which outside air is pre-cooled before passing through a space; this is done by first passing the air through a layer of wet material, from which water is evaporated, increasing the water content of the air (latent heat), and reducing its dry-bulb temperature (sensible heat).

Evaporator Fan. Fan which increases airflow over the heat exchange surface of evaporators.

Evaporator Pressure Regulator. Automatic pressure regulating valve mounted in suction line between evaporator outlet and compressor inlet; its purpose is to maintain a predetermined pressure and temperature in the evaporator.

Evaporator, Dry Type. Evaporator in which the refrigerant is in the liquid droplet form.

Evaporator, Flooded. Evaporator containing liquid refrigerant at all times.

Evaporator. Part of a refrigerating mechanism in which the refrigerant vaporizes and absorbs heat.

Eviction. The legal process of ejecting a defaulting tenant from real property; unlawful detainer action.

Evidence. Testimony, documents, and objects introduced in a judicial proceeding to support the contentions of the parties.

EVT. Equi-Viscous Temperature.

EWC. Electric Water Cooler.

EWWF. Electrically Welded Wire Fabric.

EWWM. Electrically Welded Wire Mesh.

Ex Parte Proceeding. One party to a dispute appears before a judge when the other is not present.

Excavation, Unclassified. Excavation paid for at a fixed price per yard, regardless of whether it is earth or rock.

Excavation. A cavity formed by cutting, digging or scooping.

Excelsior. Wood wool.

Exclusive Agency. A listing agreement with a real estate broker to sell or lease one's property during a specified period, excluding all other brokers, but reserving the owner's right to sell without paying a commission.

Exclusive Listing. An Exclusive Agency listing.

Exclusive Right to Sell. A listing agreement with a real estate broker to sell or lease one's property during a specified period, excluding all other brokers, and with the provision that the broker will receive a commission even if sold by the owner.

Excrete. To separate and expel waste matter in a continuous process.

Exculpatory Clause. A provision in a contract that relieves a party of liability.

Execute. 1. To carry out some direction, process, or work. 2. To validate a legal document by signing it.

Executive Ability. A behavioral aspect of management; the ability to manage, delegate responsibility, and coordinate all aspects of a business.

Executor. One appointed in a will to carry out its provisions.

Executrix. A female executor.

Exemplary Damages. Damages awarded to a private litigant and against a defendant to punish the wrongdoing of the defendant.

Exercise Bicycle. A stationary bicycle used for health and fitness purposes.

Exfiltration. Slow flow of air from the building to the outdoors.

Exfoliate. Removal from the surface in thin layers, flakes, or scales.

Exhaust Fan. An electrical powered device to withdraw fumes, dusts, or odors from an enclosure.

Exhaust Hood. Usually a square or rectangular hood housing an exhaust fan to withdraw fumes, dust, or odors from an enclosure.

Exhaust Port. That opening which carries the fluid to the downstream pressure of a fluid system.

Exhaust Valve. A movable port which provides an outlet for the cylinder gases in a compressor or engine.

Exhaust Ventilation. Mechanical removal of air from a portion of a building.

Exhibit. A document or thing that is offered as evidence in a proceeding.

Exit Device. Panic hardware.

Exit Light. A light assembly over an exit door that is independently powered to remain lit in the event of a power failure to guide persons to safety.

Exit Lock. A dead bolt assembly mounted in an exit door.

Exit Sign. A sign located to identify the way out of a room or building.

Exit. A passage or door by which one may leave a room or building; egress; a continuous and unobstructed means of egress to a public way, including intervening doorways, corridors, ramps, stairways, smokeproof enclosures, horizontal exits, exit court, and yards.

Exothermic Reaction. A process that gives off heat.

Expanded Metal Lath. Open mesh cut and drawn from solid sheet of ferrous or non-ferrous metal; made in various patterns and metal thicknesses with uneven or flattened surface; used as a metal reinforcing for plaster.

Expanding Anchor. A soft metal sleeve, commonly lead, into which a screw-type bolt is placed to provide a stable fastener.

Expansion Coefficient. The increase in length or volume per unit for a rise in temperature of 1° F.

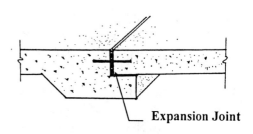

Expansion Joint

Expansion Joint. 1. A vertical joint or space to allow for movement due to volume changes; also known as a relief joint or control joint. 2. A separation between two sections of concrete which is provided to allow for free movement due to temperature changes; complete separation between parts of a concrete structure; used in locations where expansion and contraction forces are anticipated; the sections are usually divided by a strip of metal, cork or bituminous material. 3. A joint through tile, mortar, concrete, or masonry down to the substrate, intended to allow for gross movement due to thermal stress or material shrinkage. 4. A device usually formed from sheet metal and having a W shaped cross section; used to provide controlled discontinuity at locations in a plaster membrane where high stresses may be encountered. 5. A structural separation between two building elements designed to minimize the effect of the stresses and movements of a building's components and to prevent these stresses from splitting or ridging the roof membrane. 6. A joint which permits pipe to expand or contract without breaking and to allow movement of the pipe caused by the pipe's expansion and contraction.

Expansion Screed. A sheetmetal plaster screed that allows expansion and contraction of adjoining panels of exterior plaster.

Expansion Shield. Device inserted in predrilled holes, usually in concrete or masonry, which expands as a screw or bolt is tightened within it, used to fasten items to concrete or masonry.

Expansion Valve. Device in a refrigerating system which reduces the pressure from the high side to the low side and is operated by pressure.

Expansion. An increase in size when something gets hotter.

Expendable Refrigerant System. See Chemical Refrigeration.

Experimental Analysis. An analysis procedure based on the measurement of deformations and/or strains of the structure or its model; experimental analysis may be based on either elastic or inelastic behavior.

Expert Witness. A witness who, by virtue of experience, training, skill, or knowledge of a particular field or subject, is recognized as qualified to render an informed opinion on matters relating to that field or subject.

Explosion-Proof Apparatus. Apparatus enclosed in a case which is capable of withstanding an explosion of a specified gas or vapor which may occur within it and of preventing the ignition of a specified gas or vapor surrounding the enclosure by sparks, flashes, or explosion of the gas or vapor within, and which operates at such an external temperature that a surrounding flammable atmosphere will not be ignited thereby.

Explosive Limits. A range of the ratio of solvent vapor to air in which the mixture will explode if ignited; below the lower or above the higher explosive limit, the mixture is too lean or too rich to explode; the critical ratio runs from about one to twelve percent of solvent vapor by volume at atmospheric pressure.

Explosive. An explosive substance used to dislodge or loosen certain formations of earth and rock.

Exposed Aggregate Finish. A concrete surface in which the coarse aggregate is revealed; a decorative finish for concrete achieved by removing, generally before the concrete has fully hardened, the outer skin of mortar and fine aggregates and exposing the coarse aggregate.

Exposed Grid. A framework for a suspended acoustical ceiling that is visible from below after the ceiling is completed.

Exposure. 1. The part of a felt that is not overlapped by an adjacent felt in a built-up roofing membrane; the exposure then would be that part of the felt that would be covered by the flood coat. 2. The part of a wood shingle or shake that is exposed to the weather.

Expressway. Multiple lane roadway for fast moving traffic, with access and exit ramps limited to intervals, no level crossing roads, and no pedestrian traffic allowed; also known as a Freeway.

Extender. Pigment which provides very little hiding power but is useful in stabilizing suspension, improving flow, lowering gloss, and providing other desirable qualities; generally low in cost.

Extension Ladder. A flat ladder that can be extended by sliding one section on the other to gain additional height.

Extension Link. A device used to extend the backset in a bored lock.

Exterior Door. A door which is manufactured to withstand the elements and vandalism, so it can be installed on the exterior of a building.

Exterior Elevation. An architectural drawing showing the projection on a vertical plane of an exterior surface of a building.

Exterior Fixture. An electrical lighting fixture that can be installed outdoors because of its ability to withstand the elements.

Exterior Painting. The act or process of applying paint or sealer to an exterior surface.

Exterior Plywood. A general term for plywood bonded with a type of adhesive that is highly resistant to weather, water (cold, hot, and boiling), microorganisms, steam, and dry heat.

Exterior Tile. Tile installed in exterior applications.

Exterior Veneer. A veneer applied to weather-exposed surfaces.

Exterior Wall. Any outer wall serving as a vertical enclosure of a building other than a party wall.

Exterior Wood Trim

Exterior Wood Door Trim. Finish components of wood such as moldings applied around openings of exterior doors.

Exterior Wood Trim. Finish components of wood such as moldings applied around openings and intersections at exterior locations.

Exterior Wood Window Trim. Finish components of wood such as moldings applied around openings of windows.

External Drive. Term used to indicate a compressor driven directly from the shaft or by a belt using an external motor; compressor and motor are serviceable separately.

External Equalizer. Tube connected to low-pressure side of a thermostatic expansion valve diaphragm and to exit end of evaporator.

External Vibrator. Vibrating device attached to formwork to consolidate fresh concrete; used primarily in precast construction.

Extinguisher, Fire. See Fire Extinguisher.

Extra Work. Work performed by a contractor that is not included within the scope of the work defined by the contract documents.

Extract. 1. To remove with effort. 2. To separate from a mixture by a physical or chemical process. 3. To separate a metal from its ore. 4. To calculate the mathematical root of a number. 5. A concentrated product.

Extraction. Obtaining a useful substance from a raw material; obtaining of a natural resource from the earth.

Extractive Bleeding. Stains on the surface of wood caused by extractives being leached out by water or moisture.

Extractive. Substances in wood, not an integral part of the cellular structure, that can be removed by solution in hot or cold water, ether, benzene, or other solvents that do not react chemically with wood components.

Extractor. A laundry appliance that removes the free water from washed fabrics by high speed centrifugal spinning, leaving them damp and ready for hot air tumble drying.

Extrados. The exterior curve in an arch or vault.

Extraordinary Flood. A deluge of greater volume than is expected in a certain place by prudent persons based on historic experience.

Extrapolate. Calculate approximate information beyond known values by projecting the trend.

Extruded Tile. A tile or trim unit that is formed when plastic clay mixtures are forced through a pug mill die of suitable configuration, resulting in a continuous ribbon of formed clay; a wire cutter or similar cut-off device is then used to cut the ribbon into appropriate lengths and widths of tile.

Extrusion. The process of squeezing a material through a shaped orifice to produce a linear element with the desired cross section; an element produced by this process.

Eye. 1. Oculus. 2. An area in the center of a tropical cyclone marked by only light winds or complete calm with no rain. 3. A metal ring through which a rope or rod is passed. 4. A loop to receive a hook.

Eyebolt, Forged. A heavy metal bolt which contains an eye at its end which may be used as a fastening device.

Eyebrow. A small roofed projection from a building.

FA. Fresh Air.

FAA. 1. Federal Arbitration Act. 2. Federal Aviation Administration.

Fabric, Form. Welded-wire fabric used to reinforce concrete while it is setting and gaining sufficient strength to be self-supporting.

Fabric. 1. The basic structure of a building. 2. Texture and quality, of textiles. 3. A woven cloth of organic or inorganic filaments, threads, or yarns. 4. A material that resembles cloth, like steel reinforcing mesh.

Fabrication, Metal. The building, construction, or manufacture of metal structures or devices.

Fabrication. The construction, processing, or assembly of parts.

Facade. An exterior face of a building.

Face Amount. The sum of money stated on a check, bond, note or other instrument, exclusive of interest accumulations.

Face Board. An exterior trim board of a roof soffit system.

Face Brick. A brick selected on the basis of appearance and durability for use in the exposed surface of a wall.

Face Frame. In cabinetwork, the front framework from which the doors are hung and drawers are inserted.

Face Layer. The outer layer of gypsum board in multilayer applications.

Face Mounted Tile. Tile assembled into units or sheets and bonded together to facilitate handling.

Face of Weld. The exposed surface of a weld made by an arc or gas welding process on the side from which the welding was done.

Face Reinforcement. Reinforcement of weld at the side of the joint from which welding was done.

Face Seam. A carpet seam, either sewed or cemented, that is made without turning the entire carpet over or face-down; made during installation where it is not possible to make a back seam.

Face Shell. The side wall of a hollow concrete masonry or clay masonry unit.

Face. 1. The front or façade. 2. The exposed surface of a wall. 3. The long narrow side of a brick; the exposed surface of a wall or masonry unit. 4. The surface designed to be left exposed to view. 5. The more or less vertical surface of rock exposed by blasting or excavating, or the cutting end of a drill hole; an edge of rock used as a starting point in figuring drilling and blasting. 6. See Steel Square.

Faced Wall. A wall in which the masonry facing and the backing are of different materials and are so bonded as to act together under load.

Face-Mounted Tile. Tile with paper applied to the face of the tile and removed with water prior to grouting.

Face-Nailing. To nail perpendicular to the surface or to the junction of the pieces joined; also called Direct Nailing.

Facet. A small flat surface.

Facework. Pipe valves and manifold fittings connecting the filter to the circulation lines.

Facia. See Fascia.

Facial Defect. That portion of the facial surface of the tile which is readily observed to be nonconforming and which will detract from the aesthetic appearance or serviceability of the installed tile.

Facilities Planning. Planning for the long-term use of a building, which may include furnishings, equipment, operations, maintenance, renovation, expansion, and life- cycle planning.

Facility Design. The architectural and engineering design of a facility.

Facility Management. The management, maintenance, and operation of a facility.

Facility Requirements. The operational needs of a facility.

Facility Services. The operational requirements of a facility including capital, personnel, materials, transportation, and public utilities.

Facility. A property or building that is built, installed, or established to perform a particular function, such as a factory, university, or airport.

Facing Brick. Brick made especially for exterior use with special consideration of color, texture and size.

Facing Tile Institute (FTI). Box 8880, Canton, Ohio 44711, (216) 488-1211.

Facing. Any masonry forming an integral part of a wall used as a finished surface; compare with Veneer.

Factor of Safety. The ratio of ultimate strength to the working stress of a material.

Factored Load. The load imposed on a member multiplied by appropriate factors, used to design reinforced concrete members.

Factory and Shop Lumber. Lumber intended to be cut up for use in further manufacture.

Factory Edge. See Edge, 5.

Factory Finished Flooring. Flooring that has been manufactured in a factory and is ready to be installed without any further work needed on its surface.

Factory Floor. See Industrial Wood Floor.

F

Factory Mutual (FM). 1151 Boston-Providence Turnpike, Norwood, Massachusetts 02062, (617) 762-4300.

Factory Mutual. An agency of the insurance industry that sets standards for fire safety in buildings, of which compliance is a prerequisite for fire insurance.

Factory Square. 108 square feet (10 square meters) of roofing material.

Factory. A building or buildings containing plant or equipment for manufacturing of machinery or goods.

Fadeometer. Device for measuring color retention or fade resistance.

Fade-O-Meter. Mechanism used to artificially reproduce effect of sunlight on paint.

Fading. Reduction in brightness of color.

FAF. Forced air furnace.

Fahrenheit. A scale for registering temperature where freezing is 32° above zero and boiling is 212°.

FAIA. Fellow of the American Institute of Architects.

Faience Mosaic. Faience tile that are less than 6 in. by 2 in. facial area, commonly 5/16 to 3/8 in. (8 to 9.5 mm) thick, and usually pre-mounted to facilitate installation.

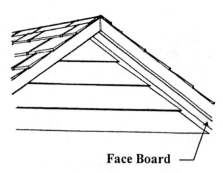

Face Board

Faience Tile. Glazed or unglazed tile, generally made by the plastic process, showing characteristic variations in the face, edges, and glaze that give a hand-crafted, non-mechanical, decorative effect.

Fail-Safe Control. Device which opens a circuit when sensing element loses its pressure.

Failure. 1. Lack of success. 2. Bankruptcy of an individual or business. 3. Unsuccessful use of a material or process. 4. Deterioration, decay, or breaking down of a material at the end of its useful life. 5. The collapse of a building or structure.

Fall. 1. The slope of land. 2. The amount of slope given to horizontal runs of pipe to provide gravity flow in the line.

Fallback. A reduction of the bitumen softening point; sometimes caused by mixing asphalt with coal-tar pitch or overheating the bitumen.

Fallout. Spray gun overspray.

False Body. Thixotropic; characteristic of paint which becomes viscous on standing but thins down on stirring.

False Set. The rapid development of rigidity in a mixed portland cement paste, mortar or concrete without the evolution of much heat; this rigidity can be dispelled and plasticity regained by further mixing without addition of water; also called Premature Stiffening and Rubber Set.

Falsework. Shoring and formwork for concrete that is removed after the concrete beams, slabs, and columns are cured.

Fan Coil Unit (FCU). A packaged unit consisting of a heating/cooling coil, fan, and filter, without ductwork, used to serve a space or a group of small spaces.

Fan Light. A decorative curve topped window above the transom over an entrance door or window.

Fan Pattern. In using spray gun, the geometry or shape of spray pattern.

Fan Truss. A standard peaked roof truss configuration.

Fan, Exhaust. See Exhaust Fan.

Fan. 1. Radial or axial flow device used for moving or producing flow of gases. 2. Spacing tile joints to widen certain areas so they will conform to a section that is not parallel; also called Fanning.

Fanning. 1. In using a spray gun, the technique of arcing; moving the spray gun away from the work. 2. See Fan, 2.

Fanny Mae. See Federal National Mortgage Association.

Far Face. Face of a wall farthest from the viewer; may be the outside or inside face depending on whether one is inside looking out or outside looking in.

Farad. The SI unit of capacitance, such that one coulomb of charge causes a potential difference of one volt.

Faraday Experiment. Silver chloride absorbs ammonia when cool and releases it when heated; this is the basis on which some absorption refrigerators operate.

Fascia, Plaster. See Plaster Fascia.

Fascia. A flat, vertical face member or band at the surface of a building or the edge beam of a bridge, or exposed eaves of a building; also spelled Facia; a flat member of a cornice or other finish; generally the board of the cornice to which the gutter is fastened.

Fast Pin Hinge. One in which the pin is fastened permanently in place; nonrising pin. See Quick Set, 2.

Fast Track. Compression of a construction schedule by over lapping some activities that otherwise would be performed sequentially.

Fastener, Insulation. See Insulation Fastener.

Fastener, Pneumatically Driven. See Pneumatically Driven Fastener.

Fastener, Powder-Actuated. See Powder-Actuated Fastener.

Fastener. Generic term for welds, bolts, rivets, screws, and other connecting devices.

Fat Mortar. Mortar containing a high percentage of cementitious components; mortar which usually does not have a sufficient amount of sand; a sticky mortar which adheres to a trowel.

Fat Mud. Mortar containing lime.

Fat Paint. Paint with too much oil.

Fat. 1. Material accumulated on the trowel during the finishing operation; often used to fill in small imperfections. 2. Describes working characteristics of a mortar containing a high proportion of cementitious material. 3. Describes working characteristics of highly plastic mortars.

Fathom. A unit of length for measuring the depth of water, equal to 6 feet.

Fatigue Resistance. The ability of asphalt pavement to withstand repeated flexing or slight bending caused by the passage of wheel loads; generally, the higher the asphalt content, the greater the fatigue resistance.

Fatigue. A structural failure which occurs as the result of a load being applied and removed, or reversed, repeatedly over a long period of time, or a large number of cycles.

Fatty Acid. Acid which is present in oils or fats in combination with glycerine.

Faucet. A valve for drawing liquid.

Fault Zone. Instead of being a single clear fracture, a zone may be hundreds or thousands of feet wide, consisting of numerous interlacing small faults.

Fault, Lateral Slip. See Lateral Slip Fault.

Fault, Normal. See Normal Fault.

Fault, Oblique Slip. See Oblique Slip Fault.

Fault, Reverse Thrust. See Reverse Thrust.

Fault, Strike Slip. See Strike Slip Fault.

Fault, Thrust. See Thrust Fault.

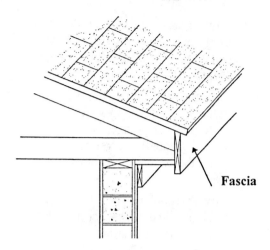

Fascia

Fault. A break in the continuity of a rock formation which is caused by the shifting of the earth's crust.

Faulting. In geology, the movement which produces relative displacement of adjacent rock masses along a fracture.

Faux Marble. Fake marble; marble simulated by painting marble graining on a painted background on wood, plaster, or metal.

Feasibility Study. An analysis performed to determine the financial, economic, technical, or other advisability of a proposed project.

Feather Edge. 1. A wood or metal tool having a beveled edge used to straighten re-entrant angles in finish plaster coat. 2. The edge of a concrete, plaster, or mortar placement such as a patch or topping that is beveled at an acute angle. 3. A tapered edge.

Feather. See Spline.

Feathered Edge. The skived edge of joint tape.

Featheredging Tile. A method of mitering tile by chipping away the body from beneath a facial edge of a tile in order to form a miter.

Feathering. A term used to describe the sanding or rubbing down of a surface to a feathery edge. where coating material gradually becomes thinner around the edge until it finally disappears.

Feature Strip. A narrow strip of decorated liner tile that has a contrasting color, texture, or design.

Featured Edge. Of gypsum board, an edge configuration of the paper bound edge that provides special design or performance.

Featured Edge. A configuration of the paper-bound edge of gypsum boards which provides special design or performance features.

Federal Emergency Management Agency (FEMA). 500 C Street, SW, Washington, DC 20472, (202) 646-4600

Federal Home Loan Mortgage Corporation (FHLMC). A U.S. Government chartered purchaser of mortgage loans, similar to FNMA; also called Freddy Mac.

Federal Housing Administration (FHA). A federal government agency that administers and regulates FHA insured real estate loans.

Federal National Mortgage Association (FNMA). A quasi-governmental corporation that issues debentures to create a secondary market to purchase FHA and VA loans at market prices.

Federal Specifications (FS). Superintendent of Documents, Government Printing Office, Washington, DC 20234.

Federal Specifications. Government specifications for products, components, and performance.

Fee Simple. An absolute and total interest in real property.

Fee, Professional. See Professional Fee.

Feeder Duct. An enclosure for a group of electrical conductors that runs power from a central, large source to one or more secondary distribution centers.

Feeder. A conductor of a wiring system between the service equipment, or the generator switchboard of an isolated plant, and the branch circuit overcurrent device.

Feel. The painter's term for the working qualities of a paint.

Feeler Gauge. A set of thin flat strips of metal of known thickness that may be used to measure small distances between surfaces.

Feldspar. A mineral aggregate consisting chiefly of microcline, albite or anorthite.

Felt Mill Ream. The mass in pounds of 480 square feet of dry, unsaturated felt; also called Point Weight.

Felt. Roofing ply sheets, consisting of a mat of organic or inorganic fibers, unsaturated, saturated, or coated with coal tar pitch or asphalt.

Felting. The process of pressing or matting together various types of hair or fibers to form a continuous fabric, known as felt.

FEMA. Federal Emergency Management Agency.

Female Thread. Inside threads in a pipe or fitting.

F

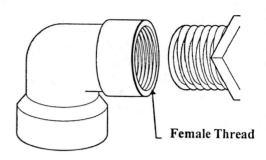

Female Thread

Fence Auger. A rotating drill with a screw thread used to drill deep, straight, and narrow holes for the installation of fence posts.

Fence Gate. 1. An opening in a fence. 2. A hinged or sliding panel in a fence.

Fence Hole. A hole in the ground where the main vertical support is inserted in the construction of a fence or wall.

Fence Post Hole. A cylindrical hole dug in the ground, for the insertion of a fence post.

Fence Reuse. The act or process of reusing old fence boards and other fencing materials in a new fence application.

Fence. 1. A barrier used to prevent escape or intrusion or to mark a boundary, usually made of posts, boards or wire. 2. In a cast plaster shop, a plaster or clay damplaced around a model before pouring material to make the mould.

Fender Pile. Outside row of piles that protects a pier or wharf from damage by ships.

Fenestration. The arrangement, proportioning, and design of windows and doors in a building.

Feng Shui. Chinese cultural system for siting, arranging, and shaping buildings.

Ferroconcrete. Reinforced concrete.

Ferrous Metal. Metal alloy containing iron; ferrous pipes include wrought iron, wrought steel, rolled steel and cast iron.

Ferrous Sulphate. A green pigment commonly known as copperas.

Ferrous. Being iron or containing iron.

Ferrule. 1. A metal band, ring, or cap, such as found on a tool handle to prevent splitting. 2. A cast-iron pipe fitting which, when installed in the bell of a cast-iron pipe, permits a threaded cleanout to close the opening.

Fertilizing. The act or process of adding a substance, such as manure or a chemical mixture, in order to make soil more fertile.

Festoon Lighting. An aerial span of conductors installed outdoors and supplying only weatherproof lampholders attached thereto.

FG. 1. Flat Grain. 2. Foundation Grade. 3. Fuel Gas. 4. Finish Grade.

FGMA. Flat Glass Marketing Association.

FHA. Federal Housing Administration.

FHMS. Flat Head Machine Screw.

FHWS. Flat Head Wood Screw.

Fiber Saturation Point. The stage in the drying or wetting of wood at which the cell walls are saturated and the cell cavities free from water; it applies to an individual cell or group of cells, not to whole boards; it is usually taken as approximately 30 percent moisture content, based on ovendry weight.

Fiber. 1. In wood, a comparatively long, narrow, tapering wood cell, closed at both ends; also called a Tracheid. 2. An additive such as glass or cellulosic fiber to improve core flexibility and gypsum board core integrity. 3. Animal hair, sisal, manila, or glass fibers of appropriate length added to plaster mortar to increase its cohesiveness.

Fiberboard Insulation. Insulation sheets made from wood or cane fibers.

Fiberboard Panel. Flat sheet material made from wood that has been reduced to fibers and bonded through either a wet or dry process to obtain a specific density.

Fiberboard Sheathing. Exterior sheathing manufactured from wood or cane fibers.

Fiberboard Wall Sheathing. Flat sheet material of fiberboard secured to exterior side of exterior wall studs used to create rigidity in building superstructure and serve as base to receive siding or veneer construction.

Fiberboard. A prefabricated sheet of compressed wood or plant fibers used for building; a homogeneous panel made from wood or cane fibers; bonding agents and other materials may be added to increase strength, resistance to moisture, fire, or decay, or to impart some other property; also called Insulation Board.

Fibered Plaster. Basecoat plaster containing animal, vegetable, or glass fiber.

Fiberglass Insulation. A type of insulation made from glass fibers having the consistency of wool.

Fiberglass Reinforced Pipe. A pipe for liquid or gas, fabricated from glass fibers and resins for strength and durability.

Fiberglass Tank. A large container for the holding of liquids, constructed of fiberglass-reinforced plastic.

Fiberglass. 1. Glass in fibrous form used in making textiles and thermal and acoustical insulation and used for reinforcing plastics. 2. The name for products made of or with glass fibers ranging from 5 to 600 hundred-thousandths inch in diameter.

Fiber-Optics. The technique of using very thin bundles of glass or plastic fibers that transmit light throughout their length by internal reflections for bending light or seeing around corners.

Fiber-Saturation Point. The point in the drying or wetting of wood at which the cell walls are saturated but the cell cavities are free from water; it is usually taken as approximately 23 to 30% moisture content, based on weight when oven-dry.

Fibrous Concrete. A light concrete made from a fibrous aggregate, like sawdust or asbestos, for increased tensile strength and making it easy to nail.

Fiddleback-Grained Woods. Wood grain figure produced by a type of fine wavy grain found, for example, in species of maple, such wood being traditionally used for the backs of violins.

Fiduciary Relationship. A relationship of trust and confidence between principal and agent.

Fiduciary. A person in a relationship or position of trust; a trusted overseer.

F

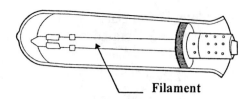

Filament

Field Concrete. Concrete delivered or mixed, placed, and cured on the job site.

Field Engineer. An engineer who works primarily at the jobsite as opposed to the home office; commonly represents the owner or agency and often empowered to make small engineering changes at the site to facilitate construction.

Field House. A building for athletic activities.

Field Painting. Painting at the job site.

Field Pole. Part of stator of motor which concentrates magnetic field of field winding.

Field Primed. The first coat of paint applied on the construction site rather than in the fabrication shop.

Field Tile. An area of tile covering a wall or floor; the field is bordered by tile trim.

Field Welded Truss. A truss fabricated and welded at a jobsite.

Field. 1. The construction site, as contrasted to offices, factories, and work shops. 2. The surface area of a single piece of gypsum board. 3. In brick masonry, the expanse or area of wall between openings and corners, composed for the most part of stretcher units.

Fiery Finish. Wood finish in which hard portion of grain stands out in relief; produced by using blowtorch and stiff bristled brush.

Fifty-Fifty. A dry or dampened mixture of one part portland cement and one part extra-fine sand; this mix is used as a filler in the joints of mounted ceramic mosaic tiles to keep them evenly spaced during installation.

Figure. The pattern produced in a wood surface by annual growth rings, rays, knots, irregular coloration, and deviations from regular grain such as interlocked and wavy grain.

Filament. A conducting wire with a high melting point in an electric bulb made incandescent by an electric current.

File. A rasp-type device which is used on either wood or metal to remove burrs and rough edges or to shape the object.

Filigree. 1. Ornamental work of gold, silver, or copper as fine wire formed into delicate tracery; fine metal openwork. 2. Anything resembling this.

Fill, Gypsum. Troweled on plaster material to make depressions level.

Fill. 1. To raise the level of land by adding earth moved in from another place or obtained by cutting. 2. An earth or broken rock structure or embankment. 3. Soil or loose rock used to raise a grade. 4. Soil that has no value except bulk. 5. Sand, gravel or other loose earth used to raise the ground level around a structure.

Filler Block. Concrete masonry unit for use in conjunction with concrete joists for concrete floor or roof construction.

Filler Metal. The metal to be added in making a welded, brazed, or soldered joint.

Filler, Joint. See Joint Filler.

Filler, Wood. See Wood Filler.

Fillet Weld. A weld at the inside intersection of two metal surfaces that meet at right angles.

Fillet Weld. A weld of approximately triangular cross section joining two surfaces approximately at right angles to each other in a lap joint, tee joint, or corner joint.

Filling In. Laying the brick or other masonry unit in the center of a wall between the face and the back.

Filling. In carpet making, yarns, usually of cotton, jute, or kraftcord, running across the fabric and used with the chain yarns to bind the pile tufts to the backing yarns; see Weft.

Film Build. In painting, dry thickness characteristics per coat.

Film Former. Substance which forms skin or membrane when dried from liquid state.

Film Integrity. In paint, degree of continuity of film.

Film Thickness Gauge. In painting, device for measuring film thickness above substrates; dry or wet film thickness gauges are available.

Filmogens. In paint, film-forming materials such as linseed oil and varnish resins.

Filter Aid. A powderlike substance such as diatomaceous earth or volcanic ash used to coat a septum-type filter; also refers to alum as an aid to sand filtration.

Filter Block. A hollow, vitrified clay masonry unit, sometimes salt-glazed designed for trickling filter floors in sewage disposal plants.

Filter Cycle. The time of filter operation between backwash procedures; also called Filter Run.

Filter Glass. A glass, usually colored, used in goggles, helmets, and hand shields to exclude harmful light rays.

Filter Media. Fine-grain material that entraps suspended materials as they pass through the material.

Filter Rate. The rate of water flow through a filter during the filtering cycle, expressed in gallons per minute per square foot of effective filter area.

Filter Rock. Graded, rounded rock, or gravel not subject to degradation by common pool chemical used to support filter media.

Filter Run. See Filter Cycle.

Filter Sand. A type of pool filter media composed of hard sharp silica, quartz, or similar particles with proper grading for size and uniformity.

Filter Septum. Part of a filter on which diatomaceous earth or similar filter media is deposited.

Filter Waste Discharge Piping. Piping that conducts waste water from a filter to a drainage system; connection to drainage system is made through an air gap or other approved methods.

Filter, Sand. See Sand Filter.

Filter. 1. Strainer; purifier. 2. A device for straining suspended particles from pool water. 3. An adjunct to air cleaning device available to serve four purposes. (1) Commercial filter - used to remove visible particles of dust, dirt, and debris, (2) Electrostatic filter - to remove microscopic particles such as smoke and haze, (3) Activated charcoal filter - to destroy odors, (4) Ultraviolet lamps or chemicals - to kill bacteria.

Filtration. A method of separating a mixture made of solid particles in a liquid; after passing through a filter, the filtrate is separated from the residue.

Final Set. See Set, 4.

Final Setting Time. The time required for a freshly mixed cement paste, mortar, or concrete to achieve final set.

Financial Analysis. Three types are common: (1) comparison of current financial data with prior years, (2) comparing current financial data with that of other business in the same industry, (3) ratio analysis.

Financial Leverage. The use of borrowed money to increase the return on the investment of the owners.

Financial Statement. A balance sheet, statement of income and expense, or any other presentation of financial data.

Fine Aggregate. Concrete aggregate smaller than 1/4-inch diameter; sand.

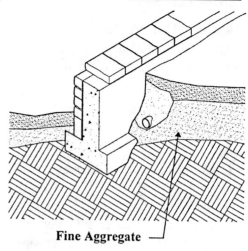

Fine Aggregate

Fine Mineral Surfacing. A water-insoluble, inorganic material, more than 50% of which passes through the No. 35 sieve, that may be used on the surface of roofing material.

Fine Textured Wood. Wood with small and closely spaced pores.

Fine-Graded Aggregate. One having a continuous grading in sizes of particles from coarse through fine with a predominance of fine sizes.

Fine-Grained Wood. See Close-Grained Wood.

Fineness Modulus. An abstract number used to compare different particles or gradations of aggregate; fineness modulus is computed by adding the cumulative percentages retained on the six standard screens (#4, #8, #16, #30, #50 and #100) and dividing the sum by 100.

Fineness Modulus. An empirical factor obtained by adding the total percentages of a sample of the aggregate retained on each of a specified series of sieves and dividing the sum by 100.

F

Fineness of Grind. Measure of particle size or roughness of liquid paint; degree of dispersion of pigment in the binder.

Fines. Term usually pertaining to small aggregate particles capable of passing through a #200 sieve.

Finger Joint. An end joint made up of several meshing wedges or fingers of wood bonded together with an adhesive; fingers are sloped and may be cut parallel to either the wide or narrow face of the piece.

Fingers. In airless spraying, a broken spray pattern.

Finial. An architectural ornament finishing off the top of a roof, gable, tower, canopy, or cupola.

Finish Carpentry. Installation of wood finish and materials, such as doors, moldings, and window trim to a building.

Finish Coat Floating. In plastering, the act of bringing the aggregate to the surface to produce a uniform texture.

Finish Coat. The final layer of any finish type material.

Finish Grade. The final earth grade required by specifications.

Finish Paving. Finish coats of concrete, asphalt, or coated macadam on streets, sidewalks, and parking lots.

Finish Plaster. The final layer of plaster coating.

Finish Plywood. The finest grade of plywood.

Finish Shotcrete. The final coat of air-blown mortar.

Finish. 1. Wood products to be used in joinery, such as doors and stairs, and other fine work required to complete a building, especially the interior. 2. The process of adding stains, filler, and other materials to protect and beautify the surface of wood.

Finisher. A craftsman with skill in the finishing of gypsum board joints.

Finishing Brushes. In plastering, brushes used to apply water to a smooth lime finish coat during final troweling.

Finishing Sheetrock. The taping and sanding of sheetrock seams to make ready for painting or finish.

Finishing Tools. Trowels, knives, and other special equipment required for the finishing of gypsum board joints.

Fink Truss. A three triangle symmetrical truss, commonly used in supporting large, sloping roofs.

Fir Decking. Softwood sheathing material for a deck or floor that is two inches or thicker.

Fir Floor. A flooring system constructed of a softwood.

Fir Siding. Exterior wall covering boards made from a softwood.

Fire Alarm Cable. A specific electrical system cable which carries electric current to a pull station, warning horn, light, or bell in the event of a fire or other catastrophe.

Fire Alarm. An warning apparatus that can be activated to warn of fire danger.

Fire Blocking. Intermittent solid cross-framing to retard the spread of flame within the framing cavity.

Fire Box. The interior of a fireplace or furnace, serving as combustion space.

Fire Brick. Brick that has been tested for fire-resistance and then graded for specific construction uses; brick whose composition and characteristics make it suitable to use for masonry fireplace fire-boxes and fire chambers.

Fire Brick. Brick made of refractory ceramic material which will resist high temperatures.

Fire Clay. An earthy or stony mineral aggregate which has as the essential constituent hydrous silicates of aluminum with or without free silica, plastic when sufficiently pulverized and wetted, rigid when subsequently dried, and suitable for use in commercial refractory products; a type of clay that can withstand great heat and is used for fire brick; also used for mortar for laying fire brick.

Fire Damper. A damper that automatically closes a duct or opening upon detection of fire.

Fire Decorating. The process of firing ceramic or metallic decorations on the surface of glazed ceramic ware.

Fire Detection Annunciator. A mechanical device which audibly announces a warning in case of a fire.

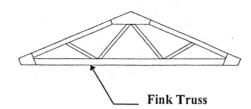

Fink Truss

Fire Division Wall. Any wall which subdivides a building so as to resist the spread of fire, but is not necessarily continuous through all stories to and above the roof.

Fire Door. A door which has been manufactured with specific materials and rated, designed to help hold back the spread of fire.

Fire Endurance. The ability of an assembly to perform to certain fire test criteria.

Fire Escape. Continuous unobstructed route of escape from building in case of fire, usually located on exterior of building and composed of stairs, ladders, and landings.

Fire Extinguisher. A small, portable apparatus which ejects fire-extinguishing chemicals to put out small fires.

Fire Hydrant. A discharge fitting or apparatus with a valve and spout at which water may be drawn from the mains of waterworks; used primarily in fire-fighting, also used to service water mains and systems; also called Fire Plug.

Fire Lines. The complete wet standpipe system of buildings, including the water service, standpipe, roof manifold, Siamese connections, and pumps.

Fire Plug. Fire Hydrant.

Fire Rated Brick. Brick that has been tested for fire-resistance and then graded for specific construction uses.

Fire Rated Door. A door which has been given a rating of how long in time it can withstand fire before failure.

Fire Rated Frame. A door frame which has been given a rating on how long in time it can withstand fire before failure.

Fire Rated Panel. A panel which has been given a rating in time on how long it can withstand fire before failure.

Fire Resistance Rating. The time, in hours, or fractions of an hour, that a material or assembly will resist fire exposure.

Fire Resistance. The ability of an assembly to maintain structural stability and act as an effective barrier to the transmission of heat for a specified period of time.

Fire Resistant Sheetrock. Sheetrock which has been manufactured with fire-resistant chemicals.

Fire Resistant. Incombustible or slow to be damaged by fire; forming a barrier to the passage of fire.

Fire Retardant Chemical. A chemical preparation used to reduce flammability or to retard the spread of flame.

Fire Retardant Paint. Paint containing substance which slows down rate of combustion of flammable material or renders material incapable of supporting flame.

Fire Retardant Roof. Roofing material which has been manufactured with fire retardant chemicals.

Fire Retardant Treated Lumber. Lumber with applied or pressurized chemical treatment of lumber to retard combustion.

Fire Retardant Treated Plywood. Plywood with applied or pressurized chemical treatment to retard combustion.

Fire Retardant. A chemical treatment to reduce the ignitability of the treated material; a material or treatment which effects a reduction in flammability and in spread of fire.

Fire Separation Wall. A wall required under the building code to divide two parts of a building as a deterrent to the spread of fire; a fire wall.

Fire Sprinkler Detector. A water pressure actuated switch that will signal when a sprinkler head has been activated.

Fire Sprinkler Head. The water spray device that is part of a fire sprinkler system.

Fire Sprinkler System. An arrangement of water pipes under continuous pressure that will release a water spray through spaced heads that are activated by increased temperature.

Fire Standpipe. Lines and connections within buildings that provide water source for fire fighting.

Fire Station. A building housing fire-fighting apparatus, and usually fire-fighters.

Fire Stop. 1. A projection of brick-work from walls between the joists to prevent fire from traveling between the plaster and the brick wall. 2. Material or member that seals open construction to inhibit spread of fire.

Fire Stopping. The sealing of contiguous framing cavities or areas to reduce the opportunity for rapid spread of fire and smoke; see Fire Blocking.

Fire Tapping. The taping of gypsumboard joints without subsequent finishing coats. A treatment method used in attic, plenum or mechanical areas where esthetics are not important.

Fire, Bisque. See Bisque Fire.

Fire, Glost. See Glost Fire.

Fire, Single. See Single Fire.

Firebox. The portion of a fireplace that contains the fire.

Firebrick. Brick that resist high temperatures and chemical disintegration, used for lining furnaces, fireboxes, and chimneys.

Firecut. A sloping end cut on a wood beam or joist where it enters a masonry wall, to allow the wood member to rotate out of the wall without prying the wall apart, if the floor or roof structure should burn through in a fire.

Fireplace. A masonry chamber within a building, to facilitate an open fire, with a flue to carry off smoke and fumes.

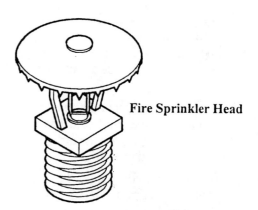

Fire Sprinkler Head

Fireproof Vault. An enclosure or room for the storing of valuables that has been constructed to withstand fire and heat.

Fireproofing. Material applied to a building element to insulate it against excessive temperatures in case of fire.

Fire-Trap. A building without proper provision for escape in case of fire.

Firewall, Brick. See Brick Firewall.

Firewall. A wall constructed to prevent or slow down the spread of fire.

Firing Range. 1. The range of firing temperature within which a ceramic composition develops properties which render it commercially useful. 2. A place where one practices the shooting of weapons.

Firing. The controlled heat treatment of ceramics in a kiln or furnace, during the process of manufacture, to develop the desired properties.

First Aid Kit. A packet of medical supplies for emergency use.

Fiscal Year. Financial year; an accounting period of 12 months.

Fish Eyes. A term used to describe small blemishes occasionally found in lime finish coats; approximately 1/4 inch in diameter, they are caused by lumpy lime.

Fish Oil. The only animal oil used to any extent in the paint industry; extracted from fish such as sardine, menhaden, and pilchard.

Fish Tape. A steel tape used by electricians to push through an electrical conduit and pull back the conductors being installed.

Fished Joint. An end butt splice strengthened by pieces nailed on the sides.

Fished. A means of installing electrical wiring in existing inaccessible hollow spaces of buildings with a minimum damage to the building finish.

Fishmouth. An opening at the exposed edge of a roofing ply sheet where the asphalt bond is lacking or the felt is wrinkled.

Fissure. A narrow crack, usually long and deep, caused by breaking or parting.

Fitting Gain. The space inside a fitting required by a pipe.

Fitting. 1. A device used for connecting pipes together. 2. An accessory such as a locknut, bushing or other part of a wiring system which is intended primarily to perform a mechanical rather than an electrical function.

Fixative. A protective coating applied to drawings in crayons, pastel, charcoal, and pencil, usually by spraying, to prevent colors from rubbing off.

Fixed Assets. Assets of a more or less permanent nature whose useful life is more than one year, such as fixtures, equipment, real estate, and trucks.

Fixed Displacement Pump. A pump in which the displacement per cycle cannot be varied.

Fixed Door. A door, of a pair, that is normally kept in the closed position.

Fixed End Beam. A beam fixed upon a support which prevents its rotation.

Fixed Fee. A stipulated lump sum fee.

Fixed Liabilities. Liabilities that are carried for over a one year period, such as mortgage, and equipment payments.

Fixed Window. Sash or glass that is immovably mounted in a wall.

Fixed. Attached or fastened in place by nails, screws, bolts, conduit, piping systems, or other means.

Fixture Branch. A pipe connecting several fixtures with the water supply.

Fixture Carrier. A mounting frame built into a wall to support a plumbing fixture.

Fixture Drain. The drain pipe from the trap of a fixture to where the pipe joins with any other drain pipe.

Fixture Unit. A design factor to determine the load-producing value from a given fixture so that the drainage piping is large enough to carry the liquids and wastes; for instance, the unit flow rate from fixtures is assumed to be one cubic foot or 7.5 gallons of liquid per minute; fixtures are rated as multiples of this unit of flow.

F

Fixture, Lighting. See Lighting Fixture.

Fixture, Plumbing. See Plumbing Fixture.

Flag. 1. See Flagstone. 2. End of hog brush bristle which divides into two or more branches like a tree; flagging provides brush with ability to hold paint.

Flagging. 1. Flagstone paving. 2. See Flag, 2.

Flagpole Outrigger. A projecting pole run out from a structure to provide a flagpole.

Flagpole. A pole used for raising and displaying flags.

Flagstone. A hard flat, usually rectangular, stone slab used for paving; also called a flag.

Flake. A small flat wood particle of predetermined dimensions, uniform thickness, with fiber direction essentially in the plane of the flake; in overall character resembling a small piece of veneer; produced by special equipment for use in the manufacture of flakeboard.

Flakeboard. A type of particleboard composed of flakes bonded together with a synthetic resin of other suitable binder.

Flaked. Irregularities left on the edge of the tile mainly due to the use of machine cutting tools.

Flaking. Detachment of small pieces of paint film.

Flame Spread Classification. A standard measurement of the relative surface burning characteristics of a building material when tested.

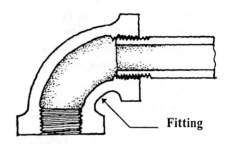

Fitting

Flame Spread Rating. A measure of the rapidity with which fire will spread across the surface of a material.

Flame Spread. The propagation of a flame away from the source of ignition across the surface of a liquid or a solid, or throughout the volume of a gaseous mixture.

Flame Test for Leaks. Tool which is principally a torch; when a halogen mixture is fed to the flame, this flame will change color in the presence of heated copper.

Flameproof Tile. Tile that has been made resistant to flame.

Flameproofing. An act or process of making an object resistant to the action of flame.

Flammable Liquids. Liquids having a flash point below 140° F. (60° C.) and a vapor pressure not exceeding 40 psia (2.81 kg/cm²) at 100° F. (38° C.).

Flammable. Capable of being easily ignited; same as inflammable; the opposite is nonflammable.

Flange. A rib or rim on an object for strength, for guiding, or for attachment to another object.

F

Flanged Pipe. A pipe that has been fitted with a projecting ring, ridge or collar, to strengthen, prevent sliding, or accommodate attachments.

Flanking Paths. Paths by which sound travels around an element that is intended to impede it.

Flapper Valves. Thin metal valve used in refrigeration compressors which allows gaseous refrigerants to flow in only one direction.

Flare Header. 1. A brick burned on one end to a darker color than the face. 2. A header of darker color than the field of the wall.

Flare Nut. Fitting used to clamp tubing flare against another fitting.

Flare. An enlargement at the end of a piece of flexible tubing by which the tubing is connected to a fitting or another piece of tubing; this enlargement is made at about a 45 degree angle; fittings grip it firmly to make the joint leakproof and strong.

Flash Cove. A detail in which a sheet of resilient flooring is turned up at the edge and finished against the wall to create an integral baseboard.

Flash Gas. Instantaneous evaporation of some liquid refrigerant in evaporator which cools remaining liquid refrigerant to desired evaporation temperature.

Flash Point. The temperature at which the material gives off flammable vapor in sufficient quantity to ignite momentarily on the application of a flame under specified conditions.

Flash Set. The rapid development of rigidity in a mixed portland cement paste, mortar, or concrete usually with the evolution of considerable heat, which rigidity cannot be dispelled nor can the plasticity be regained by further mixing without addition of water; also referred to as Quick Set or Grab Set.

Flash Weld. Resistance type weld in which mating parts are brought together under considerable pressure while a heavy electrical current is passed through the joint to be welded.

Flashback. In welding. a recession of the flame into or back of the mixing chamber of the torch.

Flashing Block. In masonry, metal flashing used to block a parapet wall and prevent roof leaks around such a wall.

Flashing Cement. A plastic mixture of cutback bitumen and mineral stabilizers, asbestos fibers or other inorganic fibers, used for reinforcing, mixed with a solvent and used as an adhesive.

Flashing, Roof. See Roof Flashing.

Flashing, Step. See Step Flashing.

Flashing, Through-Wall. See Through-Wall Flashing.

Flashing. 1. A thin, continuous sheet of impervious material such as metal, plastic, rubber, or waterproof paper used to prevent the passage of water through a joint in a wall, roof, or at a chimney; the material used and the process of making watertight the roof intersections and other exposed places on the outside of a structure. 2. Non-uniform appearance on surfaces in which coating dries with spotty differences of color or gloss.

Flat Arch. An arch or span constructed with a flat top and bottom; steel angle iron is usually used to help support this arch.

Flat Asphalt. A roofing asphalt that has a softening point of approximately 170° F. (77° C).

Flat Cut Veneer. See Plain Sliced Veneer.

Flat Finish. Dull finish, no gloss.

Flat Glass Marketing Association (FGMA). 3310 SW Harrison Street, Topeka, Kansas 66611, (913) 266-7013, FAX (913) 266-0272.

Flat Head Screwdriver. See Slotted Screwdriver.

Flat Head Wood Screw (FHWS). A wood screw with a countersunk head flush with the surface, driven with a screwdriver.

Flat Mill. A kind of grinding mill used to grind paint pigments; the mill consists of two stones, a lower stone which revolves and an upper stone which is stationary.

Flat Paint. An interior paint with a high proportion of pigment; it dries to a flat, or lusterless, finish.

Flat Position Weld. The position of welding wherein welding is performed from the upper side of the joint and the face of the weld is approximately horizontal.

Flat Seam. A sheet metal roofing seam that is formed flat against the surface of the roof.

Flat Slab. A reinforced concrete slab that is designed to span, without any beams or girders, in two directions to supporting columns; a 2-way slab supported at its corners only.

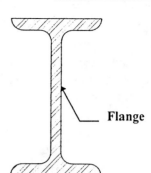

Flange

Flat Stretcher Course. A brick course of stretchers that are set on edge and expose their flat sides on the surface or face of the wall.

Flat Trowel. The flat trowel is used in conjunction with the hawk for the transferring of mortar from the mortarboard to the wall or to other vertical surfaces; frequently used for spreading pure cement on the finished floor coat; also used for spreading mortar on floor surfaces before tiles are set; commonly used to put the finish on a freshly poured concrete floor.

Flat Varnish. Varnish which dries with reduced gloss, made by adding such materials as silica, wax, or metallic soaps to the varnish.

Flat Wall Paint. A type of interior paint which is designed to produce a flat or lusterless finish.

Flat Washer. A washer which goes under a bolt head or a nut to spread the load, prevent loosening, and protect the surface.

Flat-Grained Wood. Lumber that has been sawed parallel to the pith and approximately tangent to the growth rings; lumber is considered flat-grain when the annual growth rings make an angle of less than 45 degrees with the surface of the piece.

F

Flatplate Collector. The most common and often-used collector in a solar system; the most visible portion of a solar system appearing on the roof or in the yard; device in which sunlight is converted to heat on a plane surface usually made of metal or plastic; a heat transfer fluid circulates through the collector to transport heat to be used directly or to be stored.

Flattening Agent. An ingredient, usually a metallic soap, such as calcium, aluminum or zinc stearate, used in lacquers and varnishes to reduce the gloss or to give a rubbed appearance; also called Flatting Agent.

Flattening Oil. A varnish-like composition, made of thickened oil dissolved in a thinner, used to reduce paste paint to a flat paint; also called Flatting Oil.

Flatting Agent. See Flattening Agent.

Flatting Oil. See Flattening Oil.

Flatting. Loss of gloss in coating film.

Flatwork. Concrete slabs on grade.

Flecks. See Wood Rays.

Flemish Bond. A brick bond consisting of headers and stretchers alternating in every course, so laid as always to break joints, each header being placed in the middle of the stretchers in courses above and below.

Flemish Cross Bond. A brick bond that has alternate courses of Flemish headers and stretcher courses. The headers are directly over each other and the alternate stretcher courses are crossed over each other.

Flemish Garden Bond. A brick bond where bricks are laid so that each course has a header to every 3 or 4 stretchers.

Flex. Flexible electrical conduit.

Flexibility. 1. The ability to bend or conform to new conditions. 2. The ability of an asphalt pavement structure to conform to settlement of the foundation; generally, flexibility of the asphalt paving mixture is enhanced by high asphalt content.

Flexible Conduit. Electrical conduit made of a spirally wound metallic strip.

Flexible Cord. A flexible insulated electrical cable having a plug at one or both ends used to connect an appliance or other fixture to a receptacle, manufactured to resist tangling.

Flexible Coupling. A mechanical connection that adapts to misalignment between moving parts.

Flexible Ductwork. Flexible ductwork manufactured in various diameters, made from spiral wire covered in plastic and commonly insulated, for use in the transfer of air in heating, cooling and ventilating systems.

Flexible Wiring. Electrical wiring that permits movement from expansion, contraction, vibration, or rotation.

Flexible. Limber, bendable; not stiff; the opposite of rigid.

Flexural Strength. 1. A property of a material or structural member that indicates its ability to resist failure in bending. 2. The resistance of a specified sample size of a gypsum board to failure caused by a transverse load.

Flexure Formula. A formula for determining values for the design of members or elements subjected to bending.

Flint Paper. Abrasive paper which is grayish-white in color; inexpensive but has short working life.

Flitch. A portion of a log sawn on two or more faces, commonly on opposite faces leaving two waney edges; when intended for resawing into lumber, it is resawn parallel to its original wide faces; or, it may be sliced or sawn into veneer, in which case the resulting sheets of veneer laid together in the sequence of cutting are called a flitch.

Flitch Beam. A composite beam made up of one or more wooden members along with a steel plate, the assembly usually bolted together to act as a single structural unit.

Flitch-Sliced Veneer. A thin sheet of wood cut by passing a block of wood vertically against a long, sharp knife.

Float Coat. The final mortar coat over which the neat coat, pure coat, or skim coat is applied.

Float Finish. A finish coat that is rough from aggregate material found in the plaster mortar.

Float Glass. Glass sheet manufactured by cooling a layer of liquid glass on a bath of molten tin.

Float Strip. A strip of wood about 1/4 inch thick and 1-1/4 inch wide, used as a guide to align mortar surfaces.

Float Switch. An electrical switch that is actuated by the rising or falling of a float in liquid.

Float Trap. A floating device in a plumbing fixture which opens or closes a valve to prevent sewer air gases from escaping back through the fixture.

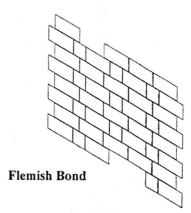

Flemish Bond

Float Valve. Type of valve which is operated by sphere or pan which floats on liquid surface and controls level of liquid.

Float. 1. A tool or apparatus for smoothing a surface, used by plasterers and concrete masons. 2. A sum of money in excess of daily needs. 3. See Slack, 1.

Floating Angle. A method of applying gypsumboard designed to allow structural movement at interior corners.

Floating Edge. In gypsumboard installation, an actory edge applied in such a manner that the edge does not lie directly over a framing member and is unsupported.

Floating Joint. In gypsum board installation, a condition where the butt joint does not lie directly over a framing member; floating joints should be back-blocked.

Floating. 1. Separation or layering of pigment in mixture of pigment. 2. A method of using a straightedge to align mortar with the float strips or screeds; this technique also is called dragging, pulling, rodding, or rodding off. 3. The rising of a swimming pool out of the ground, caused by water pressure under the pool; this only occurs when the pool is not filled with water.

Floc. Gel-like substance formed when coagulant, usually alum, combines with suspended alkaline matter in pool water and precipitates out.

Flocculate. Agglomeration of undispersed pigment particles.

Flocculating Agent. Compound that coalesces finely suspended particles.

Flocculent. A compound, usually some type of alum, used with sand-type filters to form a thin layer of gelatinous substance on the top of the sand; aids in trapping fine suspended particles which might pass through the floc.

Flock Finish. Finish obtained by spraying or sifting flock (short fibers of wool, silk, rayon) onto a surface to which the flock fibers will adhere.

Flocked Carpet. Single-level velvety pile carpet composed of short fibers embedded on an adhesive-coated backing.

Flocked Paper. Wallpaper covered with flocking.

Flocking. Very short or pulverized fiber used to form a velvety pattern or design on a surface.

Flococoating. Process of finishing by flowing finishing material on article by means of hose, allowing excess to drain into tank.

Flood Coat. The top layer of bitumen of a built-up roofing membrane; usually it is aggregate surfaced; if the flood coat is applied correctly it is poured, not mopped, to a weight of 60 pounds per square for asphalt and to a weight of 75 pounds per square for coal-tar pitch.

Flood Lamp. A strong source of artificial illumination which projects a broad beam.

Flood Level Rim. See Flood Rim.

Flood Rim. The highest point that water can reach within a plumbing fixture without overflowing; also called Flood Level Rim.

Flood. An overwhelming volume of water by overflowing of a body of water; a deluge.

Flooded System, High-Side Float. Refrigeration system which has a float operated by the level of the high-side liquid refrigerant.

Flooded System, Low-Side Float. Refrigerating system which has a low-side float refrigerant control.

Flooded System. Type of refrigerating system in which liquid refrigerant fills most of the evaporator.

Flooding. Act of allowing a liquid to flow into a part of a system.

Flooding. Same as floating.

Floodlight. Artificial lighting in a broad beam.

Floodplain. Level land that may be submerged by floodwaters.

Floor Box. A metal electrical rough-in box fed by conduits in or under the floor to provide for a floor outlet.

Floor Brick. Smooth dense brick, highly resistant to abrasion, used as finished floor surfaces.

Floor Closer. A closing device installed in the floor under a door.

Floor Deck. Sheet steel formed to fluted or ribbed profile to span between supports (usually joists or beams) to support floor system (usually concrete slab) and live loads.

Floor Drain. 1. A drain in a floor which connects to the plumbing system and removes unwanted water from an area or room. 2. An opening or receptacle located in a concrete floor, connected to a trap to receive the discharge from indirect waste.

Floor Finish. The stain, paint, wax, or polish on a floor.

Floor Finishing. The act or process of the final sanding, coloring, and sealing of a wooden floor system.

Floor Grating. Open grid of metal bars structurally formed.

Floor Hardener. A chemical applied to concrete floors to protect them from wear.

Floor Hinge. A combined pivot hinge and closing device set either in the floor or in the bottom of the door; it may be spring type only or may be combined with liquid control.

Floor Joist. A support beam, commonly installed in parallel with other beams to create a structural floor system, after which floor sheathing is fastened.

Floor Level. The elevation of any floor in a building.

Floor Masonry. Shaped or molded masonry units such as, stone, ceramic brick, tile or concrete used for finished flooring.

Floor Mat Frame. A frame, usually metal, to contain a floor mat and provide a means to anchor the mat to the floor.

Flood Lamp

Floor Mat. A piece of coarse woven or plaited fabric used as a floor covering.

Floor Molding. A piece of trim at the juncture of the wall and floor.

Floor Mounted Bench. A work surface that has its support legs mounted to the floor.

Floor Mounted Transformer. A transformer that has its supports mounted to the floor to prevent vibration and movement.

Floor Mud. Mortar which does not contain lime, used to float horizontal surfaces; also called Deck Mud.

Floor Patch. Material used to repair damage to a floor.

Floor Pedestal. A member, such as a short pier, used as a base for a floor system.

Floor Plate. Steel or aluminum plate with a non-skid design formed on one side for use in constructing industrial floors, platforms, landings, and stairs.

Floor Price. The least amount that an informed seller would accept for leasing or selling property.

Floor Removal. The act or process of the tearing up and carrying away an old floor system.

Floor Safe. A place or receptacle to keep valuables, installed in a floor.

Floor Sink. A small sink, typically 12" x 12", set in the floor to receive drainage such as condensation water.

Floor Slab. A reinforced concrete slab on grade or elevated, used as a floor.

Floor Tile, Clay. See Clay Floor Tile.

Floor Varnish. A varnish made specifically for application to floors.

Floor Wax. A substance spread onto flooring that seals, protects, and can be polished to a shine.

Floor, Access. See Access Floor.

Floor, Gym. See Gym Floor.

Floor, Industrial Wood. See Industrial Wood Floor.

Floor, Level. See Level Floor.

Floor, Terrazzo. See Terrazzo Floor.

Floor, Tile. See Tile Floor.

Floor. The lower interior surface of a room in a building.

Floorboard. One of various types of boards used for flooring.

Flooring, Marble. See Marble Flooring.

Flooring. The material with which a floor is surfaced, such as carpeting, wood strips, wood parquet, linoleum, or sheet vinyl.

Flow After Suction. Flow of mortar measured after subjecting it to a vacuum produced by a head of two inches of mercury; the suction apparatus and its use is described in Sections 27 and 28 of ASTM C 91.

Flow Gradient. A drainageway slope determined by the elevation and distance of the inlet and outlet and by required volume and velocity.

Flow Meter. Instrument used to measure velocity or volume of fluid movement.

Flow of Mortar. Measure of mortar consistency, sometimes termed the Initial Flow, determined on the flow table described in ASTM C 230; use of the flow table and method of calculating the flow is described in Section 9 of ASTM C 109.

Flow. A measure of self leveling.

Flowing Varnish. A varnish which has been designed to produce a smooth lustrous surface without rubbing or polishing.

Flue Liner. Heat-resistant firebrick or other fire clay materials that make up the lining of a chimney.

Flue. A passage for smoke and combustion products from a furnace, stove, water heater or fireplace; the opening in a chimney through which smoke passes.

Fluffing. Lint and fuzz that appears on newly installed carpet, which is merely the factory-sheared pile ends working their way to the surface, not the tufts or pile yarns themselves; this condition disappears as the carpet is used; also called Shedding.

Fluid Adjusting Screw. A screw on a spray gun which controls the amount of fluid entering the gun.

Fluid Applied Elastomer. An elastomeric material, which is fluid at ambient temperature, that dries or cures after application to form a continuous membrane.

Fluid Applied Roof Membrane. A roof membrane applied in one or more coats of a liquid that cure to form an impervious sheet.

Fluid Applied Roof. A roof coated with an asphalt-based liquid.

Fluid Flow. A measure of flow through a spray gun with atomizing air shut off.

Fluid Hose. Specially designed hose for paint materials; usually black.

Fluid Nozzle. Spray gun fluid tip with orifice; the needle and tip combination.

Fluid. Substance in either a liquid or gaseous state; substance containing particles which move and change position without separation of the mass.

Fluorescent Fixture. An illuminating device which uses a tubular electric lamp which has a coating of fluorescent material on its inner surface and contains mercury vapor.

Fluorescent Lamp Starter. See Ballast.

Fluorescent Paint. Luminous paint which glows only during activation by ultraviolet or black light.

Fluorite. Fluorspar, CaF_2, an inorganic mineral of the isometric form, used as a source of fluorine for fluxing of glasses and glazes.

Fluoropolymer Coating. A factory-applied oven-baked finish.

Flush Bolt. A door bolt so designed that when applied it is flush with the face or edge of the door.

Flush Bushing, Copper. A threaded pipe fitting used to connect two pipes with different diameters together and create a continuous smooth surface, made of copper.

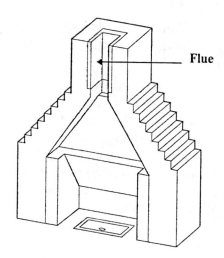

Flue

Flush Cabinet Construction. A method of building cabinets where the fronts of the doors and drawers are flush with the face frame.

Flush Door. A door with a smooth surface and no protrusions.

Flush Molding. A piece of trim which is fastened flat onto or flush with a wall and sometimes is used as a divider or chair rail.

Flush Tank. The water tank or cistern on a toilet to supply flushing water.

Flush Valve. A regulated water valve at a toilet or urinal to supply flushing water directly from a high pressure water line.

Flush. 1. Adjacent surfaces even, or in the same plane. 2. Operation to remove any material or fluids from refrigeration system parts by purging them to the atmosphere using refrigerant or other fluids. 3. To operate a water closet or urinal valve.

Flushing. Of asphalt paving, see Bleeding, 1.

Flushometer. A high pressure flush valve for a toilet or urinal.

Fluted Column. Column with vertical decorative, semi-circular channeled shafts.

Flux. 1. A substance that promotes fusion in a given ceramic mixture. 2. A chemical used when soldering, brazing, or welding to prevent oxides from forming when heated. 3. Fusible material used in welding or oxygen cutting to dissolve and facilitate removal of oxides and other undesirable substances. 4. The density of a magnetic field.

Fly Ash. Extremely fine ash from the burning of pulverized coal; when it has over 90 percent silica content it may be used as a pozzolan in concrete; it can improve the concrete by increasing the volume of cement paste.

Fly Gallery. The space above a theater stage where the scenery is suspended.

Flying Buttress. A masonry buttress springing from a separate column, forming an arch with the wall it supports.

Flying Formwork. Large sections of slab formwork that are moved by crane.

FM. 1. Facilitated Management. 2. Factory Mutual.

FNMA. Federal National Mortgage Association.

Foam Concrete. See Cellular Concrete.

Foam Extinguisher. A extinguisher which releases chemical foam to put out fires.

Foam Gasket. Joint sealing material made of rubber or plastic foam strips.

Foam Leak Detector. System of soap bubbles or special foaming liquids brushed over joints and connections to locate leaks.

Foam Plaster Base. A rigid type foamed backing which acts as a plaster base.

Foamed Concrete. Concrete made very light and cellular by the addition of a prepared foam or by generation of gas within the unhardened mixture.

Foamglass Insulation. A thermal insulation made by foaming glass with hydrogen sulfide; it is manufactured in the form of block or board and has a low fire hazard rating.

Foaming. 1. Frothing. 2. Formation of a foam in an oil-refrigerant mixture due to rapid evaporation of refrigerant dissolved in the oil; this is most likely to occur when the compressor starts and the pressure is suddenly reduced.

FOB. Free On Board; without charge for delivery to and placing on board a carrier at a specified point; FOB Jobsite means material cost includes cost of delivery to jobsite.

Focal Depth. Depth of the earthquake focus (or hypocenter) below the ground surface.

Focus. The point of origin of an earthquake; the point at which the rupture occurs; synonymous with the hypocenter.

Fog Curing. 1. Storage of concrete in a moist room in which the desired high humidity is achieved by the atomization of fresh water. 2. Application of atomized fresh water to concrete, stucco, mortar, or plaster.

Fogging. Misting.

F

Foil Back. See Foil Backed Gypsum Board.

Foil Backed Gypsumboard. Gypsumboard with aluminum foil laminated to its back surface as a vapor retardant and thermal insulator. Also called Foil Back.

Foil. 1. Metal hammered into a thin sheet. 2. A cusped element in Gothic tracery.

Folded Plate. A roof structure in which strength and stiffness derive from a pleated or folded geometry; special class of shell structure formed by joining flat, thin slabs along their edges so as to create a three dimensional spatial structure.

Folding Door. A door with hinged leaves, often provided with a ceiling or floor track.

Folding Stair. A hinged stair, attached to and concealed within the ceiling, which can be raised and lowered.

Folio. A page number referring to the disposition of source of an entry or posting in the books of the business.

Font. A receptacle in a church for baptismal water or holy water.

Food Preparation. The action or process of making food ready for cooking and consumption.

Food Service Equipment. Appliances, fixtures, and materials used in the cooking and serving of food in large quantities.

Foot Grille. Metal grating usually located in floor near building entrance to allow for scraping of dirt, ice, and snow from footwear before proceeding to building interior.

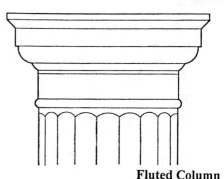

Fluted Column

Foot. A unit of length equal to 12 inches or 30.48 centimeters.

Foot-Candle. The illumination produced by one lumen of luminous flux spread uniformly over an area of one square foot.

Footing Drain. 1. Drain around the perimeter of a building, at the footings, to drain groundwater or rainwater away from the building. 2. Drain at the footing of a retaining wall to prevent accumulation of hydrostatic pressure that could collapse the wall.

Footing Excavation. The temporary removal of earth for installing a footing.

Footing Form. A wooden or steel structure, placed around the footing that will hold the concrete to the desired shape and size; also called Footing Formwork.

Footing Formwork. See Footing Form.

Footing Reinforcing. The placing of metal or steel bars in a freshly poured or placed concrete footing to strengthen it.

Footing. A masonry section, usually concrete, in a rectangular form wider than the bottom of the foundation wall or pier it supports.

Foot-Pound. A unit of work equal to the work done by a force of one pound acting through a distance of one foot in the direction of the force.

Foot-Poundal. An absolute unit of work that is equal to the work done by a force of 1 poundal in moving a body through a distance of 1 foot, equivalent to about 0.04 joules.

Footprint. The outline of a building on the ground, used in site planning and in judging compliance with planning and zoning laws.

Foots. Dregs; settlings in vegetable oils.

Force Account. Construction work on a time and material basis.

Force Majeure. An overwhelming, but unanticipated event.

Force. Energy exerted on an element tending to cause motion; magnitude, direction and point of application are all used to describe a force.

Forced Air Furnace. An appliance, with a heat exchanger and ventilating fan, that supplies hot air for space heating.

Forced Air. Air blown by a fan from a furnace or air conditioner.

Forced Convection. Movement of fluid by mechanical force such as fans or pumps.

Forced Draft. In a boiler, combustion air forced by a blower in the burner section through the boiler and up the stack.

Forced Drying. Acceleration of drying by increasing the temperature above ambient temperature using an oven, infrared lamp, or other heat source.

Force-Feed Oiling. Lubrication system which uses a pump to force oil to surfaces of moving parts.

Foreclosure. A legal procedure for selling a mortgaged property in default to satisfy the debt.

Forehand Welding. A gas-welding technique wherein the flame is directed toward the progress of welding.

Foreman. An experienced worker who works with and usually leads a crew or gang.

Forensic Architecture. A science and practice that deals with application of architecture and construction facts and scientific methods to construction and legal problems.

Forensic Engineering. A science and practice that deals with application of engineering and construction facts and scientific methods to construction and legal problems.

Forest Products Laboratory (FPL). Forest Service, U.S. Dept. of Agriculture, Madison, Wisconsin 53705.

Forge. 1. A furnace or hearth for melting or refining metal. 2. A furnace where metal is heated and wrought.

Forged Eyebolt. A heavy metal bolt which contains an eye at its end which may be used as a fastening device.

Forged Steel. Wrought steel.

Forged. Formed by heating and hammering.

Fork Lift. A vehicle with a horizontal fork in front for lifting and moving loads; a fork lift truck.

Form Deck. Thin, corrugated steel decking that serves as permanent formwork for a reinforced concrete deck.

Form Fabric. Welded-wire fabric used to reinforce concrete while it is setting and gaining sufficient strength to be self-supporting.

Form Lacquer. Thin lacquer or varnish used to coat concrete forms to prevent concrete from adhering to the forms.

Form Oil. Oil applied to the interior surface of formwork to promote easy release from the concrete when forms are removed.

Form Release. A substance applied to concrete forms to make stripping after pouring easier.

Form Tie. A steel rod with fasteners on each end, used to hold together the formwork for a concrete wall.

Form, Slab. See Slab Form.

Form. Temporary structure built to contain concrete while it sets; also called Formwork.

Form-Fit Area. See Free Form.

Formica Partition. A dividing wall finished with a layer of Formica.

Formica. Brand name and trademark for any of the various laminated plastic products, usually used for surface finish on cabinets or millwork.

Forming. 1. The act of installing wooden forms to accept reinforcing steel and concrete. 2. The shaping or molding of ceramic ware.

Formwork. See Form.

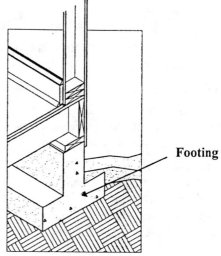

Footing

Forsterite Porcelain. A vitreous ceramic whiteware for technical applications in which forsterite, $Mg_2\ SiO_4$, is the essential crystalline phase.

Fossil Fuel. A natural fuel such as coal, gas, or petroleum formed in the geological past from the remains of living organisms.

Fossil Resin. Any of the natural or earth type resins, such as kauri and the Congo copals, which derive their characteristics through aging in the ground.

Foundation Backfill. Earth or earthen material displaced during excavation and replaced around the constructed foundation.

Foundation Block. Masonry block used for foundation work, commonly denser than regular block.

Foundation Resin. Piping around the base of a building to collect water and convey to the drainage system.

Foundation Investigation. See Geotechnical Investigation.

Foundation Mat. A grid of reinforcement steel for concrete foundations.

Foundation Reinforcing. The placing of metal or steel bars in a freshly poured or placed concrete foundation to strengthen it.

Foundation Soil. See Subgrade.

Foundation Vent. Screened opening below the floor line to provide natural ventilation to the foundation crawl spaces.

Foundation Wall. That portion of a load-bearing wall below the level of the adjacent grade, or below the first tier of floor beams or joists, which transmits the superimposed load to the footing.

Foundation. The portion of a building that has the sole purpose of transmitting structural loads from the building into the earth. That part of a building or wall which supports the superstructure.

Foundry. A workshop for casting metals.

Fountain, Wash. See Wash Fountain.

Fountain. 1. A jet or jets of water made to spout for ornamental purposes or for drinking and the structure for this purpose. 2. A natural spring of water.

Four Dimensional. Involving four dimensions, consisting of elements requiring four coordinates to determine them; the space-time continuum.

Four Pole Motor. 1800 rpm, 60 Hz electric synchronous speed motor.

Fourth Dimension. Time, in addition to height, width, and depth.

Foyer. An anteroom, lobby, entrance hallway, or vestibule.

FPL. Forest Products Laboratory.

Fraction. 1. A numerical quantity that is not a whole number, as in 1/2 or 0.5; a number expressed as one number divided by another, as in 2/3 or 4/5; the upper number is the numerator and the bottom number is the denominator. 2. A portion of a mixture separated by distillation.

Fracture. 1. A break, usually resulting in actual separation of the material; in structures, the characteristic result of tension failure. 2. See Crack.

Frame High. To construct the brickwork up to the top of the door or window frame; the lintel is then laid across the opening and rests upon the brickwork on each side of the frame.

Frame, Balloon. See Balloon Frame.

Frame, Braced. See Braced Frame.

Frame, Door. See Door Frame.

Frame, Ductile Moment Resisting Space. See Ductile Moment Resisting Space Frame.

Frame, Moment Resisting. See Moment Resisting Frame.

Frame, Platform. See Platform Frame.

Frame. 1. An enclosing border as in a picture frame. 2. The surrounding or enclosing woodwork, as around windows or doors. 3. The skeleton of a building; that is, the rough structure of a building, including interior and exterior walls, floor, roof, and ceilings. 4. To form together or construct large assemblies from smaller components. 5. A structural system consisting of relatively long, prismatic members fastened together; a rigid frame is one in which the joints can transmit moments as well as forces and which therefore does not require a braced frame for rigidity.

Framed Connection. 1. One that is capable of resisting moments. 2. A shear connection between steel members made by means of steel angles or plates connecting to the web of the beam or girder.

Framer. 1. A carpenter who constructs wood framing. 2. A carpentry contractor. 3. One who frames pictures and makes their frames.

Frames. Racks at the back of a Jacquard loom, each holding a different color of pile yarn; in Wilton carpets, 2 to 6 frames may be used and the number is a measure of quality as well as an indication of the number of colors in the pattern, unless some of the yarns are buried in the backing.

Framing Lumber. Wood members of framing systems which are manufactured by sawing, resawing, passing lengthwise through standard planing machine, crosscutting to length, and matching, but without further manufacturing.

Framing Member. The stud, plate, joist, or furring component to which the exterior and interior surfacing materials are attached; normally made of wood or metal.

Framing, Ceiling. See Ceiling Framing.

Framing, Door. See Door Framing.

Framing, Roof. See Roof Framing.

Framing, Timber. See Timber Framing.

Framing, Wall. See Wall Framing.

Framing. The rough wooden structural skeleton of a building, including interior and exterior walls, floor, roof, and ceilings.

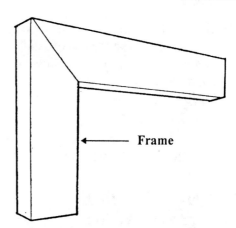

Frame

Franchise Tax Board. In California, a department of state government that collects taxes from individuals and businesses.

Fraud. A false statement of fact that is designed to deceive.

Freddy Mac. See Federal Home Loan Mortgage Corporation.

Free and Clear. Real property that has no liens or encumbrances.

Free Body Diagram. A diagram, or drawing, in which on element of structure is isolated from its surroundings, and the effect of its surroundings is shown only as forces; see Vector, 1.

Free Form. A floor area, usually in a department store or salon, not bounded by walls and of nonrectangular shape; sometimes called Form-Fit Area.

Free Water. All water contained by gypsum board, concrete, mortar, or plaster in excess of that chemically held as water of crystallization;

Freeboard. Distance between pool-water surface level and deck level.

Freeform. Freely adapted poolshape.

Freehand Floating. The application of wall mortar without the use of guide screeds; this technique is used by specialists when they are setting glass mosaic murals.

Freestone. Fine-grained sedimentary rock that has no planes of cleavage or sedimentation along which it is likely to split.

Freeway. See Expressway.

Freezer Alarm. A bell or buzzer used in many freezers which sounds an alarm when freezer temperature rises above safe limit.

Freezer Burn. Condition applied to food which has not been properly wrapped and that has become hard, dry, and discolored.

Freeze-Up. See Frozen, 2.

Freeze-Up. 1. Formation of ice in the refrigerant control device which may stop the flow of refrigerant into the evaporator. 2. Frost formation on an evaporator which may stop the air-flow through the evaporator.

Freezing Point. Temperature at which a liquid will solidify upon removal of heat. The freezing temperature for water is 32° F. (0° C.) at atmospheric pressure.

Freezing. Change of state from liquid to solid.

Freight Elevator. A hoisting device to raise or lower cargo.

Freight-In. A merchandising account that records the transportation costs on purchases paid by the buyer.

French Curve. A transparent plastic drafter's tool for drawing miscellaneous curves.

French Door. A door with rectangular glass panes extending the full length.

French Drain. A trench filled with gravel, often with a perforated pipe on the bottom, to carry off intercepted subterranean drainage.

French Polishing. High-grade wood finish obtained by applying shellac or French varnish with a cloth pad and linseed oil as a lubricant to prevent the pad from sticking.

French Process Zinc Oxide. Zinc oxide pigment made from metallic zinc; sometimes called indirect process; see American Process Zinc Oxide.

French Window. A pair of casement windows that reaches to the floor.

Freon. Trade name for a family of synthetic chemical refrigerants manufactured by E. I. duPont de Nemours & Co., Inc.

Freon-12. Dichlorodifluoromethane.

Frequency Meter. A device used to measure vibrations in cycles per second.

Frequency. 1. Commonness of occurrence. 2. The rate of recurrence of a vibration, oscillation, or cycle; the number of repetitions in a given time, such as per second; one vibration per second is one hertz.

Fresco. The art of painting on fresh plaster with water-based pigments.

Fresh Water. Those waters having a specific conductivity less than a solution containing six thousand (6,000) parts per million of sodium chloride.

Fret. A decorative pattern of interlaced designs.

Fretsaw. A finetoothed saw with its narrow blade held under tension, used for sawing curves.

Fretwork. 1. Ornamental openwork designs in woodwork cut by a fretsaw. 2. In masonry, any ornamental openwork or work in relief.

Friable Asbestos. Any materials that contain greater than one percent asbestos, and which can be crumbled, pulverized, or reduced to powder by hand pressure; this may also include previously non-friable material which becomes broken or damaged by mechanical force.

Friction Catch. Any catch which, when engaged in a strike, is held in the engaged position by friction.

Friction Connection. Two or more structural steel members clamped together by high-strength bolts with sufficient force that the loads on the members are transmitted between them by friction along their mating surfaces.

Friction Hinge. A hinge designed to hold a door at any desired degree of opening by means of friction control incorporated in the knuckle of the hinge.

Friction Loss. Loss in efficiency due to friction.

Friction Pile. A pile calculated to carry all of its load by skin friction, neglecting any contribution of direct bearing on its point.

Friction. 1. The rubbing of one material against another. 2. The tangential surface resistance between two bodies in contact which move or tend to move with respect to each other.

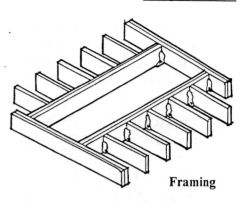

Framing

Frictional Soil. A soil, such as sand, in which there is little or no attraction between its particles, and which derives its strength from geometric interlocking of the particles; a non-cohesive soil.

Freize Carpet. A rough, nubby-textured carpet using tightly twisted yarns; a carpet with pile of uncut loops or a combination of cut and uncut loops.

Freize Yarn. A tightly twisted yarn that gives a rough, nubby appearance to the pile; in addition to use in plain colors, it is employed to form designs against plain grounds and thus gives an engraved effect.

Frieze. 1. A horizontal, often decorative, member of a cornice, set flat against a wall. 2. More broadly, any sculptured or ornamental band on a house, in a room, or on furniture.

Frigidarium. The room containing the cold bath in Roman baths.

Fringe Benefits. Benefits paid for by an employer for an employee in addition to basic wages; includes such benefits as vacations, sick time, health care, retirement, and disability insurance.

Frit. 1. The wholly or partly fused materials of which glass is made. 2. Ground-up glass used as a basis for glaze or enamel.

Fritted Glass. Float glass with crushed glass melted onto its surface, for the purpose of screening out glare, limiting solar gain, and providing various levels of opacity and color.

Fritted Glaze. A glaze in which a part or all of the fluxing constituents are pre-fused.

Froe. A cleaving tool consisting of a wedge-shaped blade mounted at right angles to the blade; used for splitting blocks of wood into shingles or barrel staves.

Frog. 1. A loop on a belt to carry a tool. 2. A mechanical device that allows the train wheels on one track to cross over an intersecting track. 3. See Panel, 8.

Front Door Lock. A lock assembly mounted in a front door.

Front End Loader. A tractor or bulldozer with a bucket which operates from the front of the vehicle.

Front End Loading. The fraudulent practice of a contractor's distorting the schedule of values so that work done early in the contract will have higher values than work done later, for the purpose of obtaining payment before it is earned; also called Unbalancing the Schedule of Values.

Front Money. The cash needed to pay all the property development costs before financing can be put in place.

Front Yard. The space between a building and the front property line.

Frost Back. Condition in which liquid refrigerant flows from the evaporator into the suction line; usually indicated by sweating or frosting of the suction line.

Frost Control, Automatic. See Automatic Frost Control.

Frost Control, Manual. See Manual Frost Control.

Frost Control Semiautomatic. See Semiautomatic Frost Control.

Frost Free Refrigerator. Refrigerated cabinet which operates with an automatic defrost during each cycle.

Frost Line. The greatest depth to which ground may be expected to freeze.

Frost Proof Tile. Tile produced for use where freezing and thawing conditions occur.

Frost. 1. A covering of minute ice crystals that form on a cold surface. 2. Frozen soil.

Frosting Type Evaporator. Refrigerating system which maintains the evaporator at frosting temperatures during all phases of cycle.

Frostline. The depth of frost penetration in soil; this depth varies in different parts of the country; footings should be placed below this depth to prevent movement.

Frothing. Foaming.

Frozen Food Case. A storage unit that freezes or maintains foods at freezing temperatures.

Frozen. 1. Water in its solid state. 2. Seized, as in machine parts, due to lack of lubrication; the term Freeze-Up is often applied to this situation.

FRP. Fiberglass Reinforced Plastic.

Fryer, Kitchen. See Kitchen Fryer.

FS. Federal Specifications.

FTI. Facing Tile Institute.

Fuel Oil. 1. Any oil that is used for fuel in engines or furnaces. 2. Kerosene or any hydrocarbon oil having a flash point not less than 100° F. (38° C.).

Fuel. A substance burned to provide heat.

Fugitive Colors. Colors which are not permanent; subject to fading.

Fulcrum. A fixed point of support about which a lever is pivoted.

Full Fillet Weld. A fillet weld whose size is equal to the thickness of the thinner member joined.

Full Floating. Mechanism construction in which a shaft is free to turn in all the parts in which it is inserted.

Full Frame. In cabinet construction, a mortised-and-tenoned frame, in which every joint is mortised and tenoned.

Full Header. A brick course composed of all headers.

Full Load Torque. In an electric motor, the maximum torque delivered without overheating.

Full Mopping. In roofing, a mopping layer that completely coats the surface with hot bitumen.

Full-Cell Process. A process for impregnating wood with preservatives or chemicals; in this process a vacuum is formed to remove air from the wood before adding the preservative.

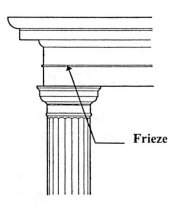

Frieze

Full-Depth Asphalt Pavement. The term Full-Depth, registered by The Asphalt Institute with the U. S. Patent Office, certifies that the pavement is one in which asphalt mixtures are employed for all courses above the subgrade or improved subgrade; a Full- Depth asphalt pavement is laid directly on the prepared subgrade.

Fuller's Earth. A type of clay mineral that lacks plasticity and is used as an adsorbent and as a filter.

Fully Adhered. A roof membrane that is fully mopped to the substrate.

Fume Hood. A ventilated laboratory enclosure for disposing of fumes.

Functional Obsolescence. The loss in value of real property due to outmoded features or poor design.

Functional. Designed to be practical; utilitarian.

Functionalism. A philosophy of design holding that form should be adapted to use, material, and structure.

Fundamental Period. The time it will take for an element or structure to stop moving and return to its original vertical position when a horizontal force is exerted on it; for example, when a horizontal force is exerted on a building causing displacement.

Fungi Resistance. The ability of any material to withstand fungi growth and their metabolic products under normal conditions of service.

Fungi. See Fungus.

Fungicide. A substance that destroys fungi or inhibits their growth.

Fungus (pl. Fungi). Any of a major group of saprophytic and parasitic lower plants that lack chlorophyll and include molds, rusts, mildews, smuts, mushrooms, and yeasts.

Funnel. A utensil consisting of a tube at the small end of a hollow cone; used for pouring liquids into a container with a small opening.

Funny Papers. A disparaging term used on the jobsite in reference to the construction drawings and specifications

Furan Grout. A two-part grout system of furan resin and furan hardener used for filling joints between quarry tile and pavers where chemical resistant properties are required; also called Furan Resin Grout.

Furan Mortar. A two-part mortar system of furan resin and furan hardener used for bonding tile to back-up material where chemical resistance of floors is important; also called Furan Resin Mortar.

Furan Plastics. Plastics based on resins in which the furan ring is an integral part of the polymer chain, made by the polymerization or polycondensation of furfural, furfuryl alcohol, or other compounds containing a furan ring, or by the reaction of these furan compounds with other compounds, the furan being in greater amount by weight.

Furan Resin Grout. See Furan Grout.

Furan Resin Mortar. See Furan Mortar.

Furan Resin. See Furan Plastics.

Furnace, Central Warm Air. Self-contained appliance designed to supply heated air through ducts to spaces remote from or adjacent to the appliance location.

Furnace. A unit which draws in cool air from an occupied space and passes the air through a heating chamber, combustion or electric, and then is returned to the occupied space; a heat system using air as the distribution fluid.

Furnish. 1. To provide what is needed. 2. To supply furniture. 3. The raw materials needed for making paper pulp.

Furnishings. Articles, especially furniture, found in the interior of a structure, generally to increase comfort or utility.

Furred Ceiling. A ceiling having spacer elements, usually furring channels, round rods, or wood strips, interposed between it and the supporting structure above.

Furring Channels. 3/4 inch cold or hot rolled steel channels used in plaster base construction.

Furring, Ceiling. See Ceiling Furring.

Furring, Channel. See Channel Furring.

Furring, Metal. See Metal Furring.

Furring, Wood. See Wood Furring.

Furring. 1. Wood or metal strips used to build out a surface such as a studded wall. 2. Narrow strips fastened to the walls and ceilings to form a straight surface upon which to lay the lath or other finish.

Furrowing. To strike a V shaped trough in a bed of mortar.

Fuse. An electrical safety device consisting of strip of fusible metal in circuit which melts when circuit is overloaded.

Fused Reducer. A pipe coupling with a larger size at one end that the other and is attached to a length of pipe by welding.

Fusel Oil. Oily liquid produced in small quantities when ethyl alcohol is produced by fermentation of grain.

Fusible Link. A safety device consisting of a metal of low melting point inserted into a machine, boiler, or linkage that will melt if the temperature rises above a certain point thereby closing or shutting off the appliance.

Fusible Plug. Plug or fitting made with a metal of a known low- melting temperature; used as safety device to release pressure.

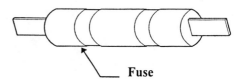

Fuse

Fusible Switch. An electric switch that has a fusible link in the wiring for an electric circuit to break the circuit by melting in the event of an overload.

Fusion. 1. The liquid or plastic state of a material caused by heat. 2. In welding, the process of melting to form a union; usually the result of interaction of two or more materials.

Fuzzing. A temporary condition on new carpet consisting of irregular fuzzing appearance caused by slack yarn twist, fibers snagging, or breaking of yarn; can be remedied by spot shearing.

G

GA. Gypsum Association.

GAAP. Generally Accepted Accounting Principles.

Gabion. A metal or wire cage filled with ballast or stone, used in large scale retaining walls.

Gable End Rafter. The last rafter system installed at the gable end of a building.

Gable End. An end wall having a gable.

Gable Louver. A louver system installed at the gable end of a building.

Gable Roof. A roof consisting of two opposite sloping planes that intersect at a level ridge.

Gable Truss. The truss installed at the gable end of a building.

Gable Vent. A screened of louvered, opening in a gable, used for venting from an attic.

Gable. The triangular wall beneath the end of a gable roof; the vertical triangular end of a building from the eaves to the apex of the roof.

Gaffer. A lighting electrician on a motion picture or television production unit.

Gage. A tool used by carpenters to strike a line parallel to the edge of a board.

Gain. 1. A notch or mortise, as in a beam or a wall, for a joist, girder or similar member. 2. The notch to set a hinge in a door or frame.

Gallery. 1. A roofed outdoor promenade or balcony. 2. A room or building for displaying art or artifacts.

Galvanic Action. See Galvanic Corrosion.

Galvanic Corrosion. Corrosion in a metal caused by a galvanic cell created by dissimilar metals in close proximity accompanied by an electrolyte, often atmospheric moisture; also called Galvanic Action.

Galvanized Accessory. A building product such as a door, window, skylight, handrail, grating, or ventilator, that has been galvanized.

Galvanized Iron. Zinc coated iron or steel.

Galvanized Mesh. Mesh screening that has been galvanized; used as wire lath, reinforcing, or fencing.

Galvanized Metal Joist Hanger. Formed steel component used to support end of load bearing joists and transmit loads to another joist or beam, galvanized.

Galvanized Pipe. A steel pipe galvanized to prevent rusting.

Galvanized. Zinc plated for corrosion protection achieved by hot dipping into molten zinc or by electrolysis.

Galvanizing Repair. Repair of damaged galvanized surfaces by application of zinc rich paint.

Gambrel Roof. A type of gable roof where the roof planes on each side of the ridge each have two pitches, the lower being steeper and longer than the upper; also called a Curb Roof.

Gang Box. Electrical rough-in box constructed of metal or hard plastic, to provide for two or more outlets or switches.

Gang Nail. A sheet metal plate that has numerous sharp tabs that act as nails to hold the sheet metal plate over a joint in a light wood truss.

Gangway. 1. A passageway. 2. A temporary passage over sometimes cleated planks on a construction site.

Gap. 1. An empty space between two objects. 2. A break in continuity.

Gap-Graded Aggregate. Aggregate containing particles of both large and small sizes, in which particles of certain intermediate sizes are wholly or substantially absent.

Garage Door Opener. An electrically driven mechanical device to raise and lower a garage door, usually activated by a radio control in the car or by a wall switch.

Garage. A room or building for storage of automobiles.

Garbage Disposal Unit. An electrically powered device, usually located in a kitchen sink, to grind household garbage and discharge it to the sewage system.

Garbage Handling. The act or process of the transport and discharge of waste materials.

Garden Wall Bond. A name that is given to any brick bond that is especially suitable for a wall two tiers thick, consisting of one header to three stretchers in every course.

Garnet Paper. Abrasive which is reddish in color, hard and sharp; comes from same source as semi-precious jewel by that name; more expensive than flint paper but lasts longer.

Garret. An attic room or attic.

Gas Checking. Fine checking; wrinkling, frosting under certain drying conditions; said to be caused by rapid oxygen absorption or by impurities in the air.

Gas Cock. A fuel gas valve.

Gas Concrete. See Cellular Concrete.

Gas Furnace. A heating system that burns gas to produce heat.

Gas Log. A hollow perforated imitation log used as a gas burner in a fireplace.

Gas Metal Arc Welding (GMAW). An arc welding process wherein coalescence is produced by heating with an arc between a continuous filler metal (consumable) electrode and the work. Shielding is obtained entirely from an externally supplied gas, or gas mixture. Some methods of this process are called MIG or CO welding.

Gas Meter. A mechanical device for measuring and recording the volume of gas passing a given point.

Gas Pocket. A weld cavity caused by entrapped gas.

Gas Pump. A mechanical device to move gas or gasoline from one location to another.

Gas Tungsten Arc Welding (GTAW). An arc-welding process wherein coalescence is produced by heating with an arc between a single tungsten (nonconsumable) electrode and the work; shielding is obtained from a gas or gas mixture; pressure may or may not be used and filler metal may or may not be used; also called TIG welding.

Gas Valve. Device in a pipeline for starting, stopping or regulating flow of gas.

Gas Welding. A welding process wherein coalescence is produced by heating with a gas flame or flames, with or without the application of pressure, and with or without the use of filler metal.

Gas, Noncondensable. See Noncondensable Gas.

Gas. 1. Any airlike substance which moves freely to fill the available space regardless of the quantity. 2. A combustible gaseous mixture; fuel gas. 3. Vapor; one of the three states of matter; compare with Liquid and Solid.

Gasholder. A large cylindrical tank for storing gas under pressure.

Gasket, Foam. See Foam Gasket.

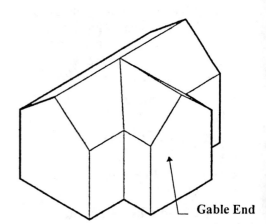

Gable End

Gasket. Sealing material at a crack; a sheet or ring of rubber, vinyl, or similar material shaped to seal the junction of metal surfaces.

Gas-Tight Manhole Cover and Frame. Air tight metal cover and frame over vertical access shaft from grade level down to underground utility or chamber.

Gasworks. A place where gas is manufactured and processed.

Gate Post. The vertical member in a fence system upon which the gate hinges are mounted.

Gate Valve. A valve utilizing a wedge-shaped gate, which allows fluid flow when the gate is lifted from the seat.

Gate, Fence. See Fence Gate.

Gate, Swing. See Swing Gate.

Gateway. 1. A gate. 2. An opening for a gate.

Gauge Port. Opening or connection provided for a service technician to install a gauge.

Gauge Pressure. The pressure indicated on a pressure gauge, measured in pounds per square inch, and indicated as PSIG; compare with Absolute Pressure, indicated as PSIA.

Gauge, Compound. See Compound Gauge.

Gauge, High Pressure. See High Pressure Gauge.

Gauge, Low Pressure. See Low Pressure Gauge.

Gauge, Manifold. See Manifold Gauge.

Gauge, Vacuum. See Vacuum Gauge.

Gauge. 1. Measurement according to some standard system. 2. The thickness of sheet metal. 3. An instrument for measuring the size or quantity. 4. The distance between tufts across the width of knitted and tufted carpets, expressed in fractions of an inch.

Gauged Brick. 1. Brick which have been sorted or otherwise produced to accurate dimensions. 2. A brick that has been rubbed on an abrasive stone to reduce it to a trapezoidal shape for use in an arch.

Gauging Plaster. A gypsum plaster formulated for use in combination with finish lime in finish coat plaster.

Gauging Trowel. The gauging trowel is larger than the pointing trowel but smaller than the buttering trowel.

Gauging. 1. Another cementitious material (usually calcined gypsum, Keene's cement, or portland cement) added to lime putty to provide and control set. 2. The act of adding gauging material.

GB. Gigabyte. 1,024 megabytes.

Gel. A semi-solid colloidal suspension or jell; a solid dispersed in a liquid; a jelly-like substance.

Gelatin. A product of the packing house; a glutinous material obtained from animal tissues by boiling; which can be cast into a semi-rigid mould; on account of its flexibility, it is particularly adaptable to moulds containing undercuts.

General and Administrative Expenses. Shown in the income statement as all items of expense of a general nature which cannot be specifically attributed to individual construction projects.

General Conditions. A written document, supplementing the specifications, which indicates and defines areas of the project relating to other than specific building trades.

General Contract. The agreement covering the work of general construction done by a prime contractor.

General Contractor. A prime contractor for general construction.

General Drawing. 1. A drawing showing elevations, plans and cross sections of the structure and the main dimensions. 2. A drawing showing the borings for substructure.

General Journal. The record used for recording business transactions not recorded in specific journals.

General Ledger. The record containing the summarization of all transactions of a business.

General Partner. A member of a partnership with unlimited management authority and liability.

General Partnership. Embodies the financial and personal resources of two or more persons who share in the owning and running of the business.

General Requirements. See Job Requirements.

Generator. A machine by which mechanical energy is changed into electrical energy.

Gentrification. The purchase and rehabilitation by the middle-classes of run-down properties.

Geodesic Dome. A dome constructed according to the patents of R. Buckminster Fuller, in which the pattern of surface divisions is always a function of an entire sphere; commonly constructed with prefabricated structural triangles linked together in a mosaic to create the domed shape.

Geodesic. The shortest line between two points on a given surface.

Geoid. The sphere that defines the earth at the location of mean sea level.

Geomancy. Divination by means of lines or geographic features; feng shui.

Geometry. A branch of mathematics that deals with the measurement, properties, and relationships of points, lines, surfaces, and solids.

Geotechnical Engineer. An engineer who specializes in rock and soil mechanics, groundwater, and foundations.

Geotechnical Investigation. The process of boring, sampling, and testing the soil at various depths to enable the geotechnical engineer to discover and analyze characteristics of the soil; also called Subsurface Investigation, Foundation Investigation, Soil Investigation, or Soil Test; the test results and engineer's recommendations are explained and summarized in a Soil Report.

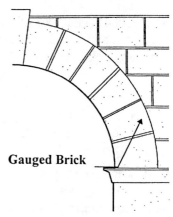

Gauged Brick

Geotextile. Synthetic fabrics used to separate backfill materials for proper drainage; used in high retaining walls and landscape design.

Gesso. A plaster surface composed of gypsum plaster, whiting, and glue, used as a base for decorative painting.

GFI Breaker. Ground Fault Interrupter Breaker; supplies power, as any breaker does, but also monitors the amount of incoming and outgoing current; whenever the entering current does not equal the leaving current, indicating current leakage, the GFI instantly opens the circuit; a faster overcurrent protection device than either a fuse or circuit breaker.

GFRC. Glass-Fiber-Reinforced Concrete.

Ghosting. In painting, patches of lighter color showing in dry coat; a coating with a skippy appearance.

GI. Galvanized Iron.

Gild. To apply a thin covering of gold.

Gilding. Process of obtaining a finish by using metal leaf.

Gillmore Needle. A device used in determining time of setting of hydraulic cement.

Gilsonite. A black, coal-like substance obtained from mines and used in the manufacture of black asphaltum varnish.

Gin Pole. A simple derrick consisting of a pole held nearly vertical by guy cables, a pulley at the top of the pole, and a hoisting rope.

Girder. A beam that supports other beams; a very large beam, especially one that is built up from smaller elements; a timber beam used to support wall beams or joists.

Girt. A beam that supports wall cladding between columns.

Gland. A device for preventing leakage of fluid past a joint in machinery, as at a bearing.

Glare. Excessive brightness in the field of view.

Glass Bead. A narrow strip of plastic, metal, or wood used to hold glass in a sash; removable trim that holds glass in place.

Glass Block. A hollow masonry unit made of glass.

Glass Cutter. A tool for cutting glass.

Glass Door. A metal or wood framed door manufactured with heat-strengthened or tempered glass.

Glass Felt. A felt sheet in which glass fibers are bonded into the felt sheet with resin; they are suitable for impregnation and coating; used in the manufacture and coating of bituminous waterproofing materials, roof membranes, and shingles.

Glass Fiber. Glass in the form of fine fibers used in fabrics.

Glass Mat. A thin mat of glass fibers with or without a binder.

Glass Mosaic Tiles. Tiles made of glass, usually in sizes not over two inches square and 1/4 inch thick, mounted on sheets of paper, usually twelve inches square.

Glass Mullion System. A method of constructing a large glazed area by stiffening the sheets of glass with perpendicular glass ribs.

Glass Pipe. Glass and glass-lined pipe used in process piping and in laboratories.

Glass Tempering Association (GTA). White Lakes Professional Building, 3310 SW Harrison, Topeka, Kansas 66611, (913) 266-7064.

Glass Wool. Glass in the form of fine fibers used in insulation.

Glass, Wire. See Wire Glass.

Glass. A hard brittle usually transparent, translucent, or shiny substance, made by fusing sand with soda and lime.

Glass-Fiber-Reinforced Concrete (GFRC). Concrete with a strengthening admixture of short alkali-resistant glass fiber.

Glassware Washer. An apparatus for the cleaning of items made of glass.

Glaze Coat. 1. The top layer of asphalt used on a smooth-surfaced built-up roof assembly. 2. A thin protective layer or coating of bitumen that is applied to the lower plies or the top ply of a built-up membrane when the top pouring and aggregate surfacing are delayed.

Glaze Fit. The stress relationship between the glaze and body of a fired ceramic product.

Glaze. A ceramic coating matured to the glassy state on a formed ceramic article; also refers to the material or mixture from which the coating is made.

Glazed Block. Concrete blocks with a surface produced by fusing it with a glazing material.

Glazed Brick. Brick or tile with a surface produced by fusing it with a glazing material.

Glazed Bullnose. A ceramic trim tile with a convex radius on one edge that has been given a glassy or glossy surface.

Glazed Ceramic Mosaic Tile. A glazed tile with a body that is suitable for interior use and which is usually nonvitreous and is not required to withstand excessive impact or be subject to freezing.

Glazed Floor. A floor covering that has had an application of a nearly transparent coating that enhances and protects the coat underneath.

Glazed Interior Tile. A glazed tile with a body that is suitable for interior use and which is usually non-vitreous, and is not required or expected to withstand excessive impact or be subject to freezing and thawing conditions.

Glazed Paver Tile. See Paver.

Glazed Quarry Tile. See Quarry Tile.

Glazed Structural Unit, GSU. A solid or hollow unit with a surface of applied smooth glossy nature, e.g. a tile with a fired glaze finish.

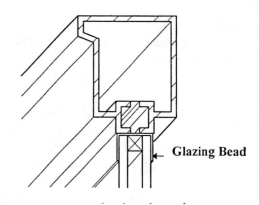
Glazing Bead

Glazed Tile, Extra Duty Glaze. Tile with a durable glaze that is suitable for floors and all other surfaces.

Glazed Tile. Tile with a fused impervious facial finish composed of ceramic materials, fused into the body of the tile which may be a non-vitreous, semi-vitreous, vitreous, or impervious body; the glazed surface may be clear, white, or colored.

Glazed Wall Tile. A wall tile with an impervious, glossy finish.

Glazier. One whose trade is setting glass.

Glazier's point. A small triangular sheetmetal nail to keep glass in place in a wood sash before puttying.

Glazing Accessory. Implements or devices needed in the glazing trade.

Glazing Bead. 1. A strip of wood or metal for holding a pane of glass in its frame or sash. 2. A sealant after application in a joint irrespective of the method of application.

Glazing Compound. Any of a number of types of mastic or putty used to bed small lights of glass in a frame.

Glazing. 1. The trade of installing glass; the trade practiced by glaziers. 2. A transparent or translucent color applied to modify the effect of a painted surface. 3. In plastering, a condition created by the fines of a machine-dash texture plaster traveling to the surface and producing a flattened texture and shine or discoloration; this may be caused by the basecoat being too wet or the acoustical mortar being too moist; glazing occurs in hand application when mortar being worked is excessively wet.

Glitter. A reflective material such as glass, diamond dust, or small pieces of variously colored aluminum foil projected into the surface of wet plaster or paint as a decorative treatment.

Globe Valve. A spherically shaped valve body which controls the flow of water with a compression disc which is opened and closed by means of a stem and mates with a ground seat to stop water flow.

Globular Transfer. A mode of metal transfer in gas metal-arc welding in which the consumable electrode is transferred across the arc in large droplets.

Gloss Enamel. A finishing material made of varnish and pigments; such an enamel forms a hard coating with a smooth surface and high gloss.

Gloss Oil. A varnish composed primarily of limed rosin and petroleum thinner.

Gloss Retention. Ability to retain original sheen.

Gloss. A shiny, lustrous finish which reflects light. The term also refers to paint or enamel that dries to a high sheen or luster, usually with a hard, smooth coat.

Glost Fire. The process of kiln-firing bisque ware to which glaze has been applied.

Glovebag. A polyethylene or polyvinyl chloride bag-like enclosure with integral gloves affixed around an asbestos-containing source so that the material may be removed while minimizing release of airborne fibers to the surrounding atmosphere.

Gloves. Protective covering for the hands to eliminate injury and improve grip.

Gluability. The ease or difficulty in bonding a material with adhesive; bondability.

Glue Block. A small piece of wood used to strengthen and support two pieces of wood joined at an angle; see Corner Block.

Glue Built-Up Members. Structural elements, the sections of which are composed of built-up lumber, wood structural panels, or wood structural panels in combination with lumber; all parts bonded together with adhesives.

Glue Gun. A hand tool for the application of bulk or cartridge- type adhesives.

Glue Joint. A joint held together with glue.

Glue Laminated Member. Assemblies of wood laminations bonded together with adhesive so that longitudinal grain of laminations are approximately parallel; Architectural Grade commonly has a better degree of external finish and would be exposed in the finished building; Industrial grade would not be finished to the same degree as Architectural Grade.

Glue Laminated Timber. A timber made up of a large number of small strips of wood glued together.

Glue. Adhesive; commonly used in joining wood parts.

Glulam. Glue laminated timber.

Glycerine. Glycerol.

Glycerol. A sweet viscous liquid alcohol, almost colorless and odorless; used extensively in the production of alkyd resins and ester gum.

Glycol. Ethylene glycol, a thick chemical alcohol, used as an antifreeze.

Glyph. 1. A sculptured character or symbol. 2. A vertical groove as on a Doric frieze.

GMP. Guaranteed Maximum Price.

Gneiss. A usually coarse grained metamorphic rock foliated by mineral layers, principally of feldspar, quartz, and ferromagnesian minerals.

Gnomon. 1. The rod, pin, or style on a sundial that casts its shadow indicating the time. 2. The L shaped remainder of a parallelogram after the removal of a smaller similar parallelogram containing one of its corners.

Goggles. Protective covering for the eyes.

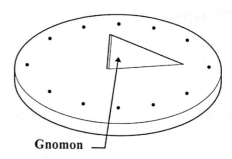

Gnomon

Going Concern Concept. The accounting assumption that a business will continue to operate for an indefinite period.

Gold Leaf. Gold beaten into a very thin sheet, ordinarily 4 to 5 millionths of an inch in thickness.

Gold Paint. Mixture of bronze powder and bronzing liquid.

Gold. A yellow malleable ductile high density precious metallic element resistant to chemical reaction, occurring naturally in quartz veins and gravel.

Golf Shelter. A structure placed on a golf course to be used for the safety of golfers in inclement weather.

Gondola. 1. A railroad car with a flat bottom, no top, and fixed sides for shipping heavy construction materials and equipment. 2. An island of shelves to display goods in a supermarket.

Goodwill. The value of a business over the book value or agreed value of its net assets based on its experience and reputation.

Goose Neck. A blacksmith's tool used to handle hot metals.

Gothic Arch. A pointed arch, with a joint at the apex rather than a keystone.

Gothic. An architectural style of Western Europe of the 12-16th centuries featuring pointed arches, vertical piers, buttresses, and vaults.

Gouge. 1. A chisel with a concave blade. 2. To overcharge.

Grab Bar. Metal or plastic bar attached to a bathroom wall, above a bathtub, near a toilet, or in a shower, to be used as a hand hold. See Flash Set.

Graben. A long, narrow trough bounded by one or more parallel normal faults; also called a Rift Valley.

Grad. A metric unit of circular measurement; 400 grads equals a full circle equals 360 degrees.

Gradall. A hydraulic, wheel-mounted backhoe often used with a wide bucket for dressing earth slopes.

Gradation. The sizing of granular materials. for concrete materials, usually expressed in terms of cumulative percentages larger or smaller than each of a series of sieve openings or the percentages between certain ranges of sieve openings.

Grade Beam Formwork. The system of support for freshly poured or placed concrete in a reinforced grade beam.

Grade Beam Reinforcing. The placing of metal or steel bars in a concrete grade beam.

Grade Beam. A reinforced concrete beam that transmits the load from a bearing wall into spaced foundations such as pile caps or caissons.

Grade Mark. The stamp on lumber indicating its grade according to the rules of the lumber grading bureau that performed the inspection.

Grade MW Brick. A grade of brick intended to be used for exterior wall surfaces where water permeability is not expected and only slight frost action; MW means moderate weather.

Grade NW Brick. A grade of brick intended to be used for back up withes or interior exposure; NW means no weather.

Grade Stake. A stake indicating the amount of cut or fill required to bring the ground to a specified level.

Grade SW Brick. A grade of brick intended to be used for exterior wall surfaces and where water permeability and frost action may be expected; SW means severe weather.

Grade. 1. The classification of lumber in regard to strength and utility in accordance with the rules of an approved lumber grading agency. 2. A gradient. 3. A predetermined degree of slope that a finished floor or ramped surface should have. 4. The horizontal ground level of a building or structure. 5. To level off to a smooth horizontal or sloping surface. 6. The slope or pitch, known as the fall, usually expressed in drainage piping as a fraction of an inch per foot. 7. The elevation of a real or planned surface or structure.

Graded Sand. A mixture of sand in which the granules consist of fine, medium, and coarse sizes.

Gradient. 1. A stretch of road or railway that slopes from the horizontal. 2. The amount of such a slope. 3. The rate of rise or fall of temperature or pressure in passing from one region to another.

Grading. See Grade.

Gradiometer. A geophysical instrument that measures the vertical component of the earth's magnetic field; a very sensitive instrument, it can detect minor variations due to buried features.

Grain Alcohol. Ethyl alcohol made from grain by distillation.

Grain Raising. Causing short fibers on surface of bare wood to stand up by applying water; liquids that do not raise the grain are known as non-grain-raising.

Grain. 1. The direction, size, arrangement, appearance, or quality of the fibers in wood or lumber. 2. The weight unit equal to 1/7000 pound; used in measuring atmospheric moisture content.

Graining. Simulating the grain of wood by using paint.

Grainness. Roughness of a protective film resembling grains of sand.

Gram Calorie. See Calorie, 1.

Gram. A metric unit of mass and weight equal to 1/1000 of a kilogram and nearly equal to one cubic centimeter of water at its maximum density.

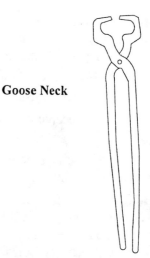

Goose Neck

Grand Master Key. A key that operates locks in several groups, each of which has its own master key.

Grandstand Seat. A seat in a stand for spectators at a racecourse or stadium.

Grandstand. A stand or bleachers for spectators, usually for a sporting event.

Granite Block. A masonry unit consisting of a very hard natural igneous rock used for its firmness and endurance.

Granite Curb. Street curbing that is manufactured from granite blocks.

Granite Paver. A type of igneous stone that is harder and more durable than regular bricks, used for driveways and patios for use with or without mortar.

Granite Veneer. A masonry facing attached to a main structure, made from granite.

Granite. Igneous rock with visible crystals of quartz and feldspar.

Granolithic Topping. A covering layer consisting of an artificial stone of crushed granite and cement.

Grant Deed. A deed by which ownership in real property is transferred.

Grantee. A person receiving an interest in property.

Grantor. A person selling, granting, or giving up an interest in property.

Granular Insulation. Insulating material made of granular vermiculite or perlite or beaded styrene, placed in hollow, usually masonry, construction.

Granular Waterproofing. Waterproofing material made of granular vermiculite or perlite or beaded styrene.

Granules. The size-graded mineral particles that are imbedded into the top surface of factory made mineral-surfaced asphalt shingles and cap sheet.

Graph. A diagram showing the relation between variable quantities, usually two variables, each measured along one of a pair of axes at right angles.

Graphic. Of or relating to the visual or descriptive arts, like writing, drawing, or photography.

Graphics. The products of the graphic arts.

Graphite. Black pigment consisting mostly of carbon obtained from natural deposits or produced from coke in an electric furnace.

Grass Cloth Tile. Tile made from woven vegetable fibers that are laminated.

Grass Cloth. A wall covering manufactured from vegetable fibers, woven and laminated to paper for backing.

Grass. Green low growing vegetation widely used as ground cover; see Lawn.

Grate Cutting. The act or process of snipping away a grate with hand tools, or cutting away a grate by acetylene torch or power tools.

Grate Painting. The act or process of applying paint to seal, prevent rust, or add color to a grating.

Grate, Metal. See Metal Grating.

Grating Frame. Frame, usually metal, to contain floor grating and provide means to anchor to floor construction.

Grating Tread. Horizontal surface of step made of steel grating.

Grating, Metal. See Metal Grating.

Grating. Open grid of metal bars structurally formed.

Gravel Fill. Gravel layered at the bottom of an excavation to insure drainage of water.

Gravel Mulch. A protective covering of rock particles, used on the ground to reduce evaporation, prevent erosion, and control weeds.

Gravel Roof. A roof composed of layers of roofing felt for waterproofing, then sealed with tar or pitch and covered with a layer of gravel to assist in protection from wear and the sun.

Gravel Stop. A metal flange or strip with a vertical lip placed around a built-up roof to prevent loose gravel from falling off the roof; it is also used to present a finished edge detail in built-up roofing.

Gravel. Small rock particles resulting from natural disintegration and weathering such as river gravel; mechanically crushed stone.

Gravity Furnace. An appliance, with a heat exchanger, that supplies hot air that is distributed by gravity.

Gravity Ventilator. A device installed in an opening in a room or building which is activated by air passing through to remove stale air and replace it with fresh air.

Gravity Wall. A retaining wall which depends solely on its weight to resist lateral forces of retained earth.

Gravity. 1. The force that attracts a body to the center of the earth. 2. The quality of having weight.

Gray Blast Cleaning. Commercial blast.

Gray Paper. The unsized, uncalendered paper used on the back side of regular gypsum board products and as the face and back paper on backing board products.

Grayish. Lacking in intensity of color.

Grease Trap. A device to trap and retain the grease content of wastewater and sewage.

Great Calorie. See Calorie, 2.

Green Architecture. Design that embraces energy conservation and minimal use of fossil fuels and nonrenewable resources.

Green Board. A gypsum board having a tinted face paper usually light green (or blue) to distinguish special board types.

Green Brickwork. Brickwork in which the mortar has not yet set.

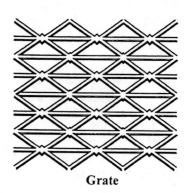

Grate

Green Concrete. Concrete which has set but which has not appreciably cured or hardened to its design strength.

Green Design. See Green Architecture.

Green Lake. A pigment; a mixture of Prussian blue and yellow lake, sold under various trade names.

Green Masonry. In masonry, a molded clay unit before it has been burned in preparation for building purposes; an uncured concrete masonry unit.

Green Plaster. Newly applied plaster that has not dried.

Green Room. A room in a theater for actors who are off-stage.

Green Wood. Freshly sawed or undried wood; twenty percent moisture or more.

Greenfield Site. A building site that has never before been built upon; compare with Brownfield Site.

Greenhouse Effect. The trapping of the sun's warmth in the lower atmosphere of the earth caused by an increase in carbon dioxide, which is more transparent to solar radiation from the earth.

Greenhouse. A glassed enclosure used for cultivation or protection of tender plants.

Gremlin. A mischievous creature said to be responsible for malfunctions of machinery and equipment.

Grid. 1. A network of uniformly spaced horizontal and perpendicular lines, as for locating and referring to points on a drawing. 2. A set of surveyor's closely spaced reference lines laid out at right angles, with elevations taken at line intersections.

Griddle, Kitchen. A flat surface upon which food is cooked by dry heat.

Grillage. A framework composed of main runner channels and furring channels to support ceilings.

Grille. Component of the ventilation system that promotes air circulation in the occupied space by providing a means to return air; a metal screen or grating that allows for the circulation of air.

Grin. Condition where the backing shows through sparsely spaced pile tufts; carpets may be grinned (bent back) deliberately to reveal the carpet construction.

Grind Gauge, Hegeman. See Hegeman Grind Gauge.

Grinder, Electric. See Electric Grinder.

Grit. An abrasive obtained from slag and other materials.

Grommet. Plastic, metal or rubber doughnut-shaped protector which lines holes where wires or tubing pass through wiring boxes or panels.

Groove Weld. A weld made in the groove between two members to be joined; the standard types of groove welds are. Square, Single-Vee, Single-Bevel, Single-U, Single-J, Double-Vee, Double-Bevel, Double-U, and Double-J.

Groove. 1. The opening provided for a groove weld. 2. See Dado. 3. See Housing. 4. See Tongue and Groove.

Gross Building Area. The total building area, without deductions, expressed in square feet.

Gross Cross Sectional Area. Area measured by overall dimension, including voids as area.

Gross Errors. Errors that are easily detected as they are large in proportion to the context in which they arise.

Gross Income. The total income produced by a real estate investment, without deduction for loan repayment or expenses.

Gross Leasable Area. The total area of a building that is leasable to tenants; the gross building area less the unrentable areas.

Gross Multiplier. A rule-of-thumb method of appraising the value of income property by multiplying the gross annual income by a multiplier; this is an unreliable method, used primarily on residential property.

Gross Profit. The excess of net sales over the cost of goods sold.

Gross. 1. An overall total exclusive of deductions. 2. A quantity of 12 dozen; 144.

Ground Acceleration. Acceleration of the ground due to earthquake forces.

Ground Bus. In the main electrical service panel, where the neutral service wire attaches and is linked to the earth by the ground wire.

Ground Coat. In painting, the coating material which is applied before the graining colors or glazing coat.

Ground Coil. Heat exchanger buried in the ground; may be used either as an evaporator or as a condenser.

Ground Color. The background color against which the top colors create the pattern or figure in the design.

Ground Cover. Low lying planting materials that cover the ground, such as grass and ivy. 2. See Soil Cover.

Ground Displacement. The distance which ground moves from its original position during an earthquake.

Ground Failure. A situation in which the ground does not hold together such as during a landslide, mud flow, or liquefaction.

Ground Fault Circuit Interrupter (GFCI). Supplies power, as any receptacle does, but also monitors the amount of incoming and outgoing current; whenever the entering current does not equal the leaving current, indicating current leakage, the GFCI instantly opens the circuit. A faster overcurrent protection device than either a fuse or circuit breaker.

Ground Hydrant. A water hydrant for the use in fighting fires, installed in the ground.

Ground Lease. A lease of the land.

Ground Movement. A general term; includes all aspects of motion. acceleration, velocity, displacement.

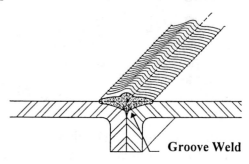

Groove Weld

Ground Paper. Wallpaper coated with an overall background color.

Ground Rent. Rent consideration under a ground lease.

Ground Rod. A rod acting as a connecting body between electrical equipment and the ground.

Ground Short Circuit. Fault in an electrical circuit allowing electricity to flow into the metal parts of a mechanism.

Ground Water. Water in the subsoil.

Ground Wire. 1. Electrical wire which will safely conduct electricity from a structure into the ground. 2. A wire attached to dissipate electrostatic charge in airless spraying.

Ground. 1. An electrical connection with the earth. 2. A conducting connection whether intentional or accidental, between an electrical circuit or equipment and earth, or to some conducting body which serves in place of the earth. 3. A strip of wood assisting the plasterer in making a straight wall and in giving a place to which the finish trim of the room may be nailed. 4. A strip attached to a wall or ceiling to establish the level to which plaster should be applied.

Grounded Conductor. A system or circuit conductor which is intentionally grounded.

Grounded. Connected to the electrical circuit grounding conductor where one exists; or in other cases, connected to a grounding electrode meeting the required characteristics of the installation.

Grounding Clip. A metal clipping device which is used to fasten a grounding wire to a pipe or fixture.

Grounding Conductor, Common Main. In a grounded system, the conductor that connects both the circuit grounded conductor and the equipment grounding conductor to the grounding electrode.

Grounding Conductor, Equipment. A conductor used to connect the equipment being grounded to the service equipment enclosure.

Grounding Conductor, Main. In an ungrounded system, the conductor connecting the equipment grounding conductor at the service to the grounding electrode.

Grounding Conductor. A conductor that is used to ground electrical equipment or a grounded circuit to a grounding electrode.

Grounding Locknut. A locknut serving as a connecting body between electrical equipment and a ground wire.

Grounding Plug. A type of electrical plug which grounds a device.

Grounding. 1. The act or process of making an electrical connection with the earth. 2. A large conduction body, as the earth, used as a common return for an electric circuit.

Groundscraper. A wide spreading building of offices or apartments; compare with Skyscraper.

Grout Bond. The adhesion to, and the interlocking of grout with the masonry units and the reinforcement.

Grout Lift. An increment of height that grout is poured.

Grout Mix. The amount of each material specified, portland cement, fine aggregate, admixture, and water.

Grout Pour. The total height of masonry wall to be grouted prior to the erection of additional masonry. A grout pour will consist of one or more grout lifts.

Grout Saw. A saw-toothed carbide steel blade mounted on a wooden handle, used to remove old tile grout, also used in patching work; care should be used, as adjacent tiles can be easily damage; the carbide steel blade is brittle and will shatter if it is dropped or abused; the spring steel tip in front of the saw blade is used for scraping grout out of corners where the saw blade cannot reach.

Grout Scrubbing Pad. A non-scratch nylon pad impregnated with abrasive used for cleaning grout from the surface of tile.

Grout, Block. See Block Grout.

Grout, Epoxy. See Epoxy Grout.

Grout, Masonry. See Masonry Grout.

Grout. A rich or strong cementitious or chemically setting mix used for filling masonry or tile joints and voids; a mixture of portland cement, aggregates, and water, which can be poured or pumped into cavities in concrete or masonry; also used for filling hollow metal door frames.

Grout-Aid. A proprietary admix to reduce the shrinkage of grout, as placed by the high-lift grouting method.

Grouted Hollow-Unit Masonry. That form of grouted masonry construction in which certain designated cells of hollow units are continuously filled with grout.

Grouted Multi-Wythe Masonry. That form of grouted masonry construction in which the space between the wythes is solidly or periodically filled with grout.

Grouting. Process of filling tile joints, masonry block or cells, or any masonry type product with grout.

Growth Ring. The layer of wood growth put on a tree during a single growing season; annual growth ring.

Grubbing. The act or process of clearing and digging up roots and stumps.

GSU. Glazed Structural Unit.

GTA. Glass Tempering Association.

GTAW. Gas Tungsten-Arc Welding.

Guarantee. 1. Written or implied assurances for a specific part of the project, or for the project as a total. 2. An undertaking or document stating that a thing will or will not happen.

Guaranteed Maximum Price (GMP). An amount stipulated in a construction contract as the maximum sum payable by the owner to the contractor for the work specified.

Guarantor. One who gives a guarantee.

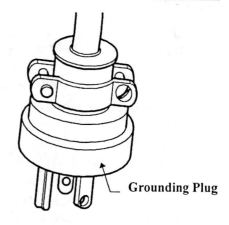

Grounding Plug

Guarded. Covered, shielded, fenced, enclosed, or otherwise protected by suitable means to remove the liability of dangerous contact or approach by persons or objects to a point of danger.

Guardrail Removal. The act or process of removing a protective or safety device.

Guardrail. 1. A safety railing used as a barrier to prevent encroachment or accidental falling from heights. 2. In street or highway construction, a barrier to keep vehicles in their lanes. 3. A device for protecting a machine part or the operator of a machine.

Guest Room. Any room or rooms occupied, or intended or designed to be occupied by a guest for sleeping purposes.

Guest. Any person hiring or occupying a room for living or sleeping purposes.

Guesstimate. An estimate based on a mixture of guesswork and calculation.

Guide Coat. In painting, a coat similar in composition to the finish or color coat, but of a different color to help obtain complete coverage.

Gum Arabic. The dry gummy exudation of Acacia Senegal, a white powdered resin, used in adhesives, cold water paint, and in show card colors.

Gum Turpentine. Oleoresinous material obtained from living pine trees; gum turpentine, when distilled, provides gum rosin and gum spirits of turpentine.

Gum. A nonvolatile viscous plant exudate which either dissolves or swells up in contact with water; many substances referred to as gums, such as pine and spruce gum, are actually oleoresins; used in making varnishes and paints.

Gun Distance. Space between tip of spray gun and work.

Gunite. A concrete material applied by pumping through a hose; also called dry-mix Shotcrete.

Gusset Plate. A flat steel plate to which the chords of a truss are connected at a joint; a stiffener plate.

Gutta-Percha. A tough plastic substance obtained from the latex of various Malaysian trees, resembles rubber but contains more resin, is used as insulation and other uses.

Gutter Nail. A long nail used for hanging rainwater gutters on light wood framed residences.

Gutter Removal. The act or process of tearing down and carrying away an old gutter system.

Gutter Strap. Metal band used to support the gutter.

Gutter. A channel to collect rainwater and snow melt at the eaves of a roof; a shallow channel constructed of steel, copper, aluminum, or plastic compounded with other materials to increase strength and wearing or rust resistance, positioned just below and along the eaves of a building for collecting and diverting water from a roof.

Guy Cable. A wire used to secure a tall exterior mast, antenna, or other structure in place.

Guy Rod. A metal rod with a cable or rope attached, leading to an object to support and stabilize it.

Gym Curtain. A vertical piece of fabric used to divide a gymnasium into separate areas.

Gym Floor. A level surface, usually made of a resilient hardwood, used for events and sports that take place in a gymnasium.

Gym Locker. A lockable compartment used to store personal effects while a person is using gym facilities.

Gym Scoreboard. A large board found in a gymnasium, which displays the score and often other information of a game or match.

Gym. Gymnasium.

Gymnasium Floor Anchor Channel. Metal channel used to anchor wood gymnasium flooring.

Gymnasium Underfloor Insulation. Insulation, usually board insulation, located between sleepers below gymnasium wood flooring to dampen the sound drum effect.

Gymnasium. A room or building used for indoor sports.

Gypsite. An earthy deposit found at or near the surface of the ground, consisting of finely crystalline gypsum mixed with loam, clay, sand, and humus; gypsum content generally ranges from 60% to more than 90%.

Gypsum Association (GA). 810 First Street, NE, #510, Washington, DC 20002, (202) 289-5440.

Gypsum Backing Board. A 1/4 (6.35 mm)to 5/8 inch (15.875 mm) thick gypsum board for use as a backing for gypsum wallboard, acoustical tile or other dry cladding.

Gypsum Block. A cast gypsum building unit; also called a Gypsum Tile.

Gypsum Concrete. A calcium gypsum mixed with wood chips, or aggregate, or both, used primarily for poured roof decks.

Gypsum Core Board. A 3/4 (19.0 mm) to 1 inch (25.4 mm) gypsum-board consisting of a single board or factory laminated multiple boards used as a gypsum stud or core in semisolid or solid gypsum board partitions.

Gypsum Fill. Troweled on plaster material to make depressions level.

Gypsum Formboard. A gypsum-board used as the permanent form for poured gypsum roof deck.

Gypsum Gauging Plaster. A plaster for mixing with lime putty to control the setting time and initial strength of the finish coat; classified either as quickset or slowset.

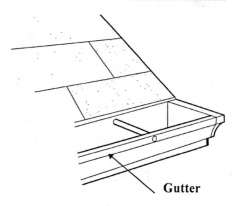

Gutter

Gypsum High Strength Basecoat Plaster. A gypsum cement for use with sand aggregate to achieve high compressive strength plaster.

Gypsum Lath. A plaster base manufactured in the form of sheets or slabs of various sizes and either 3/8 or 1/2 inch thick, having an incombustible core, essentially gypsum, and surfaced with special paper suitable for receiving gypsum plaster.

Gypsum Masonry. Molded, lightweight masonry units made from gypsum plaster, water and fiber.

Gypsum Molding Plaster. A specially formulated plaster used in casting and ornamental plasterwork; may be used neat or with lime.

Gypsum Neat Plaster. A plaster requiring the addition of aggregate on the job; it may be unfibered or fibered (vegetable, or glass fibers).

Gypsum Plaster. Plaster in which the cementing substance is gypsum.

Gypsum Ready Mixed Plaster. A plaster which is mixed at the mill with a mineral aggregate and may contain other ingredients to control time of set and working properties; only the addition and mixture of water is required on the job; also called Mill-Mixed and Pre-Mixed.

Gypsum Roof Deck. A lightweight roofing substrate made of gypsum in the form of structural boards or poured in place over a structural deck.

Gypsum Sheathing. Flat sheet material of gypsum board with water repellent paper secured to exterior side of walls, roof or floor framing used to create rigidity in building superstructure, serve as base to receive other construction, and add fire resistive characteristics; not designed for long term direct exposure to the elements.

Gypsum Tile. See Gypsum Block.

Gypsum Trowel Finish. Various proprietary ready-mixed finish coat materials consisting essentially of calcined gypsum.

Gypsum Wall Sheathing, Fire-Rated. Flat sheet material of fire rated gypsum board.

Gypsum Wallboard. A gypsum board used primarily as interior surfacing material for building structures.

Gypsum Wood-Fibered Plaster. A mill-mixed plaster containing a small percentage of wood fiber as an aggregate, used for fireproofing and high strength.

Gypsum, Alpha. See Alpha Gypsum.

Gypsum. A common naturally occurring mineral composed of hydrous calcium sulfate, $CaSO_4$ $2H2O$; the main component of sheetrock or drywall; gypsum, when heated, forms plaster-of-Paris.

Gypsumboard. An interior facing panel consisting of a gypsum core sandwiched between paper faces; also called Drywall, Plasterboard, or Sheetrock.

Gyrate. To go in a circle or spiral; revolve, whirl.

H Beam. A steel beam which in cross section resembles the letter H; commonly used in earthwork as a retaining structure or piling.

Hack. 1. To cut with irregular or unskilled blows. 2. To cut back and roughen a plastered or other surface.

Hacking. Laying masonry units so that the bottom edge is set back from the plane surface of the wall.

Hacksaw. Small toothed handsaw for sawing metals.

Hair Trap. A device to trap and retain the hair content of waste water and sewage, as in a barber shop or beauty salon.

Hairline Cracking. 1. Tiny cracks in a surface. 2. Small cracks of random pattern in an exposed concrete surface. 3. Very narrow cracks in a paint or varnish film.

Hairpin Bars. Bars, usually small sizes, bent to a hair pin shape and used for such purposes as short hooked spacer bars in columns and walls and for special dowels.

Half and Half. A dry or dampened mixture of one part portland cement and one part extra-fine sand; used as a filler in the joints of mounted ceramic mosaic tiles to keep them evenly spaced during installation.

Half Bath. A room containing a toilet and a lavatory.

Half Block. A concrete masonry unit that is half the length of a standard unit.

Half Round Gutter. A narrow rounded channel from one piece of metal, similar to a pipe cut longitudinally or along its length, used as a gutter.

Half Round Molding. An ornamental strip having one flat side and one rounded side.

Half-Timber. A form of construction composed of timber framing with the spaces filled in with brick or plaster.

Halide Fixture. A lighting fixture with a Halide Light.

Halide Light. An electric-discharge lamp that makes its light from a metal vapor such as sodium or mercury.

Halide Refrigerants. Family of refrigerants containing halogen chemicals.

Halide Torch. Type of torch used to safely detect halogen refrigerant leaks in system.

Hall. 1. Lobby or entrance room. 2. A university building for a special purpose. 3. A hallway or corridor.

Hallway. A corridor or hall.

Halogens. Any of the group of poisonous non-metallic elements, fluorine, chlorine, bromine, iodine, and astatine, which form halides by simple union with a metal, as in sodium chloride.

Halon System. A system using halon gas for the fire protection of water sensitive equipment.

Hammer Fracture. A tear in the gypsum board face and core caused by improperly hitting or over driving the nail; also called Nail Fracture.

Hammer Loss. A test to determine the impact resistance of gypsum board.

Hammer. A handtool consisting of a solid head set crosswise on a handle and used for pounding.

Hammered Effect Finish. So called because of its resemblance to hammered metal; produced by incorporating an aluminum powder in vehicle which controls leafing and nonleafing effect to create unique designs.

Hand Buggy. A hand pushed or pulled two or four wheeled cart, used to transport small amounts of concrete or material.

Hand Dryer. A mechanical/electrical device that blows hot air over wet hands to dry them.

Hand Excavation. The act or process of digging out earth using hand tools.

Hand Level. A hand-held viewing tube with a spirit level attached to fix a level line of sight.

Hand of a Lock or Door. Indicates the direction of swing or movement and/or locking security side of a door; designated as left hand, left hand reverse, right hand, or right hand reverse.

Hand Prints. Wallpapers printed by hand, usually with the silk-screen process.

Hand Seeding. The act or process of planting seed by hand.

Hand. The feel of a carpet in the hand; determined by such factors as pile height, quality and kind of fibers, type of construction, type of backing, and dimensional stability.

Handed Locks. Locks that must be used on the same handed door, as opposed to reversible locks.

Handhole. 1. Access hole used for repair and cleaning. 2. An enclosure installed in the earth, deck, floor of a building, or similar location and used as a pull or junction box for underground electrical or communication conductors; the enclosure is provided with a removable cover and so designed that the conductors may be pulled, spliced, or otherwise handled without requiring a person to enter the enclosure.

Handicapped Parking. A parking space designated as reserved for the vehicles of disabled persons.

Handicapped Partition. A dividing wall that forms a toilet cubicle area that is accessible to the disabled; commonly larger than a standard cubicle, large enough for wheelchair access.

Handicapped Plumbing. Plumbing devices and layouts specifically designed for use by disabled individuals.

Handicapped Symbol. The profile of a wheelchair and occupant placed on a sign to denote access for disabled persons.

Handling, Air. See Air Handling System.

Handling, Rubbish. See Rubbish Handling.

Handling, Trash. See Trash Handling.

Handling, Waste. See Waste Handling.

Handrail Bracket. A support for a handrail.

Handrail, Pipe. See Pipe Handrail.

Handrail. Member which is normally grasped by hand for support at stairways and other places where needed for safety of pedestrians.

Handsaw. A wood saw which is powered by hand, commonly called a Carpenter's Saw or Wood Saw.

Handsplit Shingle. A shingle made by splitting a block of wood, usually cedar or redwood, along its grain and thereby creating a shingle which may be used for roofing or siding; also called Shakes or Shake Shingles.

Handsplit Shingles. See Cedar Shake.

Hang. To install certain materials such as gypsum lath, gypsum wallboard, or wallpaper.

Hangar. A shelter for storing and repairing aircraft.

Hanger Rod. A rod for connecting pipe, gutters, or ceiling framework to a support.

Hanger Wire. Malleable wire for hanging ceilings and equipment from the structure above.

Hanger, Door Rail. See Door Rail Hanger.

Hanger, Joist. See Joist Hanger.

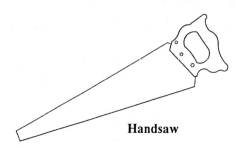

Handsaw

Hanger. 1. A tradesman who applies gypsum board products. 2. Vertical-tension member supporting a load. 3. Device attached to walls or other structure for support of pipe lines.

Hanger's Bench. See Bench, 4.

Hanger's Tee. See Tee Square.

Hansa Yellow. A family of organic yellow pigments.

Hard Edge. A special core formulation used along the paperbound edges to improve resistance to damage during the handling and application of gypsum board.

Hard Money. The actual cash paid to a seller in addition to any promissory notes.

Hard Oil Finish. A varnish giving the effect of a rubbed-in oil finish but producing a hard surface; any interior architectural varnish with a moderate luster.

Hard Tile. A term used in the tile trade to designate types of tile, such as ceramic, glass mosaic, or marble tile, over which the tile trade has jurisdiction, as compared to resilient tile.

Hard Water. Water with an excessive mineral content.

Hardboard Door. A door constructed of compressed wood fibers.

Hardboard Panel Siding. Panel siding made of hardboard.

Hardboard Panel. Flat sheet material of fibers consolidated under heat and pressure in hot press.

Hardboard Soffit. Hardboard finish material covering the underside of a part or member.

Hardboard Underlayment. Flat sheet material of hardboard placed over subflooring to provide smooth and even surface to receive finish floor covering.

Hardboard. A very dense panel product, usually with at least one smooth face, made of highly compressed wood fibers; consolidated under heat and pressure, and having a density factor of approximately 50 to 80 pounds per cubic foot.

Hard-Burned. Clay products which have been fired at high temperatures to near vitrification, generally producing relatively low absorption and high compressive strengths.

Hardener. Curing agent; promoter; and catalyst used in making and finishing concrete.

Hardening. The gain of strength of plaster, mortar, or concrete after setting; see Set.

Hardhat. 1. A safety helmet of metal or plastic worn by workers on a construction site. 2. A construction worker who wears such a hat.

Hardness. 1. A property of a material that resists indentation. 2. Of water, the quantity of dissolved minerals, such as calcium and magnesium compounds.

Hardpan. Hard, tight soil; a hard layer that may form just below plow depth on cultivated land.

Hardscape. The non-organic elements of a landscape development, such as walls, paving, fences, and lighting standards.

Hardstand. A hard surfaced area for parking an airplane.

Hardwall. 1. Base coat plaster. 2. Regionally the term differs; in some cases it refers to sanded plaster, while in others to neat.

Hardware Cloth. Steel wire mesh, usually galvanized, commonly 1/8-, 1/4-, or 1/2-inch mesh.

Hardware. The miscellany of metal or plastic fittings and fastenings used in or on a building and its parts.

Hardwood Dimension Lumber. See Dimension Stock.

Hardwood Plywood & Veneer Association (HPVA). 1825 Michael Faraday Drive, PO Box 2789, Reston, Virginia 22090-0789, (703) 435-2900.

Hardwood Plywood Manufacturer's Association (HPMA). PO Box 2789, Reston, Virginia 22090, (703) 435-2900.

Hardwood Plywood Panel. Plywood with face plies manufactured from hardwood species such as cherry, oak, birch, ash, walnut, maple, gum, mahogany, or teak.

Hardwoods. Generally one of the botanical groups of trees that have broad leaves in contrast to the conifers or softwoods. The term has no reference to the actual hardness of the wood.

Hardy Cross Method. See Moment Distribution.

Harmon Hinge. A hinge designed to swing a door into a pocket at a right angle with the frame.

Harsh Mortar. A masonry mortar that is hard to spread under the trowel, usually because of a lean ratio of cement paste to aggregates.

Hasp Assembly. A fastener for a door or lid consisting of a hinged metal strap that fits over a loop and is secured by a pin or padlock.

Hatch, Roof. See Roof Hatch.

Hatch. A hinged or removable cover in a floor or roof which permits ventilation or the passage of persons or objects.

Hatchet. A multi-purpose short handled hand tool with a head suitable for alternative use as a hammer or axe; the convex head configuration to properly dimple the nail and the hatchet blade used to cleave and adjust framing and wallboard edges to fit, or as a jacking wedge.

Hatchway. An opening giving access usually by ladder to a lower space like a cellar.

Haul Resistance. The distance measured along the most direct practical route between the center of the mass of excavation and the center of mass of the fill as finally placed; the distance material is moved.

Haul, Station Yards of. See Station Yards of Haul.

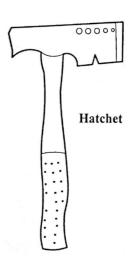

Hatchet

Hawk. A metal square with a wooden handle at the center, used by a plasterer or tilesetter, to temporarily hold mortar or plaster; commonly, a rubber pad fits over the handle and covers that portion of the metal hawk that would come in contact with the hand.

Hazardous Waste. A material or substance characterized by a propensity to be unhealthy, toxic, or dangerous.

Hazardous Wiring. All wiring except that which conforms with all applicable laws in effect at the time of installation and which has been maintained in good condition and is being used in a safe manner.

HB. Hose Bibb.

Head Flashing. The exterior flashing installed at the head of a door, window, or other opening.

Head Friction. Head required to overcome friction of the interior surface of a conductor and between fluid particles in motion.

Head Joint. See Butt Joint, 4.

Head Pressure Control. Pressure-operated control which opens electrical circuit if high-side pressure becomes too high.

Head Pressure. Pressure in the condensing side of a refrigerating system.

Head Velocity. Height of fluid equivalent to its velocity pressure in a flowing fluid.

Head, Pop-Up. See Pop-Up Head.

Head, Service. See Service Head.

Head, Static. See Static Head.

Head, Total Static. See Total Static Head.

Head. 1. The operative part of a tool. 2. The flattened part of a nail. 3. The upper horizontal part of a door, window, or other opening. 4. The pressure produced by a difference in elevation of two points in a body of fluid. 5. The top part of a column or pillar.

Header Bond. A brick bond that will show only headers on the face or surface of a wall; each header would be divided evenly over the header underneath it.

Header Course. A brick bond in which a course made up of headers only.

Header High. The height up to the top of the brick course that is directly under a header course.

Header, Ceiling. A joist perpendicular to and supporting ceiling joists.

Header. 1. Length of pipe or vessel to which two or more pipe lines are joined carries fluid from a common source to various points of use; a manifold. 2. The pipe that runs across the top or bottom of an absorber plate, gathering or distributing the heat transfer fluid from or to the grid of pipes that runs across the absorber surface. 3. A lintel. 4. A structural support over an opening. 5. A joist that supports other joists; a short joist into which the common joists are framed around or over an opening. 6. A brick that is laid on its flat surface across a wall and will show only its end on the surface or face of the wall. 7. A masonry unit which overlaps two or more adjacent wythes of masonry to tie them together; also called a Bonder.

Headlap. The minimum distance, measured at 90 degrees to the eave along the face of a shingle or felt as applied to a roof, from the upper edge of the shingle or felt, to the nearest exposed surface.

Headroom. The clear space between the floor line and ceiling, as in a stairway.

Headwall, Removal. The act or process of demolition and removal of a headwall.

Headwall. A wall, usually constructed of concrete or masonry, that is placed at the inlet side of a drain or culvert to protect fill from scouring, undermining, or to divert flow.

Healing Power. The ability of a glaze to heal surface blemishes during firing.

Hearing. A proceeding conducted by a judge or arbitrator who receives evidence about a dispute.

Hearsay. Evidence of an event that the witness did not personally perceive.

Heart Rot. Any rot characteristically confined to the heartwood; it generally originates in the living tree.

Hearth. The masonry floor of a fireplace and the floor surface in front of a fireplace.

Heartwood. The wood extending from the pith to the sapwood, the cells of which no longer participate in the life processes of the tree; heartwood may contain phenolic compounds, gums, resins, and other materials that usually make it darker and more decay resistant than sapwood.

Heat Exchanger. A device for the transfer of heat from one medium to another.

Heat Intensity. Heat concentration in a substance as indicated by the temperature of the substance through use of a thermometer.

Heat Lag. The time it takes for heat to travel through a substance heated on one side.

Heat Leakage. Flow of heat through a substance.

Heat Load. Amount of heat, measured in BTUs or watts, which is removed during a period of 24 hours.

Heat Loss. The heat lost from a building through conduction, convection, and radiation as well as through air infiltration and ventilation systems.

Heat of Compression. Mechanical energy of pressure changed into energy of heat.

Heat of Fusion. Heat released from a substance to change it from a liquid state to a solid state; the heat of fusion of ice is 144 BTUs per pound (151.9 joules).

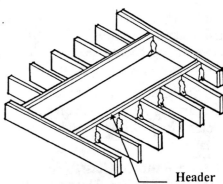

Header

Heat of Hydration. The thermal energy given off by concrete, masonry, or gypsum as it cures.

Heat of Repression. 1. Process by which oxygen and carbohydrates are assimilated by a substance. 2. When carbon dioxide and water are given off by a substance.

Heat Pipe. A method of heat recovery from exhaust air in which a coil has a partition wall dividing the coil face into two sections; this system has no moving parts and relies on the slope of the coil to operate, sloping down from the cold end to the hot end.

Heat Pump. A combination heating and cooling device. In the winter it extracts heat from air as cold as 20 degrees, and in the summer it works in reverse to become an air conditioner.

Heat Recovery. The extraction of heat from a source not primarily designed to produce heat.

Heat Sensor. A battery operated or hard wired mechanical/electrical device that either makes a loud noise or flashes a light when raised air temperature is detected; for use as a substitute for a smoke detector in locations that might have ambient smoke such as in kitchens.

Heat Sink. A substance or device for the absorption and dissipation or recovery of heat in connection with a passive solar heating system.

Heat Storage. A device or medium that absorbs collected solar heat and stores it for periods of cloudy or cold weather.

Heat Strengthened Glass. Glass that has been strengthened by heat treatment, though not to as great an extent as tempered glass.

Heat Time. In multiple-impulse welding or seam welding the time during which the current flows during any one impulse.

Heat Transfer. Movement of heat from one body or substance to another; heat may be transferred by radiation, conduction, convection, or a combination of these three methods.

Heat Wheel. A method of heat recovery from exhaust air in which both the intake and exhaust ducts are connected to the wheel.

Heat. 1. The condition of being hot. 2. A form of energy arising from the random motion of the molecules of bodies, which may be transferred by conduction, convection, or radiation.

Heat/Smoke Vent. A vent or chimney forming a passageway for expulsion of vent gases from gas-burning units to the outside air.

Heat-Affected Zone. That portion of the base metal which has been altered by the heat of welding, cutting, or brazing.

Heater Duct. The conduit in a heating system that conveys warmed air throughout a building or structure.

Heater, Electric. See Electric Heater.

Heater, Unit. See Unit Heater.

Heater, Water. See Water Heater.

Heater. Any heat-producing appliance or unit, such as a furnace.

Heating Coil. An element where heat is transferred to an air stream; the heating medium may be steam, hot water, or electrical.

Heating Control. Device which controls the temperature of a heat transfer unit.

Heating Value. Amount of heat which may be obtained by burning a fuel; usually expressed in BTUs per pound, BTUs per gallon, or calories per gram.

Heat-Set Nylon. Nylon fiber which has been heat treated to retain a desired shape.

Heavy Bodied Oil. A high viscosity oil.

Heavy Centered Pattern. Spray pattern having most paint in center, less at edges.

Heavy Duty Tile. Tile suitable for areas where heavy pedestrian traffic is prevalent.

Heavy Soil. A fine grained soil made up largely of clay or silt.

Heavy Timber. Construction requiring noncombustible exterior walls with a minimal fire-resistance rating of two-hours, laminated or solid interior members, heavy plank, or laminated wood floors and roofs.

Hectare. A unit of land area equaling 10,000 square meters or 2.471 acres.

Heel, Rafter. The end or foot that rests on the wall plate.

Heel. 1. The point at the end of a truss where the top and bottom chords intersect. 2. See Steel Square.

Heeling In. Temporary planting of trees and shrubs.

Hegeman Grind Gauge. Proprietary instrument for measuring smoothness of pigment dispersion in liquid paint.

Height. The dimension measured at right angles to the direction of the thickness and length of a masonry unit, the vertical dimension, as used in a wall.

Helical. Like a helix.

Helix. A spiral curve.

Helmet. A protective device used in arc welding for shielding the face and neck; a helmet is equipped with a suitable filter glass and is designed to be worn on the head.

Hemihydrate. A hydrate containing half a molecule of water to one of the material forming the hydrate; used to describe the form of calcined gypsum generally used for plaster.

HEPA Filter. High-Efficiency Particulate Air Filter; such filters are rated to trap at least 99.9% of all particles 0.3 microns in diameter or larger.

HEPA. High Efficiency Particulate Arrestance; see HEPA Filter.

Hermetic Compressor. Compressor which has the driving motor sealed inside the compressor housing; the motor operates in an atmosphere of the refrigerant.

Hermetic Motor. Compressor drive motor sealed within same casing which contains the compressor.

Hermetic Seal. An airtight seal.

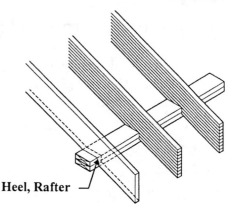

Heel, Rafter

Hermetic System. Refrigeration system which has a compressor driven by a motor contained in the compressor dome or housing.

Heroic Scale. A statue of colossal size or proportions.

Herringbone Blocking. Wood blocking between studs set at a slight angle in alternate directions in adjacent stud spaces, the angle making it possible to drive the blocks in tightly between studs.

Herringbone Bond. Bricks laid in a zigzag fashion representing a herringbone pattern.

Herringbone Floor. A floor covering made up of rows of parallel lines with adjacent rows slanting in reverse directions.

Hertz (Hz). A measurement of the frequency of vibration of sound measured in cycles per second; one cycle per second equals one Hertz.

Hessian. See Burlap.

Heteropolymer. The result of Heteropolymerization.

Heteropolymerization. See Polymerization.

Heuristic. By trial and error and experiment rather than by science.

Hew. To fell, chop, or shape with an axe.

Hex Nut. A six-sided, short metal nut with a threaded hole for receiving a rod or threaded bolt.

Hex. 1. Hexagon. 2. Hexagonal.

Hexagon. A polygon of six sides.

Hg. Mercury.

HI. Hydraulic Institute.

HID Lamp. High Intensity Discharge Lamp.

Hickey. A tool for bending pipe, conduit, and reinforcing bars.

Hiding Bond. Ability of a paint to hide or obscure the surface on which it is applied; degree of opacity of a pigment or paint.

High Boiling Solvent. A solvent with an initial boiling point above 302° F. (150° C.) such as diacetone alcohol or cellosolve acetate.

High Build. Producing thick dry films per coat.

High Calcium Lime. A type of lime containing principally calcium oxide or hydroxide and not more than 5% magnesium oxide or hydroxide.

High Chair Reinforcing. A chair-shaped device used to hold steel reinforcement off of the bottom form of a slab while the concrete is being poured.

High Early Strength Cement. Cement that develops strength more quickly than ordinary cement, typically will be as strong at 3 days and 7 days as normal cement would in 7 days and 28 days; alumina cement.

High Flash Naphtha. Aromatic solvent having a high flash point, minimum 113° F, 45° C.

High Frequency Curing. See Radio Frequency Curing.

High Joint. See Crown, 1.

High Lift Grouting. A method of construction in which concrete block units may be laid the entire height of the wall before grouting; this method of construction requires special inspection.

High Magnesium Lime. A type of lime containing more than 5% magnesium oxide or hydroxide.

High Pressure Gauge. Instrument for measuring pressures in range of 0 psi to 500 psi (0 kg/cm^2 to 3.52 kg/cm^2).

High Pressure Steam Curing. See Autoclave Curing.

High Rate Sand Filter. A sand filter designed for flows in excess of five gallons per minute per square foot, but not in excess of 20 GPM per square foot.

High Relief. Sculptural relief in which at least half of the circumference of the modeled form projects.

High Side. Parts of a refrigerating system which are under condensing or high-side pressure.

High Strength Bolt. A bolt designed to connect steel members by clamping them together with sufficient force that the load is transferred between them by friction.

High Tension. High Voltage.

High Voltage Cable. Cable manufactured to withstand high-voltage.

High Voltage Circuit. Any circuit having a difference of potential of more than six hundred volts, or seven hundred fifty volts where specified in certain regulations, between any two conductors of the circuit.

High Voltage. Electrical power of over 600 volts.

High Wall. Undisturbed soil or rock bordering a cut; a face which is being excavated, as distinguished from spoil piles.

Highbay Lighting. A lighting system located high above work or floor level.

Highest and Best Use. The use of land that will produce the highest future net income.

Highlight. A smooth area, noticeably different from the normal surface.

Highlighting. Making certain parts of finished project appear lighter than other parts.

High-Pressure Boiler. Boiler furnishing steam at pressures of 15 pounds per square inch gauge or higher (1.05 kg/cm^2).

High-Pressure Cut-Off. Electrical control switch operated by the high-side pressure which automatically opens electrical circuit if too high a pressure is reached.

High-Rise. A multi-storied building with elevators.

High-Side Float. Refrigerant control mechanism which controls the level of the liquid refrigerant in the high-pressure side of mechanism.

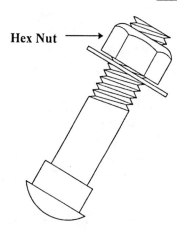

Hex Nut

High-Vacuum Pump. Mechanism which can create a vacuum in the 1000 to 1 micron range.

Highway Demolition. The destruction and removal of a public road.

Hinge Stile. The door stile to which the hinges are installed as distinguished from the lock stile.

Hinge. 1. A joint fixing the relative position of the ends of two or more structural members, but permitting their relative rotation. 2. A piece of door hardware that permits the opening and closing of a door by joining the door to the jamb with a flexible device.

Hip Rafter. A structural member of a roof forming the junction of an external roof angle or, where the planes of a hip roof meet.

Hip Roof. A roof consisting of four sloping planes that intersect to form a pyramidal or elongated pyramid shape; a roof which slopes up toward the center from all sides, necessitating a hip rafter at each corner.

Hip. The diagonal intersection of planes in a hip roof.

Hippodamian Plan. Greek plan, derived its name from Hippodamos, with grid city blocks, informal organization of public spaces and facilities within the grid matrix; better known examples are the Ionian cities of Ephesus and Miletus, 5th Century BC.

Hod. A portable trough for carrying plaster, mortar, and bricks, fixed crosswise on top of a pole and carried on the shoulder.

Hoddability. A term descriptive of the ease with which a plaster mortar may be handled with a hod or hawk, dependent upon flow characteristics and angle of repose of the mortar.

Hoe. A long handled tool with a thin flat metal blade for cultivating, weeding, and mixing.

Hog Ring. A heavy galvanized wire staple applied with a pneumatic gun which clinches it in the form of a closed ring around stud, rod, pencil rod or channel.

Hoist, Industrial. See Industrial Hoist.

Hoistway. Any shaftway, hatchway, well hole, or other vertical opening or space in which an elevator or dumbwaiter is designed to operate.

Hold Harmless. See Indemnification.

Hold Out. Ability to prevent soaking into substrate.

Holder, Door. See Door Holder.

Holding Period. In the manufacture of concrete products, the period between completion of casting and the introduction of additional heat or the steam curing period.

Holdover Tenant. A tenant that does not vacate the property at the end of a lease term.

Hole. 1. An opening or perforation in a material. 2. A pit in the ground.

Holiday. 1. In roofing, a skipped area of liquid applied roofing. 2. In painting, unintentional missing of an area of surface being painted.

Hollow Block. Concrete blocks that can be filled with insulation or reinforced and grouted.

Hollow Clay Tile. A tile building material manufactured in a variety of sizes and forms, used for both exterior walls and partitions.

Hollow Column. A vertical column constructed so as to create an air space within.

Hollow Concrete Masonry. Concrete masonry units that are manufactured with open cores, such as ordinary concrete blocks.

Hollow Core Door. A flush hardwood or plywood door glued to a skeletal framework with the interior remaining void or honeycombed.

Hollow Core Slab. A precast concrete slab element that has internal longitudinal cavities to reduce its self-weight.

Hollow Masonry Unit. A masonry unit whose net cross-sectional area in any plane parallel to the bearing surface is less than 75% of its cross-sectional area measured in the same plane.

Hollow Metal Door. A hollow core door constructed of sheet metal that has been channel-reinforced; the core is often filled with some type of lightweight material.

Hollow Metal Frame. A sheet metal door frame with reinforcing at the hinges and strikes.

Hollow Wall. A wall built of solid masonry units laid in and so constructed as to provide an air space within the wall.

Hollow-Back. Removal of a portion of the wood on the unexposed face of a wood trim member to more properly fit any irregularity in bearing surface.

Hollow-Core Construction. A panel construction with faces of plywood, hardboard, or similar material bonded to a framed-core assembly of wood lattice, paperboard rings, or the like, which support the facing at spaced intervals.

Hollow-Tube Gasket. Sealing device made of rubber or plastic with tubular cross-section.

Homestead. 1. A tract of land granted by the U.S. upon recording, improving, and cultivating the property. 2. A residence. 3. A home and adjoining land occupied by a family; this can be exempt from judgment up to a certain amount by filing a Declaration of Homestead.

Hone. A honing stone.

Honeycomb Core. A sandwich core material, for doors or building panels, constructed of thin sheet materials or ribbons formed to honeycomblike configurations.

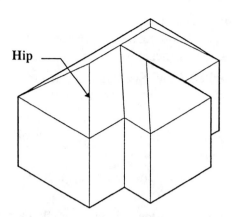

Hip

Honeycomb. Voids in concrete resulting from segregation and poor consolidation at the time concrete is placed.

Honeycombing. Checks, often not visible at the surface, that occur in the interior of a piece of wood, usually along the wood rays.

Honing Stone. A fine-grit stone for sharpening cutting tools.

Hood, Exhaust. See Exhaust Hood.

Hood, Range. See Range Hood.

Hook and Eye. A fastening device consisting of a hook that attaches to a u-shaped loop for holding a door in the open position.

Hook Knife. A special curved knife used primarily for cutting resilient floor coverings; also called a Linoleum Knife.

Hook Stick. An implement that is curved or bent for holding, catching, or pulling.

Hook Strip. Strip of wood, plastic, or metal secured to a wall surface to provide a mounting surface for hooks, such as coat hooks or key hooks.

Hook Wire. A hook implement that is curved or bent for holding, catching, or pulling.

Hook. A semicircular bend in the end of a reinforcing bar to develop bond, and thus anchorage.

Hooke's Law. Stress and strain are proportional up to the elastic limit of the material.

Hoop. Circular shaped pieces of steel reinforcing bar.

Hopped-Up Mud. Mortar mixed with an accelerator.

Hopper Window. A window whose sash pivots on an axis along the sill, and that opens by tilting toward the interior of the building.

Hopper. Deep water basin with all four sides sloping for diving.

Horizon. The line at which the earth and sky appear to meet.

Horizontal Angle. An angle in a horizontal plane.

Horizontal Application. The application of gypsum board with the long dimension at right angles to the framing members; also referred to as perpendicular.

Horizontal Branch. A branch drain in the DWV system that extends from a soil, waste stack, or building drain to the fixture trap; it may or may not have vertical sections or branches off of it which receive discharge from one or more fixture drains and carry this discharge to the soil, waste stack, or house drain; also called a Lateral Branch.

Horizontal Broken Joints. A style of laying tile with each course offset one-half its length.

Horizontal Fixed Position. The position of a pipe joint wherein the axis of the pipe is approximately horizontal and the pipe is not rotated during welding.

Horizontal Forces. Any lateral loads imposed on a structure, including those caused by wind or earthquake forces.

Horizontal Hardboard Siding. Horizontal siding made of hardboard often used as clapboards on a house.

Horizontal Rolled Position. In pipe welding, the position of a pipe joint wherein welding is performed in the flat position by rotating the pipe.

Horizontal Shaft Mixer. A mixer having a stationary cylindrical mixing compartment, with the axis of the cylinder horizontal, and one or more rotating horizontal shafts to which mixing blades or paddles are attached.

Horizontal Siding. Linear horizontal material, usually overlapping, used as an exterior surface or cladding for an exterior framed wall.

Horizontal Wood Siding. Linear horizontal wood material, usually overlapping, used as an exterior surface or cladding for an exterior framed wall.

Horizontal. Parallel to the horizon.

Horizontally Laminated Wood. Laminated wood in which the laminations are so arranged that the wider dimension of each lamination is approximately perpendicular to the direction of load.

Horizontals. Reinforcing bars running horizontally.

Horn. The extension of a stile, jamb or sill.

Horsepower. 1. A unit of electrical power in the U.S. equal to 746 watts. 2. In the English gravitational system, 550 foot-pounds of work per second.

Hose Bibb. Bibbcock. a water spigot or faucet with its nozzle threaded or coupling attached to accept a hose; also spelled Hose Bib; also called a Wall Hydrant or a Sill Cock.

Hose Reel. A round device upon which a hose is wound for easy dispensing.

Hospital Bed. A piece of furniture found in a hospital where patients recline, lie and sleep.

Hospital Cubicle. A small partitioned space or sleeping compartment in a hospital.

Hospital Divider. A movable partition in a hospital to create privacy.

Hospital Equipment. Apparatus or implements used in hospital applications.

Hospital Faucet. A water faucet that is controlled by feet to help eliminate the spread of germs by hands.

Hospital Panel. A partition in a hospital which divides areas into smaller spaces.

Hospital Stretcher. A narrow, movable bed with collapsible legs and wheels used for transporting patients, often from an ambulance to a hospital.

Hospital Wardrobe. A cabinet installed in a hospital patient room to store the patient's clothes; usually matches any other cabinet or furniture in the room.

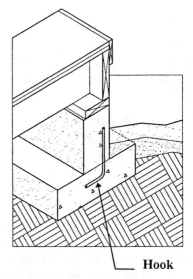

Hook

Hot Aggregate Storage Bins. Bins that store the heated and separated aggregates prior to their final proportioning into the mixer.

Hot Gas Bypass. Piping system in refrigerating unit which moves hot refrigerant gas from condenser into low-pressure side.

Hot Gas Defrost. Defrosting system in which hot refrigerant gas from the high side is directed through evaporator for short period of time and at predetermined intervals in order to remove frost from evaporator.

Hot Junction. That part of thermoelectric circuit which releases heat.

Hot Lacquer Process. Process where heat is used instead of volatile thinner to reduce the consistency of lacquer; hot lacquer can be applied with a higher percentage of solids than room-temperature lacquer.

Hot Mud. Mortar mixed with an accelerator; also called Hot Stuff.

Hot Plate, Laboratory. See Laboratory Hot Plate.

Hot Spray. Spraying material heated to reduce viscosity.

Hot Stuff. 1. Mortar mixed with an accelerator; also called Hot Mud. 2. Hot bitumen.

Hot Tub. Refers specifically to a hydrotherapy unit normally constructed of wood designed and assembled in the traditional manner of tubs or casks, with side and bottoms formed of separate boards and the whole shaped to join together by pressure of the surrounding hoops, bands or rods as distinct from spa units formed of plastic, concrete, metal or other materials.

Hot Water Circulating Loop System. A piping arrangement in which hot water is constantly circulated in a loop from which individual plumbing fixtures are fed.

Hot Water Dispenser. A device used to heat and deliver small quantities of water.

Hot Water Heating System. System in which water is circulated through heating coils.

Hot Water Storage Tank. A tank that receives heated water and holds it for future distribution.

Hot Wire Relay. See Thermal Relay.

Hot Wire. 1. The live wire, carrying electric current, as opposed to the neutral or ground. 2. Resistance wire in an electrical relay which expands when heated and contracts when cooled.

Hotel. A building containing an establishment providing lodging and meals.

Hot-Laid Plant Mixture. Plant mixes that must be spread and compacted while at an elevated temperature; to dry the aggregate and obtain sufficient fluidity of the asphalt (usually asphalt cement), both must be heated prior to mixing, hence the term hot-mix.

Hot-Water Boiler. A boiler furnishing hot water at pressures not more than 30 pounds per square inch gauge (2.12 kg/cm^2) or steam at pressures not more than 15 pounds per square inch gauge (1.06 kg/cm^2).

House Drain. The horizontal part of the drainage system which connects the DWV piping within the structure to the sanitary sewer or private sewage treatment equipment.

House Drainage System. Complete system of piping used for carrying away waste water and sewage; also called Building Drainage System.

House Paint, Outside. Paint designed for use on the exterior of buildings, fences and other surfaces exposed to the weather.

House Sewer. The piping that takes the soil and waste water from the building drain and conveys it to the public sewer or private sewage disposal system; also called Building Sewer.

House Trap. Located at the point at which the house drain leaves the building, designed to hold a quantity of water that prevents gasses from the sewer system from entering the building.

House. A building for human habitation.

Housed Joint. A joint in which a piece is grooved to receive the piece which is to form the other part of the joint.

Housing. A groove cut at any angle with the grain and partway across the piece; used for framing stair risers and treads.

Howe Truss. A standard peaked roof truss configuration where the vertical web members and the bottom chord are in tension while the sloping web members and the top chord are in compression.

HP Sodium Lamp. A high-pressure sodium vapor lamp that produces a wide-spectrum yellow light.

HP. High Pressure.

HPMA. Hardwood Plywood Manufacturer's Association.

HPVA. Hardwood Plywood & Veneer Association.

Hub Union. A pipe fitting used to join two pipes without turning either pipe.

Hub. 1. A surveyor's bench mark; used as a starting point for measuring elevations and horizontal distances. 2. The enlarged end of a hub and spigot cast-iron pipe.

Hub. See Bell.

Hue. Color.

Human Relations. A behavioral aspect of management -- handling of personnel and the consideration of each problem from the employee's point of view; the success or failure of the business often depends on the relationship between management and personnel.

Howe Truss

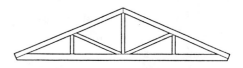

Humidification. The addition of moisture to air, thereby increasing the latent heat.

Humidified Bond. In gypsum wallboard, the ability of the surfacing paper to resist delamination from the core under extremely high humidity conditions.

Humidifier. A device for supplying or maintaining humidity in a conditioned space.

Humidistat. Electrical control which is operated by changing humidity.

Humidity Ratio. The weight of water vapor per unit weight of dry air; also called Specific Humidity.

Humidity. The amount of water vapor in a given volume of air.

Hump Joint. See Crowned Joint.

Humus. Decayed organic matter; a dark fluffy swamp soil composed chiefly of decayed vegetation; also called peat.

Hung Slab. A nonstructural concrete slab having no bearing on walls, being totally supported from above by an overhead structural element.

Hurricane. A tropical cyclone with winds over 74 miles per hour; usually accompanied by rain, thunder, and lightning.

Hut. A small or temporary dwelling of simple construction.

Hutment. A collection of huts.

HVAC Control. An apparatus that controls a heating, ventilation, and air conditioning system.

HVAC. Heating, Ventilation, and Air-Conditioning system.

HW. Hot Water.

Hydrant Removal. The act or process of removing an old hydrant.

Hydrant, Wall. See Wall Hydrant.

Hydrant. 1. A discharge pipe with a valve and threaded outlet from which water may be drawn from a water main; a faucet. 2. Fire hydrant; fire plug.

Hydrate. 1. A chemical combination of water with another compound or an element. 2. Hydrated Lime.

Hydrated Lime. Calcium hydroxide, a dry powder obtained by treating quicklime with water; also called Hydrate.

Hydration. The process of adding water to a substance.

Hydraulic Cement. A type of portland cement that will harden or set under water.

Hydraulic Elevator. A lifting device powered by pressured fluid.

Hydraulic Excavator. A digging machine powered by pressured fluid.

Hydraulic Gradient. The slope of the surface of open or underground water.

Hydraulic Institute (HI). 9 Sylvan Way, Parsnippany, New Jersey 07054, (201) 267-7772.

Hydraulic Lift. An elevating device powered by pressured fluid.

Hydraulic Spraying. Spraying by hydraulic pressure.

Hydraulic. Of a mechanism operated, moved, or effected by use of water pressure.

Hydraulics. 1. The science of the conveyance of liquids through pipes. 2. The branch of physics having to do with the mechanical properties of water and other liquids in motion.

Hydrocarbon Resins. Obtained by catalytic polymerization of petroleum fractions.

Hydrocarbon. An organic compound containing only carbon and hydrogen; fuels such as natural gas, gasoline, and coal contain hydrocarbons.

Hydrochloric Acid. A solution of the colorless gas hydrogen chloride that is a strong corrosive irritating acid, used in dilute form for cleaning portland cement paste from brick or stone work.

Hydrofluoric Acid. An aqueous solution of hydrogen fluoride that is a weak poisonous acid that attacks silica and silicates, used for finishing and etching glass.

Hydrojet Booster Pump System. A system whereby one or more hydrojets are activated by the use of a pump which is completely independent of the filtration and heating system of the spa.

Hydrojet. A fitting which blends air and water creating a high velocity, turbulent stream of air enriched water.

Hydrology. The science of the properties of the earth's water, on the surface, subsurface, and atmospheric, and its movement in relation to land.

Hydrometer. A device for measuring the density of liquids.

Hydronic Heating System. A system that circulates heated water through convectors to heat a building or space.

Hydronic. Heating system which circulates a heated fluid, usually water, through baseboard coils by means of a circulating pump which is controlled by a thermostat.

Hydronics. Systems of heating or cooling that involve transfer of heat by a circulating fluid, as water or vapor, in a closed system of pipes.

Hydrophilic. Having an affinity for water; capable of uniting with or dissolving in water.

Hydrophobic. Having antagonism for water; not capable of uniting or mixing with water.

Hydroponics. The process of growing plants in sand, gravel, or liquid, without soil and with added nutrients.

Hydrostatic Pressure. Pressure exerted by standing water.

Hydrostatic Relief Valve. A valve installed in the main drain of the swimming pool, sometimes in other areas as well; when the pool is full, the valve is closed; when pool is empty, it is opened to allow ground water to flow into the pool; this prevents the pool from floating, which happens if ground water pressure builds up.

Hydrous. Containing water.

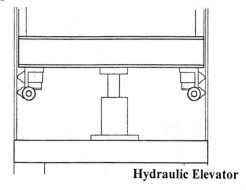

Hydraulic Elevator

Hygrometer. An instrument, consisting of a dry bulb thermometer and a wet bulb thermometer, for measuring the humidity of the atmosphere or of a gas; also called a Wet and Dry Bulb Hygrometer or a Psychrometer.

Hygroscopic. 1. Tendency to absorb water. 2. Ability of a substance to absorb and release moisture and change physical dimensions as its moisture content changes. 3. A synthetic rubber (chlorosulfonated polyethylene) often used with neoprene in elastomeric roof coverings; Hypalon is a registered trademarked name of E.I. du Pont Co.

Hyperbolic Paraboloid Shell. A concrete roof structure with a saddle shape.

Hyperbolic Paraboloid. Double curved surface which can be generated by sliding a straight line along two other lines which are skew to each other but lie in parallel planes.

Hypocenter. The point below the epicenter at which an earthquake actually begins; also called the Focus.

Hypotenuse. The side of a right triangle which is directly opposite the right angle itself.

Hypothecate. To pledge property for repayment of a debt without giving up title or possession of the property.

I

I Beam. An American Standard designation for a particular section of hot-rolled steel which in cross section is shaped like a capital "I".

I. 1. Represents the moment of inertia of a member. 2. Occupancy Importance Factor. 3. Represents Amperes.

I.D. Inside diameter

IAQ. Indoor Air Quality.

IBC. International Building Code.

ICBO. International Conference of Building Officials.

ICC. International Code Council.

ICC. Interstate Commerce Commission.

Ice Cream Cabinet. Commercial refrigerator which operates at approximately 0° F. (-18° C.); used for storage of ice cream.

Ice Dam. An ice obstruction along the eaves of a roof caused by the refreezing of water emanating from melting snow on the roof surface above.

Ice Maker. An appliance that freezes water into ice chunks or cubes.

Ice Melting Equivalent. The amount of heat absorbed by melting ice at 32° F. (0° C.) is 144 BTUs per pound of ice or 288,000 BTUs per ton; also called Ice Melting Effect.

Ice Storage Bin. A container which holds ice and maintains its solid properties.

ID. 1. Inside Diameter. 2. Inside Dimension.

Idiot Stick. See Layout Stick. See Story Pole, 2.

Idler. Pulley used on some belt drives to provide proper belt tension and to eliminate belt vibration.

Ignition Transformer. Transformer designed to provide a high- voltage current; used in many heating systems to ignite fuel; it provides a spark gap.

IIAR. International Institute of Ammonia Refrigeration.

IIC. Impact Insulation Class.

IL. Intensity Level.

ILIA. Indiana Limestone Institute of America, Inc.

Illegal. Contrary to or against the law.

Illiquid. Deficient in liquid assets.

Illuminate. Light up, make bright, usually with electrical lighting.

Ilmenite. A mineral having the theoretical composition FO TiO_2 used principally in the production of titanium oxide.

IMC. Intermediate Metal Conduit.

IME. 1. Ice Melting Equivalent. 2. Ice Melting Effect.

IMI. International Masonry Institute.

Immeasurable. Not measurable; immense.

Immiscibility. Incapability of being mixed or made homogeneous.

Immune. Not subject to legal process.

Impact Barrier. A barrier constructed to resist dynamic loading on a surface.

Impact Insulation Class (IIC). A single figure rating that provides an estimate of the sound insulating performance of a floor or ceiling assembly.

Impact. The effect that escalation or delay has on other work in a construction schedule, ripple effect.

Impasto. Thick application of pigment to canvas or other surface which makes the painting stand out in bold relief.

Impedance. The opposition to electrical flow in an alternating current electrical circuit.

Impeller. A blade of a pumping device; a propeller type fitting which produces suction when rotated.

Imperial Gallon. Unit of volume measure used in Great Britain and Canada, containing 277.42 cubic inches and weighing 10 pounds; compared with U.S. gallon of 231 cubic inches and 8.33 pounds in weight.

Imperial Units. See United States Customary Systems.

Impermeability. The resistance an asphalt pavement has to the passage of air and water into or through the pavement.

Impermeable. Does not permit the passage of fluids; impervious.

Impervious Tile. Tile designated by water absorption of 0.5 percent or less.

Impervious. Not letting water or moisture pass through or be absorbed; the degree of vitrification evidenced visually by complete resistance to dye penetration; impervious generally signifies zero absorption, except for floor and wall tile which are considered impervious up to 0.5 percent water absorption.

Imperviousness. The quality of resisting moisture penetration.

Implied Contract. A contract that is not in words.

Implied Covenant. A promise that is not expressed in a contract, but that is implied from the surrounding circumstances.

Implied Indemnity. An obligation to indemnify that arises not from the words of a contract, but from the circumstances of the parties.

Implied Warranty. A promise, usually related to a the quality or serviceability of goods, that is not in words but is implied from the circumstances of a sale.

Imported Marble. Marble shipped in from another country or area.

Impossibility. A doctrine of contract law that excuses performance that becomes physically impossible.

Impreg. Wood in which the cell walls have been impregnated with synthetic resin so as to reduce greatly its swelling and shrinking; impreg is not compressed; compare with Compreg.

Impregnate. To saturate, fill, or permeate one substance with another, as wood with chemicals.

Imprest Cash Fund. A petty cash fund maintained for minor disbursements and from time-to-time restored to its original amount by a transfer from the general cash account of an amount equal to the sum of disbursements.

Improvements. Any enhancements to value of property including buildings, paving, fencing, and landscaping.

In Situ Concrete. Poured in place concrete.

In Situ. 1. In the natural or original place. 2. Concrete poured in place.

Inactive Door. In a pair of doors, the leaf that does not contain a lock, but is bolted when closed, and to which the strike is fastened to receive the latch or bolt of the active door.

Incandescent Fixture. An electric light fixture in which the lamp or bulb has a filament that gives off light when heated to incandescence by an electric current.

Incandescent Lamp. A device that produces light through an electrically heated filament.

Inch. A unit of linear measure equal to 1/12 of a foot and equal to 2.54 centimeters.

Inchoate. Only partly formed, or just begun, as an idea or the beginnings of an operation.

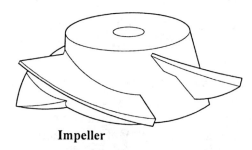

Impeller

Incinerator, Laboratory. See Laboratory Incinerator.

Incinerator, Medical. See Medical Incinerator.

Incinerator. A furnace or apparatus for reducing waste materials to ashes by burning.

Incised. 1. Cut into the surface; engraved. 2. In ceramics, decorated by cutting or indenting the ware surface.

Incline. The slope of a roof expressed either in percent or in the number of vertical units of rise per horizontal unit of run.

Incombustible. Incapable of being burned.

Income & Expense Statement. Summarizes the operations of the company over a period of time; also known as a profit and loss statement.

Income Approach. A real estate appraisal method by which the income stream is capitalized.

Incompatibility. The quality or state of two materials that are not suitable to be used together.

Incompatible. Describes material which cannot be mixed with another material without damaging original properties.

Incomplete Fusion. Fusion which is less than complete.

Increaser. A pipe fitting which increases from a smaller pipe to a larger pipe.

Increaser. A fitting in the vent stack before the stack enters the roof; installed to enlarge the stack and reduces the possibility of water vapor condensing and freezing to the point of closing the opening.

Increment Borer. 1. An augerlike instrument with a hollow bit and an extractor, used to extract thin radial cylinders of wood from trees to determine age and growth rate. 2. Also used in wood preservation to determine the depth of penetration of a preservative.

Incubator, Laboratory. See Laboratory Incubator.

Incubator. An apparatus for the maintenance of controlled conditions.

Incurable Depreciation. Decreases to value of property that cannot be rectified or are too expensive to allay.

Indemnification. A contractual obligation by which a person or entity agrees to reimburse others for loss or damage arising from specified liabilities; also called Hold Harmless.

Indemnity. A promise to hold a person harmless from liability or loss.

Indentation. In a spot, seam, or projection weld, the depression on the exterior surface or surfaces of the base metal.

Independent Contractor. A person who, in performing services for another, is responsible only for the final result, and is not subject to control as to the methods used to achieve that result.

Indeterminate. Statically indeterminate.

Indian Red. Red pigment made artificially by calcining copperas; has excellent permanency, is non-bleeding, alkali- and acid-fast.

Indiana Limestone Institute of America, Inc (ILIA). Stone City Bank Building, Suite 400, Bedford, Indiana 47421, (812) 275-4426.

Indicator Button. A device on a hotel lock to indicate whether or not the room is occupied.

Indicator Compounds. Chemical compounds, such as carbon dioxide, whose presence at certain concentrations may be used to estimate certain building conditions, for example, airflow.

Indirect Costs. All of the costs which cannot be directly attributed in their entirety to the individual unit of production or project.

Indirect Luminaire. A lighting fixture that distributes most of its light upward.

Indirect Overhead. A firm's general office overhead expenses that are not attributable to any specific job and that would continue even when there is no work on hand.

Indirect System. A solar heating or cooling system in which the solar heat is collected outside the building and transferred inside using ducts or piping and fans or pumps.

Indirect Waste Pipe. An indirect waste pipe is a pipe that does not connect directly with the drainage system but conveys liquid wastes by discharging into a plumbing fixture, interceptor, or receptacle which is directly connected to the drainage system.

Individual High Chairs. A welded wire bar support used under a support bar, to provide support for top bars in slabs, joists, or beams; also used to support upper mats of bars in slabs without support bars.

Individual Vent. A pipe which is installed to vent a fixture trap and which connects with the vent system above the fixture served or terminates in the open air.

Indoor Meter Center. A room or area inside a building or structure where the meters for measuring the usage of water, electricity, or gas are located.

Induced Draft. In a boiler, combustion air drawn through the burner or fuel bed by a power driven fan in the flue.

Induced Magnetism. Ability of a magnetic field to produce magnetism in a metal.

Inductance. A property of an electric circuit by which an electromotive force is induced in it by a variation of current either in the circuit itself or in a neighboring circuit.

Induction Baking. Using heat induced by electrostatic and electromagnetic means for baking of finishes.

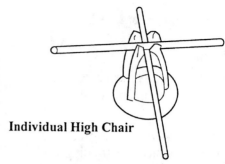

Individual High Chair

Induction Brazing. A brazing process wherein coalescence is produced by the heat obtained from resistance of the work to the flow of induced electric current and by using a nonferrous filler metal having a melting point above 800° F. but below that of the base metals; the filler metal is distributed in the joint by capillary attraction.

Induction Motor. An a-c motor which operates on principle of rotating magnetic field; rotor has no electrical connection, but receives electrical energy by transformer action from field windings.

Induction Welding. A welding process wherein coalescence is produced by the heat obtained from resistance of the work to the flow of induced electric current with or without the application of pressure.

Inductive Reactance. Electromagnetic induction in a circuit creates a counter or reverse (counter) emf (voltage) as the original current changes; it opposes the flow of alternating current.

Industrial Compactor. A machine in an industrial setting that compacts or compresses materials.

Industrial Equipment. Mechanical or non-mechanical devices used in an industrial setting.

Industrial Fluorescent. A large fluorescent light fixture used in an industrial setting.

Industrial Hoist. A mechanical device used to lift loads in an industrial setting.

Industrial Hygienist. A professional qualified by education, training, and experience to anticipate, recognize, evaluate, and develop controls for occupational health hazards.

Industrial Wastes. Liquid wastes which result from the processes employed in industrial establishments and which are free of fecal matter.

Industrial Wood Floor. A heavy duty wood floor made of 2 inch thick decking or of wooden blocks laid on end; very resistant to heavy loads and traffic; also called a Factory Floor.

Inelastic Analysis. An analysis of deformations and internal forces based on equilibrium, nonlinear stress-strain relations for concrete and reinforcement, consideration of cracking and time dependent effects and compatibility of strains; the analysis shall represent to suitable approximation the three dimensional action of the shell together with its auxiliary members.

Inelastic Behavior. Behavior of an element beyond its elastic limit.

Inert Gas. 1. A gas that is not reactive, such as helium or neon; see Noble Gas. 2. In welding, a gas which does not normally combine chemically with the base metal or filler metal.

Inert Pigment. A nonreactive pigment, filler.

Inert. Chemically unreactive.

Inertia Block. A concrete pad on which to mount mechanical equipment on vibration isolators, to prevent machinery vibration from being transmitted into the building structure.

Inertia. A property of matter by which it continues in its present state of rest or uniform motion in a straight line, unless that state is changed by an external force.

Infill. The development of underused building sites in an established neighborhood.

Infiltration. The exchange between conditioned room air and outdoor air through cracks and openings in the building enclosure.

Inflammable. Capable of being easily ignited; flammable; the opposite is nonflammable.

Inflection Point. A point on a curve at which the curvature is zero; in the elastic curve it is the point of zero moment; usually, the curvature of the curve lying to one side of the inflection point is positive, while to the other side it is negative.

Information Technology. The use of machines to handle, process, and transmit information.

Infrangible. Not capable of being broken or separated into parts.

Infrared Lamp. Electrical device which emits infrared rays; invisible rays just beyond red in the visible spectrum.

Infra-Red Sensor. A mechanical device that detects motion using infrared radiation.

Infrared. Lying beyond the visible spectrum at its red end.

Inglaze Decoration. A ceramic decoration applied on the surface of an unfired glaze and matured with the glaze.

Ingot. An oblong piece of cast metal.

Ingress. Entrance; the right to enter.

Inhibitive Pigment. One which retards the corrosion process.

Inhibitor. A substance which prevents chemical reaction such as corrosion or oxidation.

Initial Flow. See Flow of Mortar.

Initial Set. The start of the setting action in mortar, grout or concrete; see Set, 4.

Initial Setting Time. The time required for a freshly mixed cement paste, mortar or concrete to achieve initial set.

Initial Stress. In prestressed concrete, the stresses occurring in the prestressed members before any losses occur.

Injection Molding. A method of manufacturing plastic objects by injecting thermoplastics into molds.

Injunction. A court order that requires a party to do or not to do a certain act.

Inlay. A decoration in which the design is set into the surface.

Inlet Fitting. A fitting or fixture through which circulated or hydro jetted water enters the pool, spa or hot tub.

Inlet or Pool Return. Fitting through which filtered water flows into pool.

Inlet. An opening for intake.

Innovation. The introduction of something new.

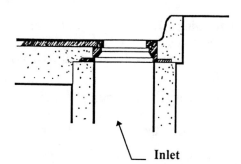

Inlet

Inorganic Coatings. Chemical colors obtained by combining two or more inorganic chemicals.

Inorganic. Not organic; usually of mineral origin.

Insect Screen. A wood or metal frame with fine wire netting to keep out insects, yet permits ventilation.

Insert. A device built into the concrete formwork so it is cast integrally into the concrete, usually for attaching something later.

Inside Corner Trowel. A finishing device used by drywallers and plasterers to finish an inside corner with one pass of a trowel.

Inside Drain. A roof drain that is not at the perimeter of the roof, requiring piping inside the building to carry roof drainage to the exterior.

Insolation. The total amount of solar radiation -- direct, diffuse and reflected -- striking a surface exposed to the sky.

Insolvency. The financial situation in which liabilities exceed assets.

Inspection Hole. A temporary or permanent opening in work to facilitate inspection.

Inspection List. List of defective or incomplete work to be completed after substantial completion; also called Punch List.

Inspection. Examination of completed work or work in process to determine its conformance to the contract requirements and the building code.

Inspector. One who examines completed work or the work under progress, such as a building inspector.

Installment Land Sale. See Conditional Sales Contract.

Instructions to Bidders. Instructions contained in the bidding documents for preparing and submitting bids for a construction project.

Instrument. Used broadly to denote a device that has measuring, recording, indicating or controlling abilities.

Insulated Block. Hollow masonry block filled with insulation.

Insulated Glass. A type of glass constructed in a manner to protect against sound, heat, heat loss, or moisture; double or triple glazed glass.

Insulated Roof. The roof of a building, constructed of materials that protect against heat, heat loss, or moisture.

Insulated Siding. Wall siding that has been manufactured with insulation installed under the finish.

Insulated. Covered and protected in an approved manner with suitable materials of the type and thickness adopted as standard requirements for the voltage and location of the particular material, conductor or part involved.

Insulating Varnish. A varnish especially designed for the electrical insulation of wires, coils, and electrical appliances.

Insulating Wallboard. Foil Backed Gypsum products; see Foil Back.

Insulation Board. 1. Insulation of various types in the form of a rigid board. 2. A low-density board made of wood, sugar cane, cornstalks, or similar material; it is dried and usually pressed to a thickness of 1/2 or 25/32 inch; also called Fiberboard.

Insulation Equipment. Mechanical devices used to distribute insulated materials in walls, attics or crawlspaces.

Insulation Fastener. A device used to fasten different types of insulation to floors, walls, roofs, or ceiling surfaces.

Insulation Fasteners. Devices used to fasten different types of insulation to floors, walls, roofs, or ceiling surfaces.

Insulation Removal. The process of ridding a building or structure of insulation.

Insulation Material. Material which is a poor conductor of heat; used to retard or slow down flow of heat through wall or partition.

Insulation, Calcium. See Calcium Insulation.

Insulation, Duct. See Duct Insulation.

Insulation, Electric. See Electric Insulation.

Insulation, Gymnasium Underfloor. See Gymnasium Underfloor Insulation.

Insulation, Masonry. See Masonry Insulation.

Insulation, Pipe. See Pipe Insulation.

Insulation, Rigid. See Rigid Insulation.

Insulation, Sill Sealer. See Sill Sealer Insulation.

Insulation, Sound Attenuating. See Sound Attenuating Insulation.

Insulation, Sprayed. See Sprayed Insulation.

Insulation. Any material used to reduce the effects of heat, cold or sound transmission and to reduce fire hazard; any material used in the prevention of the transfer of electricity, heat, cold, moisture and sound.

Insulative Panel, Foil-Face. Flat sheet material made of insulative material that has aluminum foil on one face to improve thermal resistance characteristics of the assembly.

Insulative Panel. Flat sheet material made of insulative material to improve thermal resistance characteristics of assembly.

Insulative Wall Sheathing, Foil-Faced. Flat sheet material of foil-faced insulative board material secured to the exterior side of exterior wall studs, used to create a rigid surface on the building superstructure and serve as base to receive siding or veneer construction.

Insulator Spool. A device designed and used to support a conductor and electrically separate it from other conductors or objects.

Insulator. A ceramic fitting which insulates electrical wires.

Insurance Premium. The amount paid to an insurance company for protection against property or business losses and liabilities.

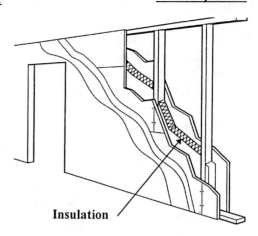

Insulation

Insurance. A contract whereby one party undertakes to indemnify or guarantee another against loss from specified perils.

Insured. Covered by insurance.

Intaglio. A carving incised in hard material.

Intangible Assets. Assets having no physical nature, but having value derived from the rights conferred upon their owner, such as goodwill.

Intangible. Something that cannot be precisely measured or assessed; impalpable; not having physical substance.

Integral Cove Base. A coved base produced by running a flooring material up the wall.

Integrated System. A solar heating or cooling system in which the solar heat is absorbed in the walls or roof of a dwelling and flows to the rooms without the aid of complex piping or ducts.

Intense Color. A strong, vivid color.

Intensity Level (IL). A measure of the acoustic power passing through a unit area expressed on a decimal scale.

Intensity. A subjective measure of the force of an earthquake at a particular place as determined by its effects on persons, structures and earth materials; intensity is a measure of effects as contrasted with magnitude which is a measure of energy.

Intercepting Drain. A drain that intercepts and diverts ground water before it reaches the area to be protected; also called a Curtain Drain.

Interceptor. Any device installed in the drainage system to prevent the passage of grease or other solid materials such as sand.

Intercoat Contamination. In painting, the presence of foreign matter between successive coats.

Intercom. An Intercommunication System.

Intercommunication System. A local telephone system within a building or a group of rooms or offices.

Interest Rate. The amount of interest, expressed as a percentage of the principal per period of time, for example, 8 percent per annum.

Interest. A charge for borrowed money; to calculate the amount of interest. Interest = Principal x Rate x Time.

Interface. 1. A surface forming a common boundary between two regions. 2. A point where interaction occurs between two systems or processes. 3. An electrical component that connects a computer to another piece of equipment.

Interference Body Steel Bolt. High strength bolt with hardened steel ribs on bolts shank to facilitate driving of bolts in misaligned holes.

Intergrown Knot. A knot whose rings of annual growth are completely intergrown with those of the surrounding wood.

Interim Financing. A construction loan, to be paid off from the permanent loan upon completion of construction; also called Interim Loan.

Interim Loan. See Interim Financing.

Interior Decorator. One who selects, and sometimes purchases, furniture, floor coverings, and fabrics, and designs interior color schemes.

Interior Designer. One who designs interiors of buildings.

Interior Elevation. An architectural drawing showing the projection on a vertical plane of an interior surface of a building.

Interior Finish. Material used to cover the interior framed areas or materials for walls and ceilings.

Interior Painting. The process of painting the inside surfaces and trim of a building or structure.

Interior Plywood. A general term for plywood manufactured for indoor use or in construction subjected to only temporary moisture; the adhesive may be interior, intermediate or exterior.

Interior Stucco. Finish plaster for walls and ceilings, finishing, smooth, or textured, consisting of a mechanically blended compound of Keene's cement, lime (Type S), and inert fine aggregate; color pigment may be added to produce integrally colored interior stucco; also called Interior Plaster; see Stucco.

Interior Trim. General term for all the molding, casing, baseboard, and other trim items applied within the building by finish carpenters.

Interior Veneer. A veneer applied to surfaces other than weather- exposed surfaces.

Interior Wood Paneling. Wood panel members used to cover and finish interior walls; often prefinished.

Interior Wood Trim. Finish components of wood such as moldings applied around openings and intersections of surfaces at interior locations.

Interlocked. So arranged that a sequence of operation or steps are electrically or mechanically assured.

Interlocked-Grained Wood. Grain in which the fibers put on for several years may slope in a right-handed direction, and then for a number of years the slope reverses to a left-handed direction, and later changes back to a right-handed pitch, and so on; such wood is exceedingly difficult to split radially, though tangentially it may split fairly easily.

Interlocking. The tiering, binding or blending of various articles with one another.

Intermediate Coat. Middle coat; guide coat.

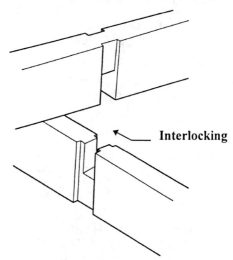

Interlocking

Intermittent Cycle. Cycle which repeats itself at varying time intervals.

Intermittent Welding. Welding wherein the continuity is broken by recurring unwelded spaces.

Intermittent. Intermittent duty is a requirement of service that demands operation for alternate intervals of (1) load and no load; or (2) load and rest; or (3) load, no load and rest.

Internal Control. The method and procedures adopted within a business to safeguard assets and control its operations.

Internal Friction. Resistance to sliding within a mass of wet or dry earth, sand, or gravel.

Internal Mix. A spray gun in which the fluid and air are combined before they leave the gun.

Internal Vibrator. Metal vibrating head immersed in fresh concrete during placement to consolidate fresh concrete; used primarily in cast in place construction.

International Building Code. Produced by International Code Council.

International Code Council. 5203 Leesburg Pike, #708, Falls Church, VA 22041, (703) 931-4533.

International Conference of Building Officials (ICBO). 5360 South Workman Mill Road, Whittier, California 90601, (310) 699-0451.

International Institute of Ammonia Refrigeration (IIAR). 1200 19th Street, NW, #300, Washington, DC 20036, (202) 223-4579.

International Masonry Institute (IMI). 823 15th Street, NW, #1001, Washington, DC 20005, (202) 783-3908

International Symbol of Accessibility. A symbol adopted by Rehabilitation International's 11th World Congress for the purpose of indicating that buildings and facilities are accessible to persons with disabilities.

Interpass Temperature. In a multiple-pass weld, the lowest temperature of the deposited weld metal before the next pass is started.

Interply. Between two layers of roofing.

Interpolate. Estimate the in between values from known ones in the same range.

Interrogatories. Written questions that a person in a lawsuit or litigation must answer in writing, under oath.

Interrupting Capacity. The maximum current at its rated voltage that a device will open without injury to any part other than replaceable fuses. The interrupting capacity shall be determined from the recognized standards for testing equipment.

Intersect. To met and cross at a point.

Intersection. A place where two or more things meet and cross at a point.

Interstate Commerce Commission. Government body which controls the design and construction of pressure containers.

Interstice. A space that is between things, like the grout space between ceramic tiles.

Intestate. A person not having made a will before death.

In-the-White. Natural or unpainted; the natural unfinished surface of the wood.

Intumesce. To foam, swell, froth, or bubble up as a result of heat, liquid, air or chemical action.

Intumescence. An enlarging, swelling or bubbling up, as under the action of heat.

Inventory. Includes all material, labor and overhead on jobs which are currently in progress, less any amounts which have been billed to customers.

Invert of a Pipe. The inside bottom elevation; the flow line.

Invert, Manhole. The lowest inside surface of a manhole; a channel in the manhole through which wastewater or stormwater flows.

Invert. The lowest portion of the inside of any horizontal pipe.

Inverted Arches. Arches that appear to have been built upside down; used to distribute the weight of the wall from pier to pier.

Inverted Roof Membrane Assembly (IRMA). A roofing system where the membrane is applied to the substrate and the insulation boards are on top, weighted down by gravel and rock ballast.

Inverted Roof. A membrane roof assembly in which the thermal insulation lies above the membrane.

Invisible Hinge. A hinge so constructed that no parts are exposed when the door is closed.

Invitation to Bid. A formal written invitation to submit a bid, usually placed in trade papers or newspapers, informing prospective bidders about a project; necessary for public work but not for private work.

Invited Bidders. A closed list of bidders decided by owner and architect as the only ones who may bid a job.

Invoice. A list of goods shipped or services rendered, with prices and charges.

Involuntary Lien. Any legal claim that is placed on a property without the permission of the owner, such as a mechanics lien.

Involute Curve. A curve traced by a thread unwound from another curve.

Iodine Number. A means of identifying and specifying qualities of oils, resins and waxes, based on fact that different qualities of these products will absorb different quantities of iodine.

Iodine. A chemical compound containing iodine; potassium or sodium iodide, when used with a suitable oxidizing agent such as chlorine, will release iodine in pool water.

Ion. Group of atoms or an atom electrically charged.

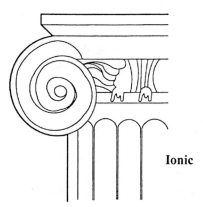

Ionic

Ionic. The order of Greek architecture characterized by a column with scroll shapes on either side of the capital.

Ionization Detector. A mechanical device indicating the presence of certain gaseous compounds and amount of ions produced when subjected to ultraviolet light or hydrogen flame.

Ionization. Breaking up of molecules into two or more oppositely charged ions; an ion is one of electrified particles into which molecules are divided by the use of water and other solvents.

IPM. Integrated Pest Management.

IPS. Iron pipe size

IR Drop. Electrical term indicating the loss in a circuit expressed in amperes times resistance (1 x R) or voltage drop.

Irish Confetti. A rock or brick used as a missile.

IRMA. Inverted Roof Membrane Assembly.

Iron Blue. Blue pigment which depends on iron content to provide blue color.

Iron Driers. Driers with high tinting strength which limits use to colored finishes.

Iron Oxide. Ferrous oxide; corrosion of iron by oxidation of the surface; rust; iron oxide is available in three forms; red, brown and yellow; it is known under a variety of names, such as Red Oxide, Jeweler's Rouge, Venetian Red, Ferric Oxide, Indian Red, Red Ochre, Mineral Rouge, Spanish Oxide, and Turkey Red.

Iron Phosphate Coating. Conversion coating; chemical deposit.

Iron. A heavy malleable ductile magnetic silver-white metallic element that readily rusts in moist air and is the most used of metals.

Iron-Phosphate Treated. Cleaning and treatment of steel deck prior to forming.

Ironstone Ware. Stone china, white granite ware, historic terms for a durable English earthenware.

Irregular Stone. Stone cut to or quarried in various shapes and sizes.

Irrigation, Lawn. To supply water to grassy areas by artificial means.

Isinglass. Mica.

Isobar. Lines on a map connecting points of equal atmospheric pressure at a particular time.

Isochronous. Occurring at the same time.

Isocyanate Resins. Urethane resins.

Isogram. A line on a map or chart connecting points of equal value, as of temperature, elevation, or precipitation.

Isohyetal Line. A line on a map or chart connecting points of equal rainfall in a specified period.

Isolated. Not readily accessible to persons unless special means for access are used.

Isolation Interface. The boundary between the upper portion of a structure, which is isolated, and the lower portion of the structure, which moves rigidly with the ground.

Isolation Joint. A separation between adjoining parts of a concrete structure, usually a vertical plane, at a designed location such as to interfere least with performance of the structure, yet such as to allow relative movement and avoid formation of cracks elsewhere in the concrete and through which all or part of the bonded reinforcement is interrupted.

Isolation System. The collection of structural elements which includes all individual isolator units, all structural elements which transfer force between elements of the isolation system, and all connections to other structural elements; the isolation system also includes the wind restraint system if such a system is used to meet the design requirements.

Isolation. An acoustical term used to describe sound privacy by reducing direct sound paths.

Isolator Unit. A horizontally flexible and vertically stiff structural element of the isolation system which permits large lateral deformations under design seismic load; an isolator unit may be used either as part of or in addition to the weight-supporting system of the building.

Isolator, Vibration. See Vibration Isolator.

Isometric Drawing. The representation of an object in isometric projection; an isometric drawing looks like a perspective but all lines parallel to the three major axes are measurable.

Isometric Projection. Axonometric projection in which all three faces are equally inclined to the drawing surface.

Isosceles Triangle. A triangle having two equal sides.

Isoseismals. Map contours drawn to define limits of estimated intensity of shaking for a given earthquake.

Isothermal Expansion and Contraction. Action which takes place without a temperature change.

Isothermal. Changes of volume or pressure under conditions of constant temperature.

Isotropic. Exhibiting the same properties in all directions.

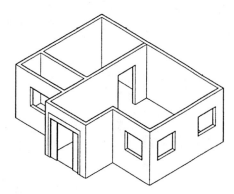

Isometric Drawing

Ivory Black. A high grade bone black pigment; formerly made by charring or burning ivory.

Ivory Paper. See Cream Paper.

Ivory Tower. 1. A state of seclusion or separation from the practicalities and realities of life. 2. Academia.

Ivory. A hard creamy white substance that makes up the tusk of an elephant, walrus, hippopotamus, or narwhal.

Jack Arch. 1. One having horizontal or nearly horizontal upper and lower surfaces; also called a Flat Arch or Straight Arch. 2. Any arch that is roughly built.

Jack Plane. A medium sized plane used in the first stages of planing.

Jack Rafter. 1. A short rafter that joins a hip rafter to the top of the wall plate. 2. A short rafter that joins a valley rafter to a ridge board.

Jack. A mechanical device for lifting heavy objects.

Jacket, Aluminum. See Aluminum Jacket.

Jackhammer. A pneumatic concrete breaker.

Jacking Equipment. In prestressed concrete, the device used to stress the tendons.

Jacking Force. The temporary force exerted by a device that introduces tension into prestressing tendons in prestressed concrete.

Jacking Pipe. Forcing pipe through the ground in a tunnel created by the pipe itself; the pipe is generally jacked horizontally in short lengths.

Jacking Stress. In prestress concrete, the maximum stress occurring in a tendon during stressing.

Jack-of-all-Trades. A person who can do many different kinds of work.

Jacquard. Mechanism for a Wilton carpet loom which produces the desired color design using a chain of perforated cardboard cards punched according to the design elements, which when brought into position activates this mechanism by causing it to select the desired color of yarn to form the design on the pile surface; the unselected colors are woven dormant through the body of the fabric.

Jagged Edges. Irregularities left on the edges of ceramic tile due to the use mainly of hand cutting tools.

Jalousie Window. A window with stationary or adjustable blinds angled to permit air and provide shade, while preventing rain from entering.

Jalousie. A blind or shutter made of a row of angled slats to keep out rain and control the influx of light.

Jamb Anchor. Steel anchor used to fasten a steel door frame to the wall or partition construction.

Jamb Block. A concrete masonry unit with a preformed slot to receive a window or door frame.

Jamb, Door. See Door Jamb.

Jamb. The vertical side of a door or window frame.

Japan Color. Colored paste made by grinding high-quality colors in hard drying varnish.

Japan Drier. Varnish gum with a large proportion of metallic salts, such as lead, cobalt, or manganese, added to hasten drying; used in paints, varnishes, and enamels.

Japanese Lacquer. Varnish made from sap of a tree which grows in Japan; becomes very hard and black as it dries.

Jaspe. Carpet surface characterized by irregular stripes produced by varying textures or shades of the same color.

J-Box. An electrical junction box.

Jib Crane. A crane which has a projecting arm off its derrick boom.

Jib. The arm of a crane.

Jig. A tool or template used to maintain mechanically the correct positional relationship between a piece of work and the tool or between parts of work during assembly.

Jiggering. Forming ceramic ware from a plastic body by differential rotation of a profile tool and mold, the mold having the contour of one surface of the ware and the profile tool that of the other surface.

Jigsaw. 1. A machine saw with a narrow reciprocating blade used for cutting curved and intricate patterns. 2. A scroll saw.

Jitterbug. A grate tamper for pushing coarse aggregate slightly below the surface of a slab to facilitate finishing.

J-Mold. A metal molding strip with a J-like section, used to edge plaster or gypsum wallboard.

Job Built Form. A temporary structure or mold constructed on a jobsite, for the support of concrete while it is setting and gaining sufficient strength to be self-supporting.

Job Lot. A miscellaneous collection of goods for sale as a lot.

Job Requirements. A list of specific, necessary, and essential tasks to bring to completion a building or structure; also called General Requirements.

Job Scheduling. An itemization in chronological order, often in chart form, of project tasks in order to start and complete a building or structure.

Job Trailer. A towed vehicle placed on a jobsite, acting as an office space.

Job. A piece of work.

Jobsite. The place where construction takes place.

Joinery. The art or trade of joinery; woodwork; finish carpentry.

Joining. The juncture of two separate plaster applications usually within a single surface plane; also called a Jointing. May also apply to asphalt paving.

Joint Cement. See Joint Compound.

Joint Check. A check made payable to more than one payee.

Joint Clip. 1. A metal fastener used vertically, sharp edge down, over the edges of two pieces of wood, and then hammered down into them. 2. In plywood sheathing, the clip fastens two abutting pieces of plywood.

Joint Compound. A cementitious material used in covering joints, corners and fasteners in the finishing of gypsum board to produce a smooth monolithic surface; also called Spackle or Joint Cement.

Joint Control. An independent escrow used to safeguard and disburse construction funds.

Joint Efficiency. The strength of a wood joint expressed as a percentage of the strength of clear straight-grained material.

Joint Factor. Joint efficiency.

Joint Filler. Compressible material used to fill a joint to prevent the infiltration of debris and to provide support for sealants.

Joint Photographing. In gypsum-board, the shadowing of the finished joint areas through the surface decoration.

Joint Reinforcement. Steel wire, bar, or prefabricated reinforcement which is placed in mortar bed joints.

Joint Reinforcing Mesh. A woven fiber screen-like material used in lieu of paper joint tape.

Joint Reinforcing Metal. Strips of expanded metal, woven or welded wire mesh used to reinforce corners and other areas of plaster and lath.

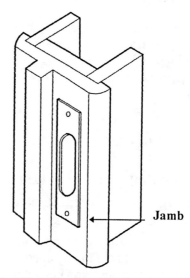

Jamb

Joint Reinforcing Tape. A type of paper, metal, fabric, glass mesh, or other material, commonly used with a cementitious compound, to reinforce the joints between adjacent gypsum boards.

Joint Reinforcing. Joint Reinforcement.

Joint Sealant. Compressible material used to exclude water and solid foreign materials from openings; a substance that prevents water and moisture from entering joints; also called Joint Sealer.

Joint Sealer. Joint Sealant.

Joint Tape System. The compound and tape system used to conceal and finish joints in gypsum board.

Joint Tape. Paper tape that is applied with gypsum-based joint compound to treat the joints in gypsum wallboard.

Joint Tenants. Two or more persons owning property together with the right of survivorship.

Joint Venture. A temporary partnership composed of individuals, partnerships, or corporations organized to accomplish a specific job or course of work.

Joint, Butt. See Butt Joint.

Joint, Contraction. See Contraction Joint.

Joint, Control. See Control Joint.

Joint, Expansion. See Expansion Joint.

Joint. 1. The location where two or more members are to be joined. 2. The point of connection between structural members. 3. The seam produced by the placement of two pieces of gypsum board together but not necessarily in the same plane. 4. The junction of two pieces, as of wood or veneer. 5. The location at which two pieces of pipe or a pipe and a fitting are connected together.

Jointer. 1. A metal hand tool used to cut a joint partly through fresh concrete. 2. In masonry, a tool used by bricklayers to form the various types of mortar joints between the courses of masonry, as the V, the concave, and weather joints. 3. An electrically powered machine for making joints.

Jointing, A. See Joining.

Jointing. 1. Smoothing and straightening the edge of a wood board. 2. Jointing is done automatically by a jointer. 2. Grinding or filing the teeth or knives of power tools to the correct height; circular saws are jointed so that there are no high or low teeth; knives of planers and jointers are jointed so that each knife makes the same depth of cut as all others. 3. The act of facing the mortar joints.

Joist Anchor. A metal rod incorporated into a masonry wall that extends out to be fastened to a joist or rafter.

Joist Bridging. 1. The bracing of joists by the fixing of lateral members between them. 2. Pieces fitted in pairs from the bottom of one floor joist to the top of adjacent joists, and crossed to distribute the floor load. 3. Wood pieces of width equal to the joists and fitted neatly between them. 3. Diagonal or longitudinal members used to keep horizontal members properly spaced, in lateral position, vertically plumb, and to distribute load.

Joist Girder. A light steel truss used to support open-web steel joists.

Joist Hanger. A metal stirrup that supports the ends of joists so that they are flush with the girder.

Joist Painting. The act or process of applying paint to seal, protect, or add color to a joist, usually refers to metal joists.

Joist, Metal. See Metal Joist.

Joist, Sister. See Sister Joist.

Joist, Wood. See Wood Joist.

Joist. A usually horizontal, structural member used as a floor, ceiling or roof framing member.

Joule. In the international system of units, the amount of energy needed to raise the temperature of 1 kilogram of water $1°$ C.

Joule-Thomson Effect. The change in the temperature of a gas on its expansion through a porous plug from a higher pressure to a lower pressure.

Journal, Crankshaft. Part of shaft which contacts the bearing on the large end of the piston rod.

Journal. 1. A record of current transactions. 2. A log. 3. The part of a rotating shaft, axle, roll, or spindle that turns in a bearing.

Journeyman. 1. A fully trained and qualified person in a craft or trade such as a carpenter, plumber, or electrician. 2. A plasterer or lather who through training and experience has become thoroughly skilled in his trade; distinguished from an apprentice or a laborer. 3. A painter who has had at least three year's experience and schooling as an apprentice.

Judgment. A final decision of a court.

Jumbo Brick. A brick that is larger than standard size and measures 8"x4"x4".

Jumpover. In piping, a double offset used to return the pipe to its original line; also called a Return Off-Set.

Junction Box. A metal or hard plastic electrical rough-in box, usually square or octagonal, housing only wire or cable connections.

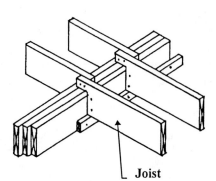

Joist

Jute Padding. A padding made of a durable yarn that comes from plant fiber used as an underlayment for carpet.

Jute. Strong, durable yarn spun from fibers of the jute plant, native to India and Far East, used in the backings of many carpets.

Juxtaposition of Colors. Placing colors side-by-side, or close together; complementary colors such as blue and orange in juxtaposition accentuate each other.

Juxtaposition. The placing of things side by side.

K

K Bracing. That form of bracing where a pair of braces located on one side of a column terminates at a single point within the clear column height.

K Series. A standard Steel Joist Institute designation for longspan steel joists.

Kalotermes. See Termite.

Kalsomine. See Calcimine.

Kaolin. 1. A refractory clay consisting essentially of minerals of the kaolin group and which fires to a white or nearly white color; also called China Clay. 2. Inert pigment which tends to impart easy brushing properties to paint products in which it is used.

Kata Thermometer. Large-bulb alcohol thermometer used to measure air speed or atmospheric conditions by means of cooling effect.

Kauri Gum. A fossil copal found in New Zealand.

Kauri Reduction. Test for solvent power of petroleum solvents.

Keene's Cement. A cement composed of finely ground, anhydrous, calcined gypsum, the set of which is accelerated by the addition of other materials, used in areas subjected to moisture; a hard, strong finishing plaster that is made from gypsum and maintains a high polish; because of its density, it excels for use in bathrooms and kitchens and is also widely used for the finish coat in auditoriums, public buildings, and other places where walls may be subjected to unusually hard wear or abuse; also called Anhydrous Calcined Gypsum.

Keeper. See Strike, 2.

Kelly Ball Test. A method by which the workability of fresh concrete can be determined.

Kelvin Scale. A temperature scale on which the unit of measurement equals the Celsius degree and according to which absolute zero is 0°, the equivalent of minus 273.16° C; water freezes at 273.16° K. and boils at 373.16° K.

Kennel. 1. A shelter for a dog. 2. A breeding or boarding establishment for dogs.

Kerf. 1. A cut made with a saw, or with a cutter, part way through a material, generally to facilitate breaking to a desired shape. 2. The space from which metal has been removed by a cutting process.

Kerfing. Longitudinal saw cuts or grooves of varying depths on the unexposed faces of millwork members to relieve stress, prevent warping, or to facilitate bending.

Kerosene. A distillate obtained in petroleum refining which evaporates slowly.

Ketones. Organic solvents containing CO grouping; commonly used ketones are acetone (dimethyl ketone); MEK (methyl ethyl ketone); and MIBK (methyl isobutyl ketone).

Kettle Bodied Oil. Oil which has been held at an elevated temperature until the oil has thickened.

Kettle, Steam. See Steam Kettle.

Kettle. Portable roofer's equipment for heating bitumen to a suitable temperature for use.

Key Changes. The different combinations of bittings or tumbler arrangements in a series of locks.

Key Lock. A lock which is activated by a key.

Key Switch. An electric on-off switch operated by a removable key.

Key. 1. A slot formed into a concrete surface for the purpose of interlocking with a subsequent pour of concrete; a slot at the edge of a precast member into which grout will be poured to lock it to an adjacent member. 2. A small piece of wood inserted in one or both parts of a joint to align it and hold it firmly together. 3. The center stone or brick of an arch. 4. The grip or mechanical bond of one coat of plaster to another coat or to a plaster base. It may be accomplished physically by the penetration of wet mortar or crystals into paper fibers, perforations, scoring irregularities, or by the embedment of the lath; see Mechanical Bond.

Keyed Joint. 1. A joint in which one structural member is keyed or notched into an adjoining member as in timber construction. 2. In masonry construction, a finished joint of mortar which has been tooled concave.

Keyhole Saw. A narrow pointed fine tooth saw for cutting small or curved holes.

Keystone. The wedge shaped central stone at the top of an arch.

Kick Hole. A hole in the roofing membrane at the base of a parapet wall, at the cant, usually caused by being stepped on or kicked.

Kick Lift. A jacking wedge used to elevate or shift the gypsum board into proper nailing position on the wall during the application procedure.

Kick Plate. A metal plate or strip that runs along the bottom edge of a door to protect against the marring of the finished surface.

Kiln Dried Lumber. Lumber that has been kiln-dried, often to a moisture content of 6 to 12%; common varieties of softwood lumber, such as framing lumber, are dried to a somewhat higher moisture content.

Kiln Dried. Lumber that has been heated in a kiln to dry and control the amounts of moisture.

Kiln Drying. Drying of wood, paint, varnish or lacquer in room or compartment with heat and humidity regulated.

Kiln Run. Brick or structural clay tile from one kiln which have not been sorted or graded for size, burning, or color variation.

Kiln. A furnace, oven, or heated enclosure for processing a substance by burning, firing, or drying.

Kilogram. A metric unit of mass and weight equal to 1000 grams.

Kilometer. A metric unit of distance equal to 1000 meters.

Kilowatt. An electrical unit of 1000 watts.

Kilowatt-Hour (KWH). A measure of electrical energy equivalent to a power consumption of 1000 watts for one hour.

Kinematic Viscosity. A method of measuring viscosity using the stoke as the basic measurement unit.

Kinetic Energy. Energy associated with motion.

Kinetic Friction. The friction between two surfaces that are moving tangentially in contact with each other; kinetic friction is always slightly less than static friction.

Kinetic. Relating to motion.

King Closer. In masonry, a closer used to fill an opening in a course larger than a half brick; about three- fourths the size of a regular sized brick; also called a king.

King Post Truss. A standard peaked roof truss configuration.

King Post. In a roof system, the member placed vertically between the center of the horizontal tie beam at the lower end of the rafters and the ridge, or apex of the inclined rafters.

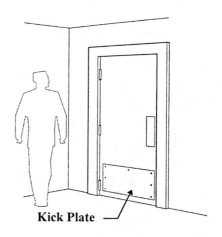

Kick Plate

King Stud. The last full length wall stud adjacent to a trimmer stud at a door or window opening.

King Valve. Liquid receiver service valve.

King. See King Closer.

Kip. A unit of weight or force equal to 1,000 pounds.

Kirk. Church.

Kitchen Cabinet. A case, box, or piece of furniture with sets of drawers or shelves, with doors, primarily used for storage, mounted on walls or floors in a kitchen area.

Kitchen Compactor. A machine in a kitchen that compresses or compacts materials by using hydraulic weight, force, or vibration.

Kitchen Faucet. A fixture found in the kitchen that is used for drawing potable water.

Kitchen Fryer. An appliance used for frying foods.

Kitchen Range. A stove with an oven and a flat top surface with burners powered by gas or electric current.

Kitchen Sink. A stationary basin that is found in the kitchen and connected to a drain.

Kitchen. A room for preparation of food.

Knee Brace. A corner brace, fastened at an angle from wall stud to rafter, stiffening a wood or steel frame to prevent angular movement.

Knee Kicker. A short tool with gripping teeth at one end and a padded cushion at the other, used in making small stretches during carpet laying.

Knee Wall. A short wall under the slope of a roof.

Kneepads. Protective foam rubber pads for workers to wear strapped on their knees when working on floors.

Knife, Board. See Board Knife.

Knife. 1. A cutting implement consisting of a blade fastened to a handle. 2. A sharp cutting blade or tool in a machine.

Knitted. A type of nonwoven carpet construction in which the backing yarn, stitching yarn and pile yarn are looped together with three sets of knitting needles, producing an uncut loop pile.

Knitted Carpet. Carpet made on a knitting machine by looping together backing, stitching and pile yarns with three sets of needles, as in hand knitting.

Knockout. A cylindrical removable section of a junction box or outlet box to accommodate installation of a conduit connector.

Knot. That portion of a tree branch or limb that has been surrounded by subsequent growth of the stem; the shape of the knot as it appears on a cut surface depends on the angle of the cut relative to the long axis of the knot.

Knotty Pine. Pine wood whose knots are exposed, often used for cabinets and interior paneling.

Knotty Score. Of gypsum board, a ragged cut edge.

Knuckle. The enlarged part of a hinge into which the pin is inserted.

Knurled. Having a surface covered with small knobs or beads, as a nail which may have such a surface for greater holding power.

Kraft Paper. A sturdy brown wrapping paper.

Kraftcord. Tightly twisted yarn made from wood pulp fiber, used as an alternate for cotton or jute in carpet backing.

Krebs Unit (KU). Arbitrary units of viscosity.

KSI. Kips per square inch.

KU. Krebs Unit.

KWH. Kilowatt Hour.

Kyanite. The most abundant of the mineral polymorphs that include andalusite and sillimanite; $Al_2O_3 \, SiO_2$; used as a source of mullite in ceramics.

L Cut. A piece of tile cut or shaped like the letter L.

Labor and Material Bond. A bond, secured by the general contractor, which guarantees that the costs for labor and materials for the project will be paid.

Labor Union. A trade union.

Laboratory Cabinet. A case, box, or piece of furniture with sets of drawers or shelves, with doors, primarily used for storage, used in a place or area for scientific studies or commercial and institutional laboratories and testing facilities.

Laboratory Counter. A level surface in a laboratory where equipment is placed and kept and where work may be performed.

Laboratory Equipment. Devices and tools used in laboratory work.

Laboratory Hot Plate. A heating device in a laboratory.

Laboratory Incinerator. A device in a laboratory in which waste materials are burned.

Laboratory Incubator. An apparatus used in a laboratory for the maintenance of controlled conditions.

Laboratory Table. A surface on which laboratory materials and devices are kept and used.

Laboratory. A room or building equipped for physical or chemical testing, experimentation, or analysis.

Lac. A natural resin secreted by certain insects which live on the sap of trees in India and other Oriental countries; marketed in various forms, such as seed lac, button lac, and shellac.

Laches. Undue delay in asserting one's legal rights, resulting in loss of the rights.

Lacquer. A sometimes colored liquid made of shellac dissolved in alcohol, or of synthetic substances, that dries to form a hard protective coating.

Lacunar. A ceiling made up of vaults or recessed panels.

Lacunaria. A recessed panel in a ceiling.

Ladder Bar. A prefabricated reinforcement designed for embedment in the horizontal mortar joints of masonry; parallel deformed side rods connected in a single plane, by cross wires, thus forming a ladder-like design.

Ladder Cage. A cage-like structure surrounding a wall-mounted ladder for safety.

Ladder Rung. A horizontal round member used as a step on a ladder.

Ladder Step. A horizontal flat member used as a step on a ladder.

Ladder, Pool. See Pool Ladder.

Ladder. A frame consisting of two parallel side pieces connected by rungs at suitable distances to form steps on which persons may climb up or down.

Lag Rod. A large diameter rod with a square or hexagonal head.

Lag Screw. A large diameter wood screw with a square or hexagonal head.

Lag. Delay in response.

Laid-Off. See Double-Up.

Laid-On. See Double-Up.

Laitance. A layer of weak and non-durable material containing cement and fines from aggregates, brought by bleeding water to the top of overwet concrete, the amount of which is generally increased by overworking or overmanipulating concrete at the surface by improper finishing or by job traffic; effluorescence.

Lake Asphalt. See Natural Asphalt.

Lake Pigment. Pigment made by putting an organic dye on a base of fine particles of inert or translucent pigment.

Lake Sand. Sand consisting predominantly of fine, rounded particles.

Lally Column. Tradename; a hollow steel column, sometimes filled with concrete.

Lamella. A unit of a surface network of closely spaced uniform ribs or beams, usually arranged in two or three intersecting diagonal lines; curved vaults and domes have been built of wood, steel, and concrete lamellas.

Lamina. The layers of material in a laminate.

Laminar Scale. Rust formation in heavy layers.

Laminate, Paper-Base. A multilayer panel made by compressing sheets of resin-impregnated paper together into a solid mass.

Laminate. 1. To form a product by bonding together two or more layers of materials. 2. The product so formed, such as a plastic laminate.

Laminated Glass. A glazing material consisting of outer layers of glass laminated to, and encasing, an inner layer of transparent plastic; used for automobile windshields and bulletproof glass.

Laminated Plastic. See Plastic Laminate.

Laminated Rubber. Several layers of rubber bonded together with adhesive under pressure.

Laminated Timber. An assembly made by bonding layers of veneer or lumber with an adhesive so that the grain of all laminations is essentially parallel.

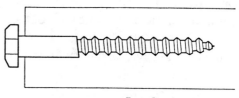

Lag Screw

Laminated Wallboard. Two or more layers of gypsum board held together with an adhesive.

Laminated Wood. A product made by bonding layers of veneer or lumber with an adhesive so that the grain of all layers is generally parallel.

Laminating Compound. A cementitious material, usually regular joint compound, used to adhere two or more layers of gypsumboard together.

Lamination. 1. The uniting of layers of material using adhesive. 2. The application of two or more layers of gypsum board.

Lamp, Incandescent. See Incandescent Lamp.

Lamp, Steri. See Steri Lamp.

Lampback. Pigment made by burning coal tar distillates without sufficient air; not quite true black.

Lampholder. A device to support an electric lamp mechanically and connect it electrically to the circuit conductors.

Lanai. Porch, veranda, or covered patio.

Land Contract. A contract for sale of land where title does not pass to the purchaser until all, or a certain number, of the payments have been made; also called Contract of Sale.

Land Plaster. Coarsely ground natural gypsum used agriculturally as a soil conditioner.

Land Residual. A real estate appraisal technique where a reasonable return on the improvements is first deducted from the income, the balance being attributable to the land.

Land Sale Contract. See Conditional Sales Contract.

Land Surveyor. A person who surveys land.

Land Title. Porous clay pipe with open butt joints.

Land. 1. The surface of the earth and all its natural resources. 2. A portion of the earth's solid surface defined by boundaries or ownership; privately or publicly owned. 3. See Root Face.

Landing Terrazzo. A staircase landing with terrazzo tile mounted on its surface.

Landing. A platform between flights of stairs or at the termination of a flight of stairs.

Landlord. The owner of property that is rented to a tenant.

Landscape Architect. One whose profession is designing the arrangement of land for human use involving vehicular and pedestrian ways and the planting of groundcover, plants, and trees.

Landscape Timber. Large, treated lengths of lumber used to decorate and act as soil erosion barriers and retaining wall members.

Landscape. 1. Natural scenery. 2. To improve a site by modification of the terrain, the planting of trees, shrubs, and ground cover, and the addition of hardscape.

Lane Joint Cracks. Longitudinal separations along the seam between two paving lanes caused by a weak seam between adjoining spreads in the courses of the pavement.

Langly. A measure of solar radiation, equal to 1 calorie per square centimeter.

Lantern. A raised structure on a roof, glazed to admit light.

Lap joint. A joint made by placing one member partly over another and bonding the overlapped portions.

Lap Siding. See Bevel Siding.

Lap Splice. See Lap, 3.

Lap. 1. The overlap of two roofing plies. 2. The length by which one bar or sheet of fabric reinforcement overlaps another. 3. The length of the overlap of two reinforcing bars; also called Lap Splice. 4. The amount of extension of one brick or any other masonry unit over another.

Lapped Joint. In wallpapering, a joint made by trimming one selvedge and overlapping the other.

Lapping. Smoothing a metal surface to high degree of refinement or accuracy using a fine abrasive.

Large Calorie. See Calorie, 2.

Laser. 1. A device that generates an intense beam of coherent monochromatic radiation in the infrared, visible, or ultraviolet region of the electromagnetic spectrum, by stimulated emission photons from an excited source; used in communications, industry, and medicine; used in construction to establish accurate sightings, and plumb and level alignments, vertical and horizontal. 2. Acronym for Light Amplification by Stimulated Emission of Radiation.

Latch Set. A fastening assembly for a door or window, operable from both sides, and lockable with a key.

Latch. 1. A bar with a catch and lever used as a fastening for a gate. 2. A spring-lock requiring a key to pass from the outside.

Latent Ambiguity. A term of a contract that appears on its face to be unambiguous, but that is made ambiguous by external circumstances.

Latent Defect. A construction defect that is not perceptible by ordinary observation.

Latent Heat Gain. The addition of heat to an enclosure by an increase in moisture content.

Latent Heat of Condensation. Amount of heat lost by a pound of a substance to change its state from a gas to a liquid.

Latent Heat of Vaporization. Amount of heat required per pound of a substance to change its state from a liquid to a gas.

Latent Heat. Heat that changes the state of material from a solid to a liquid or a liquid to a gas.

Lateral Brace. A temporary or permanent structural brace to resist lateral movement of a truss or beam.

Lateral Branch. See Horizontal Branch.

Lateral Force Coefficients. Factors applied to the weight of a structure or its parts to determine lateral force for a seismic design.

Lateral Force. A force acting generally in a horizontal direction, such as wind, earthquake, or soil pressure against a foundation wall.

Lateral Load. 1. The horizontal component of the load produced by an arch, dome, vault, or rigid frame. 2. A horizontal load applied to a structure or member such as wind or earthquake.

Lateral Pressure. Horizontal pressure such as the force of soil against the side of a high foundation wall.

Lateral Slip Fault. A fault whose relative displacement is purely horizontal.

Lateral Support. A force or structural member that prevents a structure or earthen mass from moving in a lateral or horizontal direction.

Lateral Thrust. 1. The pressure that any load or force exerts sideways or through the ends of the members. 2. The horizontal component of the force or thrust produced by an arch, dome, vault, or rigid frame.

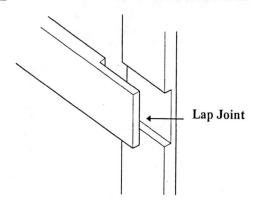

Lap Joint

Lateral. 1. Relating to the side. 2. At right angles to the long direction of the member; crosswise; transverse.

Lateral-Force-Resisting System. The portion of the structure composed of members designed to resist forces related to earthquake, wind, or other lateral effects.

Latewood. The portion of the annual growth ring that is formed after the earlywood formation has ceased; it is usually denser and stronger mechanically than earlywood; also called Summerwood.

Latex Paint. A water-based paint that can be thinned and washed from applicators with water.

Latex. A water based emulsion of a synthetic rubber or plastic obtained by polymerization and used commonly in coatings and adhesives.

Latex-Portland Cement Grout. A portland cement grout with a special latex additive which results in a less rigid, less permeable grout than regular portland cement grout.

Latex-Portland Cement Mortar. A mixture of portland cement, sand, and special latex additives which is used for bonding tile to back-up material.

Lath and Plaster Membrane. 1. A thin slab of lath and plaster including any integral supporting and stiffening members. 2. Lath and plaster as a unit of structure.

Lath, Expanded Metal. See Expanded Metal Lath.

Lath, Stucco. See Stucco Lath.

Lath. A wood strip, metal mesh, or gypsum board which acts as a backing and/or reinforcing agent for the plaster scratch coat or initial mortar coat; a plaster base.

Lathe Tool. A blacksmith's tool used to handle hot metals.

Lathe, Shop. See Shop Lathe.

Lattice Molding. Flat strip molding used in the construction of lattices.

Lattice Truss. A steel truss where the top and bottom chords are connected together by a steel lattice.

Lattice. A framework of crossed wood or metal strips.

Lauan Door. A door made of Philippine mahogany veneer.

Lauan Veneer. Very thin sheets of Philippine Mahogany, which may be combined with glue to create plywood or glued onto another surface to create a finished surface.

Lauan. A variety of Philippine mahogany, of moderate strength and durability, with wood of light yellow to reddish brown or brown.

Laundry Chute. A chute that leads from an upper floor to a lower floor by which soiled laundry is conveyed.

Laundry Equipment. Appliances and materials used for cleaning clothes.

Laundry Tray. A tub with hot and cold water supply and a drain, used in a laundry room.

Laundry. A room for washing and drying of clothes and linens.

Lav. Lavatory.

Lavatory Carrier. A horizontal structure attached to a bathroom wall to support and mount a lavatory.

Lavatory Faucet. A water dispensing device in a bathroom sink or lavatory.

Lavatory. 1. A basin with drainage and running water primarily used for washing the face and hands. 2. A room with a toilet and wash basin.

Law. A rule of conduct enforced by courts.

Lawn Irrigation. The supply of water to grassy areas by artificial means.

Lawn. An area of ground covered with mown grass.

Lawsuit. A proceeding in which the jurisdiction of a court is invoked to resolve to a dispute between two or more parties.

Lawyer. A person who is educated in and authorized to practice law.

Lay Out. The measuring and setting out of work according to the design drawings.

Layer. A stratum of weld metal consisting of one or more weld beads.

Laying Overhand. Laying brick on the farther face of a wall from a scaffold.

Layout Stick. A long strip of wood marked at the appropriate joint intervals for the tile to be used; used to check the length, width, or height of the tilework; also called Idiot Stick.

Layout. A full-sized drawing showing arrangement and structural features.

Lazy Susan. 1. A revolving circular arrangement of shelves. 2. A circular revolving cabinet shelf used in corner kitchen cabinet unit.

Lb. Symbol for pound or pounds.

Leaching Field. A land area containing a series of subterranean perforated pipes that allow septic tank effluent to percolate into the soil; also called Disposal Field.

Leaching Pit. An excavated hole in the ground that can hold solids but allows liquids to pass through and leach into the ground.

Lead Carbonate, Basic. See Basic Lead Carbonate.

Lead Drier. Almost water-white drier which works on body of paint film; various combinations of lead, cobalt, and other driers are used in formulating many modern finishes.

Lead glass. A glass of high refractive index containing lead oxide.

Lead Lined Wallboard. Gypsum wallboard with a ply of lead sheeting to be used in providing radiation protection in x-ray facilities.

Lead Oxide. Compound in several combinations of lead and oxygen, e.g., litharge and red lead.

Lead Sulphate, Basic. See Basic Lead Sulphate.

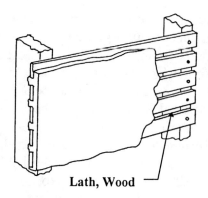

Lath, Wood

Lead Wool. A mass of lead shavings used primarily for packing or caulking.

Lead. 1. A heavy soft malleable ductile plastic but inelastic bluish white metallic element found mostly in combination, used in pipes, cable sheaths, batteries, solder, roofing, flashing, and shields against radioactivity. 2. A section of a wall built at the corners or other places wherever needed to act as a guide for the balance of the wall; the line of masonry work that is to be done is strung between the two leads which act as a guide.

Lead Colic. See Painter's Colic.

Leaded Glass. A stained glass window, with the pieces of glass, often of irregular size and shape, set in a lead framework.

Leaded Zinc Oxide. White pigment made by combining lead sulphate and zinc oxide.

Leaded Zinc. Basic lead sulphate united with zinc oxide.

Leader. See Downspout.

Leadless Glaze. A ceramic coating matured to a glassy state on a formed article, or the material or the mixture from which the coating is made, to which no lead has been deliberately added; does not imply that the glaze is nontoxic or that it contains no lead; because of plant practices and conditions, a small percentage of lead, 0.1 to 0.2% (by dry weight), expressed as lead monoxide, may be present.

Leaf. 1. The hinged or sliding part of a shutter, door, or gate. 2. One of a pair of doors.

Leafing. The overlapping arrangement of aluminum or gold bronze powders in a paint, similar to that of fallen leaves; good leafing is important in producing a metallic appearance and is caused by using treated or coated pigments along with suitable bronzing liquids.

Leak Detector. 1. Device or instrument such as halide torch, an electronic sniffer, or soap solution used to detect leaks. 2. Device used to detect and locate refrigerant leaks.

Leakage. 1. Entrance or escapage through a crack or hole, usually by a fault or mistake. 2. The loss of electricity through faulty insulation.

Lean Concrete. A concrete mix that is low in water and cement content, usually used as a stable base or fill.

Lean Mixture. Any plaster mortar containing a relatively high ratio of aggregate to cementitious material; a mortar mix that is too lean and has poor working qualities is said to be harsh.

Lean Mortar. Mortar containing a low percentage of cementitious components, not workable; see Harsh Mortar.

Leaning Edge. In gypsumboard, a factory edge formed out of square with the surface.

Lean-To. A shed on the side of a building with a sloping roof in a single plane.

Lease. An agreement whereby a tenant (lessee) obtains use of equipment, facilities, or real property for a specified period of time from the owner (lessor) usually by payment of a specified rent and under specified conditions.

Leaseback. A financial arrangement in which a tenant agrees to lease real property for a period during which the lease consideration will cover all mortgage payments and the lessor will own the property free and clear at the end of the lease term.

Lectern. A desk-like stand for holding a book or papers for a lecturer.

LED. Light-Emitting Diode.

Ledge. A narrow horizontal shelf or surface projecting from a vertical surface.

Ledger Bolt. A bolt set into a masonry wall to secure a wood or metal ledger.

Ledger. A wood or metal member bolted to a masonry wall to carry floor joists, ceiling joists, or rafters.

Ledgerboard. The support for the second-floor joists of a balloon-frame house.

Leg. 1. A tile wall running alongside a bathtub or abutment. 2. A narrow strip of tile floor.

Legal Description. 1. The description of a piece of property such that it can be positively identified for purposes such as purchase, sale, hypothecation, title insurance, and locating it on the land. 2. The lot, block, and tract numbers of a recorded subdivision. 3. A metes and bounds description.

L'Enfant Plan of Washington, DC. Adopted plan for the Capital in 1771. Daniel Burnham later worked on additions and adaptations of the plan.

Length Seam. See Side Seam.

Length. 1. The extent from end to end of an object. 2. The longer or longest of the 2 or 3 dimensions of an object. 3. The dimension measured between the ends of a masonry unit, usually the dimension of the masonry unit which is parallel to the face or length of the wall.

Leno Weave. Weave in which warp yarns, arranged in pairs, are twisted around one another between picks of weft yarn.

Lens. 1. A transparent glass or plastic substance with one or both sides curved for concentrating or dispersing light rays, as in optical devices. 2. The diffuser of a lighting fixture.

Lessee. One that holds property under a lease.

Lessor. One that conveys property by lease.

Let-In Bracing. Diagonal bracing, usually 1" x 4" or 1" x 6", nailed into notches cut in the face of the studs so as to avoid an increase in the thickness of the wall.

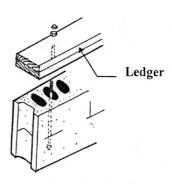

Ledger

Letter Box. A receptacle for the placement and storage of mail.

Letter of Credit. A letter written by a bank guaranteeing that drafts up to a certain amount will be honored; in some cases this will serve as a substitute for a surety bond.

Letter Sign. Sign composed of individual letters, usually of metal or plastic, mounted on a rigid background surface.

Letter Slot. An opening in a wall or door for the insertion of mail.

Level Area. A specified surface that does not have a slope in any direction exceeding 1/4 inch in one foot from the horizontal (2.083 percent gradient).

Level Floor. To bring a floor in conformance with a specified grade.

Level. 1. The position of a line or plane when parallel to the surface of still water. 2. A tool of wood, metal, or plastic used in testing for horizontal and vertical surfaces; see Spirit Level. 3. An instrument used in determining differences of elevation; see Engineer's Level. 4. A. roof is considered level up to a maximum slope of 1/2-inch per foot.

Leveler, Dock. See Dock Leveler.

Leveling Coat. A coat or layer of plaster or stucco which brings the surface to a true level plane.

Leveling Plate. A steel plate placed in grout on top of a concrete foundation to create a level bearing surface for the lower end of a steel column.

Leveling. In painting, the formation of a smooth film on either a horizontal or vertical surface, independent of the method of application; a film with good leveling characteristics is usually free of brush marks or orange peel effects.

Lever Arm. A bar operating about a fixed point or fulcrum to resist or move a load.

Lever Handle. A horizontal handle for operating the bolt of a lock.

Lever. A simple machine consisting of a rigid bar pivoted about a fulcrum (fixed point) which can be acted upon by a force in order to move a load.

LH Joist. A type of long-span high strength bar joist.

LH Series. A standard Steel Joist Institute designation for longspan steel joists.

LH. 1. Left hand. 2. Left Hand, a door handing designation.

LHR. Left Hand Reverse, a door handing designation.

Liabilities. Creditor claims on the assets of a business.

Liability Insurance. Insurance that will pay damages for which the insured becomes liable.

Liability. Legal responsibility.

Library. A room or building used for the storage and use of books.

Licensed Architect. An architect duly licensed by the state to practice.

Licensed Contractor. A contractor duly licensed by the state to construct buildings.

Lien Claim. A claim to obtain a lien against real property because of non-payment by the owner for labor, services, or material supplied for a work of improvement.

Lien Foreclosure Action. 1. A lawsuit to foreclose a Mechanics' Lien. 2. A lawsuit for foreclosures on property brought about to secure payment due to the holder of a lien and judgment against the real property involved.

Lien Release Notice. A notice showing that the lien amount has been satisfied and releasing the lien claim against the real property involved.

Lien. A legal clam by a party against another party for satisfaction of a monetary claim.

Life Estate. The right to own, use, and occupy real property during one's lifetime.

Life Expectancy. The average time a material, piece of equipment, or assembly would be expected to give satisfactory service under the conditions in which it is used.

Life-Cycle Cost. A cost that takes into account both the first cost and costs of maintenance, replacement, fuel consumed, monetary inflation, and interest over the life of the object being evaluated.

Lifeguard Chair. A raised chair, equipped with a ladder, that enables a lifeguard to view a large area.

Lift, Auto. See Auto Lift.

Lift, Hydraulic. See Hydraulic Lift.

Lift. 1. The amount of material placed at one time. 2. A layer of concrete. 3. A layer or course of paving material applied to a base or a previous layer. 4. A device to elevate loads.

Lifting. 1. In painting, the buckling of the finish coat when applied over previous coat which is not yet dry or when solvents in second coat are too strong. 2. Softening of undercoat by solvents used in coats which follow; usually caused by not allowing sufficient time for undercoat to harden before applying additional costs.

Lift-Slab Construction. A method of building multi-story sitecast concrete buildings by casting all the slabs in a stack on the ground, then lifting them up the columns with jacks and welding them in place.

Light Duty Tile. Tile suitable for limited pedestrian traffic such as entryways in single family residences.

Light Track. An electrified U-shaped member attached to a ceiling or wall and acting as a channel for sliding light fixtures.

Light Well. A court, open to the roof, to provide light and ventilation to rooms that face it.

Light, Dental. See Dental Light.

Light, Pool. See Pool Light.

Light, Surgical. See Surgical Light.

Light. 1. Daylight. 2. A lamp. 3. An appearance of brightness. 4. A pane of glass in a window. 5. Low in weight, density, or intensity.

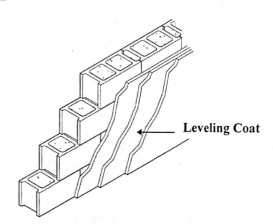

Leveling Coat

Light-Emitting Diode (LED). An electric component that emits light.

Lighting Contractor. The person in charge of installing an apparatus which supplies an artificial supply of light.

Lighting Fixture. An assembly having one or more lampholders, or a lampholder used in lieu of such an assembly.

Lighting Outlet. An outlet intended for the direct connection of a lampholder, a lighting fixture or a pendant cord terminating in a lampholder.

Lighting Standard. A pole supporting a lighting fixture as along streets or in a parking lot or athletic field.

Lighting. An artificial supply of light or the apparatus providing it.

Lightly Coated Electrode. A filler-metal electrode, used in arc welding, consisting of a metal wire with a light coating applied subsequent to the drawing operation, primarily for stabilizing the arc.

Lightning Arrester. A device connected to an electrical system to protect from lightning and voltage surges.

Lightning Rod. A metallic rod mounted on a high part of a structure, attached to a ground wire, to divert lightning into the earth.

Lightweight Aggregate. 1. Aggregate of low specific gravity, such as expanded or sintered clay, shale, slate, diatomaceous shale, perlite, vermiculite, or slag; natural pumice, scoria, volcanic cinders, tuff, and diatomite, sintered fly ash or industrial cinders; used to produce lightweight concrete. 2. Aggregate with a dry, loose weight of 70 pounds per cubic foot or less.

Lightweight Block. A concrete masonry unit constructed of lightweight materials and used to reduce the weight of walls.

Lightweight Concrete Firestop. Lightweight concrete used as firestop.

Lightweight Concrete. See Concrete, Lightweight.

Lignin. An amorphous polymeric substance related to cellulose that together with cellulose forms the woody cell walls of wood and the cementing material between them.

Lime Mortar. Mortar in which the cementing agent is lime.

Lime Plaster. Basecoat plaster consisting essentially of lime and an aggregate.

Lime Putty. Slaked or hydrated lime that is mixed with water to form a putty like mixture.

Lime Scale. The build-up of calcium carbonate on plumbing fixtures and in piping from the water supply.

Lime. 1. Various compounds of calcium oxide and magnesium oxide, obtained by heating forms of calcium carbonate, such as shells or limestone; used in mortar and plaster; also called Caustic Lime and Quicklime. 2. A dry white powder consisting essentially of the chemical compound calcium hydroxide that is made by treating caustic lime with water. 3. Various chemical and physical forms of quicklime, hydrated lime, and hydraulic hydrated lime.

Limestone Aggregate. Granular, crushed limestone.

Limestone Lintel. Limestone member placed within masonry wall or partition to support masonry or other construction over wall or partition opening.

Limestone Panel. A limestone slab, relatively thin with respect to other dimensions, and rectangular in shape.

Limestone. A sedimentary carbonate rock, composed chiefly of calcite ($CaCO_3$), but sometimes containing appreciable dolomite.

Limit Control. Control used to open or close electrical circuits as temperature or pressure limits are reached.

Limited Partnership. A partnership in which the management authority and liability of some of the partners is limited to their original investment; there must be at least one general partner who remains fully responsible for business liabilities and who runs the business.

Limiting Friction. The maximum value friction can have before motion ensues.

Line Pin. Metal pin used to attach string line used for alignment of masonry units.

Line Wire. Taut parallel horizontal wires installed on the outside of wood studs to act as a backing for weatherproof paper under the stucco netting of an exterior stucco wall assembly.

Line. 1. A length of string, thread, or rope. 2. A utility service such as a water pipe, electrical wire, or sewer pipe. 3. To cover the inside surface of an object with some material. 4. A circuit in a communication system. 5. The boundary of an area; a defining outline. 6. A straight or curved geometric element that is generated by a moving point and that has extension only along the path of the point. 7. A curve connecting all points having a specified common property. 8. A straight line is the most direct route connecting two points. 9. A string that is stretched from lead to lead to act as a guide for brick courses.

Linear Measure. Measurement along a line.

Linen Chute. An inclined plane, sloping or vertical channel, or passage down or through which soiled linen may pass.

Liner Panel. A panel used for interior finish.

Liner, Flue. See Flue Liner.

Liner, Pool. See Pool Liner.

Linerless Pool. Prefabricated component pool that retains water without additional waterproofing when assembled.

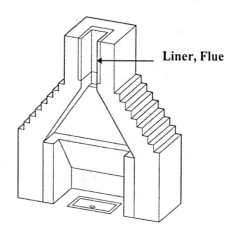

Liner, Flue

Lining Paper. Wallpaper without a ground (overall background color), used mostly for wall conditioning.

Lining, Carpet. See Carpet Padding.

Link, Beam. That part of a beam in an eccentrically braced frame which is designed to yield in shear and/or bending so that buckling of the bracing members is prevented.

Linoleum and Oilcoat Varnishes. Special highly flexible and elastic varnishes.

Linoleum Knife. See Hook Knife.

Linoleum Removal. The act or process of scraping away old linoleum flooring, commonly done by a specially designed apparatus which mechanically removes both linoleum and adhesive.

Linoleum Varnish. Special highly flexible and elastic varnish.

Linoleum. A tough wearing resilient floor covering, commonly in sheet form, consisting of a burlap back coated with a preparation of linseed oil, powdered cork, and pigments.

Linseed Oil. Yellowish drying Oil extracted from flax plant seed, widely used as a vehicle for lead-based paints; it is soluble in ether, benzene, and turpentine; metallic salts or driers are added to increase rate of drying; see also Bodied Linseed Oil and Boiled Linseed Oil.

Lint Strainer. A device mounted in the pump influent line to catch lint and other debris.

Lintel Block. A channel block.

Lintel Coarse. See String Course.

Lintel, Limestone. See Limestone Lintel.

Lintel, Masonry. See Masonry Lintel.

Lintel, Steel. Steel member placed within wall or partition to support loads over an opening.

Lintel. A horizontal structural member, usually made of stone, wood or metal, which supports the load over an opening; a Header.

Lip Molding. A molding with a lip which overlaps the piece against which the back of the molding rests.

Lip of a Strike. The projecting part of a door lock strike on which the latch bolt rides.

Lip. The chain and/or stuffer left on the edge of carpet after it has been cut.

Lipping. Laying brick so that the top edge of the unit is set in from the plane surface of the wall.

Liquefaction. Transformation of a granular material, the soil, from a solid state into a liquid state as a consequence of vibrations induced by an earthquake.

Liquefied Propane Gas. A compressed gas, such as propane or butane, used for fuel.

Liquid Absorbent. Chemical in liquid form which has the property to take on or absorb other fluids.

Liquid Applied Membrane. A roofing system applied in one or more coats, suitable for intricate shapes such as cast-in-place concrete structures.

Liquid Assets. Current assets that are, or can be, readily converted to cash.

Liquid Collector. A collector with a liquid as the heat transfer fluid.

Liquid Driers. Solution of driers in paint thinners.

Liquid Indicator. Device located in liquid line which provides a glass window through which liquid flow may be watched.

Liquid Line. Tube which carries liquid refrigerant from the condenser or liquid receiver to the refrigerant control mechanism.

Liquid Nitrogen. Nitrogen in liquid form which is used as a low- temperature refrigerant in expendable or chemical refrigerating systems.

Liquid Receiver Service Valve. Two- or three-way manual valve located at the outlet of the receiver and used for installation and service purposes; also called the King Valve.

Liquid Receiver. Cylinder (container) connected to condenser outlet for storage of liquid refrigerant in a system.

Liquid Waste. The discharge from any fixture or appliance connected to a drainage system which does not receive fecal matter.

Liquid Wood Filler. Varnishes of low viscosity, usually containing extending pigment, for use as a first coating on open- grain woods; its purpose is to afford a non-absorbent surface for succeeding coats of varnish; it is frequently colored so as to stain and fill in one operation.

Liquid. A substance that flows freely like water and has a form that has a definite volume but no fixed shape; one of the three states of matter; compare with solid and gas.

Liquidate. Wind up the affairs of a firm by ascertaining liabilities and apportioning assets; to determine by agreement or by litigation the precise amount of indebtedness, damages, or accounts.

Liquidated Damages. An amount determined by contract in advance of injury to be paid to compensate a party for an injury or damages.

Liquidity. Availability of cash, cash equivalents, and readily sold assets.

Liquid-Vapor Valve Refrigerant Cylinder. Dual hand valve on refrigerant cylinders which is used to release either gas or liquid refrigerant from the cylinder.

Liquor. Solution used in absorption refrigeration.

Lis Pendens. A document recorded in the county recorder's office that gives public notice of litigation involving title to or the possession of real property or real estate.

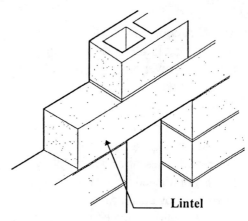

Lintel

List Price. A published price, subject to possible trade discounts.

Listed. Listed means equipment or materials included in a list published by a listing agency that maintains periodic inspection on current production of listed equipment or materials and whose listing states either that the equipment or material complies with approved standards or has been tested and found suitable for use in a specified manner.

Listing Agency. Listing agency means an agency accepted by the Administrative Authority which is in the business of listing or labeling and which maintains a periodic inspection program on current production of listed models, and which makes available a published report of such listing in which specific information is included that the product has been tested to approved standards and found safe for use in a specified manner.

Listing Mark. An independent laboratory mark or manifest indicating the material bearing this mark may be used in test certified by that lab.

Listing   L   Load Bearing Wall

Listing. A contract with a real estate broker, for a limited period of time, to produce a buyer or lessee for real property under specified rental and conditions.

Listing. Electrical materials, devices, fixtures, fittings, equipment, appliances and accessories that are shown in a list published by the enforcing agency, or by an approved testing agency, qualified and equipped for experimental testing and maintaining adequate periodic inspection of current production of listed models, and whose listing states either that the material, device, fixture, fitting, equipment, appliance and accessory complies with nationally recognized health, fire and safety requirements, or has been tested and found safe for use in a specified manner. Compliance may be evidenced by the presence of a label or identification mark of such approved testing agency.

Liter. A unit of volume equal to 1000 cubic centimeters, 1000 cubic centiliters, or 1.057 quarts.

Litmus. A dye that turns red in acidic conditions and blue in alkaline conditions; litmus paper is stained with litmus to be used as a test for acids or alkalis.

Live Load. Any load that is not permanently applied to a structure; the weight of people, furnishings, machines, and goods in or on a building; the vertical load superimposed by the use and occupancy of a building; all loads on a building except dead and lateral loads.

Live Part. Live parts are those parts which are electrically connected to points of potential different from that of the earth.

Live Space. A space with reflective surfaces so that sound can be sustained through several reflections.

Live Wire. The hot wire, carrying electric current, as opposed to the neutral or ground.

Livering. Formation of curds or gelling; coagulation of varnish finishing material into a viscous, rubber-like mass; usually caused by chemical reaction of two or more non-mixing products.

Living Room. A room in a residence for the occupants' common social pursuits.

Living Standard. Standard of living.

Living Unit. A house or apartment for the use of one family.

L-Mold. A metal molding strip with an L-like section, used to edge plaster or gypsum wallboard.

Load Bearing Metal Stud. Vertical load bearing, formed "C" channel of steel, component within framed wall, able to withstand structural loads imposed by wind loads, building loads, movement and deflection of structure.

Load Bearing Stud. A wooden stud which supports an imposed load in addition to its own weight.

Load Bearing Wall. A wall designed and built to carry superimposed vertical and shear loads as opposed to non-load-bearing walls which carry only their own weight.

Load Bearing. Supporting a superimposed weight or force.

Load Duration. The period of continuous application of a given load, or the sum of the periods of intermittent application of the same load.

Load Factor. 1. The percentage of the connected load which is likely to occur at any time. 2. The percentage of the total connected fixture-unit flow rate which is likely to occur at any point in the drainage system. 3. The percentage of the total electrical load which is likely to be in use at any time.

Load, Factored. The load, multiplied by appropriate load factors, used to proportion structural members by the strength design method.

Load, Service. See Service Load.

Load. 1. A force that is applied to a body. 2. A weight or force acting on a structure. 3. Applied external force, such as gravity and wind. 3. Any device that consumes electrical power, such as a motor, lamp, or toaster; the load on an electrical circuit.

Load-Bearing Partition. A vertical structural interior wall supporting an integral part of the construction above.

Loader. An excavating machine with a movable bucket or scoop, used to transport earth, crushed stone, or other construction materials.

Loading Dock Equipment. Powered or hand operated machinery to help load and unload freight.

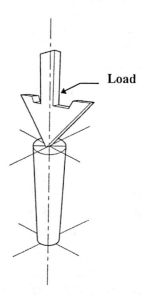

Load

Loading Dock. A raised platform adjacent to a loading door so that trucks may be loaded or unloaded with the truck bed at the same level as the warehouse floor.

Loading Pump. A hand pump for filling bulk materials into application tools.

Loam. A soft, easily worked soil containing sand, silt and clay.

Loan Fee. A fee for negotiating a loan, in addition to the interest.

Loan. Money lent at interest.

Lobby. An entrance room or anteroom.

Local Access Streets. Most of the low speed controlled access roads within sectors of a city, between collector streets, serving built-up areas, parking and pedestrians unrestricted.

Local Preheating. Preheating a specific portion of a structure.

Local Stress-Relief Heat Treatment. Stress-relief heat treatment of a specific portion of a structure.

Local Ventilating Pipe. A pipe on the fixture side of the trap through which vapor or foul air is removed from a room or fixture.

Location. A particular place or position.

Lock Plug. . Cylinder plug of a lock.

Lock Rail. The horizontal rail of a door intended to receive the lock case.

Lock Stile. The door stile to which the lock is applied as distinguished from the hinge stile.

Lock. A mechanism for fastening a door with a bolt that requires a key of a particular shape or a combination of movements to work it.

Locker Room. A room for changing clothes and storing them in lockers, especially for sporting activities.

Locker, Clothing. See Clothing Locker.

Locknut, Conduit. See Conduit Locknut.

Locknut, Grounding. See Grounding Locknut.

Locknut. 1. A nut screwed hard up against another to prevent either of them from moving. 2. A nut so constructed that it locks itself when screwed up tight.

Lockset. A device installed in a door that has both a deadbolt and doorknob assembly.

Lockwasher. A flat, split ring of metal or steel that when tightened with a nut is used to prevent loosening.

Locus. 1. A position or point. 2. A curve or line formed by all the points satisfying a particular equation.

Lodging House. Rented living quarters, usually in a private house rather than a hotel.

Loft. One of the upper floors of an unpartitioned warehouse building.

Log Lighter. A length of perforated pipe in a fireplace, connected to a valved gas line, used for lighting fires.

Log. A section of tree trunk suitable in length for sawing into commercial lumber.

Loge. 1. A small booth or compartment. 2. A box in a theater. 3. A separate forward section of a theater mezzanine or balcony.

Loggia. An opensided gallery or arcade, especially on upper floors of a building.

Logistics. The organization and details of carrying out an operation.

Long Term Creditors. Persons or companies to whom money is owed but not payable for over a year.

Long Term Debt. Mortgages and loans due in over a year.

Longitudinal Bar. Any reinforcing bar placed in the long direction of the member.

Longitudinal. 1. Running lengthwise. 2. In wood, generally, parallel to the direction of the wood fibers.

Long-Oil Varnish. Varnish with a large percentage of oil to gum resin, usually more than 25 gallons of oil to 100 pounds of resin; long-oil varnish is more elastic and more durable than short-oil varnish; spar varnish is a typical example of long-oil varnish.

Long-Term Burst. The internal pressure at which a pipe or fitting will break due to a constant internal pressure held for 100,000 hours.

Long-Term Capital Gain. The gain upon sale of capital assets that have been held for over 6 months.

Lookout. 1. The end of a rafter. 2. The construction which projects beyond the sides of a house to support the eaves. 3. The projecting timbers at the gables which support the rake boards.

Loom. Machine on which carpet is woven, as distinguished from other machines on which carpets may be tufted, flocked or punched.

Loomed Carpet. Carpet made on a modified upholstery loom with characteristic dense low-level loop pile, generally bonded to cellular rubber cushioning; see Woven Carpet.

Loop Pile. A Wilton or Velvet carpet woven with the yarn uncut; also called Round Wire.

Loop Vent. This is the same as a circuit vent; the only difference is that the vent loops back and connects with a soil stack or waste stack instead of a vent stack.

Loose Fill Insulation. Several types of thermal insulation in the form of fibers, granules, or other pieces that can be pumped, poured or placed by hand.

Loose Joint Hinge. A hinge having but two knuckles, the pin being fastened permanently to one knuckle, the other containing the pinhole, allowing the two parts of the hinge to be disengaged by lifting the door; such hinges are handed.

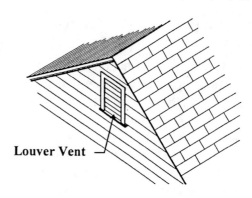

Louver Vent

Loose Knot. A knot that is not held firmly in place by growth or position and that cannot be relied upon to remain in place.

Loose Laid. A roofing membrane not attached to the substrate.

Loose Steel Lintel. Lintel made of structural steel shape (usually angle) that is set loose and built into masonry construction.

Loss on Ignition. The loss in weight expressed as a percentage, of a sample ignited at a very high temperature (1000 ± degrees C).

Lot Line. The boundary of a parcel of land.

Lots. Parcels of land in a recorded subdivision; usually numbered, and shown on a map.

Loudness. 1. The subjective human definition of the intensity of a sound; human reaction to sound is highly dependent on the sound pressure and frequency.

Louver Vent. A construction element with equally spaced slots that allow ventilation.

Louver, Door. See Door Louver.

Louver, Metal. See Metal Louver.

L

Louver, Penthouse. See Penthouse Louver.

Louver. 1. A construction of numerous sloping, closely spaced slats used to prevent the entry of rainwater into a ventilating opening. 2. A kind of window, generally in peaks of gables and the tops of towers, provided with horizontal slats which exclude rain and snow and allow ventilation. 3. An opening with a series of horizontal slats so arranged as to permit ventilation but to exclude rain, sunlight, or vision; see Attic Ventilators.

Love Wave. Transverse vibration of seismic surface wave; see Seismic Wave.

Low Boiling Solvent. Solvent with a low boiling point such as acetone or methyl alcohol.

Low Consistency Plaster. A neat (unfibered) gypsum basecoat plaster especially processed so that less mixing water is required than in standard gypsum basecoat plaster to produce workability; this type plaster is particularly adapted to machine application.

Low emissivity glass. Glass that reduces energy loss by reflecting heat back into a living space.

Low Pressure Control. Cycling device connected to low-pressure side of refrigeration system.

Low Pressure Gauge. Instrument for measuring pressures in range of 0 psi to 50 psi (0 kg/cm^2 to 3.52 kg/cm^2).

Low Relief. Bas relief.

Low Rise. A building of one or two stories and having no elevator.

Low Side. That portion of a refrigerating system which is below evaporating pressure.

Low-e Glass. Low emissivity glass.

Low-Energy Power Circuit. A circuit which is not a remote-control or signal circuit, but which has the power supply limited in accordance with the requirements of Class 2 remote control circuits. Such circuits include electric door openers and circuits used in the operation of coin-operated machines.

Lowest Responsible Bidder. The lowest bidder who is considered qualified and responsible.

Low-Iron Glass. Glass formulated with a low iron content so as to have a maximum transparency to solar energy.

Low-Lift Grouting. A method of constructing a reinforced masonry wall in which the wall is grouted in increments not higher than 4 feet.

Low-Pressure Steam Boiler. A boiler furnishing hot water at pressures not more than 30 pounds per square inch gauge (2.12 kg/cm^2) or steam at pressures not more than 15 pounds per square inch gauge (1.06 kg/cm^2).

Low-Side Float Valve. Refrigerant control valve operated by level of liquid refrigerant in low pressure side of system.

Low-Side Pressure Control. Device used to keep low-side evaporating pressure from dropping below certain pressure.

Low-Side Pressure. Pressure in cooling side of refrigerating cycle.

Low-Voltage Wiring. Wiring used for control, communication, and signal circuits, usually 12 or 24 volts.

Lozenge. A rhombus or diamond figure.

LP Fuel. Liquefied petroleum used as a fuel gas.

LPG. Liquefied Propane Gas.

LTC Ratio. In real estate loans, the Loan to Cost Ratio.

Lubrication. The use of oil or grease to make parts of machines move easily by reducing friction.

Lucite. Trademark, used for an acrylic resin or plastic consisting essentially of polymerized methyl methacrylate.

Lug Sill. A precast window or door sill for a masonry wall which is interlocked with the wall system.

Lugs. Spacers, or protuberances on the sides of ceramic tiles; these devices automatically space the tile for the grout joints.

Lumber Grading. See Grade, 1.

Lumber. Sawed parts of a log such as boards, planks, and timber; wood members which are manufactured by sawing, resawing, passing lengthwise through standard planing machine, crosscutting to length, and matching, but without further manufacturing.

Lumberjack. 1. A logger. 2. A Lumberman.

Lumberman. 1. One who processes, conveys, or merchandizes lumber. 2. A Lumberjack.

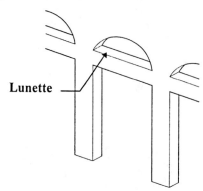

Lunette

Lumen. 1. In wood anatomy, the cell cavity. 2. A lighting unit; the amount of light emitted through an opening of 1 square foot, located at a distance of 1 foot from a light source, which emits 1 candlepower in every direction.

Luminaire. A complete lighting unit consisting of a light source, switch, globe, reflector, housing, and wiring.

Lump Lime. Quicklime as it comes from the kiln.

Lump Sum Contract. A contract in which the amount to be paid to the contractor is agreed in advance to be a stipulated sum.

Lunch Room. A room where workers or students eat their lunch.

Luncheonette. A small restaurant offering light lunches.

Lunette. An architectural element that has the shape of a crescent or half moon, like an opening in a vault for a window or an alcove for a painting or statue.

Luster Fabric. Any cut pile fabric woven with surface yarns spun from special types of staple and chemically washed, like hand-woven Oriental fabrics, to give a bright sheen or luster.

Lux. A lighting unit equal to one lumen per square meter.

Machine Direction. The direction parallel to the paper-bound edge of a sheet of gypsumboard.

Machine Room. Area where commercial and industrial refrigeration machinery, except the evaporators, is located.

Macrozones. Large zones of earthquake activity such as zones designated by the Uniform Building Code Map.

Madder. Coloring matter originally derived from the pulverized root of a plant cultivated in Europe and Asia Minor; now largely made synthetically.

Magnesium Silicate. White extender pigment which adds fluffiness to products in which it is used; provides very little opacity.

Magnet. A body having the capability of attracting iron and producing a magnetic field external to itself.

Magnetic Catch. A cabinet catch that uses a magnet to hold the door closed.

Magnetic Clutch. Device operated by magnetism to connect or disconnect a power drive.

Magnetic Compass. An instrument showing the direction of magnetic north and compass bearings.

Magnetic Core. Magnetic center of a magnetic field.

Magnetic Door Holder. A door holder using a magnet to hold it in an open position.

Magnetic Field. A region of variable force around magnets, magnetic materials, or current-carrying conductors.

Magnetic Gasket. Door-sealing material which keeps door tightly closed with small magnets inserted in gasket.

Magnetic Hammer. A special design hammer magnetically sensitized to hold a metal fastener during application.

Magnetic Pole. 1. Either of the two nonstationary areas in the north and south polar regions of the earth to which a magnetic compass needle will point 2. Either of the poles of a magnet.

Magnetic Starter. Automatic protective equipment using an electromagnet to operate; insures that a motor does not receive too high a current when starting up.

Magnetic. Having the properties of a magnet.

Magnetism. A field of force which causes a magnet to attract materials made of iron, nickel-cobalt or other ferrous material.

Magnetite. An aggregate used in heavyweight concrete, consisting primarily of ferrous metaferrite (Fe_3O_4); a black magnetic iron ore with a specific gravity of approximately 5.2 and a Moh's hardness of about 6.

Magnetron. A diode vacuum tube in which the flow of electrons is controlled by an externally applied magnetic field to generate power at microwave frequencies; the active element of a microwave oven.

Magnification Factor. An increase in lateral forces at a specific site for a specific factor or set of conditions.

Magnitude. A measure of earthquake size which describes the amount of energy released.

Mahogany Veneer. A thin layer of straight-grained medium density wood for an outer finish or decoration.

MAI. Member of the Appraisal Institute; a professional designation of a qualified real estate appraiser.

Mail Chute. An inclined or vertical channel through which mail travels from the exterior of a box or building to a container inside that box or building.

Maillechort. A silvery metal; an alloy of nickel, zinc, and copper.

Main Breaker. 1. A switch in a main electrical service panel where the service wires attach. 2. The main electrical service protective device where the power enters a building.

Main Drain. See Main Outlet.

Main Outlet. Outlet fitting(s) at the bottom of a swimming pool, spa or hot tub through which passes water to the recirculating pump; also called Main Drain or Sump Pot.

Main Runners. The heaviest integral supporting members in a suspended ceiling; main runners are supported by hangers attached to the building structure and in turn support furring channels or rods to which lath is fastened.

Main Switch. See Service Disconnect.

Main. The chief pipe, duct, or cable in any electrical, water, gas, sewer, vent or other utility system.

Maintenance Equipment. Any of a variety of implements to maintain and clean areas or equipment.

Maintenance Painting. 1. Repair painting; any painting after the initial paint job. 2. All painting except that done solely for aesthetics.

Maintenance. The systematic upkeep of property or equipment.

Maisonette. 1. A small house. 2. An apartment on two floors.

Maisoning. The process of planning and carrying out the keying and master keying of a building; especially complex systems as in hotels and office buildings.

Majolica. 1. Earthenware with an opaque luster glaze and overglaze colored decorations. 2. Any decorated earthenware having an opaque glaze.

Make-Up Air. Fresh air brought into a building from outdoors through the ventilation system and that has not been previously circulated through the system.

Make-Up Water. Fresh water used to fill or refill a swimming pool.

Male Thread. Outside threads on a pipe or fitting.

Maleic Resins. Resins based on reaction between maleic anhydride or maleic acid with glycerine and rosins.

Malice. A conscious desire to do harm.

Malicious Prosecution. Pursuing a lawsuit without probable cause.

Mall. 1. A sheltered walk or promenade. 2. A shopping area grouped around a common pedestrian way.

Malleable Iron. Iron that can be hammered or bent without breaking.

Malleable Strap. A metal fastening plate that can be hammered or bent without breaking.

Malleable. Of metals, capable of being formed into new shapes by hammering and bending.

Mallet. A hammer with a wooden head.

Malpractice Insurance. See Errors and Omissions Insurance.

Malpractice. Negligent act or omission of a professional.

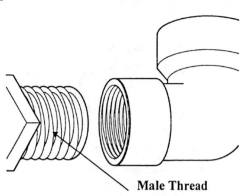

Male Thread

Managerial Competence. An aptitude for management and fulfilling of all the responsibilities inherent in the managerial functions such as planning, control, cooperation, and communication.

Mandrel Test. A physical bending test for adhesion and flexibility.

Mandrel. 1. A tapered axle inserted into a hole in a piece of work to support it during machining. 2. A metal bar used as a core around which material may be cast, molded, forged, bent, or otherwise formed. 3. The shaft and bearings on which a tool is mounted, as in a drill or circular saw.

Manganese. 1. A greyish-white, hard and brittle, metallic chemical element that resembles iron but is not magnetic. 2. A mineral that is contained in the clay used for brickmaking.

Manhole Base. The cast iron frame into which a manhole cover fits.

Manhole Cover. A heavy, round or square, steel or iron cover used to gain access to underground work through a manhole.

Manhole Removal. The act or process excavating an existing manhole and removing it.

Manhole, Communication. See Communication Manhole.

Manhole, Electric. See Electric Manhole.

Manhole, Watertight. See Watertight Manhole.

Manhole. A hole through which a person may go to gain access to an underground or enclosed structure.

Manifold Gauge. Chamber device constructed to hold both compound and high-pressure gauges; valves control flow of fluids through it.

Manifold, Service. See Service Manifold.

Manifold. A pipe fitting with several lateral outlets for connecting one pipe with others.

Manila Paper. See Cream Paper.

Manila Resins. Alcohol soluble natural resin obtained by tapping Agathis alba trees.

Manometer. A pressure gauge for gases and liquids.

Manor. The house of a landed estate.

Mansard Roof. A type of hip roof which has four sloping sides, each of which becomes steeper partway down.

Mansion. A large imposing residence.

Mantel. 1. A projecting shelf above the fireplace opening. 2. The entire finish around a fireplace.

Mantle. The main bulk of the earth that lies between the crust and the central core.

Manual Frost Control. Manual control used to change operation of refrigerating system to produce defrosting conditions.

Manual Proportioning Control. In an asphalt batching plant, a control system in which proportions of the aggregate and asphalt fractions are controlled by means of gates or valves which are opened and closed by manual means; the system may or may not include power assist devices in the actuation of gate and valve opening and closing.

Manual. 1. Done with the hands. 2. A book of instructions for operating a machine or performing some task.

Manufactory. A factory.

Manufacture. Making of articles as in a factory, especially on an industrial scale.

Manufactured Roof. A factory-finished roof system.

Manufactured Wall. A factory-finished wall system.

Manufacturer's Bond. A surety company's guarantee that it will stand behind a manufacturer's liability to finance roofing membrane repairs occasioned by ordinary wear within a period generally limited to 5, 10, 15 or 20 years.

Manufacturers Standardization Society of the Valve Fittings Industry, Inc (MSS). 127 Park Street, NE, Vienna, Virginia 22180, (703) 281-6613

Map. A two-dimensional representation of a part of the earth's surface.

Maple Flooring Manufacturers Association (MFMA). 60 Revere Drive #500, Northbrook, Illinois 60062, (708) 480-9080.

Mar. To disfigure, spoil, or impair the perfection of a surface or object.

Marble Chips. Graded aggregate of maximum hardness made from crushed marble to be thrown or blown onto a soft plaster bedding coat to produce marblecrete.

Marble Flooring. A floor system using marble as its finish material.

Marble Mosaic Tile. Tile made of small marble pieces that vary slightly in size, usually about one-half inch square.

Marble Paver. A type of igneous stone that is harder and more durable than regular bricks, used for driveways and patios for use with or without mortar.

Marble Tiles. Marble cut into tile sizes 12 inches square or less, usually 1/2 inch to 3/4 inch thick; available in several types of finishes, such as polished, honed, or split faced.

Marble, Imported. See Imported Marble.

Marble. Limestone that is more or less crystallized by metamorphism, that ranges from granular to compact in texture, that is capable of taking a high polish, and that is used in architecture and sculpture.

Marblecrete. A marble chip embedded finish for plaster.

Marbleizing. See Faux Marble.

Marblite. Plaster lining of marble granules with white cement to finish concrete pools.

Marezzo. An imitation marble produced with Keene's cement to which color pigments have been added.

Marine Plywood. Plywood panels manufactured with the same glueline requirements as other exterior-type panels but with more restrictive veneer quality requirements.

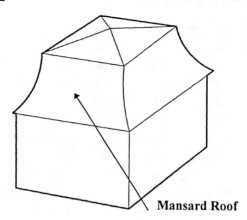

Mansard Roof

Marine Varnish. Varnishes especially designed to resist long immersion in salt or fresh water and exposure to marine atmosphere.

Marked Crossing. A crosswalk or other identified path intended for pedestrian use in crossing a vehicular way.

Market Area. The geographic area from which a real estate development expects to derive its customer base.

Market Data Approach. A real estate appraisal method by which the subject property is compared with other properties that have recently been sold or leased.

Market Study. A forecast of future demand in a particular market area for the type of real estate project proposed for development on a specific site.

Market Value. The price of something on the open market where the buyer and seller are both knowledgeable, and under no pressure to complete the transaction.

Marketable Title. Title to land that is unencumbered by any lien that would affect its value or marketability.

Marketing Management. Development of a marketing concept by a contractor, including such factors as advertising, copy for phone directory, letterhead for stationery, logo, and business cards.

Marl. A calcareous clay, containing approximately 30 to 65 percent calcium carbonate ($CaCO_3$), found normally in extinct fresh water basins, swamps, or bottoms of shallow lakes.

Marquee. 1. A canopy over an entrance. 2. A large tent set up for outdoor receptions.

Marquetry. Decorative work in which elaborate patterns are formed by the insertion of pieces of material, such as wood, shell, or ivory, into a wood veneer that is then applied to a surface, such as a piece of furniture.

Masking Power. The ability of a fired glaze to mask visually the body on which it is applied.

Masking Tape. Adhesive coated paper tape used to mask or protect parts of surface not to be finished.

Masking. 1. The presence of a background noise increased to a level to which a sound signal must be raised in order to be heard or distinguished. 2. When painting, protecting areas not to be painted. 3. Application of protective materials used in plastering machine applications.

Mason. 1. A skilled worker who builds by use of brick or stone set in mortar. 2. See Stonemason. 3. See Brickmason.

Mason's Hammer. A tool used by masons and tilesetters to score and cut brick, block, stone, or tile.

Masonite. Trademark. Fiberboard building panels made from steam-exploded wood fiber and binders.

Masonry Accessory. Any of various components necessary for masonry construction.

Masonry Anchor. A fastening device or mechanism used in a masonry wall.

Masonry Base. The lowest course of masonry in a pier, foundation, wall or footing.

Masonry Bonded Hollow Wall. A hollow wall built of masonry units in which the inner and outer wythes of the walls are bonded together with masonry units, such as in the all-rolok and rolok- bak walls.

Masonry Brick. Bricks that are shaped and molded in various sizes and shapes.

Masonry Cement. Portland cement with dry admixtures designed to increase the workability of mortar.

Masonry Chimney. A vertical non-combustible structure with a flue or flues to remove smoke and other gases constructed of shaped or molded masonry units.

Masonry Cleaning. 1. The final removal of excess grout and concrete and mortar stains, from an exterior surface of a masonry structure. 2. In older structures, the cleaning of the surface by any of several means but commonly steam, chemical, and in some rare instances sandblasting.

Masonry Control Joint. A sawed, tooled, or formed groove in a masonry joint to regulate the location of cracking resulting from the dimensional change of different parts of the structure.

Masonry Facing Panel. A masonry structure having a decorative exterior surface.

Masonry Flashing. A thin, continuous sheet of metal, plastic, rubber or waterproof paper used to prevent the passage of water through a joint in a wall, roof, or at a chimney; the material used and the process of making watertight the roof intersections and other exposed places on the outside masonry.

Masonry Floor. Shaped or molded masonry units such as, stone, brick, tile or concrete units used for finished floor coverings.

Masonry Furring. Metal or wooden strips attached to any masonry surface on which wall boards or sheathing may be conventionally attached.

Masonry Grout. A mixture of cementitious materials and aggregates and water used to fill the hollow spaces of masonry units and cavities.

Masonry Insulation. Any type of insulation for hollow masonry units.

Masonry Lintel. Masonry member placed within masonry wall or partition to support loads over an opening.

Masonry Manhole. A masonry vertical access shaft from the surface to the underground utility.

Masonry Painting. The act or process of applying paint to seal or color a masonry surface.

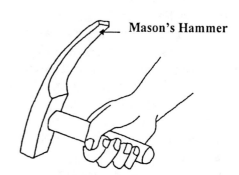
Mason's Hammer

Masonry Paver. Shaped or molded units, composed of stone, ceramic brick or tile, concrete, or cast-in place concrete used for driveways and patios.

Masonry Plaque. A commemorative or identifying inscribed tablet made of shaped or molded units, composed of stone, brick, tile, or concrete.

Masonry Plaster. A plaster surface on masonry usually made from a cement-based mixture, commonly called stucco on newer buildings, but may be some other material on older historic structures.

Masonry Plaster. A plaster surface on masonry, usually made from a cement-based mixture, commonly called stucco on newer buildings, but may be gypsum-type material on older historic structures.

Masonry Reinforcing. Lateral steel rods, wire, or mesh placed between courses of masonry units.

Masonry Research. See Masonry Institute of America.

Masonry Restoration. The act or process of the repair of a masonry structure.

Masonry Sandblast. The act or process of abrading or cutting masonry structure surfaces using sand ejected from a nozzle at high speed by compressed air; an effective form of cleaning masonry, but used sparingly because of its inclination to remove the outer (usually glazed or polished) surface.

Masonry Society, The (TMS). 3775 Iris Avenue, #6, Boulder, Colorado 80301-2043, (303) 939-9700.

Masonry Tie. A reinforcing strip, bar, or wire used to link courses of masonry together or to bond them with a wood or concrete backup wall.

Masonry Unit. Any brick, tile, stone, or block used in masonry construction.

Masonry Veneer. A nonstructural tier or layer of brick or stone attached to a structural masonry or wood framed wall.

Masonry Vent. Opening in a masonry structure to provide natural ventilation.

Masonry Wall. A wall constructed of brick, stone, or concrete block.

Masonry Waterproofing. Any of a variety of materials applied to masonry structures to resist or prevent the passage of water.

Masonry, Gypsum. See Gypsum Masonry.

Masonry. Brickwork, blockwork, and stonework.

Masonry. Construction of brick, tile, stone, or concrete block, or combination thereof, bonded together with mortar.

Mass Concrete. 1. Concrete without reinforcing; also called plain concrete. 2. Any large volume of concrete cast in place intended to resist applied loads by virtue of mass; generally a monolithic structure incorporating a low cement factor with a high proportion of large coarse aggregate.

Mass Curing. Adiabatic curing, using sealed containers.

Mass Law. In acoustics, the law relating to the transmission loss of walls, which states that in a part of a frequency range, the magnitude of the loss is controlled entirely by the mass per unit area of the panel; also, that the transmission loss increases 1 decibel for each doubling of frequency or each doubling of the panel mass per unit area.

Mass. 1. Quantity of matter held together so as to form one body. 2. A body of matter of indefinite shape. 3. The quantity of a body that is its measure of inertia.

Master Key. A key that will open several locks, each keyed separately.

Master. A fully qualified, proficient, skilled, and experienced builder, plumber, electrician, plasterer, painter, or other construction trade worker.

Masterformat. The copyrighted title of a uniform indexing system for construction specifications, as created by the Construction Specifications Institute and Construction Specifications Canada, commonly called the CSI format or numbering system.

Masterkeying. An arrangement of door lock cylinders having individual key changes which permits them all to be operated by a single key called the master key.

Mastic Grout. A chemical mixture of organic and inorganic ingredients forming a one part grouting composition that is used directly from the manufacturer; it is more flexible and stain resistant than cement grout.

Mastic. 1. High viscosity solvent-based adhesive. 2. A pasty material used as a cement, as for setting tile, or a protective coating, as for thermal insulation or waterproofing. 3. Organic tile adhesive. 4. A viscous, dough-like, adhesive substance; can be any of a large number of formulations for different purposes such as sealants, adhesives, glazing compounds, or roofing membranes; see Asphalt Mastic and Flashing Cement.

Mat Formwork. The temporary support for a concrete mat during the pouring or placing of concrete.

Mat Foundation. A concrete slab used as a building or equipment foundation.

Mat Glaze. A colorless or colored ceramic glaze having low gloss.

Mat Reinforcing. The placing of metal or steel bars in freshly poured concrete mat to strengthen it.

Mat, Concrete. See Concrete Mat.

Mat, Floor. See Floor Mat.

Mat. 1. A large footing or foundation slab used to support an entire structure. 2. A grid of reinforcing bars.

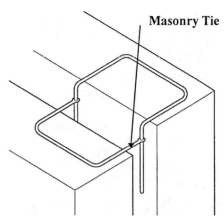

Masonry Tie

Matched Lumber. Lumber that is edge-dressed and shaped to make a close tongue-and-groove joint.

Matching. Machining boards to form tongue and groove joints.

Material Handling. The act or process of transporting materials on or to a jobsite.

Material. 1. The matter from which things are made. 2. The physical things needed for construction. 3. Important, essential, and relevant.

Materialman. An individual or organization who supplies construction materials to a project.

Materials Used. This includes the cost of all materials used on the job; usually the largest single expense item on income and expense statement.

Materiel. The materials, equipment, and supplies used in a business or on a project.

Matrix. In concrete, the material that fills the spaces between the fine and course aggregates; the cement paste.

Matter. A physical substance in general that has mass and occupies space; occurs in the state or form of solid, liquid, or gas.

Mattock. A hand implement used for digging and grubbing, with features of a pick and an adze.

Maturing Range. The time-temperature range within which a ceramic body, glaze, or other composition may be fired to yield specified properties.

Maturity Date. The date on which an obligation becomes due.

Maul. 1. A heavy hammer for driving wedges. 2. A tool like a sledge hammer with one wedge-shaped end, used to split wood.

Maulstick. A light stick with a padded leather ball at one end, held by a painter in one hand to support the other hand.

Mausoleum. A large and grand tomb.

Maximum Capable Earthquake. The maximum level of earthquake ground shaking which may ever be expected at the building site within the known geological framework; this intensity may be taken as the level of earthquake ground motion that has a 10 percent probability of being exceeded in a 100-year time period.

Maximum Size Aggregate. Aggregate whose largest particle size is present in sufficient quantity to affect the physical properties of concrete; generally designated by the sieve size on which the maximum amount permitted to be retained is 5 to 10 percent by weight.

Maximum. (Pl. maxima) The highest possible or attainable amount.

MB. Megabyte. 1,024,000 bytes.

MBH. Thousands of British Thermal Units; for example, 82 MBH = 82,000 BTU.

MBMA. Metal Building Manufacturers Association.

MC. 1. Medicine Cabinet. 2. Medium-Curing Asphalt.

MCM. Thousand circular mils, designating wire size.

MCS. Multiple Chemical Sensitivity; a condition in which a person is considered to be sensitive to a number of chemicals at very low concentrations.

MDF. Medium-Density Fiberboard.

ME. Mechanical Engineer.

Mean Effective Pressure (MEP). Average pressure on a surface when a changing pressure condition exists.

Mean Radiant Temperature. The weighted average of all radiating surface temperatures within one line of sight.

Mean. 1. The middle term in a progression of values. 2. Average.

Meat Case. A refrigerated unit for the storage and preservation of meat.

Mechanical Adhesion. Adhesion between surfaces in which the adhesive holds the parts together by interlocking action.

Mechanical Advantage. 1. The advantage gained by the use of levers or other devices to transmit force. 2. The ratio of the force that performs the useful work of a mechanism to the force that is applied.

Mechanical Application. Application of plaster mortar by mechanical means, generally pumping and spraying; distinguished from hand application with a trowel.

Mechanical Bond. 1. The physical keying of one plaster coat to another or to the plaster base. 2. Tying masonry units together with metal ties or reinforcing steel or keys.

Mechanical Cycle. Cycle which is a repetitive series of mechanical events.

Mechanical Engineer. An engineer who designs plumbing, air conditioning, and other environmental systems for buildings.

Mechanical Skimmer. See Surface Skimmer.

Mechanical Trowel. A power machine used to smooth and compact plaster finish coats; capable of producing an extremely smooth, dense surface; consists of revolving metal or rubber blades; also called a Power Trowel.

Mechanical. 1. Of or relating to machinery or tools. 2. Relating to, governed by, or in accordance the principles of mechanics. 3. Anything in the plumbing, heating, air-conditioning, or fire sprinkler trades.

Mechanics Lien. A charge against real estate for the value of work or materials incorporated thereon.

Median Strip. A space between two opposing lanes of traffic, as in a divided highway.

Median. 1. The middle value in a series of values. 2. A straight line in a triangle connecting a vertex with the midpoint of the opposite side.

Mediation. An alternative dispute resolution method by which a mediator assists and urges the parties to find a mutually acceptable resolution to their differences; unlike an arbitrator, a mediator cannot impose a solution or secession on the parties.

Medical Equipment. Apparatus or devices used in medical applications.

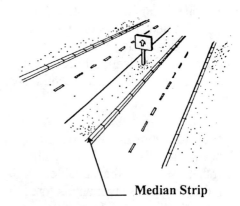

Median Strip

Medical Gases. In a hospital, the gases that are piped throughout the building for bedside and laboratory use, such as oxygen and nitrous oxide.

Medical Incinerator. A device in which medical wastes are burned.

Medical Sterilizer. An apparatus that utilizes high heat or chemicals to sterilize medical equipment.

Medical Surveillance. 1. A periodic comprehensive review of a worker's health status. 2. The required elements of an acceptable medical surveillance program for various conditions that are listed in the Occupational Safety and Health Administration standards.

Medical Utensil Washer. A mechanical washing device which sterilizes, by high heat, medical utensils for use in surgeries and other health care needs.

Medical Waste. Waste materials created by hospitals or other medical settings.

Medicine Cabinet. A cabinet, usually found in the bathroom, where articles such as toiletries and home medicines are kept.

Medieval Cities. Irregular street patterns; fortified cities were first internally subdivided to the maximum utilization of the land; often internal part of the city acted as a second fortress.

Medium Density Fiberboard. A panel product manufactured from lignocellulosic fibers combined with a synthetic resin or other suitable binder; the panels are manufactured to a density of 31 pcf (0.50 specific gravity) to 55 pcf (0.88 specific gravity) by the application of heat and pressure by a process in which the interfiber bond is substantially created by the added binder; other materials may have been added during manufacturing to improve certain properties.

Medium Duty Tile. Tile suitable for pedestrian traffic such as entryways in multiple dwellings and lobbies.

Medium Valve. A color midway between a dark color and a light color.

Medium-Curing (MC) Asphalt. Cutback asphalt composed of asphalt cement and a kerosene-type diluent of medium volatility.

Meeting Rail. See Checkrail.

Meeting Room. A room for holding meetings; a conference room.

Megahertz. One million hertz, a unit of radio frequency.

Megalopolis. 1. A large city. 2. An urban complex consisting of a metropolis and its environs.

Megawatt. 1,000,000 watts.

Megohm. A unit of measure for electrical resistance. One megohm is equal to a million ohms.

Megohmmeter. Instrument for measuring extremely high resistances, in the millions of ohms range.

MEK. Methyl Ethyl Ketone.

Melt Point. The temperature at which asphalt changes from solid to liquid.

Melt. To change a solid into a liquid by the application of heat; or the liquid resulting from such action.

Melting Point. Temperature at atmospheric pressure at which a substance will melt.

Melting Rate In welding, the weight or length of electrode melted in a unit of time; also called Burnoff Rate.

Member. A single piece in a structure, complete in itself.

Membrane Fireproofing A lath and plaster membrane having among its functions that of providing a barrier to fire and intense heat.

Membrane Roof. A roof structure with a covering of a sheet material that is impervious to water or water vapor; commonly a single sheet of material.

Membrane Waterproofing. A membrane, usually made of built-up roofing or sheet material, to provide a positive waterproof floor over the substrate, which is to receive a tile installation using a wire reinforced mortar bed.

Membrane, Neoprene. See Neoprene Membrane.

Membrane. 1. A sheet material that is impervious to water or water vapor. 2. A flexible or semi-flexible roof covering or waterproofing whose primary function is the exclusion of water.

Meniscus. The upper curved surface of a liquid in a tube, concave when the liquid wets the tube, convex when it does not.

MEP. Mean Effective Pressure.

Mercalli Scale. System used to determine the location of the epicenter of an earthquake on the basis of defining zones of intensity by observations of damage by persons experiencing the earthquake.

Mercoid Bulb. Electrical circuit switch which uses a small quantity of mercury in a sealed glass tube to make or break electrical contact with terminals within the tube.

Mercury Fixture. A light fixture that has an electric discharge lamp that produces a blue-white light by creating an arc in mercury vapor enclosed in a tube or globe.

Mercury Switch. An electrical switch that has mercury enclosed in a vial to make a silent contact.

Mercury. A silvery-white heavy liquid metallic element used in barometers, thermometers, amalgams, and electrical switches; the only metal that is liquid at ordinary room temperature; chemical symbol Hg.

Meridian Lines. Part of a governmental land surveying grid system covering the country, with base lines running east and west and meridians running north and south; townships are located with reference to a specific base line and meridian; for example, T3N, R12W, MDBM is read Township 3 North, Range 12 West, Mount Diablo Base Line and Meridian.

Mesh. The square opening of a sieve.

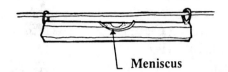

Meniscus

Mesh Tape. See Joint Reinforcing Mesh.

Mesh Tie. A wire used to hold sheets of mesh together so they will not move or spread apart when concrete is poured over the mesh.

Mesh Wire. A series of longitudinal and transverse wires arranged at right angles to each other in sheets or rolls, used to reinforce mortar and concrete; welded-wire fabric.

Mesh, Reinforcement. See Welded Wire Fanbric.

Mesh, Slab. See Slab Mesh.

Message Tube. A cylindrical tube typically installed in the walls of a building or structure, operated by the suction of air, to transport messages placed in containers.

Met. Term applied to the heat release from a human at rest. It equals 18.4 Btu/sq. ft./hr. or 50 kcal/m^2/hr.

Metal Anchor. A bolt or fastener made of metal.

Metal Beam Anchor. Formed steel component used to anchor one end of a beam to another beam, girder, or column and prevent displacement of the beam under lateral or uplift loads.

Metal Bridging. Diagonal or longitudinal metal members used to keep horizontal wood members properly spaced, in lateral position, vertically plumb, and to distribute load.

Metal Building Manufacturers Association (MBMA). 1300 Sumner Avenue, Cleveland, Ohio 44115, (216) 241-7333.

Metal Building. A building or structure constructed of a structural steel frame covered by metal roof and wall panels; commonly prefabricated in a factory and assembled at the site.

Metal Chimney. Vertical metal structure with one or more flues to carry smoke and other gases or combustion into atmosphere.

Metal Clad Cable. Electrical conduit made of a flexible steel jacket wrapped around insulated wires.

Metal Clad Door. A flush wooden door covered in sheet metal.

Metal Clip Angle. Short length of metal angle used to attach two components.

Metal Cripple Stud. Less than full height metal stud, such as under or above an opening.

Metal Decking. Light-gauge, corrugated sheets used in the construction of roofs or floors.

Metal Door Buck Anchor. Formed steel component used to anchor door bucks or jambs to concrete or masonry construction.

Metal Ductwork. The light sheet metal material out of which the ducts of an HVAC system are manufactured.

Metal Electrode. A filler or non-filler metal electrode used in arc welding consisting of a metal wire with or without a covering or coating.

Metal Fabrication. The building, construction or manufacture of metal structures or metal devices.

Metal Floor Track. Horizontal channel shaped steel member located at bottom of framed wall to receive metal studs.

Metal Framing Anchor. Fabricated metal devices used to transfer structural loads from a wooden structural member to another or other material member or supporting structure.

Metal Framing. The construction of a building or structure by using steel; the construction of frame houses and partitions by using light gauge metal studs and members.

Metal Furring. A length of metal channel attached to a masonry or concrete wall to permit the attachment of finish materials to the wall.

Metal Girder Anchor. Formed steel component used to anchor end of girder to another girder, beam, or column and prevent displacement of girder under lateral or uplift loads.

Metal Grating. Metal screening made from sets of parallel bars placed at right angles to each other to allow water to drain through, while protecting persons and vehicles from the drain opening.

Metal Halide Lamp. A lamp that uses an electric-discharge to produce light from a metal vapor such as sodium or mercury.

Metal Joist Anchor. Formed steel component used to anchor end of joists and to wall plate or beam and prevent displacement of joist under lateral or uplift loads.

Metal Joist Bridging. Diagonal or longitudinal metal members used to keep metal joists properly spaced, in lateral position, vertically plumb, and to distribute load.

Metal Joist Hanger. Formed steel component used to support end of load bearing joists and transmit loads to another joist or beam.

Metal Joist. Horizontal cold formed metal framing member of floor, ceiling or flat roof to transmit loads to bearing points; often refers to a Bar Joist.

Metal Lath Steel Framing Division of National Association of Architectural Metal Manufacturers, (MLA). 11 South LaSalle #1400, Chicago, Illinois 60603, (312) 201-0101.

Metal Lath. See Expanded Metal Lath.

Metal Lintel. A horizontal metal member spanning and carrying the load above an opening.

Metal Louver. A framed opening in a wall or door with fixed or movable flaps, manufactured from metal.

Metal Pan Stair. A stair assembly constructed to hold precast or cast-in place concrete, masonry, or stone in sheet metal pans at the treads and landings.

Metal Primer. First coating applied in finishing metal.

Metal Quarry Tile Rack. Racks that are available in many patterns and made to order for special patterns, used to maintain uniform joint widths between quarry tiles.

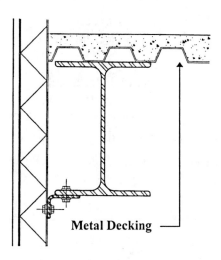

Metal Decking

Metal Rafter. Horizontal cold formed metal framing member for floor, ceiling or roof to transmit loads to bearing points.

Metal Railing Bracket. Metal wall bracket to support a railing.

Metal Railing. A guard or open fence with rails and posts made of metal.

Metal Roof Removal. The act or process of tearing off and carrying away an old metal roof.

Metal Shear Plate. Metal plates used to transfer shear loads between timber members.

Metal Shingle. 1. A roof covering unit manufactured from metal and applied in an overlapping pattern. 2. Metal material used as an exterior wall finish over sheathing.

Metal Sleeper Clip. Metal clip used to secure wood sleepers to a concrete floor.

Metal Stair Pan. A stair assembly constructed to hold precast or cast-in place concrete masonry or stone in metal sheet pans at the treads.

Metal Stair. A single or series of metal steps with framework connected at landings.

Metal Stud Bracing, Horizontal. Horizontal element used to provide stiffness and to prevent buckling or rotation of metal studs.

Metal Stud Bracing. Element used to prevent buckling or rotation of metal studs.

Metal Stud. Vertical formed steel channel within a framed wall.

Metal Tie Strap. Formed steel component used to tie one wood framing member to another.

Metal Timber Connector. Fabricated metal devices used to transfer structural loads from a timber member to another timber or other material member or supporting structure.

Metal Toilet Partition. A prefinished, manufactured dividing wall in a toilet room.

Metal Toothed Ring. Metal rings with toothed edge to embed in wood to resist shear.

Metal Track. Horizontal channel-shaped steel member located at top or bottom to receive metal studs.

Metal Trim. See Casing and Bead, 4.

Metal. 1. Any of various fusible, ductile and typically lustrous chemical elements that can conduct heat and electricity. 2. Any alloy.

Metal-Clad Cable. A fabricated assembly of insulated conductors in a flexible metallic enclosure.

Metal-Electrode-Arc Welding. An arc-welding process wherein metal electrodes are used.

Metallic Color Tile. Tile that has been coated with metal flakes to reflect light.

Metallic Grout. Grout that has been coated with metal flakes that reflect light.

Metallic Soap. A compound of metal and organic acid; used as driers, fungicides, suspending agents, and flatting agents.

Metallize. 1. To coat, treat or combine with a metal. 2. Coat with a thin layer of metal.

Metallurgy. The science concerned with the production, purification, and properties of metals and their application.

Meter Center, Indoor. See Indoor Meter Center.

Meter Center. The room or area where meters for measuring the usage of water, electricity and/or gas are located.

Meter Socket. A receptacle with electric contacts in which an electric meter is plugged.

Meter Stop. A valve used on a water main between the street and a water meter; it permits installation or removal of the meter.

Meter, Electric. See Electric Meter.

Meter, Gas. See Gas Meter.

Meter, Water. See Water Meter.

Meter. 1. An instrument for measuring consumption of gas, electrical current, or water. 2. A metric unit that equals 100 centimeters or 1000 millimeters and is equivalent to 39.37 inches.

Metering. The mechanical process of measuring the usage of water, electricity or gas.

Metes and Bounds. A type of legal description of land where each segment of the boundary line is described by length and bearing.

Methane. A colorless, odorless, highly inflammable gaseous hydrocarbon, sometimes known as marsh gas or firedamp, produced by decomposition of organic matter in marshes and mines or the carbonization of coal, used as a fuel and raw material in chemical synthesis.

Method of Joints. Mathematical procedure for determining the forces in the members of a truss or frame.

Methods of Sections. Mathematical procedure for determining the forces in the members of a truss or frame.

Methyl Acetate. An inflammable fragrant liquid used as a lacquer solvent.

Methyl Alcohol. Poisonous alcohol obtained by destructive distillation of wood.

Methyl Ethyl Ketone (MEK). A flammable liquid cleaning solvent.

Methyl Formate. Low pressure refrigerant.

Methyl Isobutyl Ketone (MIBK). A strong flammable organic solvent.

Metopes. See Tryglyphs and Metopes.

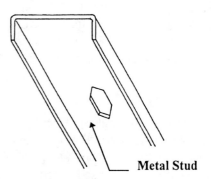

Metal Stud

Metric System. A standard measuring system based on the meter, decimally subdivided into centimeters and millimeters.

Mexican Paver Tile. Handmade terra cotta-like tile, used mainly for floors; they vary in color, texture and appearance, from tile to tile and within each tile; available in squares up to 12 inches, hexagon, octagon, elongated hexagon, fleur de lis and other shapes; coated with various types of sealers because of their soft absorptive characteristics; the coatings provide a wearing surface on the pavers which would otherwise powder away under wear.

Mezzanine. A low-ceilinged story between two main stories of a building; an intermediate story that projects in the form of a balcony.

MFMA. Maple Flooring Manufacturers Association.

Mho. The practical unit of electrical conductance equal to the reciprocal of the ohm.

MI Cable. Mineral insulated sheathed cable.

MI. Malleable Iron.

MIA. 1. Marble Institute of America, 33505 State Street, Farmington, MI 48335, (810) 476-5558. 2. Masonry Institute of America, formerly Masonry Research.

MIBK. Methyl Isobutyl Ketone.

Mica Pigment. Extender pigment made from silicates of aluminum and potassium which are split into very thin plates or sheets; used as reinforcing pigment since it tends to reduce checking and cracking.

Mica. Any of a group of silicate minerals with a layered structure.

Micro. One millionth part of unit specified.

Microbiologicals. Agents derived from or that are living organisms (for example, viruses, bacteria, fungi, and mammal and bird antigens) that can be inhaled and can cause many types of health effects including allergic reactions, respiratory disorders, hypersensitivity diseases, and infectious diseases.

Microfarad. Unit of condenser electrical capacity equal to 1/1,000,000 farad.

Microliter. One millionth of a liter.

Micrometer. 1. A gauge for accurately measuring small distances or thicknesses. 2. A unit of length equal to one millionth of a meter; also called a Micron.

Micron Gauge. Instrument for measuring vacuums very close to a perfect vacuum.

Micron. A micrometer, one millionth of a meter.

Microphone. An instrument that turns sound waves into electric current for the purposes of transmitting or recording music or voice.

Microregionalization. Breaking up of macrozones into much smaller zones of specific earthquake intensity and activity.

Microscope. An optical device found in a laboratory which has a surface for placing materials to be examined, and a lens which magnifies those materials.

Microvolt. One millionth of a volt.

Microwatt. One millionth of a watt.

Microwave Oven. An oven in which food is cooked by the heat produced as a result of short electromagnetic wave penetration of the food; the active element is the magnetron.

Middle Strip. The zone of a two-way concrete slab that lies midway between columns.

Midget Trowel. A small hand trowel used for pointing and small confined areas.

Mil. A unit of length or thickness equal to 1/1000th of an inch.

Mild Steel. Steel containing less than three-tenths of one percent carbon, not used as structural steel because of its low strength.

Mildew. 1. A discoloration caused by fungi. 2. A parasitic fungus growth occurring in insufficiently vented and damp surface areas.

Mildewcide. Substance poisonous to mildew; prevents or retards growth of mildew.

Mile. A measure of distance equal to 5280 feet.

Milieu. Environment or setting.

Military Crest. A ridge that interrupts the view between a valley and a hilltop.

Mill Scale. The oxide layer formed during the hot rolling of metals, such as that formed on hot-rolled reinforcing bars.

Mill White. White paint used to augment illumination on interior wall surface of industrial plants, office and school buildings; the vehicle is usually of the varnish type.

Mill. 1. A building used for grinding, machining, manufacturing, or woodworking. 2. A machine for grinding or crushing.

Milli. Combining form denoting one thousandth (1/1,000); for example, millivolt means one thousandth of a volt.

Milliammeter. An instrument that measures very small amounts of electrical current.

Milliampere. One thousandth of an ampere.

Millibar. One thousandth of a bar, the metric unit of atmospheric pressure equivalent to 100 pascals.

Millimeter. A metric unit of distance measurement equal to 1/10 of a centimeter or 1/1000 of a meter.

Millimicron. 1. A unit of length equal to one thousandth of a micrometer. 2. A nanometer.

Millisecond. One thousandth of a second.

Mill-Mixed Plaster. See Gypsum Ready Mixed Plaster.

Millwork. Generally, all wood materials manufactured in millwork plants and planing mills, including such items as inside and outside doors, window and door frames, blinds, mantels, panel work, stairways, moldings and interior trim, but not including flooring, ceiling, or siding.

Milori Blue. An iron blue pigment.

Mineral Aggregate. Aggregate consisting essentially of inorganic nonmetallic materials.

Mineral Black. A natural black pigment based on graphite.

Mineral Dust. The portion of the fine aggregate passing the 0.075mm (No. 200) sieve.

Mineral Fiber Felt. A building felt with mineral wool as its principle component.

Mineral Fiber Tile. A pre-formed ceiling tile that is composed of mineral fiber and a binder; has good thermal properties and acoustics.

Mineral Filler. A finely divided mineral product at least 70 percent of which will pass a 0.075mm (No. 200) sieve; pulverized limestone is the most commonly manufactured filler, although other stone dust, hydrated lime, portland cement, and certain natural deposits of finely divided mineral matter are also used.

Mineral Granules. Opaque, natural or synthetically colored aggregate commonly used to surface cap sheets, granule-surfaced sheets, and roofing shingles.

Mineral Insulated Cable. An electric cable insulated with a mineral sheathing.

Mineral Oil. An oil obtained from mineral sources as opposed to vegetable; usually petroleum or one of its distillate products.

Mineral Spirits. A clear distillate of petroleum, a solvent for asphaltic coatings.

Mineral Surfaced Roofings. Built-up roofing materials whose top ply consists of a granule-surfaced sheet; this is an asphalt saturated felt that is coated on one or both sides and is surfaced on the weather exposed side with mineral granules.

Mineral Surfaced. Factory made roofing cap sheet or shingles covered with mineral granules.

Mineral Wool Insulation. Insulation manufactured from a lightweight vitreous fibrous material.

Mineral Wool. Any of various lightweight fibrous materials used in heat and sound insulation.

Minimal. Very minute or slight.

Minimall. A small regional shopping center.

Minimum. (Pl. minima) The least possible or attainable amount.

Miniscule. Very small.

Minor Changes in the Work. Changes ordered by the architect that are within the intent of the contract documents and do not involve a change in the contract sum or the contract time.

Minus Pressure. See Negative Pressure, 2.

Minute. 1. A unit of measurement of angles, equal to 1/60 of a degree. 2. A unit of time, equal to 1/60 of an hour 3. Extremely small or tiny.

Mirror Frame. Shop fabricated or field applied perimeter trim to contain mirror and protect and conceal edges of glass, usually constructed of metal or wood.

Mirror, Plate. Plate Mirror.

Mirror. Polished surface that forms images by reflection of light rays, usually fabricated of glass with silver coating on reverse side.

Miscellaneous ACM. Interior asbestos-containing building material on structural components, structural members or fixtures, such as floor and ceiling tiles; does not include surfacing material or thermal system insulation.

Miscellaneous Taxes Payable. Estimated taxes incurred during an accounting period and owed to local and State agencies.

Miscibility. Capability of being mixed or made homogeneous.

Misconduct. Wrongful conduct.

Misses. In painting, holidays; skips; voids.

Mistake. A legal doctrine under which formation of a contract may be prevented if a party entered into the contract under a material mistake of fact.

Mist-Coat. Thin tack coat; thin adhesive coat.

Miter Box. A device for guiding a handsaw in the correct angle for making a miter joint in wood, plastic, or metal.

Miter Saw. A fine toothed saw, usually with a stiffened back, for cutting miters.

Miter. The junction of two pieces of carpet, wood, or other material at an angle; usually 45°, to form a right angle, but may be any combination of angles.

Mix Design. Devising the proportioning of water, portland cement, fine and coarse aggregates, and admixtures in a concrete mix.

Mix. The act or process of mixing; also mixture of materials, such as mortar or concrete.

Mixer Efficiency. In concrete making, the adequacy of a mixer in rendering a homogeneous product within a stated period; homogeneity is determinable by testing for relative differences in physical properties of samples extracted from different portions of a freshly mixed batch.

Mixer, Colloidal. A mixer designed to produce colloidal grout.

Mixer, Horizontal Shaft. See Horizontal Shaft Mixer.

Mixer, Kitchen. A kitchen utensil which blends or combines food ingredients.

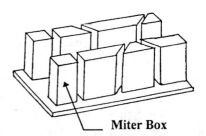

Miter Box

Mixer, Non-Tilting. A horizontally rotating drum mixer that charges, mixes, and discharges without tilting.

Mixer, Pan. See Mixer, Vertical Shaft.

Mixer, Plant. An operating installation of equipment including batchers and mixers as required for batching or for batching and mixing concrete materials; also called mixing plant when equipment is included.

Mixer, Tilting. A rotating drum mixer that discharges by tilting the drum about a fixed or movable horizontal axis at right angles to the drum axis; the drum axis may be horizontal or inclined while charging and mixing.

Mixer, Vertical Shaft. A cylindrical or annular mixing compartment having an essentially level floor and containing one or more vertical rotating shafts to which blades or paddles are attached; the mixing compartment may be stationary or rotate about a vertical axis.

Mixer. A machine used for blending the constituents of concrete, grout, mortar, cement paste, or other mixtures.

Mixing Chamber. That part of a gas-welding or oxygen-cutting torch wherein the gases are mixed prior to combustion.

Mixing Cycle. The time taken for a complete cycle in a batch mixer.

Mixing Faucet. Separate faucets having a common spout providing control of the water temperature.

Mixing Speed. Rotation rate of a mixer drum or of the paddles in an open-top, pan, or trough mixer, when mixing a batch; expressed in revolutions per minute, or in peripheral feet per minute of a point on the circumference at maximum diameter.

Mixing Time. The period during which the constituents of a batch of concrete are mixed by a mixer; for a stationary mixer, time is given in minutes from the completion of mixer charging until the beginning of discharge.

Mixing Valve. A valve that permits mixing of liquids or a liquid and a gas.

Mixing Water. The water in freshly mixed sand-cement grout, mortar, or concrete, exclusive of any previously absorbed by the aggregate.

Mixing, Open-Top. A truck-mounted mixer consisting of a trough or a segment of a cylindrical mixing compartment within which paddles or blades rotate about the horizontal axis of the trough.

Mixture. The assembled, blended, comingled ingredients of mortar, concrete, or the like; or the proportions for their assembly.

MLA. Metal Lath Steel Framing Division of National Association of Architectural Metal Manufacturers.

Mnemonic. 1. Tending to assist memory. 2. Markings on fabricated parts to assist in assembling or installing in the right place.

Mobility. The degree to which a material flows.

Mobilization. 1. The act of putting into movement or circulation. 2. The assembly and movement of equipment to a jobsite.

Modacrylics. Modified acrylics; see Acrylics.

Modal Analysis. Determination of earthquake design forces based upon the theoretical response of a structure in its several modes of vibration to excitation.

Mode. 1. A manner of style or fashion. 2. A way of doing something. 3. The most frequent value in a set of data. 4. The slope of the vibration curve.

Model. The original from which a mould or copy is made.

Modem. A device that connects a computer to the telephone system.

Modification. An agreed change to the terms of a contract.

Modified Bitumen. Bituminous membranes modified with styrene butadiene or atactic polypropylene to improve flexibility, elasticity, cohesive strength, resistance to flow and toughness.

Modified Mercalli. Modification of the original Mercalli Scale to represent construction materials and methods in the United States compared to the European construction methods and materials of the original Mercalli Scale.

M

Modular Dimensional Standards. Dimensional standards approved by the American Standards Association for all building material and equipment, based upon a common unit of measure of four inches, known as the module; this module is used as a basis for the grid which is essential for dimensional coordination of two or more different materials.

Modular Masonry Unit. One whose nominal dimensions are based on the four inch module.

Modular Ratio. The ratio of modulus of elasticity of steel (Es) to that of concrete (Ec) usually denoted by the symbol n.

Modular. Of or consisting of modules.

Modulating Refrigeration System. Refrigerating system of variable capacity.

Modulating. Type of device or control which tends to adjust by increments (minute changes) rather than by either full on or full off operation.

Module. A standardized part or independent unit used in construction.

Modulus of Deformation. A concept of modulus of elasticity expressed as a function of two time variables; strain in loaded concrete as a function of the age at which the load is initially applied and of the length of time the load is sustained.

Modulus of Elasticity. The ratio of the unit stress in a material to the corresponding unit strain; the ratio of normal stress to corresponding strain for tensile or compressive stresses below the proportional limit of the material; referred to as elastic modulus of elasticity, Young's modulus, and Young's modulus of elasticity; denoted by the symbol E.

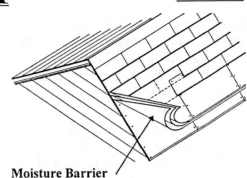

Moisture Barrier

Modulus of Rigidity. The ratio of unit shearing stress to the corresponding unit shearing strain; referred to as shear modulus and modulus of elasticity in shear; denoted by the symbol G.

Modulus of Rupture. A measure of the ultimate load-carrying capacity of a beam and sometimes referred to as rupture modulus, or rupture strength; it is calculated for apparent tensile stress in the extreme fiber of a transverse test specimen under the load which produces rupture.

Modulus of Subgrade Reaction. Ratio of load per unit area of horizontal surface (of a mass of soil) to corresponding settlement of the surface.

Mogul Base. A screw-in style base for an incandescent lamp of generally 300 watts or more.

Moist Air Curing. Curing with moist air at atmospheric pressure and a temperature of about 70° F.

Moist Cure. See Cure, 4.

Moist Room. A room in which the atmosphere is maintained at a selected temperature (usually 23.0° C ±1.7° C or 73.4° F ±3.0° F) and a relative humidity of at least 98 percent, for the purpose of curing and storing cementitious test specimens; the facilities must be sufficient to maintain free moisture continuously on the exterior of test specimen.

Moisture Barrier. A membrane used to prevent the migration of liquid water through a floor or wall.

Moisture Content of Wood. The amount of water contained in the wood, usually expressed as a percentage of the weight of ovendry wood.

Moisture Content. The amount of water in a substance; in wood, it is usually expressed as a percentage of the weight of the ovendry wood.

Moisture Expansion. An increase in dimension or bulk volume of a ceramic article caused by reaction with water or water vapor; this reaction may occur in time at atmospheric temperature and pressure, but is expedited by exposure of the article to water or water vapor at elevated temperatures and pressures.

Moisture Indicator. Instrument used to measure moisture content of a refrigerant.

Moisture Meter. An instrument that measures electrical resistance in a material and equates it to the amount of ambient moisture content; often used by painters to judge the suitability of plaster, masonry, or other surface for the application of paint.

Moisture Movement. The movement of moisture through a porous medium; the effects of such movement on efflorescence and volume change in hardened cement paste, mortar, concrete, or rock.

Moisture Protection. The act or process of retarding the seepage of moisture.

Moisture Vapor Transmission (MVT). Moisture vapor transmission rate through a membrane.

Moisture, Absorbed. See Absorbed Moisture.

Moisture. Finely divided particles of water; vapor.

Mold Oil. A mineral oil that is applied to the interior surface of a clean mold, before casting concrete or mortar therein, to facilitate removal of the mold after the concrete or mortar has hardened.

Mold, Casting. See Casting Mold.

Mold, Running. See Running Mold.

Mold. A divider containing a cavity into which neat cement, mortar, or concrete test specimens are cast; a form used in the fabrication of precast mortar or concrete units (e.g., masonry units); also spelled mould.

Molded Plywood. Plywood made to some desired shape other than perfectly flat; often this shaping is done at the time the layers are glued together; two ways of molding plywood are by applying fluid pressure and with curved forms.

Molding Plaster. A fast-setting gypsum plaster used for the manufacture of cast ornaments.

Molding, Base. See Base Molding.

Molding, Bed. See Bed Molding.

Molding, Lip. See Lip Molding.

Molding, Picture. See Picture Molding.

Molding, Rake. See Rake Molding.

Molding. A strip of wood, metal or plastic trim, used to conceal joints or provide decoration; also spelled moulding.

Mole Run. A meandering ridge in a roof membrane not associated with insulation or deck joints.

Molecule. The smallest fundamental unit of a chemical compound that can take part in a chemical reaction; a group of atoms that are linked together.

Molliers Diagram. Graph of refrigerant pressure, heat, and temperature properties.

Molybdenum. A metallic element that resembles chromium and tungsten in many properties, used in strengthening and hardening steel.

Moment Arm. A lever arm.

Moment Connection. A connection between two structural members that is resistant to rotation between the members, as differentiated from a pin connection, which allows rotation.

Moment Diagram. A graphical method of representing the value of the bending moment at any point along a beam.

Moment Distribution. A method of structural analysis for continuous beams and rigid frames whereby successive converging corrections are made to an assumed set of moments until the desired precision is obtained; also known as the Hardy Cross method.

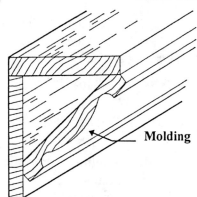

Molding

Moment of Inertia. The summation of the products obtained by multiplying each individual unit of area by the square of its distance to an axis.

Moment Resisting Frame. A structural frame composed of rigid joints in which the members and joints resist both vertical and horizontal forces.

Moment, Negative. See Negative Moment.

Moment, Positive. See Positive Moment.

Moment. A measure of the tendency to produce rotational motion, equal to the result of multiplying a the magnitude of a force) by its perpendicular distance from a particular axis or point; Bending Moment.

Momentum. The quantity of motion of a moving body, measured as a product of its mass and velocity. The impetus gained by movement.

Monastery. A house for persons under religious vows.

Monel Metal. An alloy of nickel with copper, aluminum, and iron that is resistant to corrosive liquids.

Monitor Cabinet. A cabinet whose doors have louvered panels to allow for ventilation, light, or finish design.

Monitor. A structure raised above the general roof level to provide vertical windows for light and air.

Monkey Wrench. An adjustable wrench.

Mono Pitch Truss. A truss which would develop a shed type roof.

Monochlorodifluoromethane. Popular low temperature refrigerant, Freon 12 or R-22, with a boiling point of − 41° F. at atmospheric pressure; cylinder color code is green.

Monochromatic Harmony. Color harmony formed by using shades and tints of a single color.

Monochromatic. Having or consisting of one color or hue.

Monochrome Decoration. A single color decoration.

Monocoque. A type of construction in which the outer skin carries all or a major part of the stresses.

Monocottura. A single-fired tile manufactured by a process which allows the simultaneous firing of the clay with the glaze producing a finished tile with a single firing.

Monolith. A body of plain or reinforced concrete cast or erected as a single integral mass or structure.

Monolithic Concrete. Concrete cast with no joints other than construction joints or as one piece; generally, the term is used on larger structures.

Monolithic Construction. Constructed as one piece.

Monolithic Tank. One-piece shell structure.

Monolithic Terrazzo. The application of a 5/8 inch (15 mm) terrazzo topping directly to a specially prepared concrete substrata, eliminating an underbed.

Monolithic Topping. On concrete flatwork; a higher quality, more serviceable topping course placed promptly after the base course has lost all slump and bleeding water.

Monomer. 1. A chemical compound that can undergo polymerization. 2. An organic liquid, of relatively low molecular weight, that creates a solid polymer by reacting with itself or other compounds of low molecular weight or both.

Monomolecular Film. Films that are one molecule thick; used over bleeding water at the surface of freshly placed concrete or mortar as a curing agent, as it is a means of reducing the rate of evaporation.

Monomolecular. Composed of single molecules; substances that are one molecule thick.

Month-to Month Tenancy. A lease of real property, written or oral, that provides for rent to be paid monthly, for no stated term, and would require 30 days notice by either party to terminate.

Montmorillonite. See Montmorillonoid.

Montmorillonoid. A group of clay minerals, including montmorillonite, characterized by a sheet-like internal molecular structure; consisting of extremely finely-divided hydrous aluminum or magnesium silicates that swell on wetting, and shrink on drying.

Monument. 1. A commemorative structure or building. 2. A permanent surveyor's benchmark.

Monumental. 1. Of or relating to a monument. 2. Extremely great; massive; outstanding.

Mop and Flop. An application procedure in which roofing elements, such as insulation boards, felt plies, and cap sheets, are initially placed upside down adjacent to their ultimate locations, are coated with adhesive, and are then turned over and applied to the substrate.

Mop Basin. A floor set service sink; also called a Mop Receptor.

Mop Receptor. See Mop Basin.

Mop Sink. A deep well plumbing fixture with a faucet and a drain used for collecting and dispensing water for mopping and other janitorial purposes.

Mopping. 1. A layer of hot pitch or asphalt between plies of roofing felt. 2. Swabbing, as with roofing asphalt.

Moratorium. A legally authorized or prohibited suspension of contractual obligations, such as debt repayment.

Moresque. Multicolored yarn made by twisting together two or more strands of different shades or colors.

Morphology. The study of the forms of things.

Mortar Board. A square shaped board, about 3 feet square, that is used to receive the mortar on a scaffold, for the use of the mason; also called Mud Board.

Mortar Bond. The adhesion of mortar to masonry units.

Mortar Box. The box used to mix mortar.

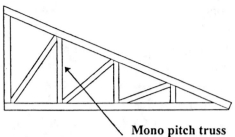

Mono pitch truss

Mortar Hoe. The mortar hoe is used for hand-mixing mortar, usually with a perforated blade and a handle about 66 inches in length.

Mortar Mix. The amount of each material specified, portland cement, fine aggregate, admixture, and water.

Mortar Mixer. A mechanical device for the mixing of mortar; most are driven by gasoline combustion engines; electrically driven mixers are used when small batches of mortar are needed.

Mortar Pumping Machine. The mortar pumping machine is used with the mortar mixer; mixed mortar is poured into the hopper, and a pneumatic gun forces the mortar through a hose; the mortar can be delivered through the hose to masons and tile-setters working high above street level; asbestos fines are added to the mortar as a bonding mechanism so that the mortar in the hose will not separate.

Mortar. 1. A mixture of cement, sand and water; when used in masonry construction, the mixture may contain masonry cement, or standard portland cement with lime or other admixtures which may produce greater degrees of plasticity and/or durability. 2. A plaster mix. 3. The cement, sand, and water mixture that fills the interstices between gravel aggregates in a concrete mix.

Mortgage. A lien against real estate that secures payment of a debt.

Mortgagee. The lender under a mortgage.

Mortgagor. The borrower under a mortgage.

Mortise Lockset. A lockset assembly with the mechanism installed in a mortised pocket within the door and frame.

Mortise. The hole which is to receive a tenon, or any hole cut into or through a piece by a chisel; generally of rectangular shape.

Mortise-and-Tenon. A joint in which a tongue-like protrusion (tenon) on the end of one piece is tightly fitted into a rectangular slot (mortise) in the side of the other piece.

Mortuary. A building in which human remains are kept until cremation or burial.

Mosaic. 1. Tile with small inlaid pieces of porcelain or natural clay materials to form decorative patterns. 2. Small tile or bits of tile, stone, or glass used to form a surface design or an intricate pattern.

Moss, Peat. See Peat Moss.

Motel. A roadside hotel providing lodging and automobile parking.

Motion. Application to a judge or arbitrator for an order or ruling.

Motor Burnout. Condition in which the insulation of an electric motor has deteriorated by overheating.

Motor Control Switches. Any type of switching device which is used in a motor.

Motor Control. Device to start or stop a motor or hermetic motor compressor at certain temperature or pressure conditions.

Motor Grader. A self-propelled machine for earth grading.

Motor Starter. High-capacity electric switches usually operated by electromagnets.

Motor Stator. Stationary part of an electric motor.

Motor, Capacitor. See Capacitor Motor.

Motor, Four Pole. See Four Pole Motor.

Motor, Two-Pole. See Two-Pole Motor.

Motor. 1. A machine, electrical or internal combustion, supplying motive power for a vehicle or some other device. 2. Rotating machine that transforms fluid or electric energy into a mechanical motion.

Motor-Generator Set. A portable gasoline motor combined with an electrical generator.

Mottling. Speckling; a non-uniform paint color.

Mould. See Mold.

Moulding. See Molding.

Mounted Tile. Tile assembled into units or sheets by suitable material to facilitate handling and installation; tile may be face-mounted, back-mounted or edge-mounted; face-mounted tile assemblies may have paper or other suitable material applied to the face of each tile, usually by water soluble adhesives so that it can be easily removed after installation but prior to grouting of the joints; back-mounted tile assemblies may have perforated paper, fiber mesh, resin or other suitable material bonded to the back and/or edges of each tile which becomes an integral part of the tile installation.

Movable Louver. An element for ventilation with equally spaced slats of wood or metal that can be opened or closed.

Movable Partition. A dividing wall that can be moved and arranged to form different walled spaces.

Moving, Shrub. See Shrub Moving.

Moving, Tree. See Tree Moving.

MPT. Male pipe thread.

MSDS. Material Safety Data Sheet.

MSHA. Mine Safety and Health Administration.

MSS Point. Minimum Stable Signal Point; the best superheat setting which will provide constant, or little change in temperature, at the thermostatic expansion valve temperature sensing element while the system is running.

MSS. Manufacturers Standardization Society of the Valve Fittings Industry, Inc.

Muck. 1. Mud rich in humus. 2. Finely blasted rock, particularly from underground.

Mud Board. See Mortar Board.

Mud Cracks. The cracks that develop in an emulsion coating that has been applied too thickly.

Mud Jacking. Raising a sunken concrete slab by pumping a slurry under the slab through a hole in the slab.

Mud Pan. A hand-held container for holding a small quantity of gypsumboard joint compound products; usually the size of a bread pan.

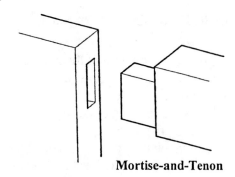

Mortise-and-Tenon

Mud. 1. A slang term for mortar. 2. A slang term for gypsumboard joint compound products. 3. Any soil containing enough water to make it soft. 4. In rotary drilling, a mixture of water with fine drill cuttings and added material which is pumped through the drill string to clean the hole and cool the bit.

Muffler, Compressor. Sound absorber chamber in refrigeration system; used to reduce sound of gas pulsations.

Mulch. A mixture, as of leaves and compost, that covers or is mixed with the earth, often to help enrich the soil; bark, crushed stone or other material used to cover planting beds, retain moisture, reduce weeds, and improve appearance.

Mullion Heater. Electrical heating element mounted in a door or window mullion; used to keep mullion from sweating or frosting.

Mullion. A vertical or horizontal bar between adjacent window or door units; the member between the openings of a window frame to accommodate two or more windows.

Mullite, Porcelain. See Porcelain Mullite.

Mullite. A rare mineral of theoretical composition, a relatively stable phase in ceramics produced by the high temperature reaction of alumina and silica or by the thermal decomposition of alumina-silica minerals such as kyanite, sillimanite, andalusite, and various clay minerals.

Multi-Color Spraying. Spraying a surface with two or more different colors at one time from one gun. The multiple colors exist separately within the material and when sprayed create an interlacing color network with each color retaining its individuality.

Multi-Family Dwelling. A dwelling that will accommodate two or more families.

Multi-Layer. Two or more layers of gypsum board used in an assembly.

Multi-Level. A carpeting texture or design created by different heights of tufts, either cut or uncut loop.

Multi-Outlet Assembly. A type of surface or flush electrical raceway designed to hold conductors and attachment plug receptacles, assembled in the field or at the factory.

Multiple Chemical Sensitivity. A condition in which a person is considered to be sensitive to a number of chemicals at very low concentrations.

Multiple Listing. A cooperative sales tool among brokers whereby they share their listings and, in the event of a sale, the listing broker and the selling broker share the commission.

Multiple Stage Compressor. See Compressor, Multiple Stage.

Multiple System. Refrigerating mechanism in which several evaporators are connected to one condensing unit.

Multiport Valve. 1. Filter control valve changing direction of water flow. 2. A valve for various pool filter operations, which combines in one unit the function of two or more single direct flow valves.

Multi-Unit Wall A wall composed of two or more wythes of masonry.

Multi-Zone Air Handling System. A system providing conditioned air similar to a single-zone system; the temperature and flow of the air supplying each zone is controlled separately; a constant supply of air is supplied to the various zones.

Muntin. 1. A small vertical or horizontal bar between small lights of glass in a sash. 2. The vertical member between two panels of the same piece of panel work.

Muntz Metal. An alloy of copper and zinc that contains 60 percent of copper, can be rolled hot, and is used for sheathing and bolts.

Mural. 1. A painting executed directly on a wall or ceiling. 2. Of a wall. 3. Tile installed in a precise area of a wall or floor to provide a decorative design or picture. 4. Glass or marble mosaic tile (tesserae) made to form a picture or design. 5. Ceramic tile, painted and fired to form a picture or design.

Muriatic Acid. Dilute hydrochloric acid (30% HCL), commercial grade, commonly used for cleaning mortar and grout stains from masonry surfaces.

Murphy Bed. A bed that may be folded or swung into a closet.

Murphy's Law. An observation that anything that can go wrong will go wrong.

Museum. A building used for storing and exhibiting objects of historical, scientific, or cultural interest.

Mushroom Capital. A flaring conical head on a concrete column.

Mutual Assent. An objective manifestation by parties that they intend to be bound by a contract.

Mutuality. The concept that a contract, to be enforceable at all, must be enforceable by both parties.

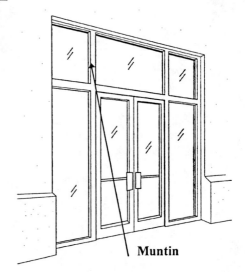

Muntin

MW. Moderate Weather; see Grade MW Brick.

Mylar. Plastic sheeting used as a drafting medium by drafters.

NAAMM. National Association of Architectural Metal Manufacturers.

Nadir. The lowest point.

NAHB. National Association of Home Builders.

Nail Apron. A leather or fabric pouch attached to a carpenter's belt to carry nails or other fasteners.

Nail Bar. A tool which pries open or pulls out driven nails.

Nail Couple. A pair of nails that will resist a rotating force.

Nail Fracture. See Hammer Fracture.

Nail Gun. A mechanical device for driving nails by compressed air.

Nail Popping. Protrusion of the nail from gypsumboard, usually attributed to the shrinkage of or use of improperly cured wood framing.

Nail Pull Resistance. The ability of gypsum board to resist nail head penetration as defined in ASTM C 473.

Nail Set. A small hand tool, like a punch, used for setting finish nails slightly below the surface of the wood.

Nail Spotter. In gypsumboard finishing, a small box-type applicator for covering dimpled nail heads with joint compound; see Ames Taping Tools.

Nail. A stiff metal wire fastening device with a point on one end and head designed for impact on the other end.

Nailable Concrete. Concrete, usually made with suitable lightweight aggregate, with or without the addition of sawdust, into which nails can be driven.

Nailable Studs. Light gauge metal studs with slots that will accept nails for attachment of wall finishes.

Nailer. 1. A wooden strip attached to a concrete, masonry, or steel deck to allow roofing materials to be mechanically fastened. 2. A wooden strip cast into a concrete member for later fastening of finishes by nailing.

Nailing Channel. Fabricated from not lighter than 25 gauge steel so as to form slots to permit attachment of lath by means of ratchet-type annular nails, or other satisfactory attachments.

Nailing Pattern. The arrangement, spacings, and dimensions of nails used to attach two or more structural members together.

NAIMA. North American Insulation Manufacturers Association.

Nameplate Plaque. A flat thin piece of metal used for the inscription of a name.

Nano-. A prefix meaning one billionth of a part, 10^{-9}.

Nanogram. One billionth of a gram.

Nanometer. One billionth of a meter.

Nanosecond. One billionth of a second.

Nanotechnology. The technology of measuring and manufacturing objects of microscopically small size.

NAP. The pile on the surface of a carpet or rug.

Napkin Dispenser. A hollow, box-like metal device installed in women's restrooms to dispense sanitary napkins and tampons.

Naptha. Any of several volatile, flammable liquids obtained by distilling certain materials containing carbon; used as a solvent or thinner in varnish and as a fuel; petroleum naptha is also known as benzine.

Narrow Carpet. Fabric woven 27 in. and 36 in. in width, as distinguished from broadloom.

Narthex. A vestibule leading to the nave of a church.

National Association of Architectural Metal Manufacturers (NAAMM). 600 South Federal Street, Chicago, Illinois 60605, (312) 922-6222.

National Association of Home Builders (NAHB). 15th and M Streets, NW, Washington, DC 20005, (202) 822-0229.

National Builders Hardware Association (NBHA). 1815 North Fort Myer Drive, Suite 412, Arlington, Virginia 22209, (703) 527-2060.

National Building Granite Quarries Association (NBGQA). North State Street, Concord, New Hampshire 03301.

National Bureau of Standards (NBS). U.S. Dept. of Commerce, Government Printing Office, Washington, DC 20234.

National Center for Earthquake Engineering Research (NCEER). State University of New York @ Buffalo, Red Jacket Quadrangle, Buffalo, New York 14261, (716) 645-3391.

National Concrete Masonry Association (NCMA). 2302 Horse Pen Road, PO Box 781, Herndon, Virginia 22070, (703) 435-9000.

National Conference of States on Building Codes and Standards, Inc (NCSBCS). 505 Huntmar Park Drive, #210, Herndon, Virginia 22070, (703) 437-0100.

National Council of Architectural Registration Boards (NCARB). 1735 New York Avenue, NW, Washington, DC 20006.

National Electrical Code (NEC). A model electrical code written by the National Fire Protection Association and adopted by various jurisdictions as their electrical code.

National Electrical Manufacturer's Association (NEMA). 2101 L Street, NW Washington, DC 20037, (202) 457-8400, FAX (202) 457-8411.

National Elevators Manufacturing Industry, Inc (NEMI). 600 Third Avenue, New York, NY 10016, (212) 986-1545.

National Fan Manufacturers' Association (NFMA). 5-157 General Motors Building, Detroit, Michigan 48202.

National Fire Protection Association (NFPA). Batterymarch Park, Quincy, Massachusetts 02269, (617) 770-3000.

National Flood Insurance Program (NFIP). PO Box 6468, Rockville, Maryland 20849-6468, (800) 638-6620.

National Glass Association (NGA). 8200 Greensboro Drive, McClean, Virginia 22102, (703) 442-4890.

National Institute for Occupational Safety and Health (NIOSH). Hubert H. Humphrey Building, 200 Independence, SW, Washington, DC 20201, (202) 401-6997. NIOSH was established by the Occupational Safety and Health Act of 1970; primary functions are to conduct research, issue technical information, and test and certify respirators.

National Institute of Building Sciences (NIBS). 1201 L Street, NW, #400, Washington, DC 20005, (202) 289-7800.

National Institute of Standards & Technology (NIST). Office of Standards Service, Gaithersburg, Maryland 20899, (301) 975-5900.

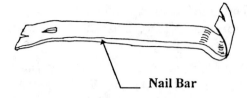

Nail Bar

National Lumber Grades Authority, The (NLGA).

National Oak Flooring Manufacturers' Association (NOFMA). PO Box 3009, Memphis, Tennessee 38173, (901) 526-5016.

National Paint & Coatings Association (NPVLA). 1500 Rhode Island Avenue, NW, Washington, DC 20005, (202) 462-6272.

National Particleboard Association (NPA). 18928 Premiere Court, Gaithersburg, Maryland 20879, (301) 670-0604.

National Ready Mixed Concrete Association NRMCA. 900 Spring Street, Silver Springs, Maryland 20910, (301) 587-1400.

National Roofing Contractors Association (NRCA). 10255 West Higgins Road, #600, Rosemont, Illinois 60018-5607, (708) 299-9070.

National Sanitation Foundation (NSF). School of Public Health, University of Michigan, NSF Building 3475 Plymouth Road, Ann Arbor, Michigan 48105, (313) 769-8010.

National Spa & Pool Institute (NSPI). 2111 Eisenhower Avenue, Alexandria, Virginia 22314-4698, (703) 838-0083.

National Technical Information Service (NTIS). Springfield, Virginia.

National Terrazzo and Mosaic Association (NTMA). 2-A West Loudoun Street, Leesburg, Virginia 22075, (703) 777-7683.

National Wood Window & Door Association (NWWDA). 1400 East Touhy Avenue, #G-54, Des Plains, Illinois 60018, (708) 299-5200.

National Woodwork Manufacturers' Association (NWMA). 205 West Touhy Avenue, Park Ridge, Illinois 60068, (312) 823-6747.

Natural Asphalt. Asphalt occurring in nature which has been derived from petroleum by natural processes of evaporation of volatile fractions leaving the asphalt fractions; the native asphalt of most importance are found in the Trinidad and Bermudez Lake deposits; asphalt from these sources is called Lake Asphalt.

Natural Clay Tile. A tile made by either the dust-pressed method or the plastic method, from clays that produce a dense body having a distinctive, slightly textured appearance.

Natural Cleft Slate. Slate which has been split into thinner pieces along its natural cleft or seam; it is rougher in appearance than machined slate.

Natural Cleft. A natural V-shaped channel, space, opening, or fissure in a material.

Natural Convection. Fluid motion of air caused by the temperature difference between the solid surface and the fluid with which it is in contact.

Natural Draft. A natural stream of air up a chimney, caused by the stack effect, which draws combustion air through the burner or fuel bed.

Natural Finish. A transparent finish, usually a drying oil, sealer or varnish, applied to wood for protection against soiling or weathering; such a finish should not seriously change the original color of the wood or obscure its grain pattern.

Natural Frequency. The constant frequency of a vibrating system in the state of natural oscillation.

Natural Gas Meter. A mechanical measuring and recording device of the volume of natural gas passing a given point.

Natural Gray Yarn. Unbleached and undyed yarn spun from a blend of black, brown or gray wools.

Natural Resins. Essentially the result of exudation of trees; divided into two large classes, dammars and copals; resins are usually named after the locality in which they are found or the port of shipment.

Natural Stone. Any type of stone which is quarried and not manufactured.

Naval Stores. A term applied to the oils, resins, tars, and pitches derived from oleoresin contained in, exuded by, or extracted from trees, chiefly species of pines; historically, these were important items in the stores of wood sailing vessels.

Nave. The main part of the interior of a church.

NBGQA. National Building Granite Quarries Association.

NBHA. National Builders Hardware Association.

NBP. A membrane typically reinforced with polyester; resists tears, punctures and weathering; remains flexible at low temperatures and has a low vapor permeability.

NBS. National Bureau of Standards.

NCARB. National Council of Architectural Registration Boards.

NCEER. National Center for Earthquake Engineering Research.

NCMA. National Concrete Masonry Association.

NCSBCS. National Conference of States on Building Codes and Standards, Inc.

Neat Cement Grout. A fluid mixture of hydraulic cement and water, with or without admixture; also the hardened equivalent of such mixture.

Neat Cement Mortar. Portland cement and water only, no aggregate.

Neat Cement Paste. A mixture of hydraulic cement and water, both before and after setting and hardening.

Neat Cement. Hydraulic cement in the unhydrated state.

Neat. 1. With no dilution and nothing added. 2. Plaster material requiring the addition of aggregate.

NEC. National Electric Code.

Needle Beam. A steel or wood beam threaded through a hole in a bearing wall and used to support the wall and its superimposed loads during underpinning of its foundation.

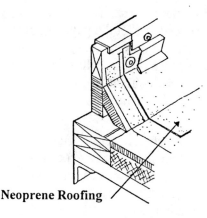

Neoprene Roofing

Needle Valve. Similar to a globe valve, it has a needle which seats into a small opening to control the flow of liquid.

Needlepoint Valve. Type of valve having a needle point plug and a small seat orifice for low flow metering.

Negative Float. The condition in a CPM schedule when there is not enough time available to perform some operation.

Negative Moment. A result of bending moment in a beam in which the upper part is in tension and the lower part is in compression; compare with Positive Moment.

Negative Pressure. 1. Condition that exists when less air is supplied to a space than is exhausted from the space, so the air pressure within that space is less than that in surrounding areas. 2. A pressure within a pipe that is less than atmospheric pressure; also called a Minus Pressure.

Neglect. 1. To pay insufficient attention to. 2. To disregard a duty, leaving it undone.

Negligence. The failure to exercise due care.

Negotiable Instrument. A written promise or request for the payment of a certain sum of money to order or bearer; includes checks, bills of exchange, letters of credit, promissory notes, trade acceptances, and certain bonds.

Negotiated Contract. A construction contract where the price has been arrived at by negotiation between owner and contractor rather than by competitive bidding.

Neighborhood Shopping Center. A small shopping center; a convenience center.

Neighborhood. In community planning, a residential area in which residents are within walking distance of each other.

NELMA. The Northeastern Lumber Manufacturers Association.

Nelson Stud. A bolt-like rod, sometimes threaded to receive a nut, welded to a steel member for attaching another member.

NEMA Switch. A electrical switch approved by the National Electrical Manufacturer's Association.

NEMA. National Electrical Manufacturer's Association.

NEMI. National Elevators Manufacturing Industry, Inc.

Neon. An inert gaseous element occurring in traces in the atmosphere and giving an orange glow when electricity is passed through it in a sealed low-pressure tube; used in advertising and commercial signs.

Neoprene Joint. A fitting of synthetic rubber, neoprene, where two or more members meet.

Neoprene Membrane. An impervious, oil-resistant synthetic rubber, manufactured to be installed in layer form; used for roofing and waterproofing.

Neoprene Roof. A roof system covered with a sheet of synthetic rubber.

Neoprene. A synthetic rubber (polychloroprene) that is used in fluid or sheet applied elastomeric roofing membranes or flashing; sometimes called chloroprene rubber; sheets are available with or without reinforcing fabric; resists weather, heat, oils, solvents and abrasion; can be used in an inverted roof assembly, a loosely laid, ballasted, or fully adhered.

Nepheline Syenite. A mineral aggregate consisting chiefly of albite, microcline, and nephelite, each in significant amount.

NESHAP. National Emission Standard for Hazardous Air Pollutants - EPA Rules under the Clean Air Act.

Net Cross Sectional Area. Of concrete masonry units, the gross cross-sectional area of a section minus the area of cores or cellular spaces; the cross- sectional area of grooves in scored units is not deducted from the gross cross-sectional area to obtain the net cross-sectional area.

Net Effective Temperature (ET). This is not an actual temperature in the sense that it can be measured by a thermometer; it is an experimentally determined index of the various combinations of dry-bulb temperature, humidity, radiant conditions, and air movement that induce the same thermal sensations.

Net Income. The remainder of revenues after direct and indirect costs are deducted; profit.

Net Lease. A lease in which the tenant pays certain specified costs, such as taxes, insurance, and maintenance, of operating the property; a net, net lease and a net, net, net lease are variations where the tenant pays more of the costs.

Net Listing. A real estate listing where the net proceeds of a sale or lease to the owner are specified and all above that figure is the broker's commission; this is an illegal arrangement in some states.

Net Profit. The excess of income after costs and expenses of a business are deducted.

Net Rentable Area. The amount of a building that is available for rental to tenants, expressed in square feet.

Net Working Capital. Current assets less current liabilities.

Net Worth. The difference between total assets and total liabilities.

Net, Safety, See Safety Net.

Net, Tennis. See Tennis Net.

Neutral Arbitrator. An arbitrator not controlled by or biased in favor of any party; also called a Neutral.

Neutral Axis. The line on a member cross-section on which the bending moment is zero.

Neutral Conductor. The neutral conductor is a grounded conductor; the insulation on a neutral conductor is colored white or a natural gray.

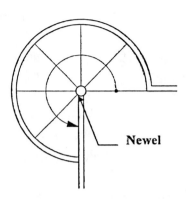
Newel

Neutral Flame. A gas flame wherein the portion of oxygen and the welding gas used is neither oxidizing nor reducing.

Neutral. 1. Without color; achromatic. 2. Being neither acid nor alkaline. 3. Not electrically live. 4. Neutral Arbitrator.

Neutralizer. Substance used to counteract acids in refrigeration system.

Neutron. That part of an atom core which has no electrical potential; electrically neutral.

New England Method. See Pick and Dip.

Newel. 1. The principal post of the foot of a staircase. 2. The central support of a winding flight of stairs.

Newton. An SI unit of force that is of such size that under its influence a body whose mass is one kilogram would experience an acceleration of one meter per second per second.

NFIP. National Flood Insurance Program.

NFMA. National Fan Manufacturers' Association.

NFoPA. Now known as AFPA.

NFPA. National Fire Protection Association.

NGA. National Glass Association.

NGR Stain. See Non-Grain-Raising Stain.

NHPMA. The Northern Hardwood and Pine Manufacturers Association.

NIBS. National Institute of Building Sciences.

NIC. Not in Contract.

Niche. Either a curved or square recess in a wall, dependent on the architecture and the use for which it is intended; used for housing statues, vases, telephones, or door chimes.

Nickel Silver. A silver-white alloy of copper, zinc, and nickel.

Nickel Steel. A steel alloy containing nickel as its principal alloying element.

Nickel. A malleable ductile silver-white metallic element, used in special steels, magnetic alloys, and as a catalyst.

Night Depository. An opening in an exterior wall of a structure or building that allows goods or materials to be delivered to a business after it has closed.

NIMBY. Not In My Back Yard, said in opposition to proposals to establish desirable developments, such as schools or hospitals, too close to one's own property.

Nineteen-Inch Selvage. A prepared roofing sheet with a 17 inch granule-surfaced exposure and a nongranule-surfaced 19 inch selvage edge; this material is also called a Split Sheet, SIS or as Wide-Selvage Asphalt Roll Roofing Surfaced with Mineral Granules.

NIOSH. National Institute for Occupational Safety and Health.

Nipple, Bushed. See Bushed Nipple.

Nipple, Offset. See Offset Nipple.

Nipple. Short length of pipe with male threads on both ends.

NIST. National Institute of Standards and Technology.

Nitrocellulose. Used extensively in making lacquer, is prepared from cellulose (cotton linters) by treatment with chemicals; see Cellulose Nitrate.

Nitrogen Dioxide. Mildly poisonous gas (NO_2) often found in smog and automobile exhaust fumes.

NKBA. National Kitchen and Bath Association.

NLGA. The National Lumber Grades Authority.

Noble Gas. Any gaseous element of a group that almost never combine with other elements, consisting of helium, neon, argon, krypton, xenon, and radon; also called the Inert Gases or the Rare Rases.

No-Fines Concrete. A concrete mixture in which only the coarse gradation (3/8 to 3/4 inch normally) of aggregate is used.

NOFMA. National Oak Flooring Manufacturers' Association.

No-Frost Freezer. Low-temperature refrigerator cabinet in which no frost or ice collects on freezer surfaces or materials stored in cabinet.

Nogging. The filling of brick between the roof rafters from the wall plate to roof boards for making the building wind tight.

No-Hub Pipe. Pipe usually manufactured of cast iron, which is fabricated without hubs and coupled together by a fastener of stainless steel and rubber.

Noil. A by-product in worsted yarn manufacture, consisting of short wool fibers, less than a determined length, which are combed out.

Noise Reduction Coefficient (NRC). A relative numerical expression of the ability of a material to absorb sound.

Noise Reduction. 1. The reduction in sound pressure level caused by making some alteration to a sound source. 2. The difference in sound pressure level measured between two adjacent rooms caused by the transmission loss of the intervening wall.

Noise. Any undesired sound.

Nomenclature. Systematic naming; the terminology of a science.

Nominal Dimension. 1. The named size; compare with Actual Size. 2. An approximate rough cut dimension assigned to a piece of material as a convenience in referring to the piece, such as 2 by 4 or 500 pound. 3. The stated size of a piece of lumber; its actual size is smaller. 4. A dimension greater than a specified masonry unit dimension by the thickness of a mortar joint. 5. This is the approximate facial size or thickness of tile, expressed in inches or fractions of an inch, for general reference, such as a "4 by 4" tile. 6. Tubing measurement which has an inside diameter the same as iron pipe of the same stated size.

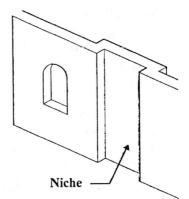

Niche

Nominal Size. The named size, for example the nominal size of a wood stud is 2- by 4- inches while the actual size is less. The nominal size of a concrete block is 4- by 8- by 16-inch while the actual size is less.

Nomogram. A graphic representation consisting of three or more lines calibrated to scale and arranged so that a straightedge connecting known values on two lines will intersect unknown values on one or more other lines.

Non-Agitating Unit. A truck mounted unit for transporting ready- mixed concrete short distances, not equipped to provide agitation (slow mixing) during delivery.

Non-Automatic. Non-automatic means that the implied action requires personal intervention for its control; see Automatic; as applied to an electric controller, non- automatic control does not necessarily imply a manual controller, but only that personal intervention is necessary.

Nonaxial. In a direction not parallel to the long axis of a structural member.

Nonbearing Partition. A partition which extends from floor to ceiling but which supports no load other than its own weight.

Non-Bearing Wall. A wall that carries no superimposed load; it could be removed without any structural effect on the remaining building.

Nonbearing. Not carrying a load.

Non-Code Installation. Functional refrigerating system installed where there are no local state or national refrigeration codes in force.

Noncombustible. Any material which will neither ignite nor actively support combustion in air at a temperature of 1200° F. when exposed to fire.

Noncondensable Gas. Gas which does not change into a liquid at operating temperatures and pressures.

Nonconforming Use. Use of land that is not in conformance with the zoning code; applies mainly to pre-existing uses when new zoning is enacted; the non- conforming use may remain as long as it is continuous; upon interruption, the non- conforming use may not continue.

Non-Decay Fungus. Mildew and mold that discolors the surface of wood but does not cause decay.

Non-Destructive Testing. Testing of building materials in a manner that does not destroy the material.

Non-Drying Oils. Oils which are unable to take up oxygen from the air and change from a liquid to a solid state; mineral oils are non-drying oils.

Non-Evaporable Water. The water in concrete which is irremovable by over drying; chemically combined during cement hydration.

Non-Ferrous Metal. A metal that does not contain iron.

Nonflammable. Incombustible, will not burn; opposite of flammable or inflammable.

Nonfrosting Evaporator. Evaporator which never collects frost or ice on its surface.

Non-Grain-Raising Stain. Wood stain which does not raise the grain of the wood; NGR Stain; made by dissolving dyes such as used in making water stains in special solvent instead of water.

Non-Load Bearing Partition. A structurally non-essential interior wall assembly for compartmentalizing floor space.

Nonmetallic Non-Shrink Grout. Nonmetallic cementitious or epoxy based mix used to fill gap created between bearing components or base plates and foundation or other supporting element.

Non-Metallic Tubing. A round sheath product, of round cross-section, fabricated from a moisture-resistant, flame-retardant material.

Non-Pressure Drainage. Means gravity flow drainage in which the sloping pipes are not filled to capacity.

Nonrecourse Mortgage. One in which the borrower has no personal liability, the lender's sole remedy being to foreclose on the property in the event of a default.

Non-Rising Pin. See Fast Pin Hinge.

Non-Shrink Grout, Nonmetallic. Nonmetallic Non-Shrink Grout.

Non-Shrink Grout. Cementitious or epoxy based mix used to fill gap created between bearing components or base plates and foundation or other supporting element.

Non-Slip Terrazzo. A mixture of marble chips and portland cement on a portland cement and sand body, which has been roughened with iron fillings, carborundum powder or indented while it is wet.

Non-Slip Tile. Tile having greater non-slip characteristics due to an abrasive admixture, abrasive particles in the surface, grooves or patterns in the surface or because of natural non-skid surface characteristics.

Nontoxic. Not poisonous.

Nonvitreous. Non-vitrified; that degree of vitrification evidenced by relatively high water absorption, usually more than 10 percent water absorption, except for floor and wall tile which are considered non-vitreous when water absorption exceeds 7 percent.

Non-Volatile. Portion of a product which does not evaporate at ordinary temperature.

Norm. 1. The standard or principle set by the average of a large group in the same trade or profession. 2. An authoritative standard or model.

Normal Charge. Thermal element charge which is part liquid and part gas under all operating conditions.

Normal Consistency. Of gypsum plaster or gypsum concrete, the number of milliliters of water per 100 grams of gypsum plaster or gypsum concrete required to produce a mortar or a slurry of specified fluidity.

Normal Fault. A fault under tension where the overlying block moves down the dip or slope of the fault plane.

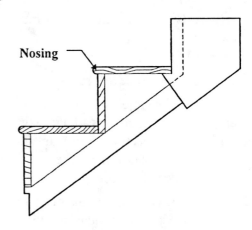

Nosing

Normal Loading. A design load that stresses a member or fastening to the full allowable stress tabulated; this loading may be applied for approximately 10 years, either continuously or cumulatively, and 90 percent of this load may be applied for the remainder of the life or the member or fastening.

Normal Weight Concrete. Concrete having a hardened density of approximately 150 pcf which is made from aggregate of approximately the same density.

Norman Brick. A brick with 2-3/4 by 4 by 12 inch dimensions.

North American Insulation Manufacturers Association (NAIMA). 44 Canal Center Plaza, #310, Alexandria, Virginia 22314, (703) 684-0084.

North Pole. 1. The northernmost point of the earth's axis of rotation. 2. The pole of a magnet that points to the north.

Northeastern Lumber Manufacturers Association, The (NELMA). 4 Fundy Road, Falmouth, Maine 04105.

Northern Hardwood and Pine Manufacturers Association, The (NHPMA).

Nosing, Rubber. See Rubber Nosing.

Nosing. 1. The projecting forward edge of a stair tread; the part of a stair tread which projects over the riser. 2. Any similar projection. 3. The rounded edge of a board.

Not in Contract (NIC). Something shown on contract drawings, usually for reference, but not to be included in the construction contract.

Notary Public. One authorized by law to acknowledge and certify documents and signatures; a notary.

Notch. A crosswise rabbet at the end of a board.

Notched Trowel. 1. A tile setter's tool consisting of a trowel with serrated edges; used to spread bonding materials; depth of application is regulated by the size and spacing of the teeth. 2. A similar tool used by applicators of resilient flooring to spread adhesive.

Note Payable. Amount owed because of a promissory note given to a creditor.

Note Receivable. A promissory note given by a customer to pay a definite sum to the business at a specific future time.

Note. A written promise to pay a certain amount of money at a certain time.

Notice of Cessation. A written notice filed by the owner or owner's agent signifying that no labor has been performed on the project for a specific period of time (such as 60 days) and establishing the time remaining to exercise lien rights by all parties concerned.

Notice of Completion. A written notice filed by the owner or owner's agent signifying completion of a project and establishing the time remaining to exercise lien rights by all parties concerned.

Notice of Nonresponsibility. A notice which, if properly recorded and posted on the premises, relieves the owner from the effects of mechanics liens for work and materials not ordered by the owner, usually ordered by lessees.

Notice to Pay Rent or Quit. A legal notice to a defaulting tenant, a prerequisite to eviction.

Notice to Proceed (NTP). A written notice to a contractor to proceed with the work of the contract.

Notice. Delivery of information to a party

Novation. The substitution of a new legal obligation for an old one.

Nozzle. An attachment at the end of a plastering machine delivery hose, which regulates the fan or spray pattern.

NPA. National Particleboard Association.

NPVLA. National Paint & Coatings Association.

NRC. Noise Reduction Coefficient.

NRCA. National Roofing Contractors Association.

NRMCA. National Ready Mixed Concrete Association.

NRP. Non-Rising Pin.

NSF. National Sanitation Foundation.

NSPI. National Spa & Pool Institute.

NTIS. National Technical Information Service.

NTMA. National Terrazzo and Mosaic Association.

NTP. Notice to Proceed.

NTS. Not to Scale.

Nuclear Meter. An instrument for detecting and measuring moisture.

Nuisance. A condition of or on real property that damages neighboring persons or property.

Nurse Station Indicator. A signaling device at a nurses station activated by patients needing assistance from nurses.

Nurse Station. An area in a hospital where nurses congregate and medical materials and records are stored.

Nursing Care Facility. Any institution which provides nursing care services to persons needing constant care and attention on a 24-hour basis; a nursing home, nursery, long term facility, convalescent home.

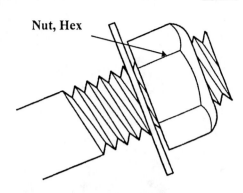

Nut, Hex

Nut, Hex. See Hex Nut.

Nut. A small square or hexagonal flat piece of metal or other material with a threaded hole through it for screwing on the end of a bolt to secure it.

NW. No Weather; see Grade NW Brick.

NWMA. National Woodwork Manufacturers' Association.

NWWDA. National Wood Window & Door Association.

Nylon Carpet. A carpet made from synthetic nylon fibers.

Nylon Plastics. Plastics based on resins composed principally on long-chain synthetic polymeric amide which has recurring amide groups as an integral part of the main polymer chain.

Nylon. Any of various synthetic polyamide fibers with tough, lightweight, elastic properties, used in industry and for textiles.

O/C. See OC.

Oak Floor. A common type of flooring, usually installed in small tongue and groove strips made of oak.

Oak Veneer. A thin layer of oak glued to an interior wood or bonded together to form a protective or ornamental facing; types available are white oak and red oak.

Oak. A strong, hard, heavy wood.

Oakum. Loosely twisted jute or hemp fiber impregnated with tar and used in caulking seams, as in wood shipbuilding or in caulking cast iron bell and spigot piping.

Oatmeal Paper. Wallpaper made by sprinkling sawdust over adhesive surface.

OB. Octave Band.

Obelisk. A tall, slender, tapering, usually 4-sided, stone monument surmounted by a pyramid, of Egyptian origin.

Oblique Slip Fault. A combination of normal and slip or thrust and slip faults whose movement is diagonal along the dip of the fault plane.

Oblique. 1. An angle; neither parallel or perpendicular. 2. Not symmetrical with its base.

Obscure Glass. Ground glass or frosted glass that transmits light but does not allow a view of objects on the other side.

Obsolescence. The loss of value due to being no longer useful.

Obtuse Angle. An angle of over 90 degrees but less than 180 degrees.

Obtuse Triangle. A triangle having one angle over 90 degrees.

OC. 1. On center; the designation of spacing for framing members in a building from the center of one member to the center of the next; also shown as O/C. 2. See Oxygen Cutting.

Occupancy Importance Factor (I). A building code term used to identify occupancies of importance.

Occupancy. A building code term referring to the use of a building, such as school, office, restaurant, warehouse, or residence.

Occupational Safety & Health Administration (OSHA). U.S. Department of Labor/OSHA, 200 Constitution Avenue, NW, Washington DC, (202) 219-7725.

Ochre. An earth pigment, usually red or yellow, made from impure iron ore, used in paint.

Octave. Frequency difference between harmonic vibrations; it is the doubling of the frequency of sound.

Octyl Alcohol-Ethyl Hexanol. Additive in absorption machines to reduce surface tension in the absorber.

Oculus. 1. A circular opening or eye in a dome. 2. Any round window.

OD. Outside Diameter.

Odor. That property of air contaminants that affect the sense of smell.

Odometer. An instrument that measures the distance traveled by a wheeled vehicle.

Off Cycle. That part of a refrigeration cycle when the system is not operating.

Offer. A promise that is enforceable if accepted.

Offeror. A person who makes an offer.

Office Building. A building containing business offices.

Office Cubicle. A small compartment unit, surrounded by fixed or movable walls on two or three sides, where an individual works.

Office Park. A planned, usually suburban, community of office buildings with common management, parking, landscaping, and other services.

Office Partition. Moveable interior wall, used to partition off an office area, assembled together to create cubicles for employees.

Office Safe. A locked receptacle that holds money or valuables in an office; usually rated by the manufacturer to designate fire-worthiness and strength.

Office Trailer. A highway vehicle parked on a job site, designed to serve wherever needed, as an office and a place to carry out business.

Office. A room or building where business activities are conducted.

Offset Cabinet Construction. A method of building cabinets where the back corners of the doors and drawer fronts are rabbeted so they overlap the face frame, thereby concealing the joint.

Offset Connector. A fitting that is a combination of elbows or bends which brings one section of pipe out of line with, but into a line parallel with, another section.

Offset Hinge. A hinge with a 90 degree offset in one or both legs, designed to shift the center of rotation of the door.

Offset Nipple. A fitting, threaded at both ends, that is a combination of elbows or bends which brings one section of pipe out of line with, but into a line parallel with, another section.

Offset. 1. A combination of elbows or bends which brings one section of pipe out of line and back in line parallel with the original sector. 2. A course of masonry units that sets in from the course that is directly below it; also called Set Back, Set In, or Set Off; the opposite of Corbel.

Off-Site. Anything not on the subject building site.

OG. Ogee.

Ogee Molding. A molding with a modified S-shaped profile.

Ogee Roof. An architectural roof design which is Eastern influenced.

Ogee. A curved section of a moulding, partly convex and partly concave.

Ogive. A pointed arch.

Ohm (R). Unit of measurement of electrical resistance; one ohm exists when one volt causes a flow of one ampere.

Ohmmeter. Instrument for measuring resistance in ohms.

Ohm's Law. A law in electricity which states the relationship between voltage, amperes, and resistance; equation. Amperes (I) x Ohms (R) = Volts (E).

Oil Absorption. A measure of the ability of pigments to absorb oil.

Oil Base Caulking. Caulking made with resins and other ingredients made of various oils, dispensed through a tube.

Oil Binding. Condition in which an oil layer on top of refrigerant liquid may prevent it from evaporating at its normal pressure-temperature.

Oil Burner Nozzle. A spraying device which atomizes fuel oil into a fine spray and increases the combustion characteristics of the fuel.

Oil Colors. Colors ground to form of paste in linseed oil.

Oil Fired Boiler. A boiler that is heated by a unit that sends oil under high pressure to a nozzle, where it is sprayed as a mist and ignited by an electric spark.

Oil Furnace. A furnace that burns heating oil.

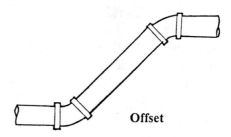

Offset

Oil Length. Oil length in varnish is measured by the oil in gallons per hundred pounds of resin; a long-oil varnish is tougher than a short-oil varnish; rubbing varnish is a typical short-oil varnish and spar varnish is a typical long-oil varnish.

Oil of Vitriol. See Sulphuric Acid.

Oil Paint. Paint in which a drying oil is the vehicle.

Oil Separator. Device used to remove oil from gaseous refrigerant.

Oil Soluble. Capable of being dissolved in oil.

Oil Stain. A wood stain consisting of oil-soluble dyes and solvents such as turpentine, naphtha, or benzol,; penetrates into pores of wood; has tendency to bleed.

Oil Switch. A switch or protector relay installed in a flue pipe, that will shut off a burner in case the stack does not come up to a predetermined temperature within 45 seconds after the motor starts.

Oil Tank. The storage canister where heating oil is stored for future use by an oil boiler, or an oil burning furnace system.

Oil Varnish. A varnish consisting of a hard resin combined with a drying oil and a drier thinned with a volatile solvent; after application, the solvent dries first by evaporation; then the oil dries by oxidation.

Oil, Refrigeration. See Refrigeration Oil.

Oil-Canning. Describing the distortion of thin-gauge metal panels that are fastened in a manner restricting normal thermal movement.

Oilcloth Varnish. Special highly flexible and elastic varnish.

Oilstone. A whetstone used with oil.

Oiticica Oil. Drying oil obtained from nut of oiticica tree.

Old English Bond. See English Bond.

Old Growth. Timber in or from a mature, naturally established forest. When the trees have grown during most if not all of their individual lives in active competition with their companions for sunlight and moisture, this timber is usually straight and relatively free of knots.

Olefin Plastics. Plastics based on resins made by the polymerization of olefins or copolymerization of olefins with other unsaturated compounds, the olefins being in greatest amount by weight; polyethylene, polypropylene and polybutylene are the most common olefin plastics encountered in pipe.

Olefin. Long-chained synthetic polymer composed of at least 85% by weight of ethylene, propylene or other olefin units; currently, only polypropylene has been produced in fiber form for carpet manufacture.

Oleoresin. A solution of resin in an essential oil that occurs in or exudes from many plants, especially softwoods; the oleoresin from pine is a solution of pine resin (rosin) in turpentine.

Oleoresinous Varnish. Varnish composed of resin or gum dissolved in a drying oil which hardens as it combines with oxygen from the air.

Olive Knuckle Hinge. A paumelle hinge with knuckles forming an oval shape.

On Center (O/C). Center to center.

On the Stump. See Standing Timber.

One Piece Toilet. A toilet where the water tank and bowl are cast in one piece of vitreous china for the purposes of quieter operation and better appearance.

One-Family Dwelling. A dwelling containing only one dwelling unit.

One-Way Action. The structural action of a slab that spans between two parallel beams or bearing walls.

One-Way Concrete Joist System. A reinforced concrete framing system in which closely spaced concrete joists span between parallel beams or bearing walls.

One-Way Glass. See Transparent Mirror.

One-Way Solid Slab. A reinforced concrete floor or roof slab that spans between parallel beams or bearing walls, with structural reinforcing in the spanning direction only.

One-Way. Constructed with reinforcing steel running in one direction only.

On-Site. Objects and activities that are on the building site.

Opacity. Hiding power, as of paint.

Opaque Glaze. A non-transparent glaze with or without color.

Opaque. Impervious to light; not transparent.

Open Circuit. Electric circuit with a physical interruption caused by a switch, a break in a conductor, or an open breaker or fuse.

Open Compressor. An external drive compressor; not hermetic.

Open Decking. A deck in which the joists on the underside are exposed.

Open Display Case. Commercial refrigerator designed to maintain its contents at refrigerating temperatures even though the contents are in an open case.

Open Front Seat. A toilet seat with an open gap at the front, usually used in public toilet rooms for improved sanitation.

Open Listing. A listing agreement with one or more real estate brokers to sell or lease one's property on a non-exclusive basis.

Open Run-Around. A method of heat recovery from exhaust air in which a fluid is inserted alternately in the supply and exhaust air stream; this system transfers both sensible and latent heat.

Open Time. The period of time during which the bond coat retains its ability to adhere to the tile and bond the tile to the substrate.

Open Type Compressor. See Compressor, Open Type.

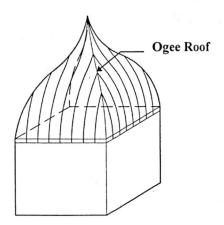

Ogee Roof

Open Type Decking. A deck in which the joists on the underside are exposed.

Open-Graded Aggregate. One containing little or no mineral filler and in which the void spaces in the compacted aggregate are relatively large.

Open-Graded Asphalt Friction Course. A pavement surface course that consists of a high-void, asphalt plant mix that permits rapid drainage of rainwater through the course and out the shoulder. The mixture is characterized by a large percentage of one-sized coarse aggregate. This course prevents tire hydroplaning and provides a skid-resistant pavement surface.

Open-Grained Wood. Woods with large pores, such as oak, ash, chestnut, and walnut; also known as Coarse-Grained or Coarse Textured.

Open-Web Steel Joist. A prefabricated, welded steel truss used at closely spaced intervals to support floor or roof decking.

Operable Door. A door, of a pair, that is normally operable.

Operable Part. A part of a piece of equipment or appliance used to insert or withdraw objects, or to activate, deactivate or adjust the equipment or appliance, such as a coin slot, push-button, or handle.

Operating Instructions. The written explanations and descriptions furnished by manufacturers and fabricators of building equipment and components for the proper operation of their products.

Operating Leverage. The ratio of the percentage change in profit to the percentage change in sales.

Operating Pressure. Actual pressure at which a refrigerating system works under normal conditions; this pressure may be positive or negative (vacuum).

Operating Profit. Revenue of the business less the total of all costs of the operation.

Operating Room. A room in a hospital for surgical operations.

Operation Expenses. Total of all expenses including cost of sales, general and administrative expenses as well as indirect operating expenses.

Opposed Blades. Two sets of blades in a damper, linked so that the adjacent blades can open and turn in opposite directions.

Option. 1. Something that may be chosen. 2. A right given for a consideration to keep an offer to purchase or lease open for a specific time.

Optionee. 1. One who receives an option. 2. The prospective purchaser or lessee in a real estate option.

Optionor. 1. One who gives an option. 2. The owner-seller or lessor in a real estate option.

Or Equal. The phrase used in specifications providing that substitutions must be equal in all relevant respects and usually subject to the approval of the architect.

Oral Agreement. A contract in words that are not reduced to writing.

Orange Mineral. Red lead prepared by roasting basic carbonate white lead; used mainly in printing ink for its characteristic color.

Orange Peel. 1. A pitted texture of a fired glaze resembling the surface of rough orange peel. 2. Spray painting defect in which the lacquer coat does not level down to a smooth surface but remains rough, like the peeling of an orange.

Orbital Sander. A hand-held platform sander that moves rapidly in a circulatory movement.

Ordinance. A law enacted by a city or a county.

Ore. Rock or earth containing workable quantities of a mineral or minerals of commercial value.

Organic Adhesive. A prepared organic material, ready to use with no further addition of liquid or powder, used for bonding tile to back-up material by the thinset method; cures or sets by evaporation.

Organic. 1. Any chemical compound containing carbon. 2. Being or composed of hydrocarbons or their derivatives or matter of plant or animal origin. 3. Pertaining to or derived from living organisms.

Organosol. Film former containing resin plasticizer and solvent; colloidal dispersion of a resin in plasticizer containing more than 5 percent volatile content.

Oriel Window. A bay window projecting from a wall and supported on corbels or brackets.

Oriented Strand Board (OSB). A type of particle panel product composed of strand-type flakes which are purposefully aligned in directions which make a panel stronger, stiffer, and with improved dimensional properties in the alignment directions than a panel with random flake orientation; also called Strandboard.

Orifice. 1. An opening through which something may pass. 2. Attachment to the nozzle on the hose of a plastering machine, of various shapes and sizes, which may be changed to help establish the pattern of the plaster as it is projected onto the surface being plastered. 3. The opening through which fuel is ejected into the burner unit of a furnace or boiler.

Original Contractor. In the law of mechanics liens, a contractor who contracts directly with an owner of real property.

O-Ring. A rubber seal used around stems of some valves to prevent water from leaking by it.

Ormolu. A gilded bronze or gold-colored alloy of copper, zinc, and tin used to decorate furniture and make ornaments.

Ornament. A thing used or serving to adorn; a decoration added to embellish.

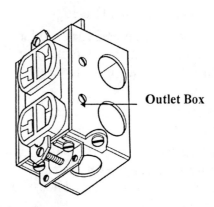

Outlet Box

Ornamental Facing. In masonry, a design formed by the laying of stone, brick, tile or other masonry units so as to produce a decorative effect.

Ornamental Metal. A detail that is added to a building constructed of metal, with the purpose of embellishment or decorating the structure.

Ornamental. Serving as an ornament; decorative.

Ornate. Elaborately adorned; highly decorated.

Orthogonal. Of or involving right angles.

Orthographic Projection. Projection of a single view of an object on a drawing surface that is perpendicular to both the view and the lines of projection; floor plans, elevations, and sections of buildings are orthographic projections.

Orthotropic. Having unique and independent properties in three mutually orthogonal (perpendicular) planes of symmetry; a special case of anisotropy.

OS&Y Valve. A type of valve, with external exposed threads supported by a yoke, indicating the open or closed position of the valve.

OSB. Oriented Strand Board.

Oscillation. A repeated back and forth movement, from one extreme to another, periodically from a maximum to a minimum.

Oscillograph. A device for recording oscillations.

Oscilloscope. A cathode ray tube that shows the changing voltage of an electrical signal as a curved line on a screen.

OSHA. Occupational Safety & Health Administration.

Osmosis. The passage of a solvent through a semi-permeable membrane into a more concentrated solution.

Ottawa Sand. A sand used as a standard in testing hydraulic cements by means of mortar test specimens; produced by processing silica rock particles obtained by hydraulic mining of the orthoquartzite situated in open pit deposits near Ottawa, Illinois; naturally rounded grains of nearly pure quartz.

Out of Phase. The state wherein a structure in motion is not at the same frequency as the ground motion; or where equipment in a building is at a different frequency from the structure.

Out of Square. The variation from true square.

Outdoor Exposure. Plastic pipe placed in service or stored so that it is not protected from the elements of normal weather conditions, the sun's rays, rain, air, and wind.

Outfall Sewer. A large sewer leading from the lower end of the collecting system to the place of disposal.

Outlet Box. A box or container which houses an electrical outlet and its connections.

Outlet. 1. An electrical receptacle into which appliances may be plugged. 2. A way out; an exit.

Outline Lighting. An arrangement of incandescent lamps or gaseous tubes to outline and call attention to certain features such as the shape of a building or the decoration of a window.

Outline Specifications. An abbreviated form of specifications normally produced with schematic design or design development drawings.

Outrigger Scaffold. A work scaffolding supported on outrigger beams cantilevered out from the building or structure being worked on.

Outrigger, Flagpole. See Flagpole Outrigger.

Outrigger. A structural member projecting from a main structure to provide additional stability or to support something, such as an outside scaffold.

Outside Four Inches. The single tier of stretcher courses on the face of the wall.

Oven, Bake. See Bake Oven.

Oven, Convection. See Convection Oven.

Oven, Wall Mounted. See Wall Mounted Oven.

Ovendry Wood. Wood dried to a relatively constant weight in a ventilated oven at 102° to 105° C.

Overatomized. Dispersed too finely by use of excessive atomizing air pressure.

O

Overcertification. The process of an architect or engineer certifying more than the proper amount to be paid to a contractor.

Overcoat. Second coat; topcoat.

Overcurrent Device. Any device that limits the current in a wire to a predetermined number of amperes.

Overcurrent. An electrical current that is in excess of the capacity of the wiring and devices.

Overflow System. Surface water collection or draw-off arrangement.

Overflow. An opening to allow rainwater to overflow after it reaches a predetermined height, commonly 2 inches in the case of roof drains.

Overglaze Decoration. A ceramic or metallic decoration applied and fired on the previously glazed surface of ceramic ware.

Overhand Work. A complete wall that is built from a scaffold that is located only on one side of the wall; the outside or farther face brickwork is laid by reaching over the wall.

Overhang Beam. A beam that is supported by two or more supports and has one or both ends projecting beyond the support.

Overhang. 1. The part of a roof that extends beyond the exterior walls of the building. 2. An overhead projection from the wall. 3. A leaning of the wall face away from the vertical.

Overhead Concealed Closer. A door closer concealed in the head frame with an arm connecting with the door at the top rail.

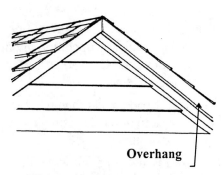

Overhang

Overhead Door. A door, commonly used in garages and warehouses, that opens upward from the ground.

Overhead. All of the indirect costs which can not be directly related to a unit of production and which must be included in the cost of the sale by some allocation method.

Overimprovement. An improvement of excessive size or cost in comparison with similar properties particularly when it is not the highest and best use of the land.

Overlap. 1. The width of fresh paint covered by next layer. 2. Protrusion of weld metal beyond the bond at the toe of the weld.

Overlay Cabinet Construction. A method of building cabinets where the doors and drawer fronts are in front of the face frame.

Overlay. 1. A drawing on a thin layer of paper superimposed over a base drawing. 2. A thin layer of paper, paper, plastic, film, metal foil, or other material bonded to one or both faces of wood panel products or to lumber to provide a protective or decorative face or a base for painting.

Overload Protector. Device, either temperature, pressure or current operated, which will stop operation of unit if dangerous conditions arise; see Overcurrent Device.

Overload. Load greater than that for which system or mechanism was intended or designed.

Oversized Brick. A brick with dimensions greater than 2-1/2"x 3-1/2"x 7-1/2".

Overspray. See Drift, 2.

Overvibration. Excessive vibration of freshly mixed concrete during placement causing segregation.

Ovolo. A quarter round molding.

Owner. 1. Any person or entity having some title or interest in a parcel of real property. 2. A person or entity who contracts for construction.

Owner-Builder. An owner of property who undertakes to construct improvements without the services of a general contractor.

Owner's Overall Budget. A budget prepared by the owner of all of the project costs including land acquisition, construction, financing, professional fees, and all other incidental costs of the project.

Owner's Right to Stop Work. The owner has the right to stop the work of a construction contract only if the contract so provides and only for the reasons stated therein.

OX, OXO, OXXO. Designations of the arrangement of sliding and fixed panels in sliding glass doors and windows; X indicates a sliding panel; O indicates a fixed panel; see also XO, XOO, XOOX.

Oxalic Acid. A poisonous strong organic acid, usually a solid white granular substance; used to dissolve iron rust stains on pool walls and floors, to clean iron rust from filter septa, and as a wood bleach.

Oxidation. The process of combining with oxygen.

Oxidize. To combine with oxygen; to become rusty.

Oxidized Sewage. Sewage which has been exposed to oxygen which causes organic substances to become stable.

Oxidizing Flame. A gas flame wherein the portion used has an oxidizing effect by an excess of oxygen.

Oxy-Acetylene. A mixture of oxygen and acetylene, used as a fuel in cutting or welding metals.

Oxygen Cutting (OC). A group of cutting processes wherein the severing or removing of metals is effected by means of the chemical reaction of oxygen with the base metal at elevated temperatures; in the case of oxidation-resistant metals the reaction is facilitated by the use of a chemical flux or metal powder.

Oxygen. A colorless tasteless odorless gaseous element, occurring naturally in air, water, and most minerals and organic substances, and essential to plant and animal life.

Oxygen-Acetylene Cutting. An oxygen-cutting process wherein the severing of metals is effected by means of the chemical reaction of oxygen with the base metal at elevated temperatures, the necessary temperature being maintained by means of gas flames obtained from the combustion of acetylene with oxygen.

O

Oxygen-Acetylene Welding. A gas welding process wherein coalescence is produced by heating with a gas flame or flames obtained from the combustion of acetylene with oxygen with or without the application of pressure and with or without the use of filler metal.

Oxygen-Arc Cutting. An oxygen-cutting process where in the severing of metals is effected by means of the chemical reaction of oxygen with the base metal at elevated temperatures the necessary temperature being maintained by means of an arc between an electrode and the base metal.

Ozalid. Tradename of a system of drawing reproduction based on the ammonia vapor process; the prints consist of purple lines on a white background; other colors of line and background are also available.

Ozone. A form of oxygen, O_3, a faintly blue unstable gas with a pungent odor and powerful oxidizing properties; used in disinfection, deodorization, oxidation, and bleaching.

P and L. Profit and Loss.

P.C. Concrete. Portland Cement Concrete; concrete composed of coarse and fine aggregates, portland cement, water, and sometimes admixtures to impart special qualities to the concrete or to aid in placing or curing.

Package Deal. An offering of design, construction, financing, and sometimes the land, for one all-inclusive price.

Package Dyeing. Placing spun and wound yarn on large perforated forms and forcing the dye through the perforations.

Package Units. Complete refrigerating system including compressor, condenser and evaporator located in the refrigerated space.

Packaged Chiller. A factory assembled piece of equipment that utilizes a refrigeration cycle to produce chilled water for circulation to a desired location or use.

Packaged Heater. A factory assembled piece of equipment that supplies heat producing units for circulation to a desired location or use.

Packaged Terminal Air Conditioner (PTAC). A self contained unit, operated in the direct expansion method, located in the space served.

Packing House Tile. Similar to quarry tile but usually of greater thickness.

Packing. Sealing device consisting of soft material or one or more mating soft elements; reshaped by manually adjustable compression to obtain or maintain a leak-proof seal.

Pad Eyes. Metal rings mounted vertically on a plate for tying small vessels.

Pad, Bearing. See Bearing Pad.

Pad, Equipment. See Equipment Pad.

Pad, Transformer. See Transformer Pad.

Padding, Carpet. See Carpet Padding.

Pagoda. A multi tiered tower of India and the Far East erected as a temple or memorial.

Paint Brush. A painter's implement composed of bristles set into a handle, used for applying paint and other coatings to a surface.

Paint Coating. Paint in position on a surface.

Paint Gauge. An instrument used to measure the thickness of paint coatings.

Paint Heater. Device for lowering viscosity of paint by heating.

Paint Preparation. The act or process of preparing a building or structure for painting, including filling, scraping, sanding, and mixing of paint.

Paint Project. Single paint job.

Paint Remover. A mixture of active solvents used to remove paint and varnish coatings.

Paint Shop. An area in a building being painted where the painter stores, mixes, cleans tools, and prepares painting materials and equipment.

Paint Spray Booth. A room or enclosed space that provides a dust-free environment and ventilation for the application of paints.

Paint System. The complete number and type of coats comprising a paint job, including surface preparation, pretreatments, dry film thickness, and manner of application.

Paint, Asphaltic. See Asphaltic Paint.

Paint, Door. See Door Paint.

Paint. A mixture of pigment, binder, and solvent as a suitable liquid to form a thin closely adherent coating when spread on a surface in a thin coat to decorate or adorn by applying lines and color.

Painter. A craft worker skilled in the application of paints, stains, and other decorative and protective treatments.

Painter's Colic. Acute abdominal pain and constipation suffered by painters, caused by long term exposure to paint products containing lead; also called Lead Colic.

Painting Preparation. The act or process preparing a building or structure for painting, including scraping, filling, sanding, and mixing of paint.

Painting. The work done by painters.

Palladian. In the neo-classical style of Andrea Palladio.

Pallet. A portable platform for storing, moving, and handling goods and materials as in a warehouse.

Pan, Metal Stair. See Metal Stair Pan.

Pan. Metal or plastic prefabricated form unit used in construction of concrete floor joist systems.

Pane. A panel of glass.

Panel Circuit Directory. A systematic listing of all of the circuits in an electric panel.

Panel Cladding. Metal sheathing panels used to provide durability, weathering and corrosion, or impact resistance.

Panel Edge Clip. H-shaped clip used to prevent differential deflection of plywood roof or floor sheathing.

Panel Hoist. A device to raise sheetrock into place while it is being nailed or screwed.

Panel Rated Siding. Panel siding made of APA proprietary siding product.

Panel Siding. Large sheets of plywood or hardboard which may serve as both sheathing and siding.

Panel, Acoustical. See Acoustical Panel.

Panel, Hardboard Siding. See Hardboard Panel Siding.

Panel, Limestone. See Limestone Panel.

Panel, Sandwich. See Sandwich Panel.

Panel. 1. A distinct, usually rectangular, section of the surface of a door, cabinet, wall, ceiling or roof. 2. A board containing instruments, controls, dials, and switches. 3. A group of people, usually three, forming an arbitration tribunal. 4. A large, thin board or sheet of lumber, plywood or other material. 5. A thin board with all its edges inserted in a groove of a surrounding frame of thick material. 6. A section of floor, wall, ceiling or roof, usually prefabricated and of large size, handled as a single unit in the operations of assembly and erection. 7. Form section consisting of sheathing and stiffeners that can be erected and stripped as a unit. 8. A recessed section on the broad face of a brick; see Frog.

Panelboard. A single panelboard or group of panel units designed for assembly in the form of a single panel including buses and with or without switches or automatic overcurrent protective devices, or both, for the control of light, heat, or power in a cabinet or enclosure placed in or against a wall or partition and accessible only from the front; also called a Switchboard.

Paneling. 1. Panels joined in a continuous surface. 2. Decorative wood panels.

Panelized Roof. A modular roof framing system so dimensioned that the sheathing and rafters can be prefabricated as a unit for efficient installation on the purlins or beams.

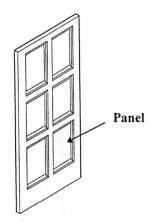

Panel

Panic Device. A mechanical device that opens a door automatically if pressure is exerted against the device from the interior of the building; a horizontal bar mounted across the full width of a door or a sort of large push-plate, acting as a latching system and operated by inside pressure on the bar or plate; also called Panic Hardware.

Panic Hardware. See Panic Device.

Pantry. A room for storing food and for serving to the table.

Paper and Wire. Asphalt impregnated paper and wire mesh, or metal lath, that are used as a backing for the installation of tile.

Paper Backed Insulation. Insulation that comes with paper facing on one side which serves as a vapor retarder.

Paper Mounted Mosaics. Ceramic mosaic tiles mounted on paper applied to face of tile; available in sheets approximately 12 by 24 inches.

Paper Rollers. Curlings of paper torn from the surface of the gypsum board; usually occurs as a result of prolonged exposure to high humidity and sliding one board across the surface of another.

Paper, Building. See Building Paper.

Paper, Curing. See Curing Paper.

Paper-Base Laminate. See Laminate, Paper-Base.

Paperboard. The distinction between paper and paperboard is not sharp, but broadly speaking, the thicker (over 0.012 inch), heavier, and more rigid grades of paper are called paperboard.

Papreg. Any of various paper products made by impregnating sheets of specially manufactured high-strength paper with synthetic resin and laminating the sheets to form a dense, moisture-resistant product.

Para Red. Pigment which is coal tar product; brilliant, opaque, non-fading, but has tendency to bleed.

Parabola. An open plane curve formed by the intersection of a cone with a plane parallel to its side.

Parabolic Troffer. A channel-like enclosure for light sources with a reflector shaped to control the light in a narrow beam.

Paraffin Oil. Light gravity mineral oil used as a lubricant in wood finishing.

Parallel Application. The long dimension of the gypsum board applied in the same direction as the framing members; also called Vertical Application or With Framing.

Parallel Bar. A pair of bars on a support adjustable in height and spacing that are parallel to each other and are used for gymnastic exercises.

Parallel Circuit. Arrangement of electrical devices in which the current divides and travels through two or more paths and then returns through a common path.

Parallel Rule. A device mounted on a drawing board to enable a drafter to produce parallel horizontal lines, and with triangles, to produce parallel vertical or sloping lines.

Parallel. Lines side by side, equidistant, and never meeting.

Parapet Wall. That portion of any exterior wall, party wall, or fire wall which extends above the roof line.

Parapet. The region of an exterior wall that projects above the level of the roof.

Parcel. A tract or plot of land under one legal description.

Parge. 1. To apply ornamental plaster to. 2. To apply waterproofing plaster to.

Parget. Parge.

Pargeting. Parging.

Parging. 1. Ornamental plaster. 2. Application of cement plaster as a part of a waterproofing system, as on foundation walls and rough masonry.

Parking Barrier. A structure either temporary or permanent, that is placed to prevent the encroachment of vehicles.

Parking Stall Painting. The act or process of painting border lines for the parking of individual motor vehicles.

Parkinson's Law. In an organization, "work expands to fill the time available for completion," and "staff increases at a fixed rate regardless of the amount of work produced."

Parol Evidence Rule. The meaning of a contract cannot be changed by reference to statements made by the parties to the contract before the contract was formed.

Parquet Floor. A floor covering laid out in a geometric design composed of small pieces of wood.

Parquetry. Wood inlay work in which geometric designs are produced by use of various colored woods.

Parterre. An ornamental garden with paths between the beds.

Parti. The basic general solution or scheme of an architectural design.

Partial Joint Penetration. Joint penetration which is less than complete.

Partial Pressures. Condition where two or more gases occupy a space and each one creates part of the total pressure.

Particleboard Underlayment. Flat sheet material of particleboard placed over subflooring to provide a smooth and even surface to receive finish floor covering.

Particleboard. A building panel composed of small particles of wood and resins bonded together under pressure; lat sheet material producing a durable and dimensionally stable product which is often used in dry conditions in place of plywood.

Particle-Size Distribution. The gradation in size of aggregate particles used in concrete, usually expressed in terms of cumulative percentages smaller or larger than each of a series of sieve openings or percentages between certain ranges of sieve openings.

Parting Bead. A narrow vertical strip in a double-hung window frame separating the upper and lower sashes; also called Parting Stop.

Parting Compound. A substance applied to concrete formwork to prevent concrete from adhering.

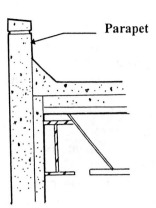

Parapet

Parting Stop. Parting Bead.

Partisan Arbitrator. An arbitrator appointed by, controlled by, or biased in favor of a party to a dispute; a Party-Appointed Arbitrator.

Partition Closure. Method of construction used to close openings formed between flutes of deck and abutting walls or partitions; resilient foam material is often used for this purpose.

Partition Door. Door located in Wire Mesh Partition.

Partition Fabric. The wire mesh fabric component of Wire Mesh Partition.

Partition Framing Member. Metal framework component of Wire Mesh Partition.

Partition System. An assembly of materials designed to perform a special function as a wall.

Partition, Clay Tile. See Clay Tile Partition.

Partition, Toilet. See Toilet Partition.

Partition, Wire Mesh. See Wire Mesh Partition.

Partition. A permanent interior wall which serves to divide a building into rooms.

Partner. One of the owners of a partnership.

Partnering. An educational procedure for obtaining the cooperation and understanding of all concerned in a construction contract by their participation at meetings or seminars attended by owner, architect, engineers, contractors, subcontractors, and suppliers to find out what is important to each participant.

Partnership. A business relationship between two or more persons, based upon a written, oral, or implied agreement, to combine their resources and skill in a joint enterprise and to share profits and losses jointly.

Parts Per Million (PPM). 1. Unit of concentration, as in solutions. 2. In swimming pools, the parts of a chemical or mineral per million parts of water by weight.

Party Wall. 1. A special purpose wall system used to divide compartments for different occupancies; may have requirements for fire and sound. 2. Partitions of brick or stone walls between buildings on two adjoining properties in which each of the respective owners of the adjacent buildings share the rights and benefits of the common wall. 3. A common wall between two tenancies; also called Demising Wall.

Party-Appointed Arbitrator. An arbitrator appointed by one of the parties; see Partisan Arbitrator.

Pascal's Law. A law of hydrostatics that states that in a perfect fluid the pressure exerted on it anywhere is transmitted undiminished in all directions.

Pass. 1. In spray painting, motion of the spray gun in one direction only. 2. A single longitudinal progression of a welding operation along a joint or weld deposit; the result of a pass is a weld bead. 3. A working trip or passage of an excavating or grading machine.

Passage Door. A door other than an exit door through which persons may traverse.

Passenger Elevator. An elevator used for people.

Passivation. Act of making inert or unreactive.

Passive Pressure. The horizontal resistance of the soil to forces against the soil.

Passive System. A solar heating or cooling system that uses no outside mechanical power to move the collected solar heat.

Pass-Through. An opening between two rooms or in an exterior wall through which things may be passed, as in serving food or returning soiled dishes.

Paste Wood Filler. A compound supplied in the form of a stiff paste for filling the open grain of hardwoods, such as oak, walnut, and mahogany.

Pastel. A light and subdued shade of a color.

Pastiche. An architectural design that imitates previous designs or styles; a stylistic imitation; a hodgepodge; see Eclectic.

Patch, Concrete. Material for the repair of small to medium holes or cracks in concrete surfaces.

Patch. A piece of material used to mend or cover a hole or as reinforcement.

Pate Dure. Hard paste, a French term designating ceramics fired at relatively high temperatures.

Pate Tendre. Soft Paste; a French term designating ceramics fired at relatively low temperatures.

Patent Ambiguity. An expression in a document that is obviously susceptible to more than one interpretation.

Patent Defect. A defect or shortcoming in a structure that is apparent to reasonable inspection.

Patent-Back Carpet. Carpet so constructed that the fabric can be cut in any direction, without raveling of edges; the edges are joined by tape and adhesives instead of being sewed.

Path of Travel. A passage that may consist of walks and sidewalks, curb ramps and pedestrian ramps, lobbies and corridors, elevators, other improved areas, or a necessary combination thereof, that provides free and unobstructed access to and egress from a particular area or location for pedestrians and/or wheelchair users.

Patina. A green film formed on copper or bronze over time.

Patio Block. Lightweight concrete paving slabs installed in lightly used foot traffic areas.

Patio. 1. An enclosed unroofed courtyard. 2. A paved area adjacent to a home, used for recreation or outdoor dining.

Pattern Length. In spray painting, length of spray pattern.

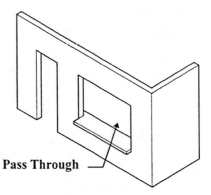

Pass Through

Pattern Width. In spray painting, width of spray at vertical center.

Pattern. A form, model, or design from which copies can be made.

Paumelle. A style of hinge embodying a single pin of the pivot type, generally of a smooth, streamlined design.

Pavement Cutting. The process of scoring or cutting through pavement surfaces with a power saw with a specific blade for that purpose.

Pavement Marking. The act or process of applying painted lines or necessary instructional signage on pavement surfaces for pedestrians or vehicle drivers.

Pavement Structure. A pavement structure with all its courses of asphalt-aggregate mixtures, or a combination of asphalt courses and untreated aggregate courses placed above the subgrade or improved subgrade.

Pavement. 1. A paved surface. 2. The material used to pave a surface such as a street or parking area.

Paver Brick. Brick units that are used in foot traffic areas; usually four inches wide, eight inches long, and 1-5/8 to 2-1/4 inches thick.

Paver, Granite. See Granite Paver.

Paver, Marble. See Marble Paver.

Paver, Masonry. See Masonry Paver.

Paver, Stone. See Stone Paver.

Paver. 1. One who paves. 2. Any durable stone, brick, or tile unit suitable for construction of a pedestrian or vehicular surface. 3. A half-thickness brick used as finish flooring. 4. Unglazed porcelain or natural clay tile formed by the dust-pressed method and similar to ceramic mosaics in composition, but thicker. 5. Glazed paver tile.

Paving Finishes. The finish coats of concrete, asphalt or coated macadam on streets, sidewalks, and parking areas.

Paving Unit. A precast masonry unit, usually 2-1/4 inches thick used for stepping stones; patios, veneering, and paving; also called a Cap Block.

Paving, Concrete. See Concrete Paving.

Paving. The surface of an outdoor area; pavement.

Payback Period. 1. The period required to recover the investment in an asset. 2. A method utilized by management to evaluate the profitability of alternative investment proposals.

Payment Bond. A guarantee by a surety that those persons who supply work and materials to a construction project will be paid for the work and materials

Payment Schedule. A provision in a construction contract that specifies the times for, and amounts of, payments for construction services.

Payment. Satisfaction, or partial satisfaction, of a debt.

Payroll. The record of wages, salaries, and fringe benefits paid by an employer to its employees.

PC. 1. Portland cement. 2. Prime coat. 3. Pull chain.

PCA. Portland Cement Association.

PCB. Polychlorinated Biphenyl.

PCC. Portland Cement Concrete.

PCF. Pounds per Cubic Foot.

PCI. 1. Pounds per Cubic Inch. 2. Prestressed Concrete Institute.

PDI. Plumbing and Drainage Institute.

PE. Professional Engineer.

Pea Gravel. That portion of concrete aggregate passing the 3/8 sieve and retained on a No. 4 sieve.

Peak. The top point of a roof, truss, spire, finial or any other similar structure.

Pearl Lacquer. Lacquer into which has been suspended guanine crystals, multi-faceted crystals found in skin attached to scales of sardine herring that thrive in cold water.

Peat Moss. Sphagnum.

Peat. See Humus.

Peavey. A lumber worker's tool for handling logs, consisting of a stout wood handle with a steel hooked lever at one end.

Pecan Veneer. A thin layer of hardwood glued over a core of sturdier less valuable solid wood, or plywood, used in flooring and furniture.

Peck. Pockets or areas of disintegrated wood caused by advanced stages of localized decay in the living tree; usually associated with cypress and incense-ceda; there is no further development of peck once the lumber is seasoned.

Peculiar Risk Doctrine. The doctrine that an owner, by employing an independent contractor, cannot escape liability to persons who may be injured during construction operations on the owner's property, since the construction operations involve a special risk of harm.

Pecuniary. Consisting of or measured in money.

Pedestal Floor. A flooring system which has short piers or legs used as a base and the flooring laid over those piers to provide a floor system; special flooring designed to prevent electrostatic buildup and sparking in a computer room; usually elevated over the existing floor to facilitate the installation of wires between the components in the room.

Pedestal. 1. An upright compression member whose height does not exceed three times its average least lateral dimension. 2. A short compression member of reinforced concrete that is placed between a column and the footing to distribute the load to the footing. 3. A support for a column or statue.

Pedestrian Barricade. An obstruction, obstacle, or barrier set up to check or control crowds of people and foot traffic.

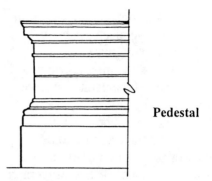

Pedestal

Pedestrian Grade Separation. A structure erected over or under an obstacle such as a freeway, roadway, street, railroad, or stream, and intended primarily for pedestrian use.

Pedestrian Ramp. A sloping path of travel intended for pedestrian traffic and as differentiated from a curb ramp.

Pedestrian Spatial Requirements. Normally 13 square feet are required per person for comfortable motion while walking.

Pedestrian Way. A route by which a pedestrian may pass.

Pedestrian. A person on foot.

Pediment. The triangular front part of a building, surmounting the columns, of a classical Grecian style.

Peel. To convert a wood log into veneer by rotary cutting.

Peeler. In gypsumboard, the clean separation of surfacing papers from the core; may be due to a variety of causes, most frequently, however, from surface calcination during the drying process or insufficient binder.

Peeling. 1. Detachment of a paint film in relatively large pieces; paint applied to a damp or greasy, surface usually peels; sometimes it is due to moisture back of the painted surface. 2. A process in which thin flakes of matrix or mortar are broken away from concrete surface, caused by adherence of surface mortar to forms as forms are removed. 3. See Shivering.

Peen. The working face of a hammer opposite the flat face, usually shaped for bending, shaping, or cutting the material being struck with the peen; see Cross Peen, Ball Peen, or Straight Peen Hammers.

Peer Review. A quality improvement system consisting of having an architect's or engineer's professional work and office procedures reviewed by a committee of peers.

Pegboard. A board with holes in which pegs or fittings may be inserted to hang tools or objects.

Pegged Flooring. Hardwood flooring with hardwood plugs set to hide recessed screws or to simulate screw covers.

PEI. Porcelain Enamel Institute.

Pein. Variant of peen.

PEL. Permissible Exposure Limits; standards set by OSHA.

Peltier Effect. When direct current is passed through two adjacent metals one junction will become cooler and the other will become warmer; this principle is the basis of thermoelectric refrigeration.

Penal Sum. The face amount of a surety bond; this is the maximum amount of the surety's liability.

Penalty and Bonus Clause. A contract provision that the contractor will pay a penalty for late completion and receive a bonus for timely or early completion.

Penalty Clause. A clause in a construction contract by which a contractor is assessed with a monetary penalty, usually on a daily basis, for delay in the completion of a project.

Penalty. A sum to be forfeited if a condition is not met.

Pencil Rods. Smooth mild steel reinforcing rods of 3/16, 1/4, or 3/8 inch diameter.

Pendentive. One of the concave triangular segments that form the transition from the supporting columns to a dome above.

Penetrating Stain Wax. Wood finish which produces color of penetrating stain with luster of wax.

Penetrating Stain. Stain made by dissolving oil-soluble dyes in oil or alcohol.

Penetration. The consistency of a bituminous material expressed as the distance in tenths of a millimeter (0.1mm) that a standard needle penetrates vertically a sample of the material under specified conditions of loading, time, and temperature.

Penetration Grading. Of asphalt cements, a classification system based on penetration in 0.1mm at 25° C (77° F); in this system, there are five standard paving grades, 40-50, 60-70, 85-100, 120-150 and 200-300.

Penny. As applied to nails, it originally indicated the price per hundred; the term now serves as a measure of nail length and is abbreviated by the letter d.

Pentaerythritol Resins. Resin made by reacting pentaerythritol, a high alcohol, with rosin.

Pentagon. A five sided plane figure.

Penthouse Louver. A louvered wall around the mechanical penthouse area of a structure, with fixed or movable flaps; this protects it from the elements and provides a visual screen around equipment.

Penthouse. A top story of a building that is smaller than the story below it.

People Competence. The aptitude for and fulfillment of necessary management of personnel to complete projects in a professional and timely manner.

Per Diem. An allowance or payment for each day.

Percentage Humidity. The degree of possible saturation of air with water vapor multiplied by 100.

Percentage Lease. A lease in which the rental consideration is an agreed percentage of sales, usually with an expressed minimum rent.

Percentage. A part of a whole expressed in hundredths.

Perched Water Table. Underground water lying over dry soil and sealed from it by an impervious layer.

Percipient Witness. A person called to testify on account of personally observing an event or being personally involved.

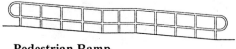

Pedestrian Ramp

Percolate. 1. Seep or ooze. 2. To cause a liquid to pass through a permeable substance.

Perforated PVC Pipe. Plastic pipe 4 inches in diameter with one or more rows of uniform holes along the length. Buried in the ground alongside building foundations or structures, to aid in drainage of groundwater and moisture.

Perforated Strap. Thin metal strips in rolls with punched holes used to hang or keep plumbing pipes in place.

Perforated Wall. Wall with relatively small openings as for ornamentation.

Performance Bond. A bond, secured by the general contractor, which guarantees that the contract will be performed; an undertaking by a surety that a contractor will perform a contract.

Performance Code. A building code that prescribes the objectives sought rather than the specific methods and materials that must be used; compare with Specification Code.

Performance Indicator. The bottom line showing profit or loss on an income and expense statement is the important indicator of the performance of a business.

P

Performance Specification. A building specification that prescribes the objectives sought rather than the specific methods and materials that must be used.

Pergola. An arbor or covered walk formed of growing plants trained over trellis work supported on parallel rows of columns.

Perilla Oil. Drying oil obtained from seeds of brush called Perilla Ocymoide, grown largely in China and Japan.

Perimeter Heating. A method of installing central heating systems so that the registers are placed on the outside walls under windows.

Perimeter Overflow System. A continuous channel formed into the sidewall entirely around the perimeter of the pool, unless interrupted by steps, into which surface pool water is continuously drawn during normal operation to provide a skimming action.

Perimeter Relief. 1. Construction detail which allows for building movement. 2. Gasketing materials which relieve stresses at the intersections of wall and ceiling surfaces.

Perimeter. The length of the circumference or outline of a figure.

Perineal Bath. A small shallow bathtub in which one bathes in a sitting position; a Sitz bath.

Period. The time for a wave crest to traverse a distance equal to one wave length or the time for two successive wave crests to pass a fixed point.

Periodic. Periodic duty is a type of intermittent duty in which the electrical load conditions are regularly recurrent.

Perjury. Wilfully lying while under oath.

Perlite Institute (PI). 600 South Federal Street, Chicago, Illinois 60605, (718) 351-5723.

Perlite Insulation. An insulation made from volcanic glass, expanded by heat.

Perlite Roof Insulation. Insulation for a roof system made from volcanic glass, expanded by heat.

Perlite. Expanded siliceous volcanic rock, expanded by heat, used as a lightweight aggregate in concrete and plaster, and as an insulating fill.

Perm. A measure of water vapor movement through a material, such as a vapor barrier; one perm equals one grain of vapor transmission per square foot, per hour, for each inch of mercury difference in vapor pressure.

Permafrost. Subsoil which remains frozen throughout the year, as in the polar regions.

Permanence. The property of a plastic which describes its resistance to appreciable changes in characteristics with time and environment.

Permanent Loan. A long term real estate loan that replaces the construction loan (Interim Loan) upon completion of the construction; also called Takeout Loan.

Permanent Magnet. 1. Material which has its molecules aligned and has its own magnetic field. 2. A bar of metal which has been permanently magnetized.

Permeability. The property of a material to permit a fluid (or gas) to pass through it; in construction, commonly refers to water vapor permeability of a sheet material or assembly and is defined as Water Vapor Permeance per unit thickness. Metric unit of measurement, metric perms per centimeter of thickness. See Water Vapor Transmission, Perm, Permeance (ASTM E 96).

Permeable. Capable of being penetrated; having pores or openings that permit liquids or gases to pass through.

Permeance. A material's resistance to water vapor transmission; the ratio of the rate of water vapor transmission through a material or assembly between its two parallel surfaces to the vapor pressure differential between the surfaces; see Water Vapor Transmission (WVT).

Permutation. An ordered arrangement or grouping of a set of numbers or things; any one of the range of possible groupings.

Perpend Bond. In masonry, a header brick or large stone extending through a wall so that one end appears on each side of the wall and acts as a binder.

Perpendicular. Standing at right angles to the plane of the horizon; vertical.

Perpends. The vertical joints in the face of a wall with all joints directly over one another.

Perpetual. Lasting forever.

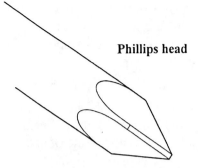
Phillips head

Personal Air Samples. An air sample taken with a sampling pump directly attached to the worker with the collecting filter and cassette placed in the worker's breathing zone; these samples are required by the OSHA asbestos standards and the EPA Worker Protection Rule.

Personal Ownership. The ownership of the business by one person; sole ownership or sole proprietor.

Personal Property. Any property that is not real property; also called Chattel.

Personnel Lift. An elevator for use by persons at a job site, a building, or structure.

Personnel. The body of people employed in a business or on a project.

Perspective. A drawing on a plane surface that represents an object as it appears to the eye; it cannot be scaled as the lines are foreshortened.

PERT. Program Evaluation Review Technique.

Pest Control. The act or process of the placement of devices or spraying of chemicals or powders to control the spread of insects and pests.

Petcock. A small valve for draining or letting out air.

Petty Cash Fund. A fund of cash that is established to make minor disbursements and avoid writing checks for small amounts.

Pew. Bench-like seating in a church.

pH Value. Measure of acidity or alkalinity; pH 7 is neutral; the pH values of acids are less than 7 and of alkalis (bases) greater than 7.

PH. Measurement of the free hydrogen ion concentration in an aqueous solution.

Pharmacy. A room or building devoted to the sale and dispensing of drugs, medical supplies, and related goods.

Phase. Distinct functional operation during a cycle.

Phased Application. The system of applying the felt plies of a built-up roofing membrane in two or more steps, separated by a delay normally of at least one day.

Phenol. A caustic poisonous derivative of benzene present in coal tar and wood tar used in dilute form as an antiseptic and disinfectant.

Phenol-Adelhyde Resins. Resins produced from phenols and formaldehyde.

Phenolic Resins. Resin based essentially on reaction between phenol and formaldehyde.

Phenolic-Resin Primer Sealer. Finish well suited for fir and other softwoods, which penetrates into pores of wood, dries and equalizes density of hard and soft grains.

Phenol-Red. A dye which is yellow at a pH of 6.8 and turns a progressively deeper red color as the pH increases to 8.4. This is the most commonly used test reagent for pH in swimming pools.

Phial. The sensing element on a thermostatic expansion valve.

Phillips Screwdriver. A cross-tipped screwdriver for installing screws.

Phloem. In a tree, the tissues of the inner bark, characterized by the presence of sieve tubes and serving for the transport of elaborate foodstuffs.

Phon. A unit of loudness for an average listener of a sound; equal to the sound level being zero at faintest, and 1000 at loudest.

Phosphatize. Form a thin inert phosphate coating on surface usually by treatment with H_3PO_4, phosphoric acid.

Phosphorescent Paint. Luminous paint which emits light after the white light has been turned off; no phosphorus is used.

Photo Equipment. Materials and devices used for photography and the processing of photographs.

Photo Processor. A piece of equipment used to develop photographs from negatives.

Photoelectric Sensor. A device that responds to light and transmits a resulting impulse.

Photoelectricity. Physical action wherein an electrical flow is generated by light waves.

P

Photogrammetry. The science of reliable measuring of aerial photographs to produce topographic maps.

Photographs. Pictures taken before a job commences to provide an accurate representation of what the site was like before construction.

Photometrics. The science of measuring the intensity of light.

Photo-Oxidation. Oxidation caused by solar rays.

Photovoltaic Cells. Semiconductor devices that convert solar energy into electrical power.

Photovoltaic Roof Tiles. Roof tiles composed of solar cells.

Photovoltaic. A process of conversion of sunlight into electricity.

Phthalic Anhydride. A white crystalline material used in making synthetic resins.

Phthalic Resins. A particular group of film formers; alkyd resins.

Phthalocyanine Blue. Organic blue pigment developed synthetically; outstanding in fade resistance.

Phthalocyanine Green. Complex copper compound pigment with bluish-green cast.

Physical Inventory. A listing of merchandise on hand, determined by actual count, weight, or measurement and the pricing or value thereof.

Physical Resources. The building activity, management and labor personnel and expertise in the operating area of the business as well as scrap materials and equipment held for resale and other assets not listed on a balance sheet.

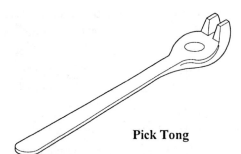

Pick Tong

Physical Therapy. The treatment of an injury or illness by physical and mechanical means, such as massage or heat.

Pi. Designated by the Greek letter P; the symbol of the ratio of the circumference of a circle to its diameter; the value of P is approximately 3.14159265; P = C/D.

PI. Perlite Institute.

Piano Hinge. A Continuous Hinge.

Pick and Dip. A method of laying brick whereby the bricklayer simultaneously picks up a brick with one hand and, with the other, enough mortar on a trowel to lay the brick; also called the Eastern or New England method.

Pick Tong. A blacksmith's tool used to handle hot metal.

Pick Up Sags. When a too-heavy coating of paint has been applied and starts to sag or run down the surface, the painter brushes up through the sagging paint to level it off.

Pick. A heavy hand implement with two pointed ends used in digging and loosening earth.

Pickling. A dipping process for cleaning steel and other metals; the pickling agent is usually an acid.

Pickup Truck. A light truck having an open body with low sides and tailboard.

Picture Framing. A rectangular pattern of ridges in a membrane over insulation or deck joints.

Picture Mold. See Picture Mold.

Picture Molding. A molding shaped to form a support for picture hooks, often placed at some distance from the ceiling upon the wall to form the lower edge of the frieze.

Picture Window. A large, often fixed, window, usually of plate or insulating glass, designed to frame an exterior view.

Piece Dyeing. Immersing an entire carpet in a dye bath to produce single- or multi-color pattern effects; see also Resist Printing.

Pier. 1. Timber, concrete, or masonry supports for girders, posts, or arches. 2. An isolated column or mass of masonry units 3. Intermediate supports for a bridge span. 4. Structure extending outward from shore into water used as a dock for ships.

Piezometer. An instrument for measuring pressure or compressibility.

Pigment Grind. Dispersion of pigment in a liquid vehicle.

Pigment Oil Stain. Consists of finely ground insoluble color pigments such as used in paints, in solution with linseed oil, varnish, mineral spirits, etc. according to formula being used; also called Wiping Stain.

Pigment Volume Concentration (PVC). Percent by volume occupied by pigment in dried film.

Pigment. 1. The coloring matter in paint or other materials. 2. Material in the form of fine powders insoluble in oils, varnishes, lacquers, thinners and the like; used to impart color, opacity, certain consistency characteristics, and other effects.

Pilaster Block. A concrete masonry unit that allows construction of a pilaster in a concrete block wall.

Pilaster. A projecting square column forming part of a wall.

Pile Butt. The large end of a pile; the small end is called the Tip.

Pile Cap Formwork. Formwork for a concrete pile cap.

Pile Cap Reinforcing. Steel reinforcing bars in a concrete pile cap.

Pile Cap, Concrete. A concrete footing resting on a group of piles.

Pile Cap. A thick slab of reinforced concrete poured across the top of a pile group to cause the group to act as a unit in supporting a column.

Pile Crushing. Of carpet, the bending of pile due to foot traffic or the pressure of furniture.

Pile Group. Several driven or poured piles in a cluster and tied together at their tops by a single poured concrete pile cap.

Pile Height. Of carpet, the height of pile measured from the top surface of the backing to the top surface of the pile; also referred to as Pile Wire Height.

Pile Setting. Of carpet, brushing after shampooing to restore the damp pile to its original height.

P

Pile Spall. A chip or piece broken from a pile by a blow from the driving hammer or by action of the elements.

Pile Testing. The act or process of measuring the resistance of a driven test pile to a pre-determined design load.

Pile Tip. The small end of a pile; the large end is called the Butt.

Pile Warp. Lengthwise pile yarns in Wilton carpets which form part of the backing.

Pile Wire Height. See Pile Height.

Pile Yarn Density. Of carpet, the weight of pile yarn per unit of volume in carpet, usually stated in ounces. per cubic. yard.

Pile Yarn. Of carpet, the yarn used to form the loops or tufts of a pile fabric.

Pile. 1. Long steel, wood, or concrete member penetrating deep into the soil to support grade-beam foundation walls or columns. 2. The raised yarn tufts of woven, tufted and knitted carpets which provide the wearing surface and desired color, design or texture; in flocked carpets, the upstanding, non-woven fibers.

Pillar. A column or post supporting a roof or for ornamentation.

Pilling. Appearance defect associated with some staple fibers where balls of tangled fibers are formed on the carpet surface which are not removed readily by vacuuming or foot traffic; pills can be removed by periodic clipping.

Pilot Circuit. Secondary circuit used to control a main circuit or a device in the main circuit.

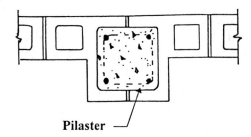

Pilaster

Pilot Hole. A small hole drilled into a piece of wood to prevent it from splitting when it receives a screw or nail.

Pilot Light. 1. A small light to indicate when a switch or other electrical device is on or in the on position. 2. A relatively small flame which may be automatic or may burn continuously; its purpose is to ignite the main supply of gas when a gas-fired heating or cooking unit is turned on.

Pin Connection. A structural connection that allows the connected members to rotate.

Pin Joint. A hinge in a structure; a structural joint that will not transmit moment.

Pin Knot. A knot that is not more than 1/2-inch in diameter.

Pin Tumblers. Small sliding pins in a lock cylinder working against coil springs which prevent the cylinder plug from rotating until the pins are raised to alignment by bitting of the proper key.

Pin, Clevis. See Clevis Pin.

Pinhole. 1. Any small hole. 2. A small perforation in the gypsum board paper or paper joint tape. 3. A small hole appearing in a cast when the water-stucco ratio has not been accurately measured; excess water causes pinholes.

P

Pinion Gear. 1. A small gear in a set of gear wheels. 2. The smaller gear of two or the smallest of three or more.

Pinnacle. A small ornamental turret usually ending in a pyramid or cone, crowning a buttress or roof.

Pinstripes. Fine stripes.

Pipe Bollard. Short pipe length, placed vertically in the ground and filled with concrete to prevent vehicular access or to protect property from damage by vehicular encroachment.

Pipe Cleanout. See Cleanout Plug.

Pipe Flange. Projecting ring, ridge or collar placed on pipe to strengthen, prevent sliding, or to accommodate attachments.

Pipe Handrail. A metal pipe used as a handrail.

Pipe Insulation. Insulation that covers pipes to help in the reduction of heat loss or gain.

Pipe Jacking. Forcing pipe through the ground in a tunnel created by the pipe itself; the pipe is generally jacked horizontally in short lengths.

Pipe Joint Compound. Putty-like material used to seal threaded pipe joints.

Pipe Painting. The act or process of painting piping to prevent rust and corrosion and also provide an identifying system in buildings and process piping plants.

Pipe Railing. A metal railing made of pipe.

Pipe Sleeve. Cylindrical insert cast into concrete wall or floor to provide for later passage or anchorage of pipe.

Pipe Wrench. A wrench with serrated jaws to grip pipe and turn it in one direction only.

Pipe, ABS. A plastic pipe made in various diameters; used for stacks and main drains in plumbing systems.

Pipe, Aluminum. A pipe for liquid or gas.

Pipe, Clay. Pipe used for drainage systems and sanitary sewers made of earthenware and glazed to eliminate porosity.

Pipe, Copper. Rigid pipe used for exterior and interior water systems, it is light and durable, resists moisture, and can be used with either mechanical or soldered connections.

Pipe, CPVC. Rigid plastic pipe used for hot and cold water supply lines.

Pipe, Fiberglass. A pipe for liquid or gas, fabricated from layers of glass fibers and resins.

Pipe, Galvanized. Zinc coated steel pipe.

Pipe, Glass. Glass and glass-lined pipe used in process piping.

Pipe, No-Hub. Pipe manufactured in cast iron, which is fabricated without hubs for coupling.

Pipe, Polyethylene. Pipe manufactured from a thermoplastic compound.

Pipe, Polypropylene. A tough plastic pipe with resistance to chemicals and heat.

Pipe, PVC. Polyvinyl chloride pipe used mainly for drain lines, particularly resistant to chemicals.

Pipe, Stainless. Pipe or tubing constructed of stainless steel which has a high resistance to corrosion.

P

Pipe, Structural. Pipe used in a structure to transfer imposed loads to the ground.

Pipe. 1. A long tube or hollow body for conducting a liquid, gas, or finely divided solid. 2. A structural column or strut.

Piping. Any system of pipes in a building.

PIR Detector. Passive Infra-Red detector. A part of a burglar alarm system.

Piscina. A basin with a drain, usually to a soak pit, near the altar in a church for disposing of water from liturgical ablutions.

Piston Displacement. Volume displaced by piston as it travels the full length of its stroke; volume obtained by multiplying area of cylinder bore by length of piston stroke.

Piston. Close-fitting part or plug which moves up and down in a cylinder.

Pit. A hole, shaft, or cavity in the earth.

Pitch Board. See Bevel Board.

Pitch Pine. Any of the pines that yield pitch; the heartwood is brownish red and resinous; the sapwood is thick and light yellow; used for lumber, fuel, and pulpwood.

Pitch Pocket. 1. An opening between growth rings of a tree which usually contains resin, bark, or both. 2. In roof construction, a flanged metal container placed around a roof penetration at roof level to receive hot bitumen or caulking and provide a roof seal; commonly found at columns or plumbing stacks.

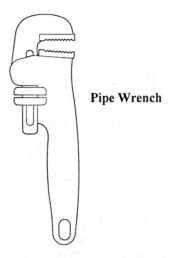

Pipe Wrench

Pitch Streak. A well-defined accumulation of pitch in a more or less regular streak in the wood of certain conifers.

Pitch. 1. Resin obtained from various conifers. 2. A black or dark viscous substance obtained as a residue in distilling tar or oil from bones; it also occurs in natural form as asphalt. 3. The slope of a stairway, in degrees. 4. The slope of a roof. 5. Degree of slope or grade given a horizontal run of pipe. 6. The spacing of rivets or bolts. 7. The frequency of sound vibrations. 8. In carpet, the number of pile ends per inch of width; actually, in practical floor covering specifications, it is taken as the number of pile ends per unit of standard 27-inch width; terms of pitch used commonly in the industry are 180, 189, 192, 216 and 256.

Pith. The small, soft core occurring near the center of a tree trunk, branch, twig, or log.

Pitot Tube. Tube used to measure air velocities.

Pits. See Pops.

P

Pitting. Formation of small, usually shallow depressions or cavities in the surface of a material.

Pivot. A shaft, axis, or pin on which something turns or oscillates.

Placement. Process of placing and consolidating concrete; also called Pour.

Plain Bar. Steel reinforcing bar without deformations.

Plain Concrete. Concrete that is either unreinforced or contains less reinforcement than the minimum amount specified in the code for reinforced concrete.

Plain Reinforcement. See Plain Bar.

Plain Sliced Veneer. Wood veneer that is sliced parallel to a line through the center of the log. Also called Flat Cut Veneer.

Plainsawed Lumber. 1. Flat-grained lumber. 2. Lumber sawed regardless of the grain, the log is simply squared and sawed to the desired thickness; also called Slash Sawed or Bastard Sawed.

Plan. A drawing or diagram made by projection on a horizontal plane; a floor plan is a horizontal section through a building.

Planar. Of or in a plane or level surface, two dimensional.

Plancher. See Plancier.

Plancier. 1. A soffit under a cornice or any projecting member; also called Plancher. 2. A plank floor or platform.

Plane of Weakness. The plane along which a structure under stress will tend to fracture; may exist because of the nature of the structure and its loading, by accident or by design.

Plane. 1. A level surface; a flat surface on which a straight line joining any two points on it would wholly lie. 2. A tool for smoothing or shaping a wood surface.

Planing Mill Products. Products worked to pattern, such as flooring, ceiling, and siding.

Plank Floor. See Random Width Flooring.

Plank. A wide piece of sawed timber, usually 1-1/2 to 4-1/2 inches thick and 6 inches or more wide.

Planned Unit Development. A cluster development that includes more land uses in the built up area (industrial, commercial) than the residential cluster system.

Planning. The process of carrying out plans for development of land or buildings.

Plant Bed Preparation. The mixing of earth with fertilizer and other soil enriching products, to create an area for the future introduction of vegetation.

Plant Screens. In a concrete batching plant, screens located between the dryer and hot bins which separate the heated aggregates into the proper hot bin sizes.

Plant. 1. To set or place plants or seeds in the ground. 2. The trees, shrubs, or seeds to be planted. 3. The land, buildings, and equipment used for processing or manufacturing.

Planter. A container in which plants are grown.

Planting. 1. A plantation. 2. A method of placing spools of different colors of surface yarn in frames back of Jacquard Wilton looms so that more colors will appear in the design than are supplied in the full solid colors used; these extra planted colors are usually arranged in groups of each shade, to give added interest to the pattern.

Plantroom. Operational pool equipment location.

Plaque. 1. A commemorative or identifying inscribed tablet. 2. A localized abnormal patch on a surface.

Plaster Accessory. Hardware or tool needed for plaster work.

Plaster Base. The lath or backing to which plaster is applied.

Plaster Bead. Built-in edging usually metal, to strengthen a plaster angle.

Plaster Bond. The state of adherence between plaster coats or between plaster and a plaster base, produced by adhesive or mechanical interlock of plaster with base or special supplementary materials.

Plaster Fascia. The exposed vertical face of a wall cornice in an interior location, finished in plaster.

Plaster of Paris. A fine white gypsum plaster used for casting and molding; pure calcined gypsum.

Plaster Patch. Plaster mixed or premixed to fill cracks and damaged areas.

Plaster Ring. A guide with a metal collar attached to a base to apply plaster to a certain thickness or provide a fastener for trim.

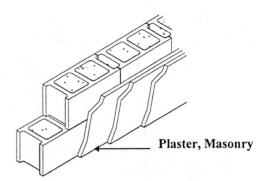

Plaster, Masonry

Plaster Screed. 1. A strip of wood or metal to regulate the thickness of plaster and furnish an edge trim. 2. A spot or strip of plaster to regulate plaster thickness.

Plaster Work. The finished product of plasterers.

Plaster, Masonry. See Masonry Plaster.

Plaster. A cementitious material, usually based on gypsum or portland cement, which is applied to lath or masonry in paste form, to harden into a finished surface; a mixture of lime, hair, and sand, or of lime, cement, and sand, used to cover exterior or interior wall surfaces; from Greek, emplastron, to daub on; Latin, emplastrum; French, platre; old English, plaister.

Plasterboard. A board used in large sheets as a backing or as a substitute for plaster in walls and consisting of fiberboard, paper, or felt, bonded to a hardened gypsum plaster core; see Gypsumboard.

Plastering Machine. A mechanical device by which plaster mortar is conveyed through a flexible hose to deposit the plaster in place; also known as a plaster pump or plastering gun; distinct from Gunite machines in which the plaster or concrete is conveyed, dry, through the flexible hose and hydrated at the nozzle.

Plastering. The work produced by plasterers.

Plastic Cement. 1. A plastic mixture of bitumen and asbestos reinforcing fibers with a solvent; see Flashing Cement. 2. Portland cement to which small amounts of plastizing agents, not more than 12% by volume, have been added at the mill.

Plastic Coated Conduit. A type of conduit for electrical wiring that is used around moist areas and highly corrosive fumes.

Plastic Consistency. Condition in which concrete, mortar, or cement paste will sustain deformation continuously in any direction without rupture.

Plastic Design. Method of structural analysis of continuous steel structures (beams and frames) based on calculating the loading which will cause collapse of the system.

Plastic Hinge. The point at which a structural member such as a beam or part of a frame is bent so that a section is stressed beyond the elastic range.

Plastic Laminate Backing Sheet. Sheet material placed on concealed side of the panel material similar to sheet material on exposed side of panel to balance the construction of the panel and give dimensional stability by minimizing moisture absorption.

Plastic Laminate Casework. Assembled plastic laminate cabinets.

Plastic Laminate Countertop. Countertop with plastic laminate finish covering substrate, usually plywood or particleboard.

Plastic Laminate. Sheet material manufactured of multiple layers of paper with top layer of plastic usually 1/16 inch (1.59 mm) thick with decorative finish; may be used in flat sheets or heat formed, bent, and adhered to single curved base material; commonly referred to by the brand name of Formica.

Plastic Pipe and Fitting Association (PPFA). 800 Roosevelt Road, Building C, #20, Glen Ellyn, Illinois 61037-5833, (708) 858-6540.

Plastic Pipe Institute (PPI). 1275 K Street, NW, #400, Washington, DC 20005, (202) 371-5200.

Plastic Pipe. Pipe manufactured from hard plastic to resist corrosion and rust.

Plastic Sheet. Plastic in which thickness is very small in relation to length or width.

Plastic Skylight. A transparent or translucent plastic molded unit that is set in a frame and mounted on a roof for use as a skylight.

Plastic Soil. A soil that can be rolled into 1/8" diameter strings without crumbling; a soft rubbery soil.

Plastic. 1. Capable of being molded; pliant; supple. 2. Any of a number of synthetic polymeric substances that can be given any required shape. 3. In stress analysis, refers to stress/strain behavior beyond the elastic range; plastic deformation usually implies some permanent shape change. 4. A condition of freshly mixed concrete, mortar, or cement paste indicating that it is workable and readily remoldable, is cohesive and has an ample content of fines and cement but is not over wet.

Plasticity. 1. Property of a material to deform under load and to retain the deformation after the load is removed. 2. A material's ability to be shaped and worked. 3. That property of plaster mortar that permits continuous and permanent deformation in any direction. 4. A plastic material is distinct from a fluid material in that it requires a measurable force (yield value) to start flow; the property exists in varying degrees in different materials and in plaster mortar is sometimes regarded as an index of working characteristics.

Plasticizer. 1. A substance added to plastics, resins, and rubbers to impart improved properties such as flexibility, workability, or stretchability. 2. A material that increases the workability or consistency of a concrete mixture, mortar or cement paste. 3. A product used to increase the flow and/or workability of plaster.

Plasticizing Agent. Plasticizer.

Plasticizing Agent. A product used to increase the flow and/or workability of plaster.

Plasticizing Wood. Softening wood by hot water, steam, or chemical treatment to make it easier to mold.

Plastisol. Film former containing resin and plasticizer with no solvents.

Plat. 1. A small piece or ground; a plot of land. 2. A drawing or map showing a plot of land.

Plate Cut. The cut at the bottom end of a rafter to allow it to fit upon the plate; the cut in a rafter which rests upon the plate; also called the Seat Cut.

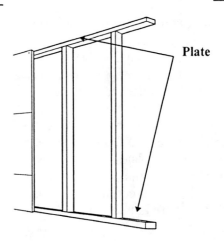

Plate

Plate Girder. A large beam made up of steel plates, sometimes in combination with steel angles, welded, bolted or riveted together.

Plate Glass. Glass of high optical quality produced by grinding and polishing both faces of a glass sheet.

Plate Mirror. Thick mirror glass manufactured to a high-quality standard.

Plate Tectonics. The theory and study of plate formation, movement, and interaction; the theory which explains seismicity in terms of plate movement.

Plate, Armor. See Armor Plate.

Plate, Duplex. See Duplex Plate.

Plate, Switch. See Switch Plate.

Plate, Ultrasonic Examined. Steel plate ultrasonically examined for structural defects.

Plate, Toe. See Toe Plate.

Plate. The horizontal framing member at the top or base of wood stud wall framing.

Platform Frame. A wooden building frame composed of closely spaced members nominally 2 inches in thickness, in which the wall members do not run past the floor framing members; typical wood stud wall framing in which the studs are one level in height and the floor framing above rests on the top plates of the wall below; the most common type of framing used in house construction; also called Western Frame; compare Balloon Frame.

Platform. Horizontal landing in stair either at the end of a flight or between flights, either at floor level or between floors.

Plating. A thin coating of metal deposited on a surface, usually by electrolysis.

Player Bench. A long seat, with or without seat back, on which athletes can sit when they are not playing.

Playground Equipment. Devices such as slides and swings, that children play on or upon in a park or playground.

Pleadings. Formal written documents, filed with a court, accusing a party of wrongdoing, or defending a party against such an accusation.

Plenum Chamber. Chamber or container for moving air or other gas under a slight positive pressure.

Plenum, Ceiling. See Ceiling Plenum.

Plenum. 1. An enclosed chamber such as the space between a suspended finished ceiling and the floor above. 2. Chamber attached directly to a furnace which receives heated air; from this largest chamber, ducts carry the air to the various registers.

Plexiglass. A plastic resilient material comparable to glass in use, usually manufactured in sheets.

Pliers. A hand tool used to grip objects.

Plinth. 1. The lower square slab at the base of a column. 2. The block of wood at the corner of a doorway where the baseboard and architrave meet.

Plot Plan. A drawing showing a plot of land with all its existing physical features and proposed improvements; a ground plan.

Plough. See Plow.

Plow. To cut a groove running in the same direction as the grain of the wood; also spelled Plough.

Plug Valve. See Core Cock.

Plug Weld. A circular weld made by either arc or gas welding through one member of a lap or tee joint joining that member to the other; the weld may or may not be made through a hole in the first member; if a hole is used, the walls may or may not be parallel and the hole may be partially or completely filled with weld metal.

Plug, Conduit. See Conduit Plug.

Plug, Grounding. See Grounding Plug.

Plug, Lock. See Lock Plug.

Plug. A pipe fitting with external threads and squared head that is used for closing the opening in another fitting.

Plumb Bob. The metal weight on the bottom of a plumb line.

Plumb Bond. In brick laying, an all stretcher bond with all joints directly over one another.

Plumb Cut. Any cut made in a vertical plane; the vertical cut at the top end of a rafter.

Plumb Rule. A narrow board having a plumb line and bob on one end, or more commonly having a bubble in a tube; it is used for establishing vertical lines and horizontal lines.

Plumb. Vertical, or perfectly straight up-and-down; at right angles to horizontal.

Plumber. A craft worker skilled in the installation, repair, and maintenance of water and waste systems in buildings.

Plumber's Friend. A tool consisting of a large rubber like cup and handle, it is used under water to force blockage through sewer lines.

Plumber's Snake. See Snake.

Plumber's Furnace. A heating source used to melt lead, heat soldering irons, or melt solder.

Plumber's Soil. A mixture of glue and lampblack used in lead work to prevent lead from sticking to selected metal parts of lead pipe and fittings.

Plumbing Access Door. A door in a floor, wall, or ceiling giving access to plumbing fittings, drains, or valves.

Plumbing and Drainage Institute (PDI). 1106 West 77th Street, South Drive, Indianapolis, Indiana 46260-3318, (317) 251-6970.

Plumbing Fixture. Plumbing equipment, usually installed last, such as sinks, water closets, bidets, and shower/bath units; devices which receive water and discharge it and/or water-borne waste into the DWV system.

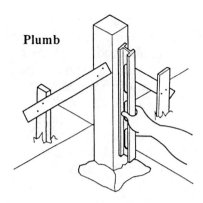

Plumb

Plumbing. 1. The act or process of installing in a building or structure the pipes, fixtures, or other apparatus for supplying potable water and removing liquid and water-borne wastes. 2. The installed fixtures and piping of a building or structure. 3. The gas piping system. 4. The storm sewer system. 5. Heating and air conditioning piping.

Plunge Router. A router in which the cutting bit enters the work from the surface of the board.

Plunge. A swimming pool.

Plunger. See Plumber's Friend.

Plus Pressure. See Positive Pressure, 2.

Plush. Of carpet, a smooth-face cut pile surface that does not show any yarn texture.

Ply Sheet. A glass fiber felt sheet coated on both sides with asphalt.

Ply. 1. A layer or thickness, as of building or roofing paper. 2. A layer of wood in plywood. 3. A layer of felt in a built-up roofing membrane; a four-ply membrane would have at least four plies of felt at any vertical cross section cut through the membrane; the exposure of any felt may be computed by dividing the felt width minus 2 inches by the number of plies, thus, the exposed surface of a 36 inch wide felt in a four-ply built-up roof membrane should be 8-1/2 inches. 4. A layer or thickness of yarns used in carpet; if the pile yarn is described as 4-ply, it means that each tuft is made of 4 yarns spun together.

Plywood Countertop. Plywood panel used as the substrate for the exposed finish surface of a countertop.

Plywood Diaphragm. Plywood sheathing on floors, roofs, or walls which provides shear strength to resist wind and earthquake loads.

Plywood Roof Sheathing. Plywood sheets secured to exterior side of roof rafters or trusses used to create rigidity in building superstructure and serve as base to receive roofing.

Plywood Shear Wall. Wall constructed of studs and plywood sheathing that in its own plane resists shear forces resulting from applied wind, earthquake or other transverse loads or provides frame stability.

Plywood Sheathing. Wall, roof, or floor sheathing of plywood.

Plywood Shelving. Horizontal mounted plywood surfaces upon which objects may be stored, supported, or displayed. OBV.

Plywood Siding. Plywood sheeting material forming the exterior surface of outside walls of frame buildings.

Plywood Soffit. Plywood finish material installed horizontally covering the underside of an assembly such as roof eaves.

Plywood Subfloor. Plywood sheets secured to the top side of floor joists.

Plywood Subfloor-Underlayment. Plywood sheets secured to top side of floor joists used to create rigidity in building superstructure and also to provide smooth and even surface to receive finish floor covering.

Plywood Underlayment. Flat sheet material of plywood placed over subflooring to provide smooth and even surface to receive finish floor covering.

Plywood Wall Sheathing. Plywood material secured to exterior side of exterior wall studs used to create rigidity in building superstructure and serve as base to receive siding.

Plywood, Finish. See Finish Plywood.

Plywood, Structural. A grade of plywood that has been tested and graded for its structural characteristics.

Plywood. A wood product made by bonding together layers of veneer or a combination of veneer layers and a lumber core; the layers are joined with an adhesive; adjoining plies are usually laid with grains at right angles to each other, and almost always an odd number of plies are used.

PM. Preventive Maintenance.

PMI. Private Mortgage Insurance.

Pneumatic Concrete. Concrete that is delivered by equipment powered by compressed air.

Pneumatic Hoist. A hoisting system powered by compressed air.

Pneumatic System. Mechanical devices powered by compressed air.

Pneumatic Tool. A tool powered by compressed air.

Pneumatic Tube System. An arrangement of tubes within a building or building complex for sending message capsules by compressed air.

Pneumatically Driven Fastener. Driven pin or threaded stud which is driven into material by use of compressed air.

Pneumatics. Study of compressible air and gases, their properties and reaction in containment.

POA Valve. See Pressure-Operated Altitude Valve.

Poché. A drafting technique in which parts of drawings are shaded in for increased readability, such as the walls on floor plans or shadows on elevations.

Pock Marks. 1. Pits; craters. 2. In gypsum board, surface imperfections appearing as a multitude of small depressions; often caused by dirt or small gravel which indented the paper surface.

Pocket, Pitch. See Pitch Pocket.

Point of Inflection. The point in a structural member where the bending moment changes from positive to negative.

Point Weight. See Felt Mill Ream.

Point. 1. A geometric element without dimension. 2. A sharpened tip as of a tool. 3. In carpet, one tuft of pile.

Pointing Trowel. A mason's tool used to point brick, stone, and block.

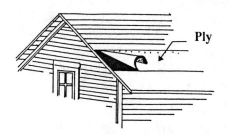

Ply

Pointing. 1. Troweling mortar into a joint after masonry units are laid. 2. The process of inserting mortar into the joints or brickwork or any other masonry units to fill open pockets or voids left when the work was originally done.

Points. A fee charged for making a loan, each point amounting to one percent of the loan amount.

Poise. A centimeter-gram-second unit of absolute viscosity, equal to the viscosity of a fluid in which a stress of one dyne per square centimeter is required to maintain a difference of velocity of one centimeter per second between two parallel planes in the fluid that lie in the direction of flow and are separated by a distance of one centimeter.

Polarity. 1. The condition of having two poles with contrary qualities. 2. The positive or negative electrical condition.

Polarized Plug. An electric plug with a configuration of prongs that prevent plugging in improperly.

Polarizing. This is the method of identifying electrical wires by colors. It is to make sure that hot wires will be connected only to hot wires and that neutral wires will run in continuous circuits without interruption back to the ground terminals.

Polder. A low lying area of land that has been reclaimed from the sea.

Pole, Closet. See Closet Pole.

Pole, Electric. See Electric Pole.

Pole, Utility. See Utility Pole.

Pole. 1. One of the two terminals of an electric cell, battery or dynamo. 2. Either extremity of the axis of a sphere. 3. The vertex of an angle coordinate. 4. A point of guidance. 5. A stake.

Pole-Gun. Spray gun equipped with an extension tube.

Police Equipment. Any of assorted devices necessary for law enforcement personnel to carry out their duties.

Police Power. The inherent power of the state to regulate its citizens and property in the interest of general security, morals, health, and welfare.

Police Station. A building housing the police and their activities in a locality.

Polish. 1. To make smooth and shiny by rubbing. 2. To make plaster finish coat smooth and glossy by troweling.

Polished Plate Glass. Glass 1/4-inch or thicker that is made by a process where both sides are ground and polished.

Polishing. Said of wall paints where shiny spots or surfaces have resulted from washing or wiping.

Pollution. Contamination of materials or the environment by harmful substances.

Polyamide. Product used in making dripless paint; see thixotropic paint.

Polybutylene Plastics. Plastics based on polymers made with butenas essentially the sole monomer.

Polychlorinated Biphenyl (PCB). Compound used in electrical transformers that is a poisonous environmental pollutant.

Polychromatic. Multi-colored.

Polychrome Finish. Finish obtained by blending together a number of colors.

Polyester Floor. Flooring material that has adhesive properties, high strength, and good chemical resistance, manufactured from a synthetic resin that polymerizes during curing.

Polyethylene Pipe. Pipe manufactured from a thermoplastic, high-molecular-weight, organic compound.

Polyethylene Vapor Barrier. A sheet form thermoplastic membrane, high molecular weight, organic compound, used as a protective cover to prevent the passage of air or moisture.

Polyethylene Wrap. A sheet form thermoplastic high-molecular-weight organic compound used to protect concrete during curing or as a temporary enclosure for construction operations.

Polyethylene. A polymer of ethylene, a thermoplastic used in packaging and insulation; plastic sheeting.

Polygon. A plane figure with 3 or more sides.

Polymer. A compound composed of one or more large molecules that are formed from repeated units of smaller molecules.

Polymeric. Composed of repeating chemical units; all plastics and polymers are polymeric.

Polymerization. A chemical reaction in which two or more small molecules combine to form larger molecules that contain repeating structural units of the original molecules; when two or more monomers are involved, the process is called copolymerization or heteropolymerization.

Polyolefin Plastics. Plastics based on polymers made with an olefin as essentially the sole monomer.

Polyphase Motor. Electrical motor designed to be used with a three or four-phase electrical circuit.

Polypropylene Pipe. A tough plastic pipe with resistance to chemicals and heat.

Polypropylene Plastics. Plastics based on polymers made with propylene as essentially the sole monomer.

Polypropylene. See Olefin.

Polystyrene. A plastic based on a resin made by polymerization of styrene as the sole monomer; polystyrene may contain minor proportions of lubricants, stabilizers, fillers, pigments and dyes; a plastic foam board used for insulation.

Polyurethane. 1. Any polymer containing urethane, used in plastics, adhesives, paints, foams, and rubbers. 2. Various polymers that are used in flexible and rigid foams, elastomers, and resins.

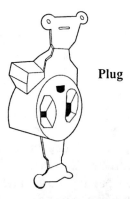

Plug

Polyvinyl Acetate (PVA). 1. A synthetic resin used extensively in emulsion (water) paints; produced by the polymerization of vinyl acetate. 2. White glue used in woodworking.

Polyvinyl Acetate Latex (PVA Latex). Latex composed primarily of vinyl acetate.

Polyvinyl Butyral. Plastic film used as an interlayer in laminated glass.

Polyvinyl Chloride Plastics (PVC). Plastics made by combining polyvinyl chloride with clorants, fillers, plasticizers, stabilizers, lubricants, other polymers, and other compounding ingredients; not all of these modifiers are used in pipe compounds.

Polyvinyl Chloride. 1. A tough transparent solid polymer, easily colored, used for pipes, fittings, flooring, and a wide variety of other plastic products. 2. A synthetic resin used in solvent type coatings and fluid bed coatings, produced by the polymerization of vinyl chloride; PVC is also used in emulsion (water) paints.

Polyvinyl Resin Emulsion Glue. White glue; wood adhesive intended for interiors; made from polyvinyl acetates which are thermoplastic and not suited for temperatures over 165° F.

Pond. 1. A small lake. 2. A roof surface which is incompletely drained.

Ponded Roof. Flat roof designed to hold a quantity of water which acts as a cooling device.

Ponding. 1. Water puddles standing on a roof due usually to improper drainage or deck deflection; also called Birdbaths. 2. Water puddles on concrete paving or asphaltic concrete paving due to insufficient sloping or imperfect finishing.

Pool Depths. The distance between the floor of pool and the maximum operating water level.

Pool Equipment. Implements and devices used for pool maintenance and operation.

Pool Ladder. A ladder used for climbing in and out of a swimming pool.

Pool Light. A light mounted in the wall of a pool for night use.

Pool Liner. Waterproof membrane for pool; usually flexible vinyl.

Pool Plumbing. All chemical, circulation, filter waste discharge piping, deck drainage and water filling system.

Pool Shell. Pool floor and walling structure; also called Pool Tank.

Pool Tank. See Pool Shell.

Pool. A body of water; or bath, hydro, hot tub, lido, pond, spa, splasher, and thermae, or covered balneum, bath house, leisure centre, natatorium, poolarium, pool enclosure, pool hall, and pool house.

Pop Rivet. Metal fastener for joining sheetmetal pieces, installed by a hand operated compressed air-assisted or spring-loaded gun; unique in that installation may be from one side of the work.

Pop-Off Valve. A safety valve which opens automatically when pressure exceeds a predetermined limit.

Pops. Ruptures in finished plaster or cement surfaces which may be caused by expansion of improperly slaked particles of lime or by foreign substances; also called Pits.

Pop-Up Drain. The part of a sink drain assembly that is operated by a linkage to open or close the drain.

Pop-Up Head. In a lawn irrigation system, a watering head that retracts when not in use, and becomes flush with the ground.

Porcelain Cleat. A ceramic electrical insulator.

Porcelain Enamel Institute (PEI). 1911 North Fort Myer Drive, Arlington, Virginia 22209, (703) 527-5257.

Porcelain Enamel. Vitreous Enamel.

Porcelain Mullite. A vitreous ceramic whiteware for technical application in which mullite ($3A_{l2}O_3\ 2SiO_2$) is the essential crystalline phase.

Porcelain Receptacle. A simple electrical lighting fixture consisting of a bulb holder, with or without a pull chain switch.

Porcelain Sink. A wash basin made from nonporous hard white glazed ceramic coated steel or cast iron.

Porcelain. Ceramic china-like coating applied to steel or cast iron surfaces; when used as a finish for metal fixtures it is called vitreous enamel.

Porch. 1. An ornamental entrance way to a house. 2. A landing before a door.

Pores. Openings on the surface of a piece of wood; these openings result when vessels in the wood are severed during sawing; see also Vessels.

Porosity. 1. Being permeable to liquids. 2. Degree of integrity or continuity. 3. Gas pockets or voids in metal. 4. The ratio of the volume of voids in the material to the total volume of the material, including the voids, usually expressed as a percentage.

Porous Fill. Soil that allows relatively free passage of water.

Porous Woods. Hardwoods having vessels or pores large enough to be seen readily without magnification.

Porphyry. A hard igneous rock quarried in ancient Egypt, composed of crystals of white or red feldspar in a red matrix.

Portable Bleacher. An uncovered stand of tiered planks that can be moved from one event to another.

Portable Stage. A movable raised platform usually used for entertainment and speaking purposes.

Portal Frame. A rigid frame; two columns and a beam attached with moment connections.

Portal. An elaborate or imposing door or entrance.

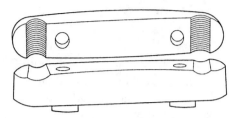

Porcelain Cleat

Portcullis. A strong heavy grating of iron or iron-bound wood sliding up and down in vertical grooves to block a gateway in a fortress.

Porte Cochere. A roofed structure adjacent to a building entrance covering a driveway, to allow shelter to automobile passengers.

Portico. A roofed colonnade adjoining a building.

Portland Blast Furnace Slag Cement. The product obtained by intimately ingrinding or intimately and uniformly blending a mixture of granulated blast furnace slag and portland cement clinker.

Portland Cement Association (PCA). 5204 Old Orchard Road, Skokie, Illinois 60077, (708) 966-6200.

Portland Cement Grout. Portland cement added to fine aggregates and water and pumped or forced into joints, cracks, and spaces as an adhesive sealer or structural fill.

Portland Cement Mortar. Mortar where the cementing agent is portland cement.

Portland Cement Plaster. A binder of portland cement mixed with plaster and used on exterior surfaces or in damp areas.

Portland Cement. A hydraulic cement produced by pulverizing clinker consisting essentially of hydraulic calcium silicates, and usually containing one or more of the forms of calcium sulfate as an interground addition; the most common type of cement used in construction; the name portland is not a brand name, it refers to a type of cement. 2. Product obtained by pulverizing and mixing limestone with other products to produce cement required for concrete mix.

Portland Cement-Lime Plaster. Portland cement and lime (either Type S hydrated lime or properly aged lime putty) combined in suitable proportions.

Portland Pozzolan Cement. The product obtained by intimately intergrinding a mixture of portland cement clinker and pozzolan, or an intimate and uniform blend of portland cement and fine pozzolan.

Positioned Weld. A weld made in a joint which has been so placed as to facilitate making the weld.

Positioner. Wire device designed to hold reinforcing steel in designated position in a masonry wall.

Positive Drainage. In roof construction, the drainage condition in which consideration has been made for all loading deflections of the deck and additional roof slope has been provided to ensure complete drainage of the roof area within 24 hours of rainfall precipitation.

Positive Moment. A result of bending moment in a beam in which the upper part is in compression and the lower part is in tension; compare with Negative Moment.

Positive Pressure. 1. Condition that exists when more air is supplied to a space than is exhausted, so the air pressure within that space is greater than that in surrounding areas. 2. A pressure within the sanitary drainage or vent piping system that is greater than atmospheric pressure; also called Plus Pressure.

Post Cap. 1. A metal connection from a wooden post to a girder. 2. A prefabricated fitting which tops a wooden post and protects it from the elements.

Post Hole. A dug out hole in the ground for the installation of a fence or gate post.

Post Light. A lighting fixture mounted on a post.

Post Line. The line which marks the outside face of a foundation, the location line of fence posts or the line of piers for a deck.

Post, Corner. See Corner Post.

Post, Treated. See Treated Post.

Post. A timber set on end to support a wall, girder, or other member of the structure.

Postal Accessories. Implements and devices used to weigh, sort, package, price, and handle mail.

Postformed Plywood. The product formed when flat plywood is reshaped into a curve configuration by steaming or plasticizing agents.

P

Post-Tensioning. A method by which concrete is compressed after it has been cast by stressing the steel reinforcing; the compressing of the concrete in a structural member by means of tensioning high-strength steel tendons against it after the concrete has cured.

Pot Hole. A small steep-sided hole usually with underground drainage.

Pot Life. See Working Life.

Potable Water. Water that is suitable for drinking.

Potassium Alum. Used as flocculent in sand filter operation.

Potential Energy. A body's ability to do work by virtue of its position relative to others, or by the arrangement of parts.

Potential Relay. Electrical switch which opens on high voltage and closes on low voltage.

Potential, Electrical. See Electrical Potential.

Potentiometer. Instrument for measuring or controlling by sensing small changes in electrical resistance.

Pouch. See Nail Apron.

Pound. 1. A unit of weight equal to 16 ounces avoirdupois or 0.4536 kilograms. 2. Crush or beat with repeated heavy blows.

Poundal. A unit of force equal to the force that would give a free mass of one pound an acceleration of one foot per second per second.

Pounds per Square Inch Absolute (PSIA). Absolute Pressure.

Pounds per Square Inch Gauge (PSIG). Gauge Pressure.

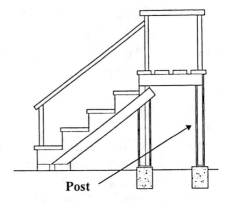

Post

Pour Coat. The top coating of bitumen on a built up roof.

Pour Point. Lowest temperature at which a liquid will pour or flow.

Pour. To place concrete; a continuous increment of concrete casting carried out without interruption.

Poured in Place. Concrete that is poured and cured on site in its final position.

Powder Room. A guest bathroom, usually containing a toilet and a lavatory; usually located near the front entry hall to be convenient to guests.

Powder Stains. Stains in form of powder which are mixed with solvents to produce wood stains.

Powder. 1. A substance in the form of fine dry particles. 2. Gunpowder.

Powder-Actuated Fastener. Driven pin or threaded stud made of special heat-treated steel to provide hard ductile fastener which is driven into material by impact of discharge of an explosive cartridge; used to fasten materials to concrete and structural steel.

Power Driven Fastener. A fastener attached to steel, concrete or masonry by a power charge cartridge or by manual impact.

Power Element. Sensitive element of a temperature-operated control.

Power Factor. 1. The ratio of the true power to the apparent power. 2. Correction coefficient for the changing current and voltage values of a-c power.

Power of Attorney. A legal instrument that authorizes one to act as the agent of another; the agent is called an attorney in fact, but does not have to be a licensed attorney.

Power Stretcher. A carpet layer's tool; an extension-type version of the knee kicker, with larger teeth arranged in a patent head which can be adjusted for depth of bite, used to stretch larger areas of carpet than can be handled by the knee kicker.

Power Tool. An apparatus or device used in construction, powered by electric current.

Power Trowel. See Mechanical Trowel.

Power. 1. A source or means of supplying energy, such as electricity. 2. Time rate at which work is done or energy emitted.

Power-Assisted Door. A door used for human passage with a mechanism that helps to open the door, or relieves the opening resistance of a door, upon the activation of a switch or a continued force applied to the door itself.

Pozzolan. Hydraulic cement obtained by grinding a mixture of fused natural material, such as volcanic ash or trass with hydrated lime; can be used to replace some of the portland cement to lower cost and in some cases improve the concrete; also spelled puzzolan.

PPFA. Plastic Pipe and Fitting Association.

PPG. Polished Plate Glass.

PPI. Plastic Pipe Institute.

PPM. Parts Per Million.

Pratt Truss. A standard peaked roof truss configuration where the vertical web members and the top chord are in compression while the sloping web members and the bottom chord are in tension.

Praxis. The practice of a skill, art, trade, or profession.

Precast Beam. A concrete horizontal structural member that is cast and cured in other than its final position, on- or off-site.

Precast Column. A concrete column that has been cast and cured in other than its final position.

Precast Concrete. Concrete parts that are cast on- or off-site and, after curing and hardening, are installed in their final position of use.

Precast Lintel. A horizontal concrete beam placed over an opening that has been cast and cured in other than its final position.

Precast Manhole. An underground enclosed structure that has been cast and cured in other than its final position, usually cast in sections.

Precast Panel. A concrete member, cast and cured in other than its final position and relatively thin with respect to other dimensions.

Precast Septic Tank. A sewage system underground tank that has been cast and cured in other than its final position.

Precast Slab. A flat, horizontal, molded layer of reinforced concrete, cast and cured in other than its final position, sometimes referred to as precast planks, often containing hollow cores.

Precast Specialty. Special shapes or ornamental objects of precast masonry or concrete.

Precast Terrazzo. Marble aggregate concrete that is cast and cured in other than its final position, and ground smooth for decorative purposes.

Precast Wall. A wall that has been cast and cured in other than its final position.

Precast. A concrete component or member cast and cured in other than its final position.

Precipitate. 1. A substance separated from a solution by chemical action or by application of heat or cold. 2. Insoluble compound formed when chlorine or alum added to pool water reacts with other chemicals or minerals.

Precipitation. Rain, hail, or snow falling to the ground.

Precoat. The initial coating of filter aid on the septum of a diatomaceous earth filter.

Preconstruction Jobsite Conference. A conference on the jobsite, before start of construction, attended by owners, architects, engineers, contractors, and subcontractors for the purpose of coordination and cooperation.

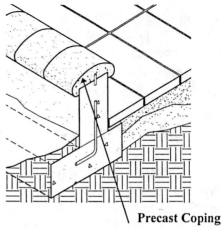

Precast Coping

Predecorated Wallboard. A gypsumboard with a finished surface applied before installation; gypsumboard may be predecorated with paint, texture, vinyl film, or printed paper coverings in a variety of patterns, styles, and colors.

Prefabricated Lined Stack. Chimney or vent fabricated in a shop into sections with multiple components to provide thermal insulation and allow for low clearance to wooden or combustible elements.

Prefabricated. Constructed and assembled in a workshop and later brought to the jobsite for incorporation into the building.

Pre-fill. An application method used in the preparation of tapered/beveled gypsum board to receive tape and joint treatment; aids in reducing the possibility of ridging and beading.

Prefinished Gypsumboard. Gypsumboard finished at the factory with a decorative layer of paint, paper, or plastic.

Pre-Function Area. An anteroom in a hotel or conference center leading to a large dining room, used for receptions prior to dinners or meetings.

Preheating. The application of heat to the base metal prior to a welding or cutting operation.

Prehung Door. A door that is hinged to its frame in a factory or shop.

Preliminary Drawings. Drawings prepared in a preliminary stage of a project.

Preliminary Notice. A notice sent to the owner, general contractor, and lender by subcontractors or materialmen within 20 days of first supplying labor, services or materials to a project informing them of what is being supplied or performed and establishing lien or stop-notice rights.

Premature Stiffening. See False Set.

Pre-Mixed Plaster. See Gypsum Ready Mixed Plaster.

Prepacked Concrete. Concrete manufactured by placing clean, graded coarse aggregate in a form and later injecting a portland cement sand grout, under pressure, to fill the voids.

Prepaid Expenses. Materials or services that a company buys and pays for before use, such as insurance premiums, or office supplies.

Preparation, Painting. See Painting Preparation.

Prepayment Penalty. A monetary penalty for paying off a note prior to its due date.

Present Value. The value now of a sum of money to be paid or received in the future, calculated by discounting future cash flows by an appropriate discount rate.

Preservation. 1. The process of treating materials to prolong their useful life. 2. Conservation.

Preservative Treated. Applied or pressurized chemical treatment of wood or plywood to make it resistant to deterioration from moisture and insects.

Preservative. 1. A substance for preserving. 2. In wood, any substance that, for a reasonable length of time, is effective in preventing the development and action of wood-rotting fungi, borers of various kinds, and harmful insects that deteriorate wood.

Press. A machine for punching, notching, and shearing steel.

Pressed Bricks. Bricks that are pressed out in a mold before they are baked or burned.

Presser, Clothing. See Clothing Presser.

Pressure Balance. In spray painting, relationship of pot pressure to atomizing air pressure.

Pressure Connector, Solderless. See Solderless Pressure Connector.

Pressure Connector. An electrical connecting device which squeezes two or more conductors together.

Pressure Differential. Pressure difference across hydraulic system.

Pressure Drop. 1. Loss in pressure due usually to length or diameter of line or hose. 2. Pressure difference at two ends of a circuit, or part of a circuit, the two sides of a filter.

P

Pressure Feed Paint Tank. In spray painting equipment, fluid container in which fluid flow is caused by air pressure; also called Pressure Pot.

Pressure Feed. In spray painting equipment, fluid flow caused by application of air or hydraulic pressure to paint.

Pressure Gauge. 1. A gauge for measuring the pressure exerted by a liquid or a gas. 2. A device to measure the pressure of an explosive.

Pressure Head. Force caused by the weight of a column or body of fluids; expressed in feet, inches or psi.

Pressure Head. The amount of force created by a depth of one foot of water.

Pressure Limiter. Device which remains closed until a certain pressure is reached and then opens and releases fluid to another part of system or breaks an electric circuit.

Pressure Motor Control. High or low-pressure control connected into the electrical circuit and used to start and stop motor; it is activated by demand for refrigeration or for safety.

Pressure Pot. See Pressure Feed Paint Tank.

Pressure Reducing Valve. A valve which maintains fluid pressure uniformly on its outlet side as long as pressure on the inlet side is at or above a design pressure.

Pressure Regulator, Evaporator. See Evaporator Pressure Regulator.

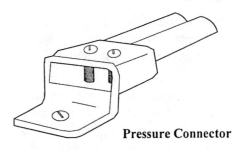

Pressure Connector

Pressure Regulator. Automatic valve located between a compressor and evaporator outlet that is responsive to its own inlet pressure or to the evaporator or refrigerator temperature; it throttles the vapor flow when necessary to prevent the evaporator pressure from falling below a selected value.

Pressure Regulator. A valve which controls water pressure in the supply line.

Pressure Switch. An electric switch that is activated by a rise or drop in air or fluid pressure.

Pressure Water Valve. Device used to control water flow; it is responsive to head pressure of refrigerating system.

Pressure, Absolute. See Absolute Pressure.

Pressure, Atmospheric. See Atmospheric Pressure.

Pressure, Back. See Back Pressure.

Pressure, Gauge. See Gauge Pressure.

Pressure, Operating. Pressure at which a system is operating.

Pressure, Suction. Pressure in low-pressure side of a refrigerating system.

Pressure. Energy impact on a unit area; force or thrust on a surface.

Pressure-Heat Diagram. Graph of refrigerant pressure, heat, and temperature properties; see Mollier's diagram.

Pressure-Operated Altitude Valve (POA Valve). Device which maintains a constant low-side pressure independent of altitude of operation.

Pressure-Sensitive Adhesive. An adhesive that will adhere to a surface at room temperature by briefly applied pressure alone.

Pressure-Treated Lumber. Lumber that has been impregnated with chemicals under pressure, for the purpose of retarding either decay or fire.

Prestressed Beam. A horizontal structural member which has had a load applied to it to increase its effectiveness in resisting working loads.

Prestressed Concrete Institute (PCI). 175 West Jackson Boulevard., Chicago, Illinois 60604, (312) 786-0300.

Prestressed Concrete. Concrete in which the steel is stretched and anchored to compress the concrete.

Prestressing. Applying compressive stress to a concrete structural member to increase its strength.

Pretax Income. Net income or profit shown on income and expense statement before deduction for Federal Taxes.

Pretensioning. A method by which the design tensile force is applied to the steel reinforcing before the concrete is set.

Pretreatment. Chemical alteration of the surface to make it suitable for painting.

Prevailing Wage Law. A law that establishes minimum wages for job classifications in the construction trades.

Prevalent Level Samples. Air samples taken under normal conditions; also called Ambient Background Samples.

Preventive Maintenance Painting. Periodic touch-up painting or application of full coats of paint before deterioration starts.

Price. The amount of money that must be paid for something.

Primary Branch. This is a single branch that slopes from the base of a soil or waste stack and joins the main building drain or another branch of the main building drain.

Primary Color. One of the colors from which all other colors may be produced by mixing; in pigments, red, yellow, and blue.

Primary Control. Device which directly controls operation of heating system.

Prime Coat. 1. The first coating of any material placed on a surface to seal and to provide a proper base for additional finish coats. 2. The first coat of paint.

Prime Contractor. A general contractor who contracts with a property owner and, in turn, employs a subcontractor or subcontractors to perform some or all of the work.

Prime In the Spots. Apply a priming coat to those spots that have been scraped, wire brushed, shellacked, have had the old paint burned off ,or consist of newly patched plaster.

Prime Professional. Any professional having a contract directly with the owner.

Prime Rate. The rate of interest paid by the most credit-worthy customers of a bank.

Primed Door. A door which has received its first coat of paint prior to installation.

Primed Siding. Exterior siding material that has received its first coat of paint.

Primer Paint. A specially mixed paint that is applied first to a surface to seal and provide a base for additional finish coats.

Primer. 1. A tinted base coat of paint to seal the surface and equalize suction differences. 2. A liquid composed of bituminous solvent that is applied to a surface to improve the adhesion of heavier subsequent applications of bitumen.

Principal. 1. Under a surety bond, the contractor is the principal. 2. The amount of a loan or the loan balance at any particular time. 3. In a debt repayment, the principal portion goes to reducing the debt as opposed to the interest portion.

Principle. 1. A fundamental truth or law as the basis of reasoning or action. 2. A general law in physics, chemistry, or engineering. 3. A law of nature forming the basis for the construction or working of a machine or mechanical device.

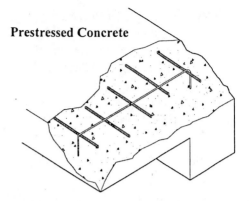

Prestressed Concrete

Print Dyeing. Screen printing a pattern on carpet by successive applications of premetalized dyes, which are driven into the pile construction by an electromagnetic charge.

Print Free. Paint sufficiently dry so that no imprint is left when something is pressed against it.

Prism Testing of Masonry. Compression testing of a sample section of wall approximately 16 inches high and 16 inches long of the thickness and type of construction similar to the wall under construction.

Prism. 1. A solid geometric figure whose two ends are similar, equal, and parallel rectilinear figures, and whose sides are parallelograms. 2. A transparent body in this form, usually triangular, with refracting surfaces at an acute angle with each other, which separates white light into a spectrum of colors.

Prismatic. 1. Similar to a prism. 2. Brilliantly colored. 3. Refracted or dispersed as if by a prism.

Privacy Lock. A lockset, without a key, that can be locked from the room side only; used mainly on bedroom and bathroom doors; also called a Bathroom Lock.

P

Private Mortgage Insurance. Life insurance carried by a home buyer to pay off the mortgage in the event of buyer's death; the insured amount declines as the loan is paid off; this type of insurance is usually required by the lender when the loan to value ratio exceeds 90%.

Private Sewer. A sewer privately owned.

Privity. Persons who have contractual relationships with each other are in privity.

Pro Forma Balance Sheet. A statement representing a future financial position of a business showing anticipated assets, liabilities and net worth.

Pro Forma. As a matter of form; provided in advance to prescribe form.

Pro Rata. Proportionate or pro-rated.

Probability. The extent to which an event is likely to occur, measured by the ratio of the favorable cases to the whole number of cases possible.

Probate. The official court process of proving the validity of a will.

Procedure. Rules governing court operations.

Process Tube. Length of tubing fastened to hermetic unit dome, used for servicing unit.

Process. 1. A procedure for manufacturing products. 2. To handle or deal with a procedure.

Processed Lime. Pulverized quick lime.

Processed Quicklime. See Lime.

Processor, Photo. See Photo Processor.

Produce Case. A bin or container, usually open, that holds market produce, such as lettuce or tomatoes.

Product Data. Illustrations, standard schedules, performance charts, instructions, brochures, diagrams, and other information furnished by the contractor to illustrate materials or equipment for some portion of the work.

Product Liability. A manufacturer's legal liability for claims arising out of damages sustained by users of the product.

Production Rate. Measurement of surface cleaned or coated in one working day by one person, usually in units of square feet per day.

Products Standards Section, U.S. Dept. of Commerce (PS). 14th and Constitution Avenue, NW, Washington, DC 20230.

Professional Engineer. A person qualified as an engineer and duly licensed by the state to practice as a civil, structural, electrical, mechanical, or other engineer.

Professional Fee. The amount of money charged by a person hired to perform a professional service.

Professional Indemnity Insurance. See Errors and Omissions Insurance.

Professional Liability. A professional's legal liability for claims arising out of damages sustained by others allegedly as a result of negligent acts, errors, or omissions in the performance of professional services.

Professional Standard of Care. The duty of an architect, engineer, or other professional to exercise such care, skill, and diligence as other professionals in the same discipline would in the same or similar circumstances.

Profile Depth. Average distance between top of peaks and bottom of valleys on the surface.

Profile. 1. The outline of something. 2. A drawing of something in section. 3. Surface contour of a blast-cleaned surface as viewed from the edge.

Profit and Loss Statement. A financial statement showing the income and expenses of a business for the period, and profit or loss resulting therefrom.

Profit and Loss. A summary account used at the end of an accounting period in which income and expenditure and the resulting net profit or loss are shown.

Profit. The excess of income over expenditure; net income.

Program Evaluation Review Technique (PERT). An early form of critical path analysis for construction scheduling.

Program. A detailed outline of all of the features to be included in a building.

Progressive Blocking. Wood blocking between joists or rafters progressively nailed as each joist or rafter is installed.

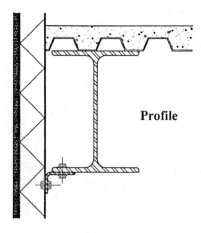

Profile

Progressive Kiln. A kiln in which the total charge of lumber is not dried as a single unit but as several units, such as kiln truckloads, that move progressively through the kiln; the kiln is designed so that the temperature is lower and the relative humidity higher at the end where the lumber enters than at the discharge end.

Progressive Loss. An injury to person or property that occurs over a period of time, such as cracking caused by gradual settlement of soils or illness caused by inhalation of dust.

Project Delivery System. Method by which owners can contract for construction, such as design-bid-build, design/build, turnkey, or other of numerous variants.

Project Manual. The volume usually assembled for the work which may include the bidding requirements, sample forms, conditions of the contract, and specifications.

Project. 1. A scheme or design for an undertaking or development of real property. 2. To protrude or extend outward.

Projecting Sash. A window with sashes that swing either inward or outward.

Projecting Window. A window that swings either inward or outward.

Projection. 1. A part that protrudes. 2. The presentation of films or slides by projecting their image from a projector onto a screen. 3. A forecast or estimate based on a present trend. 4. The forming of a plan or scheme of development. 5. A graphic representation of a building or architectural element. 6. The graphic representation of any portion of the earth's spherical surface on a flat plane surface.

Projector, Nonprofessional. The nonprofessional projector employs film other than 7/8 inch wide.

Projector, Professional. The professional projector employs a film which is more than 7/8 inch wide.

Promise. An undertaking that something will or will not happen in the future.

Promisee. A person to whom a promise is made.

Promisor. A person who makes a promise.

Promissory Estoppel. A promise that is enforceable even though it is not supported by consideration, since the promisor is estopped to withdraw it.

Promissory Note. A written promise to pay a sum on or before a particular date.

Proof of Service. Evidence that a legal document has been delivered to a specified person.

Propane. 1. A heavy gaseous hydrocarbon of the alkane series used as bottled fuel. 2. Low temperature application refrigerant.

Propeller Mixer. A type of mixing device which is inserted into a liquid, usually paint, and powered by a propeller at the end of a thin shaft.

Property Damage. Injury to tangible property.

Property Insurance. Insurance against damage to, or loss of, property that belongs to the insured.

Property. A tangible thing that is, or may be, possessed by a person.

Proportional Limit. The maximum unit stress for which Hooke's Law is valid.

Proportioning. Selection of proportions of material for concrete to make the most economical use of available materials to manufacture concrete of the required strength, placeability, and durability.

Proposal. The document submitted by the contractor to the owner for construction of the project; also called the Bid.

Propylaeum. (Pl. propylaea) An architecturally important vestibule or entrance to a building or enclosure; usually used in the plural.

Propylene Plastic. Plastics based on resins made by the polymerization of propylene with one or more other unsaturated compounds, the propylene being in greatest amount by weight.

Proration. The pro rata distribution of expenses in proportion to time of transaction, as taxes and insurance in a property sale escrow or as in shared expenses in proportion to space leased.

Proscenium Arch. The wall that separates the stage from the auditorium and provides the arch that frames it.

P

Protected Membrane Roof. A membrane roof assembly in which the thermal insulation lies above the waterproof membrane.

Protective Board. A board or sheet of material that is installed next to a waterproofing membrane and then backfilled against thus protecting the membrane from puncture or damage.

Protective Life. Interval of time during which a paint system protects substrate from deterioration.

Protector, Circuit. See Circuit Protector.

Proton. Particle of an atom with a positive charge.

Prototype. 1. An operational model of a new product. 2. An original design that is the basis of all future improvements and variations; an Archetype.

Protractor. A device for measuring angles on drawings or in constructing angles on drawings.

Provisional Remedies. Orders or proceedings that protect persons or property while legal proceedings are pending.

Prussian Blue. Ferric Ferrocyanide; pigment which is deep in color and has great strength; not affected by acids but easily affected by alkali.

Prybar. An iron or steel bar used for prying, such as a crowbar.

PS. Products Standards Section, U.S. Dept. of Commerce.

PSF. Pounds per square foot.

PSI. Pounds per square inch.

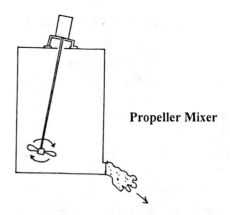

Propeller Mixer

PSIA. Pounds per Square Inch Absolute.

PSIG. Pounds per Square Inch Gauge.

Psychological Factors. Psychological, organizational, and personal stresses that could produce symptoms similar to poor indoor air quality.

Psychrometer. See Hygrometer.

Psychrometric Chart. Chart that shows relationship between the temperature, pressure, and moisture content of the air.

Psychrometric Measurement. Measurement of temperature, pressure, and humidity using a psychrometric chart.

Psychrometrics. A branch of physics dealing with the study of atmospheric conditions, specifically the relationships between moisture and air.

P-Trap, ABS. A P-trap used mostly at sinks and lavatories; manufactured of acrylonitrile butadiene styrene, a rigid plastic pipe.

P-Trap, No-Hub. A P-trap used mostly at sinks and lavatories; made to join with no-hub cast iron drain piping.

P-Trap. A P-shaped drain trap that prevents sewer gas from escaping from a plumbing fixture.

P

Public Sewer. A sewer that is publicly owned.

Public Utility Regulatory Policies Act. A law requiring public utility companies to buy on-site generated electricity from private power producers.

Public Works. Construction and projects undertaken on behalf of governmental entities.

Publication. The process of making writings or drawings available for distribution to or inspection by the public.

Puckering. A condition in a carpet seam, due to poor layout or unequal stretching, wherein the carpet on one side of the seam is longer or shorter than that on the other side, causing the long side to wrinkle or develop a pleated effect.

PUD. Planned Unit Development.

Puddle Welded. Spot weld, made by arc welding process, in which coalescence proceeds from surface of one member into other; weld is made without preparing a hole in either member.

Puddle. To compact loose soil by soaking it and allowing it to dry.

Puddling. A condition of mechanical dash textures resulting in glazing, texture deviation or discoloration caused by holding the plastering machine nozzle too long in one area.

Puffing Agent. A synthetic organic product used to produce increased viscosity in varnishes and paints.

Pull Box. An electrical rough-in box placed in a length of conduit, through which cables can be pulled.

Pull Chain. An electric switch at a fixture operated by a pull cord.

Pull Saw. A handsaw that cuts on the pull rather than on the push stroke.

Pull Wire. A wire installed in a conduit by an electrician by which additional wires are pulled through.

Pulley Stile. The members of a window frame which contain the counterweights and pulleys, and between which the edges of the sash slide.

Pulley. A grooved wheel for a cord, chain, or belt to pass over, used for changing the direction of a force or for mechanical advantage.

Pulling. See Floating.

Pulp. Wood fiber added as an aggregate to neat calcined gypsum.

Pulvinar. A cushioned seat.

Pumice Stone. A stone of volcanic origin, which is pulverized to produce a soft abrasive used extensively in rubbing finishing coats of fine wood finishes.

Pumice. A light porous volcanic rock often used as an abrasive and as plaster aggregate.

Pump Down. The act of using a compressor or a pump to reduce the pressure in a container or a system.

Pump, Centrifugal. See Centrifugal Pump.

Pump, Fixed Displacement. See Fixed Displacement Pump.

Pump, Gas. See Gas Pump.

Pump, Reciprocating Single Piston. See Reciprocating Single Piston Pump.

Pump, Screw. See Screw Pump.

Pump, Sump. See Sump Pump.

Pump. 1. A machine usually with a rotary action or the reciprocal action of a piston, for raising or moving liquids, compressing gases, spray painting, and inflating tires. 2. See Plastering Machine.

Pumped Concrete. Concrete that is pumped through a hose or pipe.

Pumping Agent. A product used to increase the flow of plaster through hoses during machine applications.

Pumping of a Slab. The ejection of mud and water up through joints or cracks when a vehicle passes over an on-grade concrete slab.

Punch list. See Inspection List.

Punch Out. A close fitting hole in gypsum board to allow penetration of plumbing lines; a pipe hole.

Punch. A small hand tool formed from a short steel bar with one end in various shapes to suit the work, such as forming, perforating, embossing, or cutting, when struck by a hammer.

Punched Carpet. Carpet made by punching loose, unspun fibers through a woven sheet which results in a pileless carpet similar to a heavy felt; usually consists entirely of synthetic fibers.

Punitive Damages. Damages awarded to a private person against a wrong-doer by way of punishment, and to deter future misconduct.

Punkiness Soft or spongy gypsum-board core.

Purchase Money Mortgage. A mortgage or trust deed given to a seller as part of the purchase consideration.

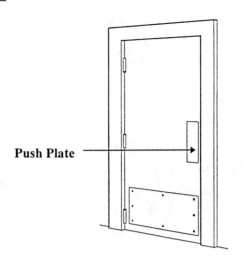

Push Plate

Purchases Account. The account in which merchandise purchased for resale is recorded.

Purger, Air. See Air Purger.

Purging. Releasing compressed gas to atmosphere through some part or parts for the purpose of removing contaminants from that part or parts.

Purlin. 1. A timber supporting several rafters at one or more points. 2. Beams or struts that span across a roof to support the roof framing system.

Push Plate. A metal plate on a door installed at hand level for pushing the door open, used for protection from damage and easier cleaning.

Push Stick. A pole or strip used to push a workpiece when cutting with power saws, jointers and other power tools; pushing a board by hand is unsafe with power equipment.

Putty Coat. Final smooth coat of plaster.

Putty Knife. A broad-bladed hand tool which is used to dispense and apply putty to surfaces.

Putty. A cement composed of whiting and raw linseed oil, used for setting panes of glass and filling small holes and cracks.

PV Cells. Photovoltaic cells.

PV. Photovoltaic.

PVA. Polyvinyl Acetate.

PVB. Polyvinyl Butyral.

PVC Caulking. Caulking made of polyvinyl chloride.

PVC Cement. A solvent cement specifically used to join PVC pipe and fittings.

PVC Chamfer Strip. A triangular or curved insert constructed of PVC placed in an inside form corner to produce a rounded or flat, beveled edge.

PVC Conduit. Lengths of rigid plastic pipe made of polyvinyl chloride.

PVC Control Joint. Control joint made of polyvinyl chloride.

PVC Fitting. A fitting made of polyvinyl chloride used in joining PVC piping.

PVC Pipe. Polyvinyl chloride pipe used mainly for drain lines and particularly resistant to chemicals.

PVC Sewer Pipe. Large diameter pipe used for the dispersal of waste material, constructed of polyvinyl chloride, a rigid plastic pipe.

PVC Waterstop. A waterstop made of polyvinyl chloride.

PVC. 1. Polyvinyl Chloride. 2. Pigment Volume Concentration.

P-Wave. The primary or fastest wave traveling away from a seismic event through the earth's crust, and consisting of a train of compressions and dilations of the material; see Seismic Wave.

Pyramid. A polyhedron having for its base a polygon and for its faces triangles, meeting at a common apex.

Pyroligneous. Obtained by destructive distillation of wood.

Pyrolysis. Chemical change brought about by the action of heat.

Pyrometer. A thermometer for measuring very high temperatures.

Pyroxylin. See Cellulose Nitrate.

Pythagorean Theorem. The square of the length of the hypotenuse of a right-angled triangle is equal to the sum of the squares of the lengths of the other two sides.

Quad. 1. A unit of energy equal to one quadrillion (1,000,000,000,000,000) BTUs. 2. Quadrangle. 3. Quadriphonic.

Quadrangle. 1. A four sided enclosure surrounded by buildings. 2. A quadrilateral. 3. One of the survey map sheets published by the USGS.

Quadrant. 1. A quarter of a circle. 2. An instrument for measuring altitudes. 3. A hardware device to fasten together the upper and lower leaves of a Dutch door.

Quadrel. A square tile; a quarrel.

Quadrilateral. A polygon of four sides.

Quadripartite. Having four parts; quadripartite vaulting is the commonest form of groined and also of ribbed vaulting.

Quadriphonic. Sound recorded or transmitted utilizing four transmission channels.

Quadrisect. To divide into 4, usually equal, parts.

Quadruplex Cable. Four wire cable.

Quag. A marsh.

Quagmire. A soft boggy or marshy area that gives way under foot.

Quake. An earthquake.

Qualified Historical Building (Or Structure). Any structure or collection of structures deemed important to the history, architecture or culture of an area by an appropriate local or state governmental jurisdiction. This shall include structures of national, state or local historical registers or official inventories, such as the National Register of Historic Places, State Historical Landmarks, State Points of Historical Interest, and city or county registers or inventories of historical or architecturally significant sites and places or landmarks.

Quality Control. All the activities undertaken to ensure adequate quality in manufactured products and in on-site construction.

Quality. The degree of excellence of a thing.

Quantity Survey. Detailed listing of all materials, equipment, and services needed to complete a project.

Quantity Surveyor. A person who measures and prices building work; see Quantity Survey.

Quantity. The size, extent, weight, amount, number, or volume of a thing.

Quantum. A desired or allowed amount.

Quarrel. 1. A square or lozenge-shaped piece of material, especially a piece of glass, set diagonally, as in a latticed sash; also called a quarrel pane. 2. The opening in a sash prepared for such glazing.

Quarry Bed. The side of a piece of building stone that is parallel with the natural strata or veins.

Quarry Tile. A large, fired, clay floor tile, usually unglazed.

Quarry. An open excavation for removing building stone, slate, or limestone; a rock pit.

Quarry-Faced. Stone having a rough face as if just split from the quarry.

Quarter Bend. A drainage pipe fitting which makes a 90 degree angle.

Quarter Drop Match. See Set Match.

Quarter Hollow. A concave molding or cavetto of which the transverse section is an arc of about 90 degrees, the converse of a quarter round, or ovolo.

Quarter Round. Wood molding with a cross section in the shape of a quarter circle.

Quarter Width. The unit of yard measure (1/4) used in referring to carpet or loom widths; early European carpet was woven in widths of 27 inches or 3/4 yards; hence, 4/4 = 1-yard width or 3'; 12/4 width = 9' and 16/4 width = 12'.

Quartersawed Lumber. Edge-grained lumber.

Quartz. Crystalline silica; a hard brittle mineral breaking with a glasslike fracture, and usually transparent to translucent, and colorless, or of a white, pink, and amethystine hue.

Quartzite. A strong sandstone cemented by quartz, about 98 percent silica.

Quasi Contract. Application of contract rules of law to parties that do not have a contractual relationship.

Quatrefoil. A four cusped circular element in Gothic tracery.

Quay. A structure built along the bank of a waterway for use as a landing place.

Queen Anne. The architecture and furniture style of England, under Dutch influence, during the short reign of Anne, 1702-1714.

Queen Closer. A brick cut in half lengthwise.

Queen Truss. A truss framed with queen posts.

Queen. 1. A half brick made by cutting a whole brick in two lengthwise as for a corner in a soldier course. 2. A half brick used to prevent vertical joints falling above one another.

Quenching. 1. Cooling of steel in cold water to increase its hardness. 2. Rapid cooling of metal in a heat treating process.

Quick Assets. 1. Assets easily converted to cash. 2. Current assets without inventory.

Quick Burst. The inter-pressure required to burst a pipe or fitting due to an internal pressure build-up, usually within 60 to 90 seconds.

Quick Drying. A material with a relatively short drying time.

Quick Fix. An expedient often inadequate solution to a problem; a Band-Aid Approach.

Quick Set. 1. See Flash Set. 2. In gypsum wallboard joints, a joint compound that chemically hydrates prior to drying; also called Fast Set.

Quick-Connect Coupling. A device which permits easy and fast connecting of two fluid lines.

Quicklime. The first solid product CaO that is obtained by calcining limestone and that develops great heat and becomes crumbly when treated with water; lime; before it can be used in construction, quicklime must be slaked in water and aged for at least 2 weeks.

Quicksand. 1. Fine sand or silt that is prevented from settling firmly together by upward movement of ground water. 2. Any wet inorganic soil so unsubstantial that it will not support any load.

Quicksilver. Mercury.

Quiet Title Action. A lawsuit brought to remove a claim on the title of real property.

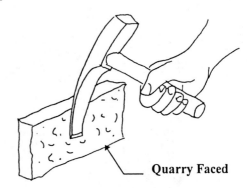

Quarry Faced

Quincunx. An arrangement of five objects in which four are at the corners and one is at the center at the crossed diagonals, in the manner of a dice or a five playing card, used in planting trees.

Quintefoil. A five cusped element in Gothic tracery.

Quirk. A groove separating two parts of a molding.

Quitclaim Deed. A form of deed in the nature of a release containing both words of grant and of release; it does not provide a guarantee or warranty of title.

Quoin. Cornerstones forming the external angle of a building; bricks or block masonry that project out from the walls at the corners of a building for ornamental effect.

Quonset. A prefabricated metal building with a semicylindrical corrugated roof.

R

R&D. Research and Development.

R. 1. Radius of gyration. 2. Electrical Resistance, in Ohms.

R-12. Dichlorodifluoromethane.

RA. 1. Registered Architect. 2. Return Air.

Rabbet. A corner cut out along the corner edge of a piece of wood.

Raceway. Any channel courses supporting and protecting electrical conductors, including conduits, wireways, surface metal raceway, cable trays, floor and ceiling raceways, busways, and cable bus.

Rack and Pinion. A mechanical device consisting of a toothed metal bar (the rack) and a small meshing gear wheel (the pinion).

Rack. 1. A framework or stand to support materials or objects. 2. A toothed bar that meshes with a gear wheel as in a Rack and Pinion.

Racking. 1. Lateral stresses exerted on an assembly. 2. In bricklaying, the method of laying the end of a wall in a series of steps so when work is resumed at a later date, the bond will continue with the old construction easily.

Rack-Type Dishwasher. A dishwasher that accepts dishes which are loaded on metal or plastic racks.

Rad. Radius.

Radial Commutator. Electrical contact surfaces on a rotor which is perpendicular or at right angles to the shaft center line.

Radial Wall Form. A temporary structure or mold for the support of a curved concrete wall while the concrete is curing.

Radial. Moving along lines diverging from a center, like spokes.

Radian. A unit of circular angle, equal to an angle at the center of a circle the arc of which is equal in length to the radius; one radian equals 57.2958 degrees; one degree equals 0.017454 radians.

Radiant Heat Transfer. Radiant heat transfer occurs when there is a large difference between the temperatures of two surfaces that are exposed to each other, but are not touching.

Radiant Heat. Heat transmitted by radiation.

Radiant Heating. A heating system in which warm or hot surfaces are used to radiate heat into the space to be conditioned.

Radiant Panels. Panels with integral passages for the flow of warm air or liquids; heat is transferred to a room by thermal radiation.

Radiation Unit. A determinate quantity adopted as a standard of measurement of radiant heating.

Radiation. 1. Heat transformed by electromagnetic waves. 2. A net exchange of radiant energy between two bodies or objects, across an open space.

Radiator. A heating device that transfers heat from the heated fluid within to the air outside.

Radio Frequency Curing. Of wood adhesives, curing of bondlines by the application of radio frequency energy; also called High Frequency Curing.

Radiology Equipment. Apparatus or devices that use high energy radiation in medicine.

Radius of Gyration. A characteristic of the cross-section of the member used in the determination of its structural characteristics.

Radius. A straight line from the center to the circumference of a circle or sphere; one- half of the diameter.

Radon. A heavy radioactive gaseous element formed by disintegration of radium; it seeps out of the earth and accumulates under buildings, in some cases causing cancer in humans; easily removed by passive ventilation.

Raft. A footing or foundation, usually a large thick concrete mat.

Rafter Anchor. A bolt or metal fastening device which attaches the rafters wall plate.

Rafter Tie. A wood or steel horizontal tension member connecting two opposite rafters to resist their spreading.

Rafter, Common. See Common Rafter.

Rafter, Cripple. See Cripple Rafter.

Rafter, Hip. See Hip Rafter.

Rafter, Jack. See Jack Rafter.

Rafter, Valley. See Valley Rafter.

Rafter. A framing member that runs up and down the slope of a pitched roof; the parallel beams that slope from the ridge of a roof to the eaves and make up the main body of the roof framework.

Raggle Block. A concrete masonry unit with an integral flashing reglet.

Raggle. A groove cut in masonry; any groove provided to receive roofing or flashing.

Rail Standard, Steel. See Steel Rail Standard.

Rail Transit Boarding Platform. A horizontal, generally level surface, whether raised above, recessed below, or level with a transit rail, from which persons embark/disembark a fixed rail vehicle.

Rail, Chair. See Chair Rail.

Rail. 1. A horizontal bar or timber extending from one post or support to another, such as a guard or barrier in a fence or staircase. 2. The horizontal members of the framework of a sash, door, blind, or any paneled assembly. 3. A horizontal support member to support a curtain.

Railing Expansion Joint. Control joint in metal railing to allow differential movement of railing components from thermal expansion and contraction.

Railing Fitting. Elbows, crosses, tees, caps, flanges made of metal, slip type, threaded, or flush welded to railing components to facilitate transitions and terminations.

Railing Flange. Flanges made of metal, slip type, threaded, or flush welded to railing components to facilitate attachment to mounting surfaces.

Railing, Metal. See Metal Railing.

Railing, Pipe. See Pipe Railing.

Railing, Structural Tubing. See Structural Tubing Railing.

Railing. 1. An open fence or guard for safety, made of rails and posts. 2. A banding in cabinetwork. 3. On plywood, the solid wood band around one or more edges.

Railroad Crossing. An grade level intersection at which railroad tracks meet vehicular, pedestrian or other kinds of traffic.

Railroad Tie. A timber support for railroad tracks.

Railroad Work. Construction or repair performed on railroad tracks.

Rain Spots. Defects on paint film caused by rain.

Raintight Enclosure. A building, a component of a building, or a structure that is impervious to the outside elements such as rain or snow.

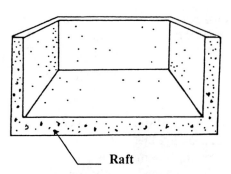

Raft

Raintight Time Clock. A weatherproof time recording clock that has been manufactured to withstand the outside elements, such as rain or snow.

Raintight. So constructed or protected that exposure to a beating rain will not result in the entrance of water.

Raised Grain. A roughened condition of the surface of dressed lumber in which the hard latewood is raised above the softer earlywood but not torn loose from it.

Rake Molding. The cornice on the gable edge of a pitch roof, the members of which are made to fit those of the molding of the horizontal eaves.

Rake. 1. A hand implement with spaced teeth for gathering or loosening material like grass, gravel, or earth. 2. The sloping edge of a pitched roof. 3. The trim of a building extending in an oblique line, as rake dado or molding. 4. The end of a wall that slopes or racks back.

Raked Joint. In brick masonry, a type of joint which has the mortar raked out to a specified depth while the mortar is still green.

Raking Bond. Bricks that are laid in a zigzag fashion.

Ram Air. Air forced through the condenser due to the rapid movement of the vehicle along the highway.

RAM. Random Access Memory.

Ramp. A sloping passageway.

Ranch Molding. An architectural style of molding that is installed as finish work or ornamentation; a type or style of trim which is gently curved and devoid of ornate design.

Random Access Memory. A memory in a computer system that is used for temporary storage of data and allows that data to be accessed at random and to be changed; compare Read Only Memory.

Random Rubble. Masonry wall built of unsquared or rudely squared stones, irregular in size and shape.

Random Width Flooring. Flooring materials of wood with varying widths, often called a plank floor.

Range Hood. An exhaust system installed over a stove to remove fumes by means of a motorized fan and a duct to outside air.

Range, Electric. See Electric Range.

Range, Firing. See Firing Range, 2.

Range, Kitchen. See Kitchen Range.

Range. 1. Pressure or temperature settings of a control. 2. Change within limits. 3. A Kitchen Range.

Rankine. A scale for registering temperature in which 0 degrees represents absolute zero; Rankine corresponds to the Fahrenheit scale, $32° \text{ F} = 491.67°$ Rankine.

Rapid Sand Filter. A swimming pool filter designed to be used with sand as the filter media and for flows not to exceed 3 GPM per square foot in commercial pools and 5 GPM in residential pools.

Rapid Start Lamp. A fluorescent lamp operated by a ballast that provides a low-voltage winding to preheat electrodes and initiate the arc without a starting switch or the application of high voltage.

Rapid Curing (RC) Asphalt. Cutback asphalt composed of asphalt cement and a naphtha or gasoline-type diluent of high volatility.

Rare Gas. See Noble Gas.

Rasp. A coarse file with cutting points instead of lines.

Ratchet. A mechanical device, consisting of a toothed bar and pawl, that allows motion in one direction only.

Rate of Flow. Volume flow per unit of time.

Rate of Growth. The rate at which a tree has increased its amount of wood; this is measured radially in the trunk or in lumber cut from the trunk; the rate is determined from the number of annual growth rings per inch.

Rational Number. Expressible as a ratio of whole numbers, that is, not involving roots or non-terminating decimals; involving only addition, subtraction, multiplication and division and only a finite number of times.

Ratproofing. Various measures taken to keep rodents out of the interior of a building, particularly food preparation or storage areas, consisting mainly of steel mesh barriers above and below grade and packing of oversized pipe and conduit holes with cement mortar or steel wool.

Ravelling. In asphalt paving, the progressive separation of aggregate particles from the surface downward or from the edges inward; ravelling is caused by lack of compaction, construction of a thin lift during cold weather, dirty or disintegrating aggregate, too little asphalt in the mix, or overheating of the asphalt mix.

Raw Glaze. A ceramic glaze compounded primarily from raw constituents; it contains no pre-fused materials.

Raw Linseed Oil. The crude product processed from flaxseed, usually without much further treatment.

Raw Oil. Oil as received from the press or separated from the solvent in the solvent extraction process.

Raw Sienna. 1. An earthy substance containing oxides of iron and usually of manganese. 2. Brownish yellow pigment, used in paint. 3. See Sienna.

Raw Umber. 1. A brown earthy substance containing oxides of iron and manganese. 2. A pigment, darker than ochre and sienna, used in paint.

Raw. Raw linseed oil.

Rayleigh Wave. Forward and vertical vibration of seismic surface wave; see Seismic Wave.

Rays, Wood. See Wood Rays.

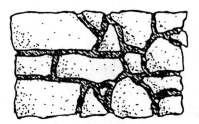

Random Rubble

RBM. Reinforced Brick Masonry.

RC Asphalt. See Rapid Curing Asphalt.

RCP. Reinforced Concrete Pipe.

RCSSB. Red Cedar Shake and Shingle Bureau.

RDF. Refuse Driven Fuel.

Reaching. In using a spray gun, extending a spray stroke too far.

Reactance. That part of the impedance of an alternating current circuit due to capacitance or inductance or both.

Reaction Wood. Wood with more or less distinctive anatomical characters, formed typically in parts of leaning or crooked stems and in branches; in hardwoods this consists of tension wood and in softwoods of compression wood.

Reaction. 1. In structures, the response of the structure to the loads. 2. The force that acts at a support of a structure.

Reactive Aggregate. See Alkali-Aggregate Reaction.

Read Only Memory. A computer memory read only at high speeds but not capable of being changed; compare Random Access Memory.

Ready-Mixed Compound. Factory mixed gypsumboard joint compound in ready-to-use form.

Ready-Mixed Concrete. Concrete manufactured at batch plants and delivered by truck to the job site in a plastic state.

Ready-Mixed. 1. Any material that is mixed prior to delivery to the jobsite. 2. Plaster which is mixed at the mill with mineral aggregate and other ingredients which control time of set; generally used in conjunction with gypsum plasters; also called mill-mixed, pre-mixed.

Real Estate Broker. A broker who deals in real property.

Real Estate. Land that is subject to ownership, with its permanent improvements and appurtenances; also called Real Property.

Real Property. See Real Estate.

Realtor. A real estate agent who is a member of the National Association of Realtors.

Realty. Real property; real estate.

Reamer. A small hand tool for enlarging or shaping holes in wood, metal, or other materials.

Reaming. Removing the burr from inside of a pipe.

Rear Yard. The space between a building and the rear property line.

Reasonable Use. A legal doctrine that allocates water rights.

Rebar. Steel reinforcing bar.

Rebate. A continuous rectangular notch cut out of the corner of a piece of wood; a rabbet.

Receiver Heating Element. Electrical resistance mounted in or around liquid receiver. It is used to maintain head pressures when ambient temperature is low.

Receiver. A person appointed by the court to administer property in foreclosure or a business in reorganization.

Receiver-Drier. Cylinder (container) in a refrigerating system for storing liquid refrigerant and which also holds a quantity of desiccant.

Receptacle, Cable. See Cable Receptacle.

Receptacle, Combination. See Combination Receptacle.

Receptacle. An electric outlet.

Receptor. 1. A metallic or nonmetallic waterproof base for a shower stall. 2. A plumbing fixture, such as a floor sink, that receives the discharge from indirect waste piping.

Recess Filler Strip. A strip of material inserted in a fissure to make flush with the surrounding surface.

Recess. A shallow depth or space in a wall; an alcove or niche.

Recessed Edge. In gypsum board, a sharply reduced caliper along the factory edge; has been replaced by the more gradually reduced caliper of the Tapered Edge. See Taper.

Recessed Light Fixture. A light fixture installed in a suspended ceiling or a recess in a plaster or gypsum board ceiling.

Recessed. Set into a niche or recess in a wall, floor, or ceiling surface.

Reciprocal. 1. Inversely related; opposite; mutually corresponding. 2. One of a pair of numbers whose product is one (8 and 1/8 or 3/4 and 4/3).

Reciprocating Chiller. A compressor (single-acting) using pistons that are driven by a connecting rod from a crankshaft.

Reciprocating Compressor. Compressor which uses a piston and cylinder mechanism to provide pumping action.

Reciprocating Single Piston Pump. A pump having a single reciprocating (moving up and down or back and forth) piston.

Reciprocating. Back and forth motion in a straight line.

Recirculating Overflow. See Surface Skimmer.

Recirculation System. In swimming pool piping, the interconnected system traversed by the recirculated water from the pool until it is returned to the pool.

Recision. The cancellation of a contract.

Reclamation Tank. See Separation Tank.

Recoat Time. In painting, the time interval required between application of successive coats.

Reconstruction. The state of being constructed again.

Reconveyance Deed. A written instrument wherein the trustee of a trust deed conveys property back to the trustor after payments have been paid in full to the beneficiary.

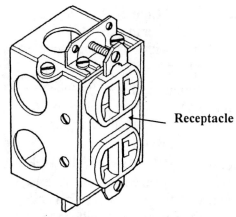

Receptacle

Record Drawings. A set of drawings prepared by the general contractor, which includes any revisions in the working drawings and specifications during construction, indicating how the project was actually constructed; sometimes called As-Built Drawings or As-Builts.

Recording Ammeter. Electrical instrument which uses a pen to record amount of current flow on a moving paper chart.

Recording Thermometer. Temperature measuring instrument which has a pen marking a moving chart.

Recording Voltmeter. A device which records the continuity, power, grounding, voltage and resistance of electrical circuits.

Recording. The process of placing a document on public record by filing in the office of the county recorder; recording a document constitutes constructive notice.

Recourse. 1. The legal right to hold someone liable for payment such as the maker of a negotiable instrument. 2. The provision of some mortgages or trust deeds to hold the borrower personally responsible for the deficit when foreclosure sale proceeds are insufficient to cover the indebtedness.

Recreational Courts. A quadrangular space walled or marked off for playing games with a ball, such as basketball or tennis.

Recreational Facilities. Establishments and structures created for sporting and entertainment purposes, such as courts and pools.

Recreational Use Statutes. Laws that have been adopted in most states to protect property owners from liability to persons who are permitted to use the property for recreational purposes.

Rectangle. A four sided plane figure whose angles are equal; a parallelogram.

Rectifier. A device for converting alternating current to direct current.

Rectilinear. Characterized by straight lines; perpendicular.

Red Cedar Shake and Shingle Bureau (RCSSB). 515 116th Avenue NE, Suite 275, Bellevue, Washington 98004, (206) 543-1323, Fax (206) 455-1314.

Red Fir. Douglas Fir.

Red Label Goods. Flammable or explosive materials with flash points below 80° F. (26.7° C.).

Red Lead. A compound formed by roasting lead or litharge; it used extensively in paints for protecting iron and steel against corrosion.

Red Ochre. An earthy pigment made from impure iron ore, used in paint.

Redan. A part of a medieval fortification consisting of two walls or parapets projecting at an acute angle and open to the rear.

Redemption. The repurchase of one's property after it has been sold in foreclosure; this right to redeem lapses after a specified time.

Reducer. 1. A pipe fitting which connects pipes of different sizes. 2. A tile trim unit used to reduce the radius of a bullnose or a cove to another radius or to a square. 3. Volatile ingredients used to thin or lower viscosity of a finishing material.

Reducing Female Adapter. A pipe fitting which is threaded on the inside of at least one end to receive a different size pipe, to make a connection.

Reducing Flame. An oxygen-fuel gas flame wherein the portion used has a reducing effect; same as Carbonizing Flame.

Reducing Tee. A pipe fitting which joins three pipes of different sizes at 90 degree angles.

Redundant Member. Any member of a truss not required for truss stability.

Redwood Inspection Service (RIS). One Lombard Street, San Francisco, California 94111, (415) 392-7880.

Redwood. A coniferous timber tree with red heartwood and yellow sapwood.

Reel. A round device upon which a hose or wire is wound for easy use and storage.

Re-Entrainment. Situation that occurs when the air being exhausted from a building is immediately brought back into the system through the air intake and other openings in the building envelope.

Reentrant Angle. An angle pointing inward; the opposite of a salient angle.

Re-Entrant Corner. An inside corner of a surface, producing stress concentrations in the roofing or waterproofing membrane.

Refectory. A dining hall.

Refinancing. The reorganization of the loans on a property usually by finding new loans at lower interest rates or larger amounts and paying off the old loans.

Refined Shellac. A grade of orange or white shellac from which the wax has been removed.

Reflectance. The ratio of light reflected by a surface to the incident light falling on the surface.

Reflected Ceiling Plan. An architectural drawing of a ceiling seen as though it is reflected from the floor.

Reflection Cracks. Cracks in asphalt overlays that reflect the crack pattern in the pavement structure underneath; caused by vertical or horizontal movement in the pavement beneath the overlay, brought on by expansion and contraction with temperature or moisture changes.

Reflection Pool. An ornamental, usually shallow, pool of water designed to reflect an important architectural element beyond.

Reflection. The return of light and sound waves from a surface.

Reflective Coated Glass. Glass onto which a thin layer of metal or metal oxide has been deposited to reflect light and/or heat.

Reflective Paint. Paint that has been formulated to reflect light away.

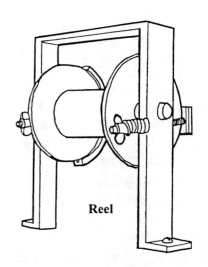

Reel

Reflector Lamp. Flood lamp or spot with bulb coated with a reflecting surface.

Reflector Lighting. A lighting fixture with a polished surface to redirect its light.

Reflectorized Sign. A sign made with highly reflective material.

Refraction. The bending of a light ray as it passes obliquely through a material.

Refractive Index. Ratio of velocity of light in a certain medium compared with its velocity in air under same conditions.

Refractories. Heat-resistant non-metallic ceramic materials.

Refractory Concrete. Concrete having refractory properties, and suitable for use at high temperatures (generally about 315 to 1315° C), in which the binding agent is a hydraulic cement.

Refractory. Capable of enduring high temperature.

Refrigerant Charge. Quantity of refrigerant in a system.

Refrigerant Control. Device which meters flow of refrigerant between two areas of a refrigerating system; can be a capillary tube, expansion valves, or high and low-side valves; maintains pressure difference between high-pressure and low-pressure side of the mechanical refrigerating system while unit is running.

Refrigerant Cylinder. Cylinder in which refrigerant is stored and dispensed; color code painted on cylinder indicates kind of refrigerant.

Refrigerant. Substance used in refrigerating mechanism; it absorbs heat in evaporator by change of state from a liquid to a gas, and releases its heat in a condenser as the substance returns from gaseous state back to a liquid state.

Refrigerated Case. A storage unit using refrigeration to keep food or materials cool.

Refrigerated Truck. Commercial vehicle equipped to maintain below-ambient temperatures in its storage compartment.

Refrigeration Oil. Specially prepared oil used in refrigerator mechanism which circulates, to some extent, with refrigerant.

Refrigeration Tubing. The hollow, cylindrical piping which holds the refrigerant fluid in a refrigeration system.

Refrigeration. The cooling of a material or space.

Refrigerator, Blood. See Blood Refrigerator.

Refuse Driven Fuel (RDF). A method of recovering heat, in an incinerator, from solid waste.

Refuse Hopper. A funnel-shaped receptacle having a device for releasing waste materials through a pipe.

Regional Shopping Center. A large shopping center serving a region.

Register, Wall. A wall mounted grille or damper for the passage of conditioned air.

Register. Combination grille and damper assembly covering an air opening or the end of an air duct.

Registered Architect. An architect who is registered with the state board of architectural examiners of one of the states; a licensed architect.

Reglet. A horizontal or sloping slot in masonry construction, or a sheetmetal slot in plastered wood construction, to accept the upper edge of counterflashing.

Regular Core Gypsumboard. Gypsumboard for general construction purposes without special core additives.

Regulations. Laws that are enacted by a public administrative agency, rather than by an elected legislative body.

Regulator. A device for controlling the delivery of gas at some substantially constant pressure regardless of variation in the higher pressure at the source.

Rehabilitation. The remodeling of a building to render it suitable for modern uses.

Rehydrate. To provide water to something that is dehydrated.

Reinforced Brick Masonry (RBM). Brickwork into which steel bars have been embedded in the mortar to impart tensile strength to the construction.

Reinforced Concrete Pipe. Factory-made precast piping

Reinforced Concrete. Concrete containing adequate reinforcement (prestressed or not prestressed) and designed on the assumption that concrete and steel act together in resisting forces; concrete work into which steel bars have been embedded to impart tensile strength to the member.

Reinforced Grouted Masonry. Wall construction consisting of brick or block that is grouted solid throughout its entire height and has both vertical and horizontal reinforcing.

Reinforced Hollow Unit Masonry. Wall construction consisting of hollow masonry units in which part of all of the cells are continuously grouted and has both horizontal and vertical reinforcing.

Reinforced Masonry. Unit masonry in which reinforcement is embedded in such a manner that the two materials act together in resisting forces.

Reinforced Membrane. A roofing or waterproofing membrane reinforced with felts, mats, fabrics or chopped fibers.

Reinforced Plastics. Plastics reinforced with glass or carbon fibers.

Reinforcement, Mesh. See Welded Wire Fabric.

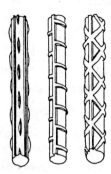

Reinforcement

Reinforcement. 1. The action or state of strengthening by additional assistance, material, or support. 2. The action or state of strengthening or increasing by fresh additions. 3. Steel bars or wires embedded in concrete and located in such a manner that the metal and the concrete act together to resist loads.

Reinforcing Accessory. Items used to install reinforcing in concrete, including, but not limited to, chairs, couplings, and tie wire.

Reinforcing Bar. A manufactured, usually deformed, steel bar, used in concrete and masonry construction to resist tensile stresses.

Reinforcing, Masonry. See Masonry Reinforcing.

Reinforcing. 1. The act of making stronger. 2. A structure or member that is used to add strength. 3. Steel rods or metal fabric placed in concrete slabs, beams, or columns to increase their strength.

REL. Recommended Exposure Limits; recommendations made by NIOSH.

Relative Humidity. The ratio of the amount of water vapor actually present in the air to the amount present when the air is saturated with water vapor at the same temperature, expressed as a percentage.

Relay, AC. See AC Relay.

Relay. An electromagnetic mechanism, moved by a small electrical current in a control circuit, which operates a valve or switch in an operating circuit.

Release Agent. Material used to prevent bonding of concrete to a form surface.

Release Clause. A provision in a mortgage or trust deed which allows a portion of the encumbered property to be released from the mortgage upon the payment of a specified sum of money to the lender.

Reliction. An increase in land area caused by the gradual recession of sea or river water.

Relief Joint. See Control Joint and Expansion Joint.

Relief Valve. A pressure safety valve installed on water heaters, hot water and boiler tanks to relieve pressure when it exceeds a preset level.

Relief Vent. In a DWV system, a vent coming from a vent stack and connecting to a horizontal branch at a point between the first fixture connection on that branch and the soil stack or waste stack; its purpose is to allow the circulation of air between the vent stack and the soil or waste stack.

Relief Ventilator. An auxiliary vent whose main purpose is to provide supplementary circulation of air.

Relief. 1. Damages or court orders awarded in litigation. 2. Ornamented prominence of parts of figures above a plane surface.

Relieving Arch. See Discharging Arch.

Reline Pipe. To install new linings in pipes; usually includes the cleaning of built-up scale or debris from the existing pipe and relining with a compatible material.

Reluctance. A force working against the passage of magnetic lines of force (flux) through a magnetic substance.

Remainder. The estate in real property that remains after a life estate terminates upon the death of the holder.

Remainderman. The person entitled to the estate that remains after termination of a life estate.

Remaining Economic Life. The time period from the present to the point where a real property improvement becomes valueless.

Remote Control Circuit. Any electrical circuit which controls any other circuit through a relay or an equivalent device.

Remote Power Element Control. Device with sensing element located apart from the operating mechanism it controls.

Remote System. Refrigerating system in which condensing unit is away from space to be cooled.

Removable Railing Section. Section of a railing designed and constructed to be removed easily to facilitate movement through the railing.

Relieving Arch

Removable Window Well Cover. Removable steel bar or expanded metal cover for a window well.

Remove. The act or process of demolishing, dismantling, and carrying away an old fitting or component.

Removers. Compositions designed to soften old varnish or paint coats so that they may be easily removed by scraping or washing.

Renaissance Cities. Formal plans of a symmetric nature; hierarchical articulation of spaces between more prominent and less prominent parts of the cities, such as palaces, monuments, and other focal points.

Renaissance. 1. Rebirth; revival. 2. The art and architecture developed during the period of 14-17th century; the transition between medieval and modern times, marked by the revival of classical styles and the beginnings of modern science.

Rendering. 1. An architectural drawing, often in color and in perspective. 2. Application of plaster directly to a masonry wall.

Rental Value. A measure of damages for deprivation of the use of property.

Repair. The reconstruction or renewal of any part of an existing building for the purpose of its maintenance.

Repeat. In carpet, fabric, and wallpaper, the distance from a point in a pattern figure to the same point where it occurs again, measuring lengthwise of the material.

Replacement Reserves. A portion of the annual income set aside for replacement of wasting assets, such as roofing, mechanical and electrical equipment, and other impermanent building components.

Replevin. A legal process whereby personal property that has been unlawfully taken from one may be recovered.

Replication. A copy or reproduction.

Reprographics. Systems for reproducing graphic materials, including architectural and engineering drawings and diagrams.

Repulsion-Start Induction Motor. An electric motor type which has an electrical winding on the rotor for starting purposes.

Required Strength. The strength of a member or cross section required to resist factored loads or related internal moments and forces.

Reredos. 1. An ornamental wall behind an altar in a church. 2. The back of a fireplace or open hearth.

Reproofing. The practice of applying new roofing materials over existing roofing materials.

Res Judicata. The doctrine that courts will not re-litigate the same issues between the same parties.

Resawing. Sawing lumber again after the first sawing; specifically, sawing into boards or dimension lumber.

Rescission. A legal remedy for a material breach of contract under which the law pretends that a contract never existed and attempts to put the parties in the positions they occupied before the contract was executed; a cancellation.

Reserves. Portions of current income set aside and accumulated at interest to pay for future maintenance, repairs, and replacement of equipment.

Residence. A building for human habitation; a dwelling.

Residential Carpet. A grade or style of carpet installed in low foot-traffic applications.

Residential Equipment. An appliance or piece of equipment found in the home, that is used by the homeowner.

Residential Occupancy. A building or any portion of a building containing one or more dwelling units.

Residual Stress. Stress remaining in a structure or member as a result of thermal or mechanical treatment or both.

Residual, Building. See Building Residual.

Residual, Land. See Land Residual.

Residual. 1. The remainder or residue after a chemical reaction or a transaction. 2. Free acting disinfectant remaining in poolwater after treating and breaking down pollution.

Resilience. 1. Ability of a material to resume its original size and shape after deformation, such as stretching, twisting, compression, or indentation. 2. The measurement of the absorption of dynamic energy by a structure without permanent deformation or fracture. 3. The property whereby a strained body gives up its stored energy on the removal of the deforming force. 4. The ability of a carpet fabric or padding to spring back to its original shape or thickness after being crushed or walked upon.

Resilient Channel. A metal furring member designed to absorb sound or noise impact which strikes the surfacing membrane.

Resilient Floor Covering Institute (RFCI). 1030 15th Street, NW, Suite 350, Washington, DC 20005, (202) 833-2635.

Resilient Flooring. Flooring materials such as asphalt, vinyl, linoleum, rubber, cork, and similar resilient materials; available in tile or sheet form.

Resin Hardness. Method of indicating hardness of resins; usually from No. 1 (hardest) to No. 6.

Resin. A sticky material obtained from the sap of certain trees and plants (natural resin) or made synthetically from coal-tar products and other organic substances (synthetic resin); resins are widely used in making varnishes and paints.

Resist Printing. Placing a dye-resist agent on carpet prior to piece dyeing so that the pile will absorb color according to a predetermined design.

Resistance or R-Value. The tendency of a material to retard the flow of heat.

Resistance. 1. An opposing or retardant force. 2. An opposition to flow or movement such as friction. 3. The property of hindering the conduction of electricity or heat.

Resistivity (r). The property of hindering the conduction of heat or electricity; the reciprocal of conductivity, $1/k$.

Resistor. Electrical device having resistance to the passage of electrical current.

Resolving Forces. Replacing a force or forces with two or more other forces that yield the same effect on a structure as the original forces.

Resonance. Movement produced in an element as a result of movement added to its natural period of movement, which is the same.

Resorcinol Glue. A glue that is high in both wet and dry strength and resistant to high temperatures; used for gluing lumber or assembly joints that must withstand severe service conditions.

Respirator Mask. A device worn over the mouth to aid in breathing.

Respirator. A device for maintaining artificial respiration.

Respondiat Superior. The doctrine that a principal is liable for the acts of an agent, and an employer for the acts of an employee.

Response. Effect produced on a structure by earthquake ground motion.

Restaurant. A room or building for preparing and serving meals to the public.

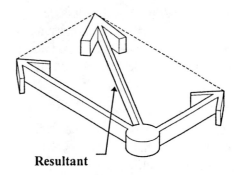

Resultant

Restoration. The act of restoring to an original condition; the act or process of bringing a structure back to its former position or condition.

Restretch. A term applied to the remedial steps necessary for the correction of improperly laid carpet resulting from application of wrong stretching techniques, carpet defects, or undetermined causes.

Restrictor. A device for producing a deliberate pressure drop or resistance in a line by reducing the cross-sectional flow area.

Restroom. A room in a public place equipped with toilets, urinals, and lavatories, usually segregated by gender.

Resultant. A force that will produce the same effect as two or more combined forces.

Retainage. Amount of construction sum held back by the owner from the contractor to be paid after construction is satisfactorily completed; also called Retention.

Retained Earnings. The accumulated earnings that are retained in a corporation since it was founded less the total of dividends declared to the stockholders.

Retaining Wall. A wall that is designed to resist the lateral pressures of retained soil; a wall that holds back a hillside or is backfilled to create a level surface.

Retardant, Fire. See Fire Retardant.

Retardation. Delaying the hardening or strength gain of fresh concrete, mortar or grout.

Retarder. 1. A concrete admixture used to slow down the natural curing process. 2. An admixture used to delay the setting action of plaster; generally used only with gypsum plasters or finish coat plaster containing calcined gypsum gauging. 3. Slow drying solvents or extenders added to lacquer to delay drying of the lacquer.

Retemper. Addition of water to portland cement plaster, mortar, or concrete after mixing but before setting process has started.

Retention. Retainage.

Retrofitting. To install new materials or equipment in an existing building.

Return Air Duct. The duct by which air is returned from a room or space to the heating or cooling system to be reconditioned and re-used.

Return Lines. 1. Pipework returning water to be recooled or reheated. 2. Pipework returning filtered water to the pool.

Return Off-Set. See Jumpover.

Return on Investment. The percentage of earnings on the total assets of a corporation.

Return Period of Earthquakes. The time period in which the probability that an earthquake of a certain magnitude will reoccur.

Return Piping. That part of the piping between the filter and the pool, spa or hot tub through which passes the filtered water.

Return. 1. Any surface turned back from the face of a principal surface. 2. The ending of a small splash wall or a wainscot at right angle to the major wall. 3. The continuation of a molding or finish of any kind in a different direction. 4. In HVAC, a term for the return-air duct of a forced air heating/cooling system. 5. The continuation in a different direction of the face of a building or any member.

Rev. 1. Revised. 2. Revision. 3. Reverse. 4. Reversed.

Reveal. The side of an opening, as a window or door jamb.

Revenues. The gross income of a business.

Reverberation Time. The time is takes sound to become inaudible; a function of a space, people, furnishings, and the absorptivity of the surfaces.

Reverberation. The persistence or echoing of previously generated sound caused by reflection of acoustic waves from the surfaces of enclosed spaces; the support of sound by successive reflections.

Reverse Coloring. The changing of yarn frames in Jacquard weaves to cause the interchanging of ground and top colors, according to customers' preferences.

Reverse Cycle Defrost. 1. Method of heating evaporator for defrosting purposes. 2. Valves to move hot gas from compressor into evaporator.

Reverse Engineering. The study and analysis of a product in order to gain information about its design, construction, and use, for designing a new improved version.

Reverse Thrust Fault. A geological fault under compression where the overlying block moves up the dip of the fault plane.

Reversible Lock. A lock that will function on doors of either hand, sometimes with slight adjustments.

Reversing Valve. Device used to reverse direction of the refrigerant flow depending upon whether heating or cooling is desired.

Reversionary Clause. A provision in a deed that provides for the property to revert to the grantor in the event that any of the restrictions are violated.

Revibration. The delayed vibration of concrete that has already been placed and consolidated; most effective when done at the latest time a running vibrator will sink of its own weight into the concrete and again make it plastic.

Revolving Door. Typically a four panel door attached at 90 degrees to each other that turns on a center axis; some are 3 panel attached at 120 degrees to each other.

Reynolds Numbers. A numerical ratio of the dynamic forces of mass flow to the shear stress due to viscosity.

RF Curing. Radio Frequency curing.

RFCI. Resilient Floor Covering Institute.

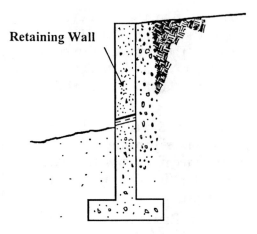

Retaining Wall

RFP. Request for Proposal.

RH. 1. Right hand. 2. Right Hand, a door handing designation.

Rheology. A science dealing with the deformation and flow of matter.

Rheostat. An electrical instrument used to control a current by varying the resistance.

RHMS. Round Head Machine Screw.

Rhombus. A polygon with four equal sides that do not meet at right angles.

RHR. Right Hand Reverse, a door handing designation.

RHWS. Round Head Wood Screw.

Rib Lath. Metal lath with ribs formed by folding to provide stiffness for greater span.

Rib. 1. A plain or molded arched member which forms a support for an arch or vault. 2. A decorative feature resembling a rib on the surface of a vault or ceiling; a projecting piece of molding upon the interior of a vault or used to form tracery in ornamental work.

Ribband. 1. The support for the second-floor joists of a balloon-frame house. 2. The horizontal member of a wood frame wall which supports floor joists or roof rafters; also called a Ribbon.

Ribbed Shells. Spatial structures with material placed primarily along certain preferred rib lines, with the area between the ribs filled with thin slabs or left open.

Ribbon. See Ribband.

Richter Scale. A logarithmic scale that is used to measure the magnitude of earthquakes.

Riddle. A coarse sieve, as for sand or gravel.

Ride the Brush. To bear down on the brush to the extent that the paint is applied with the sides of the bristles instead of the flat ends; this shortens the life of the brush.

Ridge Board. The board against which the tips of rafters are fastened; the top line of a roof; the ridge.

Ridge Cut. Any cut made in a vertical plane; the vertical cut at the top end of a rafter.

Ridge Rib. A rib forming the ridge in a vaulted roof.

Ridge Vent. A vent mounted along the ridge of a roof to aid in ventilating an attic space.

Ridge, Roof. See Roof Ridge.

Ridge. 1. The top edge or corner formed by the intersection of two roof surfaces. 2. The board against which the tips of rafters are fastened.

Ridging. 1. An upward, tenting displacement of a roofing membrane, frequently occurring over insulation joints, deck joints, and base sheet edges. 2. See Beading.

Riffler. A small handtool for filing, scribing, or scraping.

Rifle Range. A place where rifle shooting is practiced.

Rift Valley. A graben.

Right Angle. An angle of 90 degrees.

Right Triangle. A triangle having one right angle.

Right-of-Way. The legal right, usually by easement, to pass over land owned by another.

Rigid Copper Tubing. Hard copper pipe used when installing water lines, particularly where they can be seen.

Rigid Frame. Two columns and a beam or beams attached with moment connections; a moment-resisting building frame; the connections between the beam and columns are constructed so as to transmit moments.

Rigid Insulation. Thermal insulating material made of polystyrene, polyurethane, polyisocyanurate, cellular glass, or glass fiber, sometimes offered with a skin surfacing, formed to flat board shape of constant thickness.

Rigidity. Relative stiffness of a structure or element.

Rim. A designation of articles of hardware designed for application to the surface of doors and windows, such as rim locks.

Ring Nail. See Annular Ring Nail.

Ring, Split. See Split Ring.

Ring, Toothed. See Toothed Ring.

Ringelmann Sale. Device for measuring smoke density.

Rip Fence. A guide on a table saw to assist in cutting lumber in the direction parallel to the grain.

Rip. To cut lumber lengthwise, parallel to the grain.

Riparian Rights. The rights of an owner whose land adjoins water, such as a river, stream, lake, or sea.

Ripper. Narrow strips of gypsum board used for soffits, window returns and finished openings.

Ripping. Sawing wood along the grain.

Ripple Effect. The effect that escalation or delay has on other work in a construction schedule; impact.

Riprap. A foundation or sustaining wall of stones placed together without order, as in deep water or on an embankment, to prevent erosion.

Ripsaw. A saw designed for cutting lumber length wise.

RIS. Redwood Inspection Service.

Rise. The vertical distance through which anything rises, as the rise of a roof or stair.

Riser Band. A flat, horizontal member usually ornamental, of relatively slight projection, mounted on a riser between two stair treads.

Riser Hanger. The horizontal member of a stair structure which supports the riser.

Riser Valve. Device used to manually control flow of refrigerant in vertical piping.

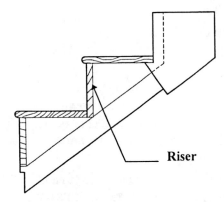

Riser

Riser. 1. A single vertical increment of a stair; the vertical face between two treads in a stair. 2. A vertical run of plumbing, wiring or ductwork. 3. A spacer between shipping units to allow entry of fork lift blades. 4. The flow channel or pipe that distribute the heat transfer liquid across the face of an absorber.

Risk Management. The taking of precautions to avoid unexpected property and casualty losses which can severely damage the business's prospects; insurance is one of the best means of reducing the company's exposure to risk.

Rising Damp. Moisture in concrete or masonry construction that is absorbed from the earth and rises through capillarity.

Risk. Chance or possibility of danger, loss, injury, or other adverse consequences.

Rivet Pattern. The arrangement, spacings, and dimensions of rivets used to attach two or more structural members together.

Rivet. A structural fastener on which a second head is formed after the fastener is in place.

RMA. Rubber Manufacturers Association.

Ro. Rough.

Road Oil. A heavy petroleum oil, usually one of the Slow-Curing (SC) grades.

Roadway Excavation. The act or process of removing or moving obstructions such as boulders, rocks, earth to prepare a path for a road or highway.

Robot. A mechanical device that is operated by automatic controls and is used in manufacturing processes to perform functions previously carried out by human workers.

Robotics. The science concerned with the design, manufacture, and utilization of robots.

Rock Anchor. A post-tensioned rod or cable inserted into a rock formation for the purpose of anchoring a structure to it.

Rock Asphalt. See Asphalt Rock.

Rock Drilling. The act or process of boring holes into rock.

Rock Flour. Crushed rock in fine particles like powder.

Rock Gun. A device for throwing aggregate onto a soft bedding coat in applying marblecrete.

Rock Pocket. A concrete defect in which a group of aggregate is not adequately surrounded by cement paste.

Rock. The hard, firm, and stable parts of earth's crust; any material which requires blasting before it can be dug by available equipment.

Rockwool Insulation. Insulation made by forming fibers from molten rock and slag; used as insulation in ceilings and walls.

Rococo. An artistic style of the 18th century that is characterized by fanciful, excessive, intricate, and elaborate ornamentation; baroque.

Rod Saw. A saw consisting of a steel rod approximately 1/8" in diameter used to cut circles or irregular curves in tile; the rod has tungsten carbide particles embedded in the surface.

Rod, Ground. See Ground Rod.

Rod, Guy. See Guy Rod.

Rod, Hanger. See Hanger Rod.

Rod, Tamping. See Tamping Rod.

Rod, Traverse. See Traverse Rod.

Rod. 1. A slender stick, pole, or bar of wood or metal. 2. A unit of length, used in land measuring, of 16.5 feet. 3. A square rod, 273.9 square feet.

Rodding Off. See Floating.

Roll Roof. A roof that has been covered with an asphaltic material that comes in rolls.

Roll Roofing. Asphaltic roofing, coated felts, either smooth or mineral surfaced, that comes in rolls;

Rolled Steel Section. Any hot-rolled steel member such as a wide flange beam, I-beam, angle, channel, tee, or similar sections.

Rolled. A brick that is laid with an overhanging face.

Roller Blind. A window covering that retracts to the top of the window by means of a spring-loaded roller.

Roller Coating. 1. A paint or other material applied to the surface by a paint roller. 2. Process of finishing an article by means of hard rubber or steel rollers.

Roller Latch. A friction door latch employing a roller under spring tension which engages a recessed strike plate formed to receive the roller.

Roller Marks. Surface defects caused during the manufacturing process as the gypsum board is moved through the various production operations.

Roller Shade. A window shade on spring-loaded a roller that allows the shade to go up and down.

Roller Strike. A door latch strike having a rolling member at the point of latch bolt contact to minimize friction.

Roll-Up Door. A door which raises and rolls into a coiled configuration and lowers on tracks on either jamb.

Rolok Wall. An economy wall and a substitute for a solid brick wall, it is a combination of several shells of brickwork laid flat or on edge to produce a cellular type of open construction.

ROM. Read Only Memory.

Roman Arch. A round topped arch.

Roman Brick. Brick whose nominal dimensions are 3-1/2 x 2 x 12 inches.

Roman Cities. Influenced by the plan used for military camps, grid iron city blocks; more formal organization of public spaces in the nucleus of city.

Roman Numeral. 1. Any of the Roman letters representing numbers, I = 1, V = 5, X = 10, L = 50, D = 500, M = 1000. 2. A numeral in the system of notation that is based on the ancient Roman system.

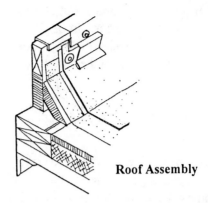

Roof Assembly

Romanesque. A style of architecture prevalent in Europe in the 9-12th century, with massive vaulting, round arches, decorative use of arcades, and profuse ornament.

Rood Screen. A decorative screen separating the nave from the chancel.

Roof Adhesive. A bonding agent used to cement roof materials.

Roof Assembly. An assembly of interacting roof components, including the roof deck, designed to weatherproof and, normally, to insulate a building's top surface.

Roof Ballast. Crushed rock or gravel which is spread on a roof surface to form its final surface.

Roof Bond. A legal guarantee that a roof installed is in accordance with specifications and will be repaired or replaced if it fails within a certain period of time due to normal weathering.

Roof Cricket. Relatively small elevated area of wood roof constructed to divert water around chimney, curb, or other projection.

Roof Decking. 1. Two inch or thicker lumber installed on a roof system. 2. The structural skin of a roof.

Roof Drain. 1. A drainage outlet on a roof. 2. A drainage conduit or pipe that collects water runoff from a roof and leads it away from the structure.

Roof Expansion Joint. A break or space between roofing materials to allow for thermal expansion and contraction.

Roof Flashing. A weather shield commonly used at penetrations of a roof by pipes or columns to insure no passage of the elements.

Roof Framing. The skeleton structure of a roof system.

Roof Hatch. Fabricated, usually of metal, horizontal access door mounted on curb to permit access to flat or low sloped roofs from the building interior.

Roof Insulation. 1. Materials used between rafters or roof supports for the protection from heat and cold. 2. Solid sheets of insulating material installed on a flat roof.

Roof Panel. Roof panel construction from boards, plywood, or metal sheets that can be installed as a pre-assembled unit.

Roof Ridge. The uppermost peak of a roof where the roof planes meet.

Roof Scupper. See Scupper, 1.

Roof Sheathing. The first layer of covering on a roof, fastened to the rafter boards, used to support the roofing material.

Roof Shingle. Pieces of wedge shaped wood or other material used in overlapping courses to cover a roof, usually made of wood, asphalt, clay tile, or metal.

Roof Slab. The flat section of a reinforced concrete roof, supported by beams, columns, or other framework.

Roof Specialties. A fitting or piece of trim used in the installation of a roof, such as gravel stop, flashing, or vent strips.

Roof Spinner. An exhaust ventilation apparatus mounted on a roof which uses the wind to create an upward draft and thus create ventilation.

Roof System. A general term for the complete assembly of waterproof covering, insulation, vapor barrier, ballast, and substrate.

Roof Vent. 1. A usually screened opening on a gable end or roof structure to allow a free flow of ventilation air into the underroof area. 2. The unit mounted on a roof that is the terminus of a ventilation pipe.

Roof Walkway. A permanent aisle for safe access across a roof, also serves as a protection for the roofing material when maintenance is being done.

Roof, Asphalt. See Asphalt Roof.

Roof, Built-Up. See Built-Up Roof.

Roof, Curb. See Gambrel Roof.

Roof, Fluid-Applied. See Fluid-Applied Roof.

Roof, Hip. See Hip Roof.

Roof. 1. The top cover of a building or structure. 2. The top limit on price, a Ceiling Price.

Roofer. A person who constructs or repairs roofs.

Roofing Cement. A general term for a variety of trowelable mastics, asphalt, or tar that are used in roof construction and repair.

Roofing Consultant. An expert in the design, specification, construction, and repair of roofs.

Roofing Felt. A felt sheet impregnated with a bituminous waterproofing material.

Roofing Nail. A short, often galvanized, nail with a broad head.

Roofing Removal. The act or process of stripping off an old roof system.

Roofing Square. 1. An area of 10 by 10 feet of roofing surface. 2. 100 square feet of roofing surface.

Roofing System. A group of interacting roof parts designed to weatherproof and to insulate a roof.

Roofing. 1. Materials for a roof. 2. The process of constructing a roof. 3. The skilled trade of applying roofing; a Roofer.

Rooftop AC. Air-conditioning system or equipment installed on a roof.

Rooftop Heater. A heating unit installed or placed on a roof.

Rooftop Unit. A system of conditioning air in a self contained unit, installed directly above the conditioned space, in a weatherproof enclosure.

Room Resonance. The sound resonance of an enclosed space, which depends in the relationship between the sound wavelength, the space dimensions, and the reflective surfaces.

Rooming House. A hotel.

Root Ball. The soil and earth that clings to roots of a dug up tree that is later to be planted in a different location.

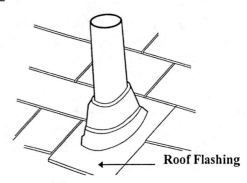

Roof Flashing

Root Buttress. A root that is above ground where it joins the trunk of a tree.

Root Face. In welding, that portion of the groove face adjacent to the root of the joint.

Root Hook. A very heavy hook designed to catch and tear out big roots when it is dragged along the ground.

Root of Joint. That portion of a joint to be welded where the members are closest to each other.

Root of Weld. The point at which the bottom of the weld intersects the base-metal surfaces and is the farthest from application of weld heat and/or filler metal side.

Root Opening. In welding, the separation between the members to be joined at the root of the joint.

Rope. A stout cord made by twisting or braiding together strands of hemp, sisal, flax, cotton, nylon, wire, or similar material.

Rose. A trim plate attached to the door hardware behind the knob.

Rosette. 1. A circular decorative ornament made of wood, metal, or plaster. 2. A metal trim at a ceiling mounted lighting fixture or at a blanked-off lighting fixture outlet.

Rose-Window. A round window filled with tracery and, usually, stained glass; characteristic of Gothic churches.

Rosewood Veneer. A thin layer of cabinet wood with a dark red or purplish color streaked and variegated with black, applied to give a superior and decorative surface.

Rosin. A hard resin obtained from pine trees containing principally isomers of abietic acid; wood rosin is obtained from stump or dead wood, using steam distillation; gum rosin is obtained from the living treerosin is used in making varnish, soap, and solder flux.

Rot. Decay.

Rotary Blade Compressor. Mechanism for pumping fluid by revolving blades inside a cylindrical housing.

Rotary Compressor. See Compressor, Rotary.

Rotary-Cut Veneer. Veneer cut in a lathe which rotates a log or bolt, chucked in the center, against a knife.

Rotate. To turn or revolve about an axis or center.

Rotation. In building structures, the torsional movement of a diaphragm about a vertical axis.

Rotisserie. 1. A restaurant specializing in roasted and barbecued meats. 2. A cooking device with a revolving spit for roasting and barbecuing meat.

Rotor. A rotary part of a machine.

Rottenstone. A siliceous limestone which when finely pulverized, is used in wood finishing; it has negligible cutting action but is fine for polishing; also called Tripoli.

Rough Horse. See Carriage.

Rough Lumber. Lumber that has not been dressed (surfaced) but which has been sawed, edged , and trimmed.

Rough Opening. The clear dimensions of the framed opening that must be provided in a wall to accept a given door, window unit, or other unit.

Rough-In. All work that must be performed before plastering.

Round Edge. A preformed factory rounded edge design.

Round Head Wood Screw (RHWS). A wood screw with a rounded head protruding above the surface; driven with a screw driver; usually installed with a washer under the head.

Round Pattern. Circular spray pattern.

Round Wire. See Loop Pile.

Round, Quarter. See Quarter Round.

Round. 1. Circular. 2. A wooden dowel.

Rout. To gouge out or make a furrow in, as in wood.

Router. Woodworking tool with a high speed bit for cutting rebates, dadoes, insetting butts, and forming decorative or rounded edges.

Row House. One of a series of houses connected by common sidewalls and forming a continuous group.

Rowlock Course. A course of headers laid on their edges instead of flat side; rows of brick laid on edge especially used for the ring of an arch.

Rows. The number of lengthwise yarn tufts in one inch of Axminster or Chenille carpet; compare with Wires.

RPM. Revolutions Per Minute.

RTU. Roof Top Unit.

Rub Joint. A glue joint made by carefully fitting the edges together, spreading glue between them, and rubbing the pieces back and forth until the pieces are well rubbed together.

Rubber Emulsion Paint. Paint with a vehicle of fine droplets of natural or synthetic rubber dispersed in water.

Rubber Manufacturers Association (RMA). 1400 K Street, NW, Washington, DC 20005, (202) 682-4800

Rubber Nosing. The projection of a rubberized stair tread over a riser in a stair system.

Rubber Set. See False Set.

Rubber Spacers. Cross and tee-shaped objects used to space tile on floors or walls; manufactured in thicknesses of 1/16", 1/8", 1/4", 3/8", and 1/2".

Rubber Trowel. A non-porous synthetic-rubber-faced float mounted on an aluminum back with a wood handle, used in tile grouting to force material deep into tile joints and to remove excess grout.

Rubber, Laminated. See Laminated Rubber.

Rubber. A tough elastic material made from the latex of rubber trees, a natural polymer of a hydrocarbon.

Rubbing Compound. An abrasive material used to produce a smooth finished wood surface.

Rubbing Oil. Neutral, medium-heavy mineral oil used as a lubricant for pumice stone in rubbing varnish and lacquer.

Rubbing Stone. A carborundum stone that is used to smooth the rough edges of tile.

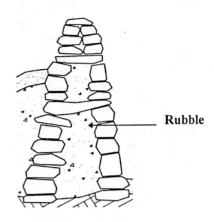

Rubble

Rubbing Varnish. A hard-drying varnish which may be rubbed with an abrasive and water or oil to a uniform leveled surface.

Rubbish Handling. The act or process of transporting or removing waste materials.

Rubbish Removal. The process of removing waste or garbage from an area.

Rubble Masonry. Uncut stone, used for rough work, foundations, backfilling, and the like.

Rubble. 1. Roughly broken quarry stone. 2. Rough broken stones or bricks used to fill in courses of walls or for other filling;

Rug. A term used to designate soft floor coverings laid on the floor but not fastened to it; usually a rug does not cover the entire floor.

Rule-of-Thumb. A rule for general guidance based on experience rather than on precision or scientific theory.

Ruler. A straight, usually graduated, piece of wood, plastic, or metal, used for drawing straight lines or for measuring.

Run. 1. In a stairway, the width of a step, measured from the face of one riser to the face of the next, and not including the nosing. 2. The horizontal distance covered by a flight of steps. 3. The length of the horizontal projection of a sloping member such as a rafter when in position. 4. A temporary planked scaffold for walking or wheeling materials. 5. That portion of a pipe or fitting continuing in a straight line in the direction of flow in the pipe to which it is connected. 6. An appreciable length of straight or nearly straight piping.

Rung. A step on a ladder.

Runner Channel. A steel channel member from which furring channels and lath are supported in a suspended plaster ceiling.

Runner. Metal or wood track or strips placed at floor and ceiling to receive framing members, such as metal or wood studs.

Runners. See Main Runners.

Running Bond. Brickwork consisting entirely of stretchers, lapping of units in successive courses so that the vertical head joints lap.

Running Mold. Constructed by plasterers in place by use of a metal pattern to form a molding of fresh plaster.

Running Time. Amount of time a condensing unit is run per hour or per 24 hours.

Running Track. An oval-shaped track for use by athletes in various running events.

Running Trap. A plumbing trap in which the inlet and outlet are in a straight horizontal line between which the water way is depressed to below the bottom of either the inlet or outlet.

Running Winding. Electrical winding of motor which has current flowing through it during normal operation of motor.

Runoff. The excess of rainwater or snowmelt that is not absorbed into the earth and drains to streams, rivers, or storm water collection devices.

Runs. In painting, curtains, sags, and other irregularities due to uneven flow.

Rural Letter Box. A roadside receptacle usually found in rural areas for delivery of mail.

Rust. A reddish or yellowish-brown coating formed on iron or steel caused by oxidation, usually as a result of moisture; ferrous oxide.

Rusticated Concrete. Beveled edges of concrete making the joints conspicuous.

Rustification. In building and masonry, the use of squared or hewn stone blocks with roughened surfaces and edges deeply beveled or grooved to make the joints conspicuous.

Rust-Inhibitive Washes. Solutions which etch the metal and form a dull gray coating of uniformly fine texture, thus producing rust-inhibitive surface receptive to priming coat.

Rutile. A mineral form of titanium oxide, TiO_2, tetragonal crystallization, but usually produced chemically for use in ceramics and other products.

R-Value. The number of minutes (seconds) required for 1 Btu (joule) to penetrate one square foot (square meter) of a material for each degree of temperature difference between the two sides of the material. The resistance of a material to the passage of heat. The reciprocal of conduction.

RW. Redwood.

Rwd. Redwood.

S

S&P. Shelf and pole.

S1E. Lumber, surfaced one edge.

S1S. Lumber, surfaced one side.

S1S1E. Lumber, surfaced one side and one edge.

S1S2E. Lumber, surfaced one side and two edges.

S2E. Lumber, surfaced two edges.

S2S. Lumber, surfaced two sides.

S4S. Lumber, surfaced four sides.

SA. Supply Air.

Sabin. The unit of acoustic absorption; one sabin is the absorption of one square foot of perfect sound-absorbing material.

Sack Joint. Joint that has been wiped or rubbed with a rag or object such as a rubber heel.

Sack. A quantity of cement, 94 pounds, I cubic foot, in the United States for portland or air entraining portland cement or as indicated on the sack for other kinds of cement.

Sacking. Removing or alleviating defects on a concrete surface by applying a mixture of sand and cement to the moistened surface and rubbing with a coarse material such as burlap.

Saddle Board. The finish of the ridge of a pitched roof house; also called Comb Board.

Saddle Fitting. A fitting used to install a branch line from an existing run of pipe; first a hole is made in the pipe, then the saddle is clamped around the hole.

Saddle Tie. A specific method of wrapping hanger wire around main runners or of wrapping tie wire around the juncture of main runner and cross furring.

Saddle Valve. Valve body shaped so it may be silver brazed or clamped on to a refrigerant tubing surface; also called Tap-a-Line.

Saddle. 1. A built up section of the roof substrate to divert water toward the drains. 2. See Threshold, 2.

Safe, Office. See Office Safe.

Safety Arch. See Discharging Arch.

Safety Belt. A solid belt that may be attached to a secured lifeline.

Safety Can. Approved container of not more than 5-gallon capacity, with a spring-closing lid and spout cover; designed to relieve internal pressure safely when exposed to fire.

Safety Chain. Chain installed horizontally in railing assembly to provide for ease in providing temporary opening in railing.

Safety Control. Device which will stop the refrigerating unit if unsafe pressures, temperatures, or dangerous conditions are reached.

Safety Factor. See Factor of Safety.

Safety Glass. Specific type of glass having the ability to withstand breaking into large jagged pieces, usually tempered and laminated.

Safety Glasses. Plastic glasses that give protection to the wearer doing work that might send off particles of matter, such as when grinding and chipping.

Safety Motor Control. Electrical device used to open circuit to motor if temperature, pressure, or current flow exceed safe conditions.

Safety Net. A woven meshed fabric that is suspended below a construction activity to protect materials and people that may fall from dangerous heights.

Safety Nosing. Stair nosing with abrasive non-slip strip surface flush with tread surface.

Safety Plug. Device which will release the contents of a container before rupture pressures are reached.

Safety Relief Valve. A valve designed to release water or steam if the temperature and pressure have risen to the point that an explosion could occur.

Safety Switch. A mechanism on a machine or device that prevents that machine from being operated until the mechanism is deactivated.

Safety Tread. Stair tread which is covered on the top surface with abrasive or non-slip material.

Safety Valve. Pressure release valve preset to be released when pressure exceeds safe operating limit.

Safeflower Oil. Oil from seed of thistle-like plant grown mostly in Egypt and India.

Safing. Fire-resistant material inserted into a space between a curtain wall and a spandrel beam or column, to retard the passage of fire through the space.

Sag Pond. A pond occupying depression along a geological fault; the depression is normally due to uneven settling of the ground or other causes.

Sag. 1. A fresh plaster wall surface has developed a slide. 2. An unevenness or irregularity in a coat of paint, varnish, or lacquer, resulting from too much of the liquid collecting in one spot or area.

Salamander. A portable source of heat in a building under construction, customarily kerosene or oil-burning, used to temporarily heat an enclosure; often used around newly placed concrete to prevent freezing.

Sale-Leaseback. A financing arrangement in which an owner sells real property to an investor and leases it back, usually for the purpose of freeing invested capital.

Sales Tax. A tax levied on sales of goods and services which is calculated as a percentage of the purchase price and collected by the seller.

Salic. Sialic.

Salient Angle. Projecting outward from a line, surface or level; the opposite of a reentrant angle.

Saline Water. Those waters having a specific conductivity in excess of a solution containing 6,000 parts per million of sodium chloride.

Saline. A solution of salt in water.

Sally Port. A gate or passage in a fortified place for use by troops making a sortie.

Salmon Brick. Relatively soft, under-burned brick, so named because of its color; used mostly where it is protected from the weather and where it does not have to support a great weight.

Salt Glaze. A glaze produced by the reaction, at elevated temperature, between the ceramic body surface and salt fumes produced in the kiln atmosphere.

Salt Spray. A salt fog test environment.

Salt. 1. Sodium chloride, NaCl, used for preserving the freshness of food. 2. Substance that results from reaction between acid and base.

Saltpetering. See Efflorescence.

Salvage Value. The estimated value of an asset at the end of its economic life.

Salvage. The saving and utilization of waste materials leftover from a remodeling job or a fire.

Sample Soil. A representative specimen of soil from a site.

Samples. Material samples requested by the architect of the general contractor or materials specified for the project.

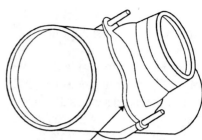

Saddle Fitting

Sampling. The method of obtaining small amounts of material for testing from an agreed-upon lot.

Sand Asphalt. A mixture of sand and asphalt cement or cutback or emulsified asphalt; it may be prepared with or without special control of aggregate grading and may or may not contain mineral filler; either mixed-in-place or plant mix construction may be employed; sand asphalt is used in construction of both base and surface courses.

San Blast. A system of abrading a surface such as concrete by a stream of sand, or other abrasive, ejected from a nozzle at high speed by water and/or compressed air.

Sand Cushion Terrazzo. Terrazzo with an underbed that is separated from the structural floor deck by a layer of sand.

Sand Down. Remove the gloss of an old surface finish and smooth it prior to refinishing.

Sand Filter. Pool filter using sand as filtering medium.

Sand Finish. Rough finish plaster wall.

Sand Holes. Tiny pits in the surface of ceramic tile.

Sand Mold Brick. See Soft Mud Brick.

Sand. Loose granular material resulting from the natural disintegration of rock or from the crushing of friable sandstone, passing through a #4 sieve but predominantly retained on a #200 sieve; manufactured sand is the fine material resulting from the crushing and classification by screening, or otherwise, of rock, gravel or blast furnace slag.

Sandbag. A bag filled with soil or sand to act as a temporary barricade or dam.

Sandblast Masonry. The act or process of abrading or cutting masonry structure surfaces, using sand ejected from a nozzle at high speed by compressed air; effective form of cleaning masonry, but used sparingly because of its inclination to remove the outer, weather resistant, usually glazed or polished, surface.

Sandblast. A system of cutting or abrading a surface such as concrete by a stream of sand ejected from a nozzle at high speed; compressed air is used to propel a stream of wet or dry sand onto the surface; often used for cleanup of horizontal construction joints or for exposure of aggregate in architectural concrete; a method of scarifying the surface of concrete or masonry to provide a bondable surface; used to clean metal before painting.

Sander. A machine that smoothes or polishes by means of abrasive material usually in the form of a disk or belt; Disk sander or Belt Sander.

Sanding Pole. In gypsumboard joint finishing, a sandpaper holder affixed to the end of a handle with a swivel to aid in the sanding process.

Sanding Sealer. A lacquer used as a seal coat over a filler; usually given some filling action by adding inert substances.

Sanding. Rubbing sandpaper or similar abrasive over a surface before applying a finish.

Sandpaper. Paper with sand or other abrasive stuck to it for smoothing or polishing.

Sandstone. A sedimentary rock formed from sand.

Sandwich Glass. A pane of glass that contains an inner layer of material between two outer layers of glass to provide additional insulating properties or resistance to breakage.

Sandwich Panel. A panel consisting of two outer faces of wood or metal, bonded to a core of insulating foam.

Sandwiching. In manufacturing gypsumboard, forming the gypsum core between two plys of paper.

Sandy Finish. A surface condition having the appearance of sandpaper; may result from overspray; or a finish with sand or walnut shells.

Sanitary Facility. Any single unit or a combination of water closets, urinals, lavatories, bathtubs or showers, together with the room or space in which they are housed.

Sanitary Piping. Drain, waste, and vent plumbing systems.

Sanitary Sewer. A sewer used only for carrying the liquid or water-borne wastes from plumbing fixtures; does not include storm, surface, or ground water.

Sanitary Tee. A fitting in a DWV plumbing system.

Sanitizer. One of three groups of antimicrobials registered by EPA for public health uses; EPA considers an antimicrobial to be a sanitizer when it reduces but does not necessarily eliminate all the microorganisms on a treated surface; to be a registered sanitizer, the test results for a product must show a reduction of at least 99.9% in the number of each test microorganism over the parallel control.

Sap. Most of the fluids in a tree; certain secretions and excretions, such as oleoresin, are excepted.

Saponification Number. Number of milligrams of potassium hydroxide needed to neutralize the acid in one gram of substance after it has been saponified.

Saponify. To convert an oil or a fat into soap by action of an alkali; when linseed oil paint comes in contact with a surface that contains strong alkali and water, like new concrete floor, the oil is saponified and loses its bonding qualities.

Sapwood. The wood of pale color near the outside of the log; under most conditions the sapwood is more susceptible to decay than heartwood.

Sarcophagus. A stone coffin.

Sash Balance. A spring-loaded device for counterbalancing the weight of a vertical sliding sash as on a double-hung window.

Sash Bead. A strip with one edge molded, against which a sash slides.

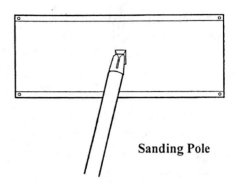

Sanding Pole

Sash Chain. A metal chain adapted for use with vertical sliding sash, attached to the sash and to the counterbalancing sash weight.

Sash Cord. Cord or rope used similarly to sash chain.

Sash Pole. A wood or metal pole to which a sash pole hook is attached, used to raise or lower a transom or sash beyond hand reach.

Sash Pulley. A pulley mortised into a double-hung sash frame over which the sash cord or sash chain passes.

Sash Socket. A metal plate, attached to a sash, containing a hole or cup to receive a sash pole hook.

Sash Tool. A beveled paint brush for use on windows and other precise cutting in.

Sash Weight. A cast iron weight used for counterbalancing the weight of a vertical sliding sash as on a double-hung window.

Sash, Transom. See Transom Sash.

Sash. The framework which holds the glass in a window.

Satin Finish. The dried film of paint or other finishing material which does not have a full luster, but a dull luster like that of satin.

Satisfaction of Mortgage. A written instrument wherein the mortgagee conveys the property back to the mortgagor when the underlying loan has been entirely repaid

Saturated Air. Air that contains all of the moisture vapor it can hold; in saturated air, the dry-bulb, wet-bulb, and dew point temperatures are all equal.

Saturated Felt. A felt that is partially impregnated with low softening point bitumen.

Saturated Surface Dry (SSD). A condition of an aggregate which holds as much water as it can without having any water within the pores between the aggregate particles.

Saturated Vapor. Vapor condition which will result in condensation into droplets of liquid if vapor temperature is reduced.

Saturation Coefficient. See C/B Ratio.

Saturation. Condition existing when substance contains all of another substance it can hold for that temperature and pressure.

Savanna. A grassy plain in tropical and subtropical regions, with few or no trees.

Saw Cut, Concrete. See Concrete Saw Cut.

Saw Cutting, Concrete. See Concrete Saw Cutting.

Saw Fence. A device on a table saw for guiding lumber being sawed.

Saw Kerf. 1. A groove in a piece of wood made by a saw blade. 2. That portion of a log, timber, or other piece of wood removed by the saw in parting the material into two pieces.

Saw, Band. See Band Saw.

Sawed Control Joints. A joint cut in hardened concrete, generally not to the full depth of the member, by means of special equipment.

Sawed Veneer. Veneer produced by sawing.

Sawhorse. A frame, usually four-legged, for supporting wood while sawing.

Saw-Tooth Roof. A roof shape that is saw-toothed in section; the vertical planes are usually to accommodate windows.

SBA. Small Business Administration.

SBCCI. Southern Building Code Congress International, Inc.

SBS. Sick Building Syndrome.

SC Asphalt. See Slow-Curing Asphalt.

SC Grade. See Road Oil.

Scab. 1. A short piece of lumber used to splice two other pieces together. 2. One who refuses to join a labor union or who participate in practices in opposition to the union.

Scaffold. 1. A temporary structure or platform for workmen to stand or sit on when working in high places; also called Staging. 2. A temporary framework structure for the support of concrete forms and other shoring; adjustable metal scaffold is frequently adapted for shoring of concrete formwork.

Scaffolding. The materials used in constructing scaffolds.

Scagliola. An imitation marble made by the plasterer; composed of a combination of Keene's cement, glue, isinglass, and coloring material; it takes a high durable polish.

Scale Drawing. A drawing in which the measurements represented are drawn to a predetermined scale, such as 1/4 inch equals one foot, so that all elements and dimensions in the drawing are proportional in length and width to the actual room, floor, or building depicted.

Scale. 1. A proportion between two sets of dimensions, as between those of a drawing and its original; for example, the scale of a drawing may be expressed as 1/4 inch = one foot. 2. A measuring tool used by architects and engineers in preparing drawings to a proportionate scale. 3. To measure a drawing with a scale. 4. Either pan or tray of a balance. 5. To climb, as a ladder. 6. A series of graduated marked spaces for measuring something, as on a thermometer. 7. Rust occurring in thin layers. 8. Hard deposit of minerals on heater coils and pool surfaces.

Scalene Triangle. A triangle having three unequal sides.

Scaler. A hand cleaning chisel.

Scaling. 1. Finish condition in which pieces of the dried finishing material come off exposing the surface below. 2. Prying or chipping loose pieces of rock off a building face or roof to avoid danger of their falling unexpectedly. 3. The breaking away of a hardened concrete surface, usually to a depth of 1/16 to 3/16. 4. Measuring a drawing with a scale.

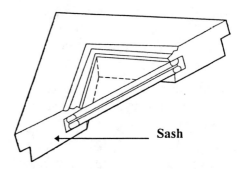

Sash

Scallops. The up-and-down uneven effect along the edge of carpet caused by indentations where tacks are driven.

Scantling. 1. A small piece of lumber. 2. Lumber with a cross-section ranging from 2 by 4 inches to 4 by 4 inches.

Scarf Joint. 1. A glued end connection between two pieces of wood, using a sloping cut to create a larger surface for the glue bond, to allow it to develop the full tensile strength of the wood it connects. 2. A timber spliced by cutting various shapes of shoulders, or jogs, which fit each other. 3. A joint between two pieces of wood which allows them to be spliced lengthwise.

Scarfing. A joint between two pieces of wood which allows them to be spliced lengthwise.

Scarifier. A piece of thin metal with teeth or serrations cut in the edge, used to roughen fresh mortar surfaces to achieve a good bond for the tile; a scarifier also can be used to roughen the surface of concrete.

Scarlet Lake. Pigment made by precipitation of aniline color upon base of alumina hydrate and barium sulphate.

Scarp. 1. A cliff, escarpment, or steep slope of some extent formed by a geological fault. 2. A cliff or steep slope along the margin of a plateau, mesa, or terrace.

Scarred Faces. Surface blemishes caused by scraping or other marring of ceramic tile.

Scavenger Pump. Mechanism used to remove fluid from sump or container.

Schedule. 1. A list, catalog, or inventory. 2. A time table of things to be done. 3. A pipe size system originated by the iron industry, giving outside diameters and wall thickness of iron and plastic pipes.

Scheduling Software. Computer software used to prepare a construction schedule.

Schematic Design Phase. One of the standard phases of architectural service (Schematic Design Phase, Design Development Phase, Construction Documents Phase, Bidding or Negotiation Phase, and Construction Phase-Administration of the Construction Contract).

Schematic Diagram. A drawing showing general layout and relationships of the elements, not necessarily to scale.

Schematic Drawings. Drawings prepared by architects for the Schematic Design Phase.

Schrader Valve. Spring-loaded device which permits fluid flow in one direction when a center pin is depressed; in other direction when a pressure difference exists.

Scissors Lift. A low-rise lift utilizing a scissorlike mechanism to raise the platform.

SCL. Structural Composite Lumber.

Sconce. A wall bracket candle holder or lighting fixture.

SCORE. The Service Core of Retired Executives.

Score. To cut a surface with a sharp blade before breaking.

Scoreboard, Gym. See Gym Scoreboard.

Scotch Yoke. Mechanism used to change reciprocating motion into rotary motion or vice-versa; used to connect crank-shaft to piston in refrigeration compressor.

Scotch. See Scutch.

Scotia. A concave molding.

SCPI. Structural Clay Products Institute.

SCR. 1. Silicon-Controlled Rectifier. 2. Structural Clay Research, trademark of the Structural Clay Products Research Foundation.

Scratch Coat. The first coat of plaster applied to a surface in three coat work; the term originates from the practice of cross-raking or scratching the surface of this coat with a comb-like tool to provide a mechanical key to aid bond with the brown coat.

Scratch. A defect in a surface made by scraping with a sharp pointed object.

Scratcher. Any serrated or sharply tined object that is used to roughen the surface of one coat of mortar to provide a mechanical key or bond for the next coat.

Scratching. The application of a scratch coat and its combing with a scratcher.

Screed Guide. Firmly established grade strips or side forms for unformed concrete which will guide the strikeoff in producing the desired plane or shape.

Screed. 1. A strip of wood, metal, or plaster that establishes the level to which concrete or plaster will be placed. 2. To strike or plane off wet mortar or concrete which is above the desired plane or shape.

Screen Door Latch. A small locking or latching device used on screen doors and operated by a knob or a lever handle.

Screen Door. A door consisting of metal or plastic insect screen stretched taut on a light wood or metal frame.

Screen. 1. A fixed or movable partition for separating, concealing, or sheltering from drafts, excessive heat or light, or from observation. 2. A white or silver surface for the projection of films. 3. A wood or metal frame with fine wire netting to keep out insects. 4. See Sieve.

Screw Conveyor. A helical conveyor.

Screw Cover, Box. See Box, Screw Cover.

Screw Gun. A hand held device resembling an electrical drill designed or adapted to mechanically drive and set screws.

Screw Jack. A lifting mechanism consisting of a threaded shaft for raising heavy loads.

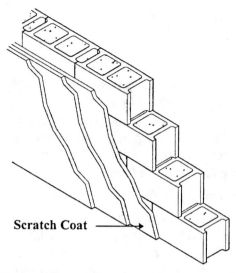

Scratch Coat

Screw Pump. Compressor constructed of two mated revolving screws.

Screw Ring. A metal or plastic ring which is threaded on its inside and used as a connector.

Screw, Self-Tapping, Drywall. See Self-Tapping Drywall Screw.

Screw. A spirally grooved metal fastener installed with a screwdriver.

Screwable Type Stud. Metal stud fabricated from not lighter than 26 gauge metal with knurled flanges to facilitate easier penetration of self-tapping screws or divergent point staples.

Screwdriver. A hand tool for installing screws.

Screwless Rose. A door knob rose that has a concealed method of attachment.

Screws, Self-Tapping, Drywall. See Self-Tapping Drywall Screws.

Scribe. To score or mark along a cutting line.

Scribing. 1. Transferring the exact irregularities of a wall or other surface onto a piece of carpet, wood or paper, which is then cut to fit those irregularities. 2. Fitting woodwork to an irregular surface; in moldings, cutting the end of one piece to fit the molded face of the other at an interior angle to replace a miter joint; such a joint is called a Coped Joint.

Scrim Back. See Double Back.

Scrim. 1. A fabric reinforcing that is used between moppings to strengthen parts of the roofing membrane. 2. A durable plain-woven fabric used for carpet backing. 3. Rough textured woven cloth worked into brown or finish coat to add crack resistance to the plaster; its use is now very rare.

Scroll. A convoluted or spiral ornamental design in metal or wood work.

Scrollsaw. A thin-bladed handsaw in a frame to keep the blade in tension, for cutting curves and irregular designs; a fretsaw.

Scrollwork. A decoration consisting of spiral lines.

Scuffed Paper. A surface scrape usually caused by sliding one gypsum board over the face of another, or by sanding beyond the finished joint.

Sculptor. An artist who makes sculptures.

Sculpture. The art of creating forms, often representational, in the round or in relief by chiseling stone, carving wood, or casting metal.

Sculptured Carpet. A pattern formed when certain tufts are eliminated or pile yarns drawn tightly to the back to form a specific design in the face of the carpet; the result simulates the effect of hand carving; also called Carved Carpet.

Sculptured Tile. Tile with a decorative design of high and low areas molded into the finished face.

Scum Channel. Perimeter overflow inset into pool wall as surface water collection for filtration.

Scupper. 1. An opening through a parapet wall through which water can drain from the edge of a flat roof. 2. An opening through an exterior wall through which water can drain.

Scutch. In masonry, a bricklayer's cutting tool used for dressing and trimming brick to a special shape; it resembles a small pick; also called a Scotch.

Scuttle. A small access opening in a wall, ceiling, or roof fitted with a lid.

SDI. 1. Steel Deck Institute. 2. Steel Door Institute.

SE. Structural Engineer.

Seal Coating. Coating used to prevent excessive absorption of the first coat of paint by the substrate; a primer.

Seal Leak. Escape of oil or refrigerant at the junction where a shaft enters a housing.

Seal, Shaft. See Shaft Seal.

Seal. 1. Something that confirms, ratifies, or makes secure. 2. An amount of water standing in a drain trap to prevent foul air from rising; see Trap Seal. 3. Close securely or hermetically. 4. Apply a non-porous coating to a surface to make it impervious. 5. A counterflashing closure strip, narrow in shape, made of bituminous materials. 5. To secure a roof from the entry of moisture.

Sealable Equipment. Electrical equipment enclosed in a case or cabinet that is provided with means of sealing or locking so that live parts cannot be made accessible without opening the enclosure.

Sealant. 1. A material used for sealing spaces between materials. 2. An elastomeric material that is used to fill and seal cracks and joints; at expansion joints, this material prevents the passage of moisture and allows horizontal and lateral movement. 3. A caulking material used to fill voids or seams for acoustical, air infiltration, or thermal purposes. 4. A mixture of polymers, fillers, and pigments used to fill and seal joints where moderate movement is expected; it cures to a resilient solid.

Sealed Refrigeration Compressor. See Hermetic Compressor.

Sealed Unit. See Hermetic System.

Sealer. A base coating of paint to seal and equalize suction differences and prevent absorption of subsequent coats.

Sealing Compound. 1. Material used to exclude water and solid foreign materials from openings. 2. A substance that prevents water and moisture from entering joints.

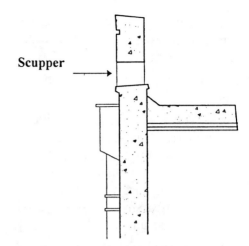

Scupper

Seam Weld. A weld consisting of a series of overlapping spot welds made by seam welding or spot welding.

Seam. 1. A line, groove, or ridge formed at the meeting of two edges of a material. 2. A layer of rock, coal or ore. 3. Treated gypsum board joints. 4. A joint between sections of carpet; see Back Seam, Face Seam, Cross-Seam, Length Seam, and Side Seam.

Seamless Gutter. Rainwater gutter without seams; usually made on a truck containing a forming machine that processes long coils of sheet metal on the building site.

SEAOC. Structural Engineers Association of California.

Seasoned Loan. A loan that has had a good payment record over an extended period of time

Seasoned Lumber. Air- or kiln-dried undefective lumber that is properly stored and is ready for use.

Seasoning. Removing moisture from green wood to improve its serviceability.

Seat Cut. The cut at the bottom end of a rafter to allow it to fit upon the plate.

Seat of a Rafter. The horizontal cut on the bottom end of a rafter which rests upon the top of the plate.

Seat, Grandstand. See Grandstand Seat.

Seat. 1. A piece of furniture for sitting on, such as a chair, stool, bench, or pew. 2. The portion of a chair where one sits. 3. That portion of a valve mechanism against which the valve presses to effect shutoff. 4. A special chair of one in authority.

Seated Connection. A connection in which a steel beam rests on top of a steel angle fastened to a column or girder.

Second Grade Ceramic Tile. Ceramic tile with appearance defects not affecting wearing or sanitary qualities.

Second Growth. New timber that has grown after the removal of all or a large part of the previous stand.

Second Law of Thermodynamics. Heat will flow only from material at higher temperature to material at a lower temperature.

Second Mortgage. A mortgage recorded after a first mortgage.

Second Trust Deed. A trust deed recorded after a first trust deed.

Second. 1. A unit of measurement of angles, equal to 1/60 of a minute. 2. A unit of measurement of time, equal to 1/60 of a minute. 3. A Second Mortgage. 4. A Second Trust Deed.

Secondary Branch. The second or more drain branch of the building drain.

Secondary Clay. See Sedimentary Clay.

Secondary Color. The result of mixing two primary colors; red and yellow yields orange; yellow and blue yields green; blue and red yields violet.

Secondary Financing. Any additional loan or financing that is junior to a mortgage or deed of trust; a Second Mortgage or Second Truct Deed.

Secondary Network. A distribution system wherein the secondary means of an alternating current system are interconnected and supplied through transformers connected in parallel on the secondary side through fuses or automatic switching devices arranged to prevent the feeding of fault current on the primary side of the transformers through the secondary mains; such a system is also called an alternating current automatic low-voltage secondary network.

Secondary Refrigerating System. Refrigerating system in which the condenser is cooled by the evaporator of another or primary refrigerating system.

Section Modulus. The ratio of the moment of inertia of a member to the distance from its neutral axis to the outermost fiber.

Section of Land. A land area one mile square, equal to 640 acres; there are 36 sections in a township;

Section Properties. The physical attributes of structural members including dimensions, area, section modulus, moment of inertia, location of centroid and any other geometric information that affects strength.

Section. 1. A drawing showing the kind, arrangement, and proportions, of the various parts of a structure; it shows how the structure would appear if cut through by a plane. 2. Any of several components that could be re-assembled to construct an object.

Security Glass. A glazing sheet with multiple laminations of glass and plastic; designed to stop bullets.

Security Grille. Steel grille to prevent intrusion or entry through openings.

Security Interest. An interest in property, real or personal, held by a creditor to secure the performance of an obligation.

Security System. An interacting or interdirected group of devices, such as alarms and electronic monitors, used to detect breaches of security.

Sedimentary Clay. A clay which has been geologically transported from its place of formation; also called Secondary Clay.

Sedimentary Rock. Rock formed from materials deposited as sediments, such as sand or sea shells, which form sandstone, shale, or limestone.

Seebeck Effect. When two different adjacent metals are heated, an electric current is generated between the metals.

Seepage. Movement of water through soil without formation of definite channels.

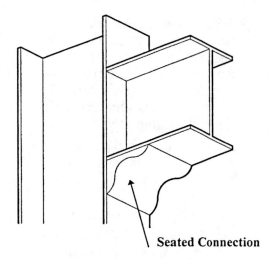

Seated Connection

Segregation. 1. Varying concentration of concrete ingredients resulting in nonuniform proportions in the concrete mix. 2. The tendency for the coarse particles to separate from the finer particles in handling; in concrete, the coarse aggregate and drier material remains behind and the mortar and wetter material flows ahead; this also occurs in a vertical direction when wet concrete is overvibrated or dropped vertically into the forms, the mortar and wetter material rising to the top. 3. In aggregate, the coarse particles roll to the outside edges of the stockpile.

SEIA. Solar Energy Industries Association.

Seiche. A wave on the surface of water in an enclosed or semi-enclosed area, such as a lake, bay or harbor.

Seismic Creep. Very slow periodic or episodic movement along a fault trace unaccompanied by earthquakes.

Seismic Design. The structural design of a structure as it is affected by horizontal and vertical stresses caused by earthquakes.

Seismic Hook. In steel reinforcing of reinforced concrete, a 135-degree bend with a six-bar diameter, but not less than 3-inch (76mm) extension that engages the longitudinal reinforcement and projects into the interior of the stirrup or hoop.

Seismic Load. A load on a structure caused by movement of the earth relative to the structure during an earthquake.

Seismic Wave. One of four distinct waves generated by an earthquake; P-wave, S-Wave, Love wave, and Rayleigh wave.

Seismic. Relating to earthquakes.

Seismicity. The world-wide or local distribution of earthquakes in space and time; a general term for the number of earthquakes in a unit of time, or for relative earthquake activity.

Seismograph. An instrument which produces a continuous record of earth motion.

Seismology. The science of earthquakes.

Selective Surface. An absorber coating that absorbs most of the sunlight hitting it but emits very little thermal radiation.

Selects. These are the accepted or suitable bricks after culling.

Self Furring Lath. Metal lath with dimples that space the lath away from the sheathing behind it to allow plaster to penetrate the lath.

Self Furring. Metal lath or welded wire fabric formed in the manufacturing process to include means by which the material is held away from the supporting surface, thus creating a space for "keying" of the insulating concrete, plaster, or stucco.

Self Priming. Use of same paint for primer and for subsequent coats; the paint may be thinned differently for the different coats.

Self Spacing Tile. Tile with lugs, spacers, or protuberances on the sides that automatically space the tile for the grout joints.

Self Tapping. Creates its own screw threads on the inside of a hole.

Self-Cleaning. Paint in which the rate of chalking is controlled so dirt on surface will be washed away with accumulated chalk.

Self-Healing. Any a material that melts with solar heat and seals cracks that were formed earlier from other causes.

Self-Inductance. Magnetic field induced in conductor carrying the current.

Self-Leveling. Any viscous material that is applied by pouring and will spread out prior to curing.

Self-Siphonage. The loss of the seal of a trap as a result of removing the water from the trap that is caused by the discharge of the fixture to which the trap is connected.

Self-Tapping Drywall Screws. Special screws with a drilling type tip configuration for use with metal framing.

Selvage. Also spelled selvedge. 1. The edge of a sheet material that is finished differently and is expected to be covered in application or cut off. 2. The unsurfaced edge of roll roofing which is covered at the lap when installed. 3. The finished lengthwise edge of woven carpet that will not unravel and will not require binding or serging. 4. Edges of wallpaper without printing.

Semiautomatic Frost Control. Control which starts the defrost part of a cycle manually and then returns the system to normal operation automatically.

Semiconductor. Any of a class of solids, such as silicon or germanium, that conduct no electricity when pure or at low temperatures but are highly conductive when containing suitable impurities or at higher temperatures; used for integrated circuits, transistors, and diodes.

Semi-Drying Oils. Oils which dry to a soft, tacky film; the principal semi-drying oil used in the paint industry is soybean oil.

Semigloss Enamel. An enamel made so that its coating, when dry, has some luster but is not very glossy.

Semigloss Paint. A paint made so that its coating, when dry, has some luster but is not very glossy.

Semigloss. Sheen on dry finish which is about half way between dead flat finish and full gloss.

Semihermetic Compressor. Hermetic compressor with service valves.

Semi-Mat Glaze. 1. A colorless or colored glaze having moderate gloss. 2. A medium-gloss ceramic glaze with or without color.

Semi-Vitreous. Three percent to 7 percent water absorption.

Sensible Heat Gain. The addition of heat to an enclosure by conduction, convection or radiation.

Sensible Heat. Heat which causes a change in temperature of a substance; the degree of molecular excitation of a given mass.

Sensitive Bulb. Part of a sealed fluid device which reacts to temperature; used to measure temperature or to control a mechanism.

Sensor, Photoelectric. A device that responds to light and transmits a resulting electrical impulse.

Sensor. Material or device which goes through a physical change or an electronic characteristic change as surrounding conditions change.

Separate Contractor. A prime contractor; any contractor who has a direct agreement with the owner.

Separation Tank. In a swimming pool, a device used to clarify filter rinse or waste water; also called a Reclamation Tank.

Separation. Division into components or layers by natural causes.

Separator, Oil. See Oil Separator.

Separator. 1. Device to separate one substance from another. 2. See Interceptor.

Septic Tank. A subterranean, usually concrete, tank in which organic sewage matter is decomposed by anaerobic bacterial action.

Septum. That part of the filter element consisting of cloth, wire screen or other porous material on which the filter cake is deposited.

Sequence Controls. Group of devices which act in sequential order.

Serging. A method of finishing a lengthwise cut edge of carpet to prevent unraveling; distinguished from finishing a cut end which may require binding.

Series Circuit. In electrical wiring, the path in which electricity to operate a second lamp or device must pass through the first; current flow travels, serially, through all devices connected together.

Serrate. To punch-perforate in a line to allow for shearing by snapping at the perforation.

Serrated. Notched or toothed on the edge.

Service Cable. Electrical service conductors made up in the form of a cable.

Service Conductors. The electrical supply conductors which extend from the supply main, duct, or transformers of the serving agency to the service equipment of the premises supplied; for overhead conductors this includes the conductors from the last line pole to the service equipment.

Service Core of Retired Executives (SCORE). A source for limited managerial expertise provided by the Small Business Administration.

Service Disconnect. A switch for, or a means of disconnecting an entire building from electrical service; also called Main Switch.

Service Drop. That portion of overhead service conductors between the pole and the first point of attachment to the building or other structure.

Service Ell. See Street Ell.

Service Entrance Conductors. 1. When overhead, that portion of the service conductors which connect the service drop to the service equipment. 2. When underground, that portion of the service conductors between the terminal box located either inside or outside the building wall, or the point of entrance in the building if no terminal box is installed, and the service equipment.

Service Equipment. The necessary equipment, usually consisting of circuit breaker or switch and fuses, and their accessories, intended to constitute the main control and means of cutoff for the electrical supply to a building or structure.

Service Head. The external fitting used at the juncture of the electrical service to a building and the mast from the breaker panel.

Service Lateral. The underground electrical service conductors between the street main, including any risers at a pole or other structure or from transformers, and the first point of connection to the service entrance conductors in a terminal box inside or outside the building wall; where there is no terminal box, the point of connection shall be considered to be the point of entrance of the service conductors into the building.

Service Load. The live and dead structural loads, without load factors.

Service Manifold. Chamber equipped with gauges and manual valves, used by service technicians to service refrigerating systems.

Service Panel. The main electric panel which houses the main switches and distributes electricity to the branch panels and circuits.

Service Pipe. The water pipe coming into the building from the street main.

Service Raceway. The rigid conduit, electrical metallic tubing, or other raceway, that encloses the electrical service entrance conductors.

Service Sink. A sink with a deep basin to accommodate a scrub pail; used for the filling and emptying of scrub pails, the rinsing of mops, and the disposal of cleaning water; also called a Slop Sink.

Service Station. An establishment that services motor vehicles and sells gasoline and oil.

Service Valve. A type of valve which provides an opening in a liquid piping system to service and monitor the system.

Service Window. An opening in a wall or partition through which business is transacted.

Service Wye. A pipe fitting which joins three pipes at a 45 degree angle; a drainage fitting in a plumbing system.

Service. The conductors and equipment for delivering energy from the electricity supply system to the wiring system of the premises served.

Serviceable Hermetic. Hermetic unit housing containing motor and compressor assembled by use of bolts or cap screws.

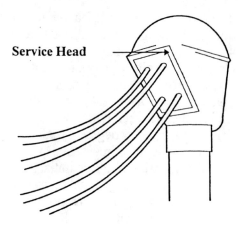

Service Head

Servo. 1. A servomotor. 2. A servomechanism.

Servomechanism. A low power device, electrical, hydraulic, or pneumatic, that controls a more powerful mechanism.

Servomotor. The motive element in a servo mechanism.

Set Back. 1. The distance a building is located from the front property line. 2. See Offset.

Set In. The distance a brick is set back from the brick directly below it.

Set Match. In a Set Match carpet pattern, the figure matches straight across on each side of the narrow carpet width; in a Drop Match, the figure matches midway of the design; in a Quarter Drop Match, the figure matches one-quarter of the length of the repeat on the opposite side.

Set Off. See Offset.

Set Screw. 1. A headless screw used to secure two separate parts in a relative position to one another, preventing the independent motion of either part. 2. A screw to adjust the tension of a spring.

Set. 1. To harden by chemical hydration. 2. A change from a plastic to a hard state. 3. The change in mortar or plaster from a plastic, workable state to a solid, rigid state. 4. The condition reached by a cement paste, mortar, or concrete when it has lost plasticity to an arbitrary degree, usually measured in terms of resistance to penetration or deformation; initial set refers to first stiffening; final set refers to attainment of significant rigidity. 5. A chisel used for cutting brick.

Setoff. An offset against a claim.

Sett. A rectangular paving block of stone or wood.

Setting Bed. 1. The layer of mortar on which the tile is set. 2. The final coat of mortar on a wall or ceiling.

Setting of a Circuit Breaker. The value of the current at which the circuit breaker is set to trip.

Setting Time. The time required for a freshly mixed cement paste, mortar, concrete, or plaster to achieve initial or final set.

Setting Type Joint Compound. A gypsumboard joint compound that hardens by chemical reaction prior to drying; used for patching and completing joint finishing in a shorter period of time.

Setting Up. Initial drying of coating to a point where it is no longer able to flow.

Settlings. Dregs; sediment.

Severability. A rule under which a contract may be divided into its component parts.

Sewage Ejector. A sealed chamber for storing sewage and periodically pumping it up to a sewer at a higher level.

Sewage. Liquid waste containing animal, vegetable, or chemical matter in suspension or solution.

Sewer Brick. Low absorption, abrasive-resistant brick intended for use in drainage structures.

Sewer Connection. The juncture where the house drain connects to the public sewer.

Sewer Tile. Glazed waterproof clay pipe with bell joints.

Sewer. A subterranean conduit for conveying sewage or storm water.

Sewing Pole. Any piece of wood or other material, more or less rounded, over which carpet may be laid prior to opening up the fabric in order to facilitate sewing and other related operations; most carpet layers prefer a wooden pole about 4 in. in diameter that has been slightly flattened on one side.

Sex Bolts. Pairs of bolts, male and female threaded, used for bolting through doors, for example, to attach door closers or surface hinges.

Sexagesimal Measure. The division of a circle into 360 degrees, each degree into 60 minutes, and each minute into 60 seconds.

Sgraffito. A decorative and artistic medium generally consisting of two layers of differently colored plaster; while still soft, the uppermost layer is scratched away, exposing the base or ground layer; countless variations on the process are possible by modulation or pigments and in combination with fresco techniques; sgraffito in Italian means scratched.

Shade. 1. The gradation of color. 2. A color produced by a pigment or mixture with some black in it. 3. A window blind. 4. A device to reduce the glare from a lamp. 5. A screen to provide shelter from light, heat, or sun. 6. A minute variation.

Shaded-Pole Motor. Small a-c motor designed to start under light loads.

Shading Coefficient. The ratio of total solar heat passing through a given sheet of glass to that passing through a sheet of clear double-strength glass.

Shading Lacquer. Transparent colored lacquer used in shading; applied with a spray gun.

Shading. In carpet, bending or crushing the pile surface so that the fibers reflect light unevenly; not a defect but an inherent characteristic of some pile fabrics.

Shadowing. 1. See Joint Photographing. 2. When preceding coats of paint show through the last coat, the finish is said to be shadowing.

Shaft Seal. Device used to prevent leakage between shaft and housing.

Shaft Wall. 1. A wall surrounding a shaft. 2. A fire-rated drywall system which is used to enclose a shaft.

Shaft. An unbroken vertical passage through a multistory building, used for elevators, wiring, plumbing, or ductwork.

Shag Carpet. A carpet made of long coarse or matted fiber; surface consisting of long twisted loops.

Shake Roof. A roof covering system made up of thicker, hand-cut cedar wood shingles.

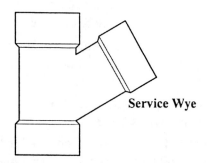

Service Wye

Shake. 1. In wood, a separation along the grain, the greater part of which occurs between the rings of annual growth; usually considered to have occurred in the standing tree or during felling. 2. A Handsplit Shingle.

Shale. A rock formed of consolidated mud.

Shall. A mandatory term as used in most building laws, construction specifications, and building contracts.

Shared Savings Provision. In a construction contract, when the actual costs including profit and overhead are less than the guaranteed maximum price (GMP), the difference will be shared by contractor and owner on some stipulated percentage basis.

Shark Fin. An upward-curled, felt sidelap or endlap.

Sharp Luster. A very high gloss.

Sharp Sand. Coarse sand of which the particles are of angular shape.

Shear Connector. Steel stud or strap, embedded within concrete slab, to transfer horizontal shear loads from slab to supporting structure.

Shear Diagram. A graphic representation of the value of the shear stresses at any point along a beam.

Shear Panel. A vertical plane that resists lateral forces; see Shear Wall.

Shear Plate. In heavy timber construction, a round steel plate used for connecting wood to non-wood materials.

Shear Strength. The strength of an element to resist shear.

Shear Stud. 1. A piece of steel welded to the top of a steel beam or girder so as to become embedded in the concrete fill over the beam and cause the beam and the concrete to act as a single structural unit. 2. Forged steel headed stud, fusion welded to horizontal structural decking, embedded within concrete slab, to transfer horizontal shear loads from slab to supporting structure.

Shear Wall. 1. A wall designed to resist lateral forces parallel to the wall. 2. A wall portion of a structural frame intended to resist lateral forces, such as earthquake, wind, and blast, acting in or parallel to the plane of the wall; see Shear Panel.

Shear. A force effect that is lateral, perpendicular to the axis of a structure.

Shearing. A carpet finishing operation that removes stray fibers and fuzz from loop pile, and produces a smooth level surface on cut pile.

Sheathing Paper. The water resistant paper installed over sheathing and under siding or shingles to insulate in the house; building paper.

Sheathing, Roof. See Roof Sheathing.

Sheathing, Wall. See Wall Sheathing.

Sheathing. 1. The rough covering applied to the outside of the roof, wall, or floor framing of a structure. 2. See Gypsum Sheathing.

Shed Roof. A single-pitched roof.

Shed. A building or dormer with a single pitched roof.

Shedding. See Fluffing.

Sheen. Special type of gloss measured in terms of reflected light at an angle of 30 degrees or less.

Sheepsfoot Roller. A towed roller consisting of a steel drum with numerous protruding bulbous steel rods, used for compacting clay soil.

Sheet Asphalt. A hot mixture of asphalt cement with clean angular, graded sand and mineral filler; its use ordinarily is confined to a surface course, usually laid on an intermediate or leveling course.

Sheet Flashing. Sheet material, commonly metal, used in construction to direct the flow of water or to prevent water penetration.

Sheet Floor. Any type of resilient flooring material, such as linoleum, that is manufactured and installed in sheets or rolls.

Sheet Flow. Rainwater runoff from land or paving in a continuous sheet in contrast with channeled flow.

Sheet Glass. Window glass.

Sheet Lath. A type of metal lath formed by punching geometrical perforations in steel sheets; made from heavier gauge steel than expanded laths, they consequently have greater stiffness.

Sheet Lead. Cold-rolled lead in a sheet whose thickness is expressed by the weight of one square foot of the finished product.

Sheet Metal & Air Conditioning Contractors' National Association, (SMACNA). 8224 Old Courthouse Road, Vienna, Virginia 22180, (703) 790-9890.

Sheet Metal Brake. A machine for flanging, bending, or folding sheet metal.

Sheet Metal Collar. Sheet metal that covers an object or connection.

Sheet Metal Roof. A roof covering of aluminum, copper, stainless steel, or galvanized metal sheets; in high-finish applications, referred to as an architectural roof.

Sheet Metal Screw. A screw with a fully threaded shank for attaching two thicknesses of sheet metal together; has a pointed tip that will start in a drilled hole and is self-tapping.

Sheet Metal. Flat rolled metal less than 1/4 inches in thickness.

Sheet Piling. Planking or sheeting made of concrete, timber, or steel that is driven in, interlocked or tongue and grooved together to provide a tight wall to resist the lateral pressure of water, adjacent earth or other materials.

Sheet Plastic. Plastic in which thickness is very small in relation to length or width.

Sheeting. 7/8" tongue and groove board.

Sheetrock Beam. A horizontal member constructed of plasterboard.

Shear

Sheetrock Finishing. The final sanding and coating of gypsum board seams to make ready for painting.

Sheetrock. See Gypsum Board.

Shelf Angle. A steel angle attached to the spandrel of a building to support a masonry facing.

Shelf Life. The period of time during which various materials, such as adhesives and sealants, can be stored under specific temperature conditions and remain suitable for use; also called Storage Life.

Shelf Pin. A pin for supporting a shelf, inserted in a hole in a vertical member of a cabinet or bookcase; also called a Shelf Support or Shelf Rest.

Shelf Rest. See Shelf Pin.

Shelf Support. See Shelf Pin.

Shelf, Checkroom. See Checkroom Shelf.

Shelf. A horizontal mounted surface upon which objects may be stored, supported, or displayed.

Shell and Tube Flooded Evaporator. Device which flows water through tubes built into a cylindrical evaporator or vice versa.

Shell Concrete. Concrete outside the transverse reinforcement confining the concrete.

Shell Masonry. The outer portion of a masonry unit as placed in masonry.

Shell of Chimney. The outside wall of a chimney.

Shell Type Condenser. Cylinder or receiver which contains condensing water coils or tubes.

Shell. A curved, stiff, surface which can carry normal forces by the normal components of tension, compression, or shear forces which can exist within the thickness of the shell.

Shellac. Lac resin melted into thin flakes and used for making varnish; lac is a resinous substance secreted as a protective covering by the lac insect (an Asian scale insect living in trees) and used to make varnish and shellac.

Shelter, Golf. See Golf Shelter.

Shelter, Truck. See Truck Shelter.

Shelving. Horizontal mounted surface upon which objects may be stored, supported, or displayed.

Shepherd Violet Toner. A complex manganese phosphate pigment having extreme acid resistance and light fastness.

Sherardizing. A protective coating on metal similar to galvanizing.

SHGF. Solar heat gain through fenestration.

Shield, Expansion. See Expansion Shield.

Shielded Metal-Arc Weld. Welding process where fusion is produced by heating with arc between covered metal electrode and work. Shielding is obtained from decomposition of electrode covering.

Shielding. In welding, a method of protecting adjacent work by positioning temporary protective sheets of rigid material; particularly used for machine applications.

Shim. 1. A thin piece of material placed between two components of a building to adjust their relative positions as they are assembled. 2. To build up low areas; to level or adjust height. 3. To insert shims.

Shingle Butt. The lower exposed edge of a shingle.

Shingle Principle. The overlapping of materials, shingle style, so that impinging water, like rainwater, will run harmlessly down and out.

Shingle Roof. 1. Pieces of wedge shaped wood or other material used in overlapping courses to cover a roof. 2. Roof covering of asphalt, asbestos, wood, tile, or slate.

Shingle, Cedar. See Cedar Shingle.

Shingle, Metal. See Metal Shingle.

Shingle, Wood. See Wood Shingle.

Shingle. 1. A small thin piece of building material often with one end thicker than the other for laying in overlapping rows as a covering for the roof or sides of a building or structure. 2. A small unit of prepared roofing material designed to be installed with similar units in overlapping rows on inclines normally exceeding 25 degrees. 3. To cover with shingles. 4. To apply any sheet material in overlapping rows like shingles.

Shingling Hatchet. A roofer's or sider's shingle hammer which has a small hatchet at its opposite end to shape wood shingles.

Shingling. 1. The pattern that is formed by laying parallel felt rolls on a roof with lapped joints, in which one longitudinal edge overlaps the longitudinal edge of one adjacent felt, and where the other longitudinal edge underlays the other adjacent felt. 2. The application of shingles on a sloped roof.

Ship and Galley Tile. A special quarry tile having an indented pattern on the face of the tile to produce an antislip effect.

Ship Bottom Paint. Special product designed to prevent corrosion and fouling with marine life on the bottom of ships.

Shiplap. 1. A board with edges rabbeted so as to overlap flush from one board to the next. 2. An offset lamination of two layers of gypsumboard.

Shiplapped Lumber. Lumber that is edge-dressed to make a lap joint; see Ship Lap.

Shipping-Dry Lumber. Lumber that is partially dried to prevent stain and mold in transit.

Ships Ladder. Ladder fabricated of metal or wood and inclined between 50° and 77° with steps in place of rungs and provided with railings; usually a standard manufactured item.

Shivering. The splintering which occurs in fired glazes or other ceramic coatings due to critical compressive stress; also called Peeling.

Shoe Mold. See Base Shoe.

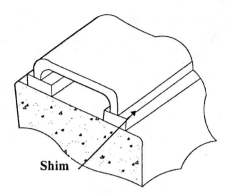

Shim

Shoe. 1. A formed metal section used in attaching metal studs to floor and ceiling tracks. 2. The end section of a channel turned to an angle, usually 90 degrees, to permit attachment, generally to other channels.

Shooting. See Sprouting.

Shop Coated. Shop Primed.

Shop Drawing Stamp. A rubber imprinting stamp used by architects, engineers, and contractors to indicate their approval, disapproval, or other administrative action on shop drawings.

Shop Drawings. Detailed fabrication or construction drawings of specific items of a project provided by subcontractors.

Shop Lathe. A machine that rotates an object about a horizontal axis which then can be shaped by a fixed cutter or tool.

Shop Painted, Asphalt-Base Type. Coating of asphalt-based paint applied in shop, usually for use in high humidity areas.

Shop Painted, Baked Enamel. Coating of baked enamel applied in shop, usually over a primer on underside of deck where exposed to view.

Shop Painted. Coating of paint applied in shop, usually a primer coat to protect metal from corrosion, which may or may not receive additional coats in the field.

Shop Primed. Shop painted.

Shopping Center. One or more sales establishments or stores grouped with associated parking facilities.

Shopping Mall. A large suburban building or group of buildings containing various shops with associated passageways.

Shore Hardness. The reading of a material's hardness on a durometer similar to the Shore A durometer, the scale of which is 0-100, used on rigid and semi-rigid materials such as polystyrene; consists of a pinpoint depression into the material; both the Shore A and Shore D instruments are made by the Shore Instrument Manufacturing Company, Inc., Jamaica, New York.

Shore. Temporary support for formwork and fresh concrete that has not developed full strength.

Shoring. 1. Temporary bracing to hold the sides of an excavation and prevent it from caving. 2. The timbers used as bracing against a wall or under a beam for temporary support.

Short Circuiting. 1. In electrical, an interrupted circuit. 2. In HVAC, a situation that occurs when the supply air flows to exhaust registers before entering the breathing zone; to avoid short-circuiting, the supply air must be delivered at a temperature and velocity that results in mixing throughout the space.

Short Cycling. Refrigerating system that starts and stops more frequently than it should.

Short Term Debt. Mortgages and loans due in less than a year.

Short Time. Short time duty is a requirement of service that demands operation at a substantially constant load for a short and definitely specified time.

Short. Paint that does not have uniform appearance; usually due to the absence of easy brushing liquids.

Shot Blasting. Blast cleaning using steel shot as the abrasive.

Shot. In carpet making, the number of weft yarns in relation to each row of pile tufts crosswise on the loom; a 2-shot fabric is one having two weft yarns for each row of pile tufts; a 3-shot fabric has three weft yarns for each row of tufts; see Weft.

Shotcrete. A low-slump concrete mixture deposited by being blown from a nozzle at high speed with a stream of compressed air, also called Gunite.

Should. As used in most building laws, construction specifications, and building contracts, it means recommended but not mandatory as when shall is used.

Shoulder Eye Bolt. A bolt type device which has an eye connection as its head.

Shoulder. Area between the tapered edge and the face of a gypsum board.

Shove Joints. Vertical joints filled by shoving a brick against the mortar on the next brick when it is being laid in a bed of mortar.

Shovel. A hand implement with a handle and a scoop used for digging, throwing, or lifting material.

Shoving. In asphalt paving, a form of plastic movement resulting in localized bulging of the pavement surface; these distortions usually occur at points where traffic starts and stops, on hills where vehicles brake on the downgrade, on sharp curves, or where vehicles hit a bump and bounce up and down; they occur in asphalt layers that lack stability, usually caused by a mixture that is too rich in asphalt, has too high a proportion of fine aggregate, has coarse or fine aggregate that is too round or too smooth, or has asphalt cement that is too soft; it may also be due to excessive moisture, contamination due to oil spillage, or lack of aeration when placing mixes using liquid asphalts; see Corrugations.

Show Window. Any window used or designed to be used for the display of goods or advertising material, whether it is fully or partly enclosed or entirely open at the rear, and whether or not it has a platform raised higher than the street floor level.

Shower Compartment. An enclosure in which water is showered on a person.

Shower Diverter Valve. A valve at the bath tub filler tap to divert water to the shower over the tub.

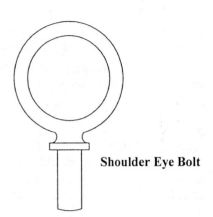

Shoulder Eye Bolt

Shower Door. A door, usually made of unbreakable plastic or shatterproof glass, used in a shower stall or combination shower/bathtub.

Shower Faucet. A water faucet specifically manufactured for use in a shower.

Shower Floor, Waterproof Membrane. A membrane, usually made of built-up roofing, to provide a positive waterproof floor over the substrate, which is to receive a tile installation using a wire reinforced mortar bed.

Shower Pan. 1. A prefabricated assembly to provide a bottom for a shower, may be made of sheet metal, plastic, or masonry. 2. Terminology used in some areas for a waterproof membrane.

Shower Receptor Liner. Waterproof membrane under the shower floor; also called Shower Receptor Lining.

Shower Receptor Lining. See Shower Receptor Liner.

Shower Receptor. The floor and side walls of the shower up to and including the curb of the shower.

Shower Rod. A horizontal pole onto which a curtain is hung to prevent water from splashing out of a shower.

Shower Stall. A compartment where water is showered on a person.

Shower. A bath in which water is showered on the person.

Shrinkage Cracking. Cracking of a structure or member due to failure in tension caused by external or internal restraints as reduction in moisture content develops.

Shrinkage Cracks. In asphalt paving, interconnected cracks forming a series of large blocks usually with sharp corners or angles; frequently they are caused by volume change in either the asphalt mix or in the base or sub-grade.

Shrinkage Temperature Steel. Reinforcing bars laid at right angles to the principal bars in a member, for the purpose of preventing excessive cracking caused by drying shrinkage or temperature stresses in concrete.

Shrinkage. 1. In gypsumboard joints, a slight concave depression in the joint treatment usually due to using too thin a mix of joint compound; also called a Starved Joint. 2. The decrease in volume, or contraction, of a material by the escape of any volatile substance, or by a chemical or physical change in the material. 3. Decrease in volume on drying. 4. Loss of bulk of soil when compacted in a fill; usually is computed on the basis of bank measure.

Shroud. Housing over condenser, evaporator or fan.

Shrub Moving. The digging out of a shrub with its root system intact, to relocate it to a different place.

Shunt. A conductor joining two points of a circuit, through which more or less of a current may be diverted.

Shut-off Valve. A valve used in a supply line to either turn on the liquid or turn it off, seldom used to reduce flow.

SI. French. System International d' Unites; International System of Units.

Sialic. Of minerals that have a high content of silica and alumina, typical of the outer layers of the earth.

Sick Building Syndrome (SBS). Describes situations in which building occupants experience acute health or discomfort effects that appear to be linked to time spent in a particular building, but where no specific illness or cause can be identified; the complaints may be localized in a particular room or zone, or may be spread throughout the building.

Side and End Matched. Wood strip flooring that is tongued and grooved on sides and ends.

Side Hill. A slope that crosses the line of work.

Side Jamb. The vertical pieces to a window or door opening on which the door or window rest.

Side Lumber. A board from the outer portion of the log. ordinarily one produced when squaring off a log for a tie or timber.

Side Seams. Seam running the length of the carpet; also called a Length Seam.

Side Vent. A vent that connects to the drain pipe through a 45 degree Y or less.

Side Yard. The space between a building and the side property line.

Side-Grained Wood. Flat-grained lumber.

Sidelap. The shortest distance in inches in which horizontally adjacent elements of roofing overlap each other.

Sidelight. A tall, narrow window alongside a door.

Sidewalk Doors. Steel or aluminum doors in the plane of a sidewalk leading to a basement area.

Sidewalk. A walk for pedestrians at the side of a street.

Siding, Steel. See Steel Siding.

Siding. The outside finish of an exterior wall.

Sienna. A pigment obtained from the earth which is brownish yellow when raw; orange red or reddish brown when burnt.

Sieve Analysis. Determination of the proportions of particles of the granular material lying within certain size ranges on sieves of different size openings; see Gradation.

Sieve. A metallic sheet or plate, woven wire cloth, or similar device, with regularly spaced openings of uniform size, mounted in a suitable frame or holder for use in separating material according to size; also called a Screen; in laboratory work an apparatus in which the apertures are square for separating sizes of material.

Sight Gauge. See Sight Glass.

Sight Glass. A glass section in piping to enable visual monitoring of the condition of the fluid or its height; also called a Sight Gauge.

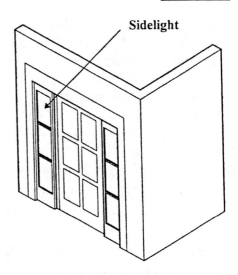
Sidelight

Sign, Electric. See Electric Sign.

Sign. A posted command, warning, or direction.

Signage. Any of a group of posted commands, warnings, or directions.

Signal Circuit. Any electrical circuit which supplies energy to an appliance which gives a recognizable signal; such circuits include circuits for door bells, buzzers, code-calling systems, and signal lights.

Signboard. A sign.

Silex. A form of silica used extensively in making paste wood fillers; it is chemically inert and does not absorb moisture or shrink.

Silica Gel. Hydrated silica in a hard granular form used as a desiccant; when heated, moisture is released and compound may be reused.

Silica Sand. 1. A white fine sand composed of silica. 2. A mineral usually contained in the clay that is used for the making of bricks. 3. An inert pigment made from quartz rock, which is highly resistant to acids, alkalis, heat and light.

Silica. SiO_2, the common oxide of silicon usually found naturally as quartz or in complex combination with other elements as silicates; various polymorphs and natural occurrences of silica include cristobalite, tridymite, cryptocrystalline chert, flint, chalcedony, and hydrated opal.

Silicate of Soda. See Water Glass, 3.

Silicate Paints. Those employing silicates as binders; used primarily in inorganic zinc rich coating.

Siliceous. Containing silicon or silicate.

Silicon Carbide. An abrasive, shiny black; very hard and brittle crystals made by fusing silica sand and coke in electric furnace.

Silicon. An abundant non-metallic element occurring in silica and silicates, used in the manufacture of glass, in alloys, and in electronic devices.

Silicon-Controlled Rectifier (SCR). Electronic semiconductor which contains silicon.

Silicone. A polymeric substance with high resistance to cold, heat, water, and passage of electricity; used in sealants, lubricants, varnishes, binders, and electric insulators.

Silk Screen Finishing. Process of finishing where paint is forced through open meshes of a fabric screen; parts of the screen are blocked off and do not print, thus producing the design.

Silking. A surface defect characterized by parallel hair-like striations in coated films.

Sill Bolt. Steel bolt placed in the foundation wall or slab to secure the wood sill.

Sill Cock. An outdoor water faucet, usually at sill height, used as a hose connection; also called a Hose Bibb or Wall Hydrant.

Sill Course. See String Course.

Sill High. The height of the window sills.

Sill Pan. A sheet metal deflector installed at each end of a wood window or door sill to carry intruded water that has run down the jamb to the exterior.

Sill Sealer Insulation. Insulation placed between sill plate and supporting concrete or masonry to prevent air leaks.

Sill. 1. The lowest member of the frame of a structure, resting on the foundation and supporting the uprights of the frame. 2. The member forming the lower side of an opening, as a door sill or a window sill. 3. See Threshold, 3.

Silt Trap. A settling hole or basin that prevents water-borne soil from entering a pond or drainage system.

Silt. 1. Sediment deposited by water in a water course. 2. A soil composed of particles less than 1/20 mm in diameter. 3. A heavy soil intermediate between clay and sand. 4. Particles so fine that they are scarcely visible to the naked eye, unless placed in a glass tube and examined before proper light.

Silver Brazing. Brazing process in which brazing alloy contains some silver as part of joining alloy.

Silver Leaf. Thin leaf made of silver, used mostly for lettering on glass; aluminum leaf is used where silver color is required on wood or metal surfaces as silver would tarnish.

Silver. A greyish-white lustrous malleable ductile precious metallic element, used chiefly in alloys; it has the highest thermal and electrical conductivity of any substance.

Simple Beam. A beam that is supported on two supports and where no bending is transferred from the beam to the support.

Simplified Practice Recommendation (SPR). U.S. Department of Commerce, 14th & Constitution Avenue, NW, Washington, DC 20230.

Sine Wave, A-C Current. Wave form of single frequency alternating current; wave whose displacement is sine of angle proportional to time or distance.

Single Family Dwelling. A one-family dwelling.

Single Fire. The process of maturing an unfired ceramic body and its glaze in one firing operation.

Single Hung Window. A window with two overlapping sashes, the lower of which can slide vertically in tracks, and the upper of which is fixed.

Single Layer. One layer of gypsum board as the finished wall membrane.

Single Lever Faucet. Any of several types of washerless faucets using a single control and springs, balls or cartridges to control flow and temperature.

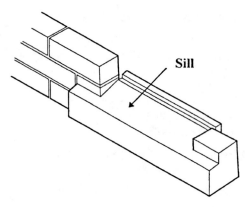

Sill

Single Ply. A roofing system consisting of only one layer of material.

Single Pole Switch. An electrical switch with one positive contact and two positions, off and on.

Single Roll. A single roll of American-made wallpaper is a roll containing 36 square feet of paper; wallpaper usually comes in bolts which contain two or three single rolls.

Single Spread. See Spread.

Single Stage Compressor. See Compressor, Single Stage.

Single Tee. A precast concrete plank with one integral stiffening rib.

Single Zone Air Handling System. A system supplying conditioned air to an entire building or to a portion of a building that can be controlled as one zone; a constant volume of air is supplied to the space served.

Single Accommodation Sanitary Facility. A room that has not more than one of each type of sanitary fixture, is intended for use by only one person at a time, has no partition around the toilet, and has a door that can be locked on the inside by the room occupant.

S

Single-Bevel Groove Weld. See Groove Weld.

Single-J Groove Weld. See Groove Weld.

Single Phase Motor. Electric motor which operates on single- phase alternating current.

Single-Ply Roofing Institute (SPRI). 175 Highland Avenue, Needham, Massachusetts 02196, (617) 444-0242.

Single-Pole, Double Throw Switch, (SPDT). Electric switch with one blade and two contact points.

Single-Pole, Single Throw Switch, (SPST). Electric switch with one blade and one contact point.

Single-U Groove Weld. See Groove Weld.

Single-Vee Groove Weld. See Groove Weld.

Sinistral. Left handed.

Sink, Darkroom. See Darkroom Sink.

Sink. 1. A basin with a drainage system and water supply, used for washing and drainage. 2. A pit or pool used for the deposit of waste.

Sinking Fund. A fund of money accumulated, and augmented by interest, over time to retire a debt or provide reserves for replacement of wasting assets.

Sintering. Heating a powder until it begins to melt, adding to its strength without destroying its porosity, used in manufacturing lightweight aggregates.

Siphon. A tube used to transfer liquids from a higher to a lower level by means of atmospheric pressure.

Siphonage. A particle vacuum created by the flow of liquids in pipes.

Siren. A device, often electrically operated, for producing a penetrating warning sound.

Sisal. A strong durable white fiber from the fleshy leaves of the agave plant; used in making rope and twine.

Sister Joist. The reinforcement of a joist by nailing, or attaching alongside the existing joist, another joist or reinforcing member.

Site Demolition. The act or process of demolishing an old building or structure to make way for new construction.

Site Development. On-site and off-site work, including, but not limited to, walks, sidewalks, ramps, curbs, curb ramps, parking facilities, stairs, planting areas, pools, promenades, exterior gathering or assembly areas and raised or depressed paved areas.

Site Plan. A plan drawing showing the site and buildings.

Site. Area of land to be used for a building location.

Sitecast Concrete. Concrete that is cast-in-place.

Sitz Bath. See Prerineal Bath.

Size. Solution of gelatin-type material, such as resin, glue or starch, used to fill or seal pores of surface and prevent absorption of finishing materials.

Sizing. 1. Working material to the desired size. 2. A coating of glue, shellac, or other substance applied to a surface to prepare it for painting or other method of finish.

SJI. Steel Joist Institute.

Skarf Joint. An end joint in wood formed by joining with adhesive the ends of two pieces that have been tapered or beveled to form sloping plane surfaces, usually to a feather-edge, and with the same slope of the plane with respect to the length of both pieces.

Skein Dyeing. Immersing skeins of yarn in vats of hot dye.

Skein. A batch of yarn.

Sketch. A rough drawing representing the chief features of a site, a building, or an object and often made as a preliminary study.

Skewback. A sloping masonry surface from which a segmental arch springs.

Skid Resistance. 1. A measure of the frictional characteristics of a surface. 2. The ability of an asphalt paving surface, particularly when wet, to offer resistance to slipping or skidding; the factors for obtaining high skid resistance are generally the same as those for obtaining high stability; proper asphalt content and aggregate with a rough surface texture are the greatest contributors; the aggregate must not only have a rough surface texture, but also resist polishing; aggregates containing non-polishing minerals with different wear or abrasion characteristics provide continuous renewal of the pavement's texture, maintaining a skid-resistant surface.

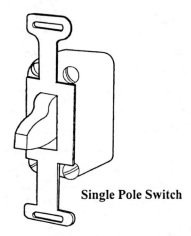

Single Pole Switch

Skim Coat. 1. A thin coat of plaster over any base system; may be the final or finish coat on plaster base. 2. In gypsumboard finishing, a thin coat of joint treatment over the entire surface to reduce surface texture and suction variations. 3. The method of treating plastered areas to receive a gloss paint such as in kitchen and bath areas.

Skim Filter. A swimming pool surface skimmer combined with a vacuum filter.

Skimmer Weir. Part of swimming pool skimmer that adjusts to small changes in water level and assures continuous flow of water into skimmer.

Skimmer. Device that continuously draws swimming pool surface water and surface debris into the filtration system.

Skin Condenser. Condenser using the outer surface of the cabinet as the heat radiating medium.

Skin Friction. The resistance of the soil surrounding a pile to vertical movement of the pile.

Skinning. A tough layer or skin formed on the surface of a paint or varnish in the container; caused by exposure to air.

Skip Trowel. A method of plaster texturing resulting in a rough Spanish Stucco effect.

Skippy. Said of paint that causes the brush to skip on the surface, leaving some spots uncoated and others too thickly coated; this condition can be caused by lack of sufficient vehicle, to permit easy, uniform application or by liquids that pull.

Skips. 1. In painting, holidays, misses, and uncoated spots on finished surface. 2. In roofing, areas that have been missed by the roofer in applying liquid roofing material.

Skirting. Baseboard at a stairway.

Skive Edge. In gypsumboard finishing, the outside edges of the paper joint tape that have been sanded to improve adhesion and reduce waviness.

Skiving. To pare or slice off in thin layers.

Skylight. A transparent opening in a roof for the admittance of light.

Skyscrape. To design or build a very tall building.

Skyscraper. A very tall building; compare with Groundscraper.

Slab Bolster. Continuous, individual support used to hold reinforcing bars in the proper position.

Slab Form. 1. The formwork used for the pouring or placing of a concrete slab. 2. A type of manufactured metal decking which is made expressly to receive a final layer of poured concrete.

Slab Mesh. Welded-wire fabric in sheets or rolls used to reinforce concrete slabs.

Slab, Precast. A flat, horizontal, molded layer of reinforced concrete, cast and cured in other than its final position.

Slab. 1. A cast concrete floor. 2. Flat section of floor or roof either on the ground or supported by beams or walls.

Slack. 1. Extra time in a CPM schedule; also called Float. 2. Hanging loose without tension. 3. Insufficiently diligent; negligent. 4. Slow in business activity.

Slag Inclusion. Non-metallic solid material entrapped in weld metal or between weld metal and base metal.

Slag. A by-product smelting iron, lead, or copper ore; used for construction aggregate; cinder.

Slake. A term denoting the process whereby lime putty is produced from quicklime; slaking consists of adding quicklime to water and allowing the resulting slurry to age for at least two weeks.

Slaked Lime. Hydrated lime.

Slander of Title. Intentionally and wrongfully placing a cloud on title to real estate.

Slash-Grained Wood. Flat-grained lumber.

Slate Flour. Filler used to considerable extent in asphalt mixtures and in roofing mastic.

Slate. A form of geologically hardened clay, easily split into thin sheets.

Sledge. See Sledgehammer.

Sledgehammer. A large heavy hammer that is wielded with both hands; also called a Sledge.

Sleeper. 1. A timber laid on the ground to support a floor joist. 2. A wood strip, usually for a wood floor system, which is fastened directly to a concrete floor thus facilitating the installation of the finished floor.

Sleeping Accommodations. Rooms in which people sleep.

Slenderness Ratio. The ratio of the unbraced length of a column to the radius of gyration.

Sliced Veneer. Wood veneer that is sliced off a log, bolt, or flitch with a knife.

Slicker. A tool often used by the plasterer in place of the darby. It is made of a thin board beveled on both sides, about 4' long and 6 to 8 inches wide, held by the thicker edge.

Slide, Water. See Water Slide.

Slide. 1. To move in continuous contact with a smooth surface, 2. A fresh tile wall that has buckled or sagged; this condition may be caused by excessive mortar, insufficient lime in the mortar, or excessive moisture in the scratch coat; a slide also may result if the surface is slick or the mortar is too soft. 3. A small landslide.

Sliding Glass Door. An exterior glass door mounted above and below on tracks for ease in movement.

Sliding Window. A window that moves horizontally in tracks.

Sling Psychrometer. Measuring device with wet and dry bulb thermometers; moved rapidly through air it measures humidity; see Hygrometer.

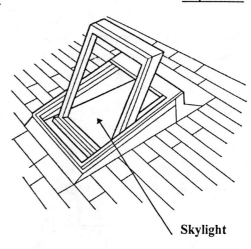

Skylight

Slip Brick. See Soft Mud Brick.

Slip Coating. A ceramic material or mixture other than a glaze, applied to a ceramic body and fired to the maturity required to develop specified characteristics.

Slip Covering. A pipe coupling which has no stop to prevent it from slipping over a pipe; used to make water tight joints in plastic or copper pipe during a repair or alteration.

Slip Form. A form which is raised or pulled as concrete is placed; may move vertically to form walls, stacks, bins or silos, usually of uniform cross section from bottom to top; or a generally horizontal direction to lay concrete evenly for highways, on slopes and inverts of canals, tunnels and siphons.

Slip Forming. Building multistory sitecast concrete walls with forms that rise up the wall as construction progresses.

Slip Glaze. A glaze consisting primarily of a readily fusible clay or silt.

Slip Joint. 1. A contraction joint in a concrete or masonry wall that allows lateral movement. 2. A connection in which one pipe slides inside another allowing for expansion and contraction without breaking.

Slip Nut. A nut used on P straps; a gasket is compressed around the joint by the slip nut to form a watertight seal.

Slip Sheet. A sheet of paper used in a built up roof installation to allow the roofing to move over the substrate.

Slip Sill. A masonry sill which fits directly into a masonry opening.

Slip. 1. An inadvertent or trivial mistake. 2. A suspension of ceramic material in liquid. 3. A small geological fault. 4. A long, thin, strip of wood.

Slip-Critical Connection. A steel connection in which high-strength bolts clamp the members together with sufficient force that the load is transferred between them by friction.

Slippage Cracks. In asphalt paving, crescent-shaped cracks that are open in the direction of the thrust of wheels on the pavement surface; they result when there is a lack of good bond between the surface layer and the course beneath.

Slippage. The lateral movement between adjacent plies of roofing felt along the bitumen lines resulting in a randomly wrinkled appearance and sometimes exposing the lower plies or even the base sheet to the weather.

Slip-Resistant Tile. Tile having greater slip-resistant characteristics due to an abrasive admixture, abrasive particles in the surface or grooves or patterns in the surface.

Sliver. A Splinter.

Slop Sink (Service Sink). A deeper fixture than an ordinary sink. Custodian's type sink See Service Sink.

Slope. An inclined position or direction; the rate of incline.

Slot Cut. Description of a tile that has been cut to fit around pipes or switch boxes. This tile is usually in the shape of the letter H or the letter L.

Slot Weld. A weld made in an elongated hole in one member of a lap or tee joint joining that member to that portion of the surface of the other member which is exposed through the hole; the hole may be open at one end and may be partially or completely filled with weld metal; a fillet-welded slot should not be construed as conforming to this definition.

Slot, Anchor. A groove in an object into which a fastener or connector is inserted to attach objects together.

Slot. An opening in a member to receive a connection with another part.

Slotted Screwdriver. The most common of screwdriver types, has a flat square blade; also called a Flat Head Screwdriver.

Slough. 1. A swamp or backwater. 2. A creek in a marsh or tidal flat.

Slow Drying. Requiring 24 hours or longer before recoating is possible.

Slow-Curing (SC) Asphalt. Cutback asphalt composed of asphalt cement and oils of low volatility.

Sludge. 1. A muddy, greasy deposit or sediment. 2. Precipitated solid matter produced by water and sewage treatment processes.

Slug. 1. In the English system (feet, pounds, seconds), the slug is that mass which when acted on by a 1 pound force acquires an acceleration of 1 foot per second per second. 2. Detached mass of liquid or oil which causes an impact or hammer in a circulating system.

Slugging. 1. Condition in which a mass of liquid enters a compressor causing hammering. 2. The act of adding a separate piece or pieces of material in a joint before or during welding resulting in a welded joint which does not comply with design, drawing, or specification requirements.

Sluice. 1. A sliding gate or other contrivance for regulating the flow of water in a channel. 2. An artificial water channel. 3. A steep, narrow waterway.

Slump Cone. A mold in the form of the lateral surface of the frustum of a cone with a base diameter of 8 inches (203 mm), top diameter 4 inches (102 mm), and height 12 inches (305 mm), used to hold a specimen of freshly mixed concrete for the slump test; a cone 6 inches (152 mm) high is used for tests of freshly mixed mortar and stucco.

Slump Test. The procedure for measuring slump of concrete, mortar, or plaster with a slump cone.

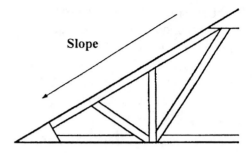

Slope

Slump. A measure of consistency of freshly mixed concrete, mortar, or stucco; the decrease in height of wet concrete when a supporting mold is removed; a measure of the consistency of plastic concrete, equal to the number of inches of subsidence of a truncated cone of concrete released immediately after molding in a standard slump cone.

Slurry Coat. A brushed application of slurry, generally applied to back of adhered veneer units and to backing.

Slurry. 1. A watery mixture of water and any finely divided insoluble material, such as portland cement, slag, or clay in suspension. 2. Cement grout. 3. In making gypsumboard, the gypsum core mixture in its fluid state prior to setting.

Slurry. Cement grout.

Slush Coat. A pure coat of a very soft consistency; also called a Slurry coat.

Slush Joint. See Slushed Joint.

Slushed Joint. Vertical joint filled, after units are laid, by throwing mortar in with the edge of a trowel, used in fireplace and chimney construction; also called a Slush Joint.

SMACNA. Sheet Metal & Air Conditioning Contractors' National Association.

Small Business Administration. This agency funds small business investment companies and minority small business investment companies that make loans to provide financing for contractors; usually restricted to those with a proven history.

Small Calorie. See Calorie, 1.

Smalt. Deep blue pigment prepared by fusing together potash, silica and oxide of cobalt, and reducing to powder the glass thus formed; smalt is sometimes applied to freshly coated surfaces to provide unusual decorating effect.

Smart Building. A modern office building that is pre-wired for all the telecommunication services and computer links.

Smelt. To melt or fuse ore and to separate metal.

Smelter. A furnace in which the raw materials are melted.

Smoke Alarm. A device that emits a warning sound if it detects smoke.

Smoke Chamber. The space in a fireplace immediately above the throat where the smoke gathers before passing into the flue and narrowed by corbeling to the size of the flue lining above.

Smoke Detector. See Smoke Alarm.

Smoke Developed Rating. An index of the toxic fumes generated by a material as it burns.

Smoke Shelf. The horizontal area behind the damper of a fireplace.

Smoke Test. Test made to determine completeness of combustion.

Smoke Vent. A vent or chimney forming a passageway for expulsion of vent gases from gas-burning units to the outside air.

Smooth-Surfaced Roof. A built-up roofing membrane that is surfaced with a layer of hot-mopped asphalt or cold-applied asphalt-clay emulsion or asphalt cutback; sometimes consisting of an unmopped inorganic felt.

Snake. A flexible tape used by plumbers for clearing stoppages in drain and sewer piping; also called Plumber's Snake.

Snap Header. A false header composed of half bricks with the good end showing on the surface.

Snap Line. See Chalk Line.

Snow Load. The live load of snow accumulating on a roof structure.

Soap Brick. A small brick for filling out a course.

Soap Dish. A concave vessel, sometimes carved into a sink, which holds a bar of soap when not in use.

Soaping Tile. The method of applying a soapy film to newly tiled walls to protect them from paint and plaster during construction.

Soapstone. See Talc.

Social Obsolescence. Depreciation; economic obsolescence.

Social Security. A U.S. government program established in 1935 whereby a portion of one's income goes toward old-age and unemployment insurance and assistance; a provision for the economic security and social welfare of individuals and their families.

Society of the Plastic Industry, Inc (SPI). 1275 K Street, NW, Washington, DC 20005, (202) 371-5200

Socket Wrench. A wrench with interchangeable sockets for different sizes of nuts and bolts.

Socket. 1. An opening to accept a mechanical device. 2. An electrical device to accept a light bulb.

Sod. The grass covered surface of the ground; turf.

Soda Ash. Sodium carbonate is used in swimming pools to raise the pH and increase total alkalinity; it is also used with alum in sand-type filters to produce floc; when chlorine gas is used as a disinfectant, this chemical is used to neutralize hydrochloric acid.

Sodium Bicarbonate. Baking soda used to raise total alkalinity of pool water with little change in pH.

Sodium Bisulfate. Dry acid that, mixed with water, lowers pH and total alkalinity of pool water.

Sodium Fixture. An electric lamp that contains sodium vapor and electrodes between which a luminous discharge takes place, commonly used outdoors.

Sodium Hypchlorite. Liquid containing 12 to 15 percent available chlorine used to disinfect pool water.

Sodium Thiosulfate. A chemical compound used to remove all chlorine from a test sample to avoid false pH test readings or false bacteria test results.

Soffit Board. A special designed and formulated gypsum board for use in exterior over hangs, carport ceilings and other weather protected areas.

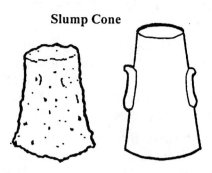

Slump Cone

Soffit Vent. An opening under the eaves of a roof used to allow air to flow into the attic or the space below the roof sheathing.

Soffit. 1. The underside part of a member of a structure, such as a beam, stairway, roof, or arch. 2. The undersurface of a horizontal element of a building, especially the underside of a stair or a roof overhang; relatively small in area as compared with a ceiling.

Soft Dollars. That portion of equity investment in a real estate development that is tax deductible in the first year, such as prepaid interest.

Soft Mud Brick. Brick produced by molding relatively wet clay, often a hand process; when insides of molds are sanded to prevent sticking of clay, the product is Sand Mold Brick; when molds are wetted to prevent sticking, the product is Water-Struck Brick or Slip Brick.

Soft Rot. A special type of decay developing under very wet conditions (as in cooling towers and boat timbers) in the outer wood layers, caused by cellulose-destroying microfungi that attack the secondary cell walls and not the intercellular layer.

Soft Water. Water with a minimal mineral content.

Softball Backstop. A three-sided enclosure with an overhang, often made of chain link fencing, that prevents the ball from leaving the field and from hitting spectators.

Soft-Burned. Clay products which have been fired at low temperature ranges, generally producing relatively high absorptions and low compressive strengths.

Softener, Water. See Water Softener.

Softening Point Drift. A lowering of the softening point of bitumen, usually caused by long overheating or mixing asphalt and coal-tar pitch; also called Dropback.

Softening Point. 1. The temperature point where a hard material becomes soft or viscous. 2. The softening point of bitumen; the asphalt softening point is measured by the ring and ball test; the softening point of coal-tar pitch is measured by the cube in water test.

Software, Scheduling. See Scheduling Software.

Softwood Plywood Panel. Plywood manufactured from softwood species as Douglas fir, western larch, western hemlock, Sitka spruce, white cedar, or redwood.

Softwoods. 1. Generally, one of the botanical groups of trees that in most cases have needlelike or scalelike leaves, the conifers. 2. The wood produced by such trees. 3. The term has no reference to the actual hardness of the wood.

Soil and Waste Pipe. Plastic, copper, cast iron or DWV drainage, water and vent.

Soil Bend. A piece of short, curved pipe, like an elbow, used to connect two straight links of pipe in a sewage system.

Soil Cover. A light covering of plastic film, roll roofing, or similar material used over the soil in crawl spaces of buildings to minimize moisture permeation of the area; also called Ground Cover.

Soil Gases. Gases that enter a building from the surrounding ground, such as radon, volatile organics, and pesticides.

Soil Investigation. See Geotechnical Investigation.

Soil Pipe. A plumbing line that carries waste water from water closets and urinals.

Soil Report. A geological report by a geological engineer providing information on subsurface soil conditions.

Soil Stack. A vertical pipe used to carry human waste from toilets; the vertical main piping of a DWV system.

Soil Structure Interaction. The effects of the properties of both soil and structure upon response of the structure.

Soil Test. See Geotechnical Investigation.

Soil, Heavy. See Heavy Soil.

Soil. The loose surface material of the earth's crust; the unconsolidated natural surface material above bedrock.

Soils Engineer. A professional engineer whose area of expertise is in soil analysis, foundation design, drainage, and the effects of other site conditions on the design of structures.

Sol-Air Temperature. The combined solar and convective sources of heat at the outside surface of a wall or roof.

Solar Array. An array of panels for collecting heat from the sun for use in a solar heating system.

Solar Collector. See Collector, 2.

Solar Heat Gain through Fenestration. The amount of heat gain through a particular cladding system or window.

Solar Heat. Heat created by visible and invisible energy waves from the sun.

Solar Heating. A heating system utilizing heat from the sun.

Solar Panel. A battery of solar cells.

Solar Screen Title. Tile manufactured for masonry screen construction.

Solar System. The sun and the celestial bodies whose motion it governs; the sun, with the planets, moons, asteroids, and comets that orbit it.

Solar Systems. Heating and cooling systems that are based on solar energy.

Solarium. A glass enclosed porch or room with exposures to the sun, used for relaxation or therapy.

Solder. A fusible alloy, usually of lead and tin, used to join less fusible metals or wires.

Soldering Iron. A pointed or wedge-shaped tool that is heated electrically or by other means, used in soldering.

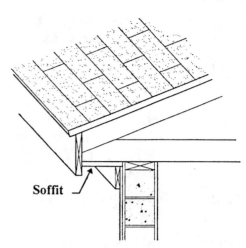

Soffit

Solderless Pressure Connector. A device which establishes the connection between two or more conductors or between one or more conductors and a terminal by means of mechanical pressure and without the use of solder.

Soldier Course. Oblong tile or brick laid with the long side vertical and all joints in alignment.

Soldier. A brick laid on its end, with its narrow face toward the outside of the wall.

Sole Ownership. A business owned entirely by one person.

Sole Plate. The horizontal piece of dimension lumber to which the bottom of the studs are attached in a wall of a light frame building.

Sole Proprietor. The sole owner of a business.

Sole Proprietorship. Same as sole ownership.

Solenoid Valve. A valve that is actuated by an electric solenoid.

Solenoid. A cylindrical coil of wire acting as an electromagnet when a current flows.

Solid Block. A concrete masonry block with small or no internal cavities.

Solid Casting. Forming castings by introducing a body slip into a porous mold which usually consists of two major sections, one section forming the contour of the inside of the object and allowing a solid cast to form between the two mold faces.

Solid Core Door. A flush door with no internal cavities.

Solid Core. An object with no internal cavities.

Solid Door. A door which is constructed with solid materials.

Solid Masonry Unit. A unit whose net cross-sectional area in every plane parallel to the bearing surface is 75% or more of its gross cross-sectional area measured in the same plane.

Solid Mopping. The mopping of a continuous roof surface area with no areas left unmopped.

Solid Slab. A concrete slab, without ribs or voids, that spans between beams or bearing walls.

Solid. 1. Firm and stable in shape and volume. 2. Having three dimensions. 3. Of the same substance throughout. 4. Without cavities. 5. One of the three states of matter; compare with Liquid and Gas. 6. Nonvolatile portion of paint.

Solids Volume. In paint, percentage of total volume occupied by nonvolatiles.

Solids. In paint, the dry ingredients remaining after evaporation of all volatile solvent or water; not a fluid and not flowable.

Solubility. 1. Degree to which a substance may be dissolved. 2. A measure of the purity of an asphalt cement; the portion of the asphalt cement that is soluble in a specified solvent such as trichloroethylene; inert matter, such as salts, free carbon, or non-organic contaminants are insoluble.

Soluble Salts. A mineral that is usually present in the clay used for the making of bricks.

Soluble. Describes the property of a substance to dissolve in another and form a solution; for example, sugar is soluble in water.

Solution Dyeing. Adding dye or colored pigments to synthetic material while in liquid solution before extrusion into fiber.

Solution. The process by which a substance (solid, liquid, or gas) is homogeneously mixed with a liquid, called the solvent, and the mixture being incapable of mechanical separation into its components; alloys and amalgams are solutions of metals in metal; brines are solutions of a salt in water; syrups are solutions of sugars in water; solution should not be confused or used interchangeably with such terms as dispersion, suspension, or emulsion.

Solvent Adhesive. An adhesive having a volatile organic liquid as a vehicle, not including water-based adhesives.

Solvent Balance. Ratio of amounts of different solvents in a mixture of solvents.

Solvent Cement. A solvent adhesive that contains a solvent that dissolves or softens the surfaces of plastic pipe being bonded so that the pipe assembly becomes essentially one piece of the same pipe of plastic.

Solvent Release. Ability to permit solvents to evaporate.

Solvent Wash. Cleaning with a solvent.

Solvent. 1. Having the money to pay all legal debts. 2. Able to dissolve or form a solution.

Sone. Calculated sound loudness rating.

SOP. Standard Operating Procedure.

Sound Absorbing Material. Acoustic insulation, usually made of glass fiber, placed in flutes of deck to provide increased sound absorption through perforations in deck.

Sound Absorption Coefficient. The ratio of sound energy absorbed to the sound energy hitting a surface.

Sound Absorption. 1. The process of dissipating or removing sound energy. 2. The property possessed by materials, objects, and structures such as rooms, of absorbing sound energy.

Sound Attenuating Insulation. A special width unfaced mineral fiber product used to improve the sound transmission loss.

Sound Attenuation. A process in which sound is reduced as its energy is converted to motion or heat.

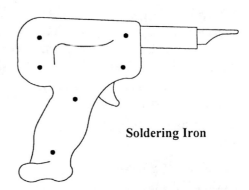

Soldering Iron

Sound Knot. A knot that is solid across its face, at least as hard as the surrounding wood, and shows no indication of decay.

Sound Transmission Class (STC). A single number rating of the sound insulation value of a partition or wall. It is derived from a curve of its insulation value as a function of frequency; the higher the number, the more effective the sound insulation.

Sound Transmission Coefficient. A measure of reduction in sound transmission through certain assemblies of materials used in floors, walls, and ceilings.

Sound Transmission Loss (STL). The decrease or attenuation in sound energy, in decibels, of airborne sound as it passes through a building material; in general, transmission loss increases with frequency, the higher the frequency, the greater the sound transmission loss.

Sound. 1. Deformation waves that are traveling in the air or other elastic materials. 2. Sound is a rapid fluctuation of air pressure.

South Pole. 1. The southernmost point of the earth's axis of rotation. 2. The pole of a magnet that points to the south.

Southern Building Code Congress International, Inc (SBCCI). 900 Montclair Road, Birmingham, Alabama 35213, (205) 591-1853.

Southern Forest Products Association (SPA). PO Box 641700, Kenner, Louisiana 70064, (504) 443-4464.

SOV. Shut Off Valve.

Sovereign Immunity. The doctrine that the sovereign is immune to litigation, originally applied to monarchs and now applied to governments.

Soybean Oil. Oil made from seed of soybean, a leguminous annual plant.

Spa Pool. A pool, not under medical supervision that incorporates water jets and/or an aeration system used for hydro massage.

Spa. 1. A resort with mineral springs. 2. A unit designed primarily for therapeutic use which is not drained, cleaned, or refilled for each individual; it may include, but is not limited to, hydrojet circulation, hot water, cold water, mineral baths, air induction bubbles, or any combination thereof; industry terminology for spa includes, but is not limited to, therapeutic pool, hydrotherapy pool, whirlpool, or hot spa.

SPA. Southern Forest Products Association.

Space Dyeing. In carpet making, alternating bands of color applied to yarn by rollers at predetermined intervals prior to tufting.

Space Frame. A truss that spans with two-way action; also called a Space Truss.

Space Truss. See Space Frame.

Space. 1. An area. 2. A distance. 3. A volume. 4. The region beyond the earth's atmosphere and beyond. 5. An amount of area set aside for some purpose, such as storage space or dining space. 6. A definable area, such as a room, toilet room, hall, assembly area, entrance, storage room, alcove, courtyard, or lobby.

Spacer Strip. A metal strip or bar inserted in the root of a joint prepared for a groove weld to serve as a backing and to maintain root opening during welding.

Spacers. T-shaped and Y-shaped, they are used in installation to separate tile on walls and floors. They are manufactured in various thicknesses from 1/16" to 1/2".

Spacing Mix. A dry or dampened mixture of one part portland cement and one part extra-fine sand; this mix is used as a filler in the joints of mounted ceramic mosaic tiles to keep them evenly spaced during installation.

Spackle. Trademark; a white powder which, when mixed with water, forms a paste used in filling or repairing cracks in plaster or gypsumboard.

Spackling Compound. Kind of plaster which is used to fill surface irregularities and cracks in plaster; this compound when mixed with paste paint makes what is known as Swedish Putty.

Spading Tool. A thin-bladed shovel which is used to place and tamp concrete.

Spall, Pile. See Pile Spall.

Spall. A fragment, usually in the shape of a flake, detached from a larger mass by a blow, by the action of weather, by pressure, or by expansion within the larger mass.

Spalling. The cracking, breaking, or splintering of materials, usually due to heat.

Span. The distance between supports for a beam, girder, truss, vault, arch or other horizontal structural device.

Spandrel Beam. A beam that runs along the outside edge of a floor or roof.

Spandrel Glass. Opaque glass manufactured especially for use in spandrel panels.

Spandrel Panel. A curtain wall panel used in a spandrel.

Spandrel. 1. The wall area between the head of a window on one story and the sill of a window on the floor above. 2. The area of a wall between adjacent arches.

Spar Varnish. A very durable varnish designed for severe service on exterior surfaces; resistant to rain, sunlight and heat; so named because of its suitability for the spars of ships.

Spark Arrestor. A mesh grill at the top of a chimney flue to prevent emission of hazardous sparks.

Spark-Proof Tools. Tools made of bronze beryllium.

Spatter Finish. Finish which provides a spattered or spackled effect.

Spatter. In arc and gas welding, the metal particles expelled during welding and which do not form a part of the weld.

Special Conditions. A contract document that revises or clarifies the general conditions, and describes specific conditions of the project or site.

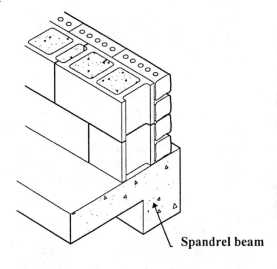

Spandrel beam

Special Door. A door that has a unique use, such as a bank vault door.

Special Steep Asphalt. A roofing asphalt that has a softening point of approximately 220°F (104°C) and that conforms to the requirements of ASTM Standard D 312, Type IV.

Special-Purpose Tile. A tile, either glazed or unglazed, made to meet or to have specific physical design or appearance characteristics such as size, thickness, shape, color, or decoration; keys or lugs on backs or sides; special resistance to staining, frost, alkalis, acids, thermal shock, physical impact; high coefficient of friction, or electrical properties.

Specialties. A designation of construction materials or components required to finish off the structure; denotes incidental items which are needed to complete construction of the building, such as roof specialties or concrete specialties.

Specialty Contractor. A contractor who follows a recognized trade; trade contractor or subcontractor; a specialty contractor commonly installs certain specific items such as flooring, windows, or terrazzo.

Specific Gravity. The specific gravity of a substance is its mass (in grams) per cubic centimeter, or it is the weight of a given volume of the substance divided by the weight of an equal volume of water at a temperature of 39.1° F. Hence, one may calculate the specific gravity of any substance by dividing the weight of a cubic foot of the substance (in pounds) by 62.425; as applied to wood, the ratio of the ovendry weight of a sample to the weight of a volume of water equal to the volume of the sample at a specified moisture content (green, airdry, or ovendry).

Specific Heat. The quantity of heat, in BTUs, needed to raise the temperature of 1 pound of material 1°F.

Specific Humidity. See Humidity Ratio.

Specific Latent Heat of Fusion. The quantity of heat, in joules, required to change unit mass (e.g. 1 kilogram) of a substance from the solid to the liquid state at the melting point.

Specific Latent Heat of Vaporization. The quantity of heat, in joules, required to change unit mass (e.g. 1 kilogram) of a substance from the liquid to the vapor state at the boiling point.

Specific Latent Heat. Latent heat per unit mass.

Specific Performance. 1. The performance of a legal contract substantially in accordance with its terms. 2. A lawsuit to require complete performance of a contract.

Specific Volume. Volume per unit mass of a substance.

Specification Code. A building code where all allowable and required materials and methods are specified in detail; compare with Performance Code.

Specified Lateral Forces. Lateral forces corresponding to the appropriate distribution of the design base shear force prescribed by the governing code for earthquake-resistant design.

Speckled Glaze. A glaze containing granules of oxides or ceramic stains that are of contrasting colors.

Specks. Any dark dots on the tile less than 1/64 inch in diameter, and noticeable at a distance of more than 3 feet.

Spectral Colors. Band of colors produced when ray of sunshine is bent by glass prism.

Spectrum. The band of colors, as seen in a rainbow, arranged in a progressive series according to their refrangibility or wavelength.

Specular Gloss. Mirror-like reflectance.

Speculation. Assumption of unusual business risk in hopes of obtaining commensurate gain.

Speculum. A reflector in an optical instrument.

Spelter. Zinc, with impurities, cast into slabs for commercial use.

Sphagnum. Atypical mosses that grow only in wet acid areas, combined with other decomposed plant debris to form peat moss.

Sphere. A solid figure, or its surface, with every point on its surface equidistant from its center; a ball shape.

Spheroid. A shape resembling a sphere.

Sphinx. Any of several ancient Egyptian stone figures having a lion's body and a human or animal head.

SPI. The Society of the Plastic Industry, Inc.

Spike Knot. A knot cut approximately parallel to its long axis so that the exposed section is definitely elongated.

Spike. A very large nail.

Spillway. An overflow channel for a pond or a terrace channel.

Spinner, Roof. See Roof Spinner.

Spiral Balance. A spring-loaded metallic spiral that provides the counterweight for a doublehung window.

Spiral Column. A concrete column with a continuous spiral of wire around the longitudinal steel.

Spiral Reinforcing. Continuously wound reinforcing in the form of a cylindrical helix.

Spiral Spacers. Usually made of channels or angles, punched to form hooks, which are bent over the coiled spiral to maintain it to a definite pitch.

Spiral, Column. See Column Spiral.

Spiral. A continuously coiled bar or wire, used for reinforcing in a concrete column.

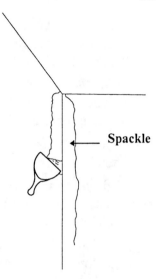

Spackle

Spiral-Grained Wood. Wood in which the fibers take a spiral course about the trunk of the tree instead of the normal vertical course. The spiral may extend in a right- handed or left-handed direction around the tree trunk; spiral grain is a form of cross grain.

Spire. A tapering cone shaped structure surmounting a church roof or tower; a steeple.

Spirit Level. A device for establishing a horizontal line or plane by means of centering a bubble in a slightly arched glass tube.

Spirit Stain. A stain made by dissolving a dye in an alcohol.

Spirit Varnish. A varnish made by dissolving a resin in a solvent; it dries primarily by evaporation rather than by oxidation.

Spitout. A glaze defect of the pinhole type developed in the decorating kiln, due to evolution of minute gas bubbles from body or glaze.

Splash Block. A small precast block of concrete or plastic used to divert water at the bottom of a downspout.

Splash System, Oiling. Method of lubricating moving parts by agitating or splashing oil in the crankcase.

Splash Wall. The wall of a tile drainboard or bathtub.

Splash. 1. A daub or a spot from a splashed liquid. 2. To dash a liquid on a wall or floor. 3. A Splash Wall.

Splay Angle. Where two surfaces come together forming an angle of more than 90 degrees.

Splay. 1. An inclined surface, as the slope or bevel at the sides of a door or window. 2. To make a beveled surface or to spread out or make oblique.

Splice, Compression. See Compression Splice.

Splice. 1. Joining of two members by overlapping, soldering, or by adhesives. 2. Joining of two similar members in a straight line. 3. Joining two lengths of rope by interweaving the strands. 4. Connection of one reinforcing bar to another by lapping, welding, mechanical couplings or other means. 5. The lap between sheet or rolls of welded wire fabric.

Spline Joint. A type of connecting strip fitted into slots in each member, used in cabinetmaking.

Spline. A thin strip of wood used to reinforce joints; also known as a Feather or Tongue.

Splinter. A long, thin, sharp, broken fragment of a material such as wood, metal, or glass; a Sliver.

Split Astragal. An astragal that is split through the middle, allowing each door leaf to operate independently.

Split Face Block. A concrete block with a roughened face.

Split Jamb. A door frame fabricated in two interlocking halves, to be installed from the opposite sides of an opening.

Split L Cut. An improper L cut that is made by splitting a tile instead of cutting it.

Split Receptacle. An electrical outlet where each of the two plugs are on different circuits.

Split Ring. A shear-resisting connector for timbers consisting of a metal ring set in circular grooves in two pieces; the assembly is held together by bolts.

Split Sheet. See Nineteen-Inch Selvage.

Split System. An air conditioning system consisting of two pieces of equipment, an indoor and an outdoor unit; the indoor unit consists of an evaporative coil, provision for heating, and an air handler; the outdoor unit consists of a compressor and an air cooled condenser.

Split. A tear in the membrane, usually resulting from tensile stress of movement of the substrate; see Crack, 3.

Split-Phase Motor. Motor with two stator windings; both windings are in use while starting; one is disconnected by centrifugal switch after motor attains speed; motor then operates on other winding only.

Splitter. A cleavage or lateral fracture in the gypsumboard core; see Core Separation.

Splitting. Tearing or cracking completely through a roof membrane, usually caused by expansion, contraction, deflection, and shearing stresses in the underlying deck construction.

Spodumene (Alpha Spodumene). A lithium mineral of the theoretical composition $Li_2O\ A_{l2}O_3\ 4SiO_2$ (monoclinic crystallization) which on heating inverts to beta spodumene, a form having very low thermal expansion.

Spoil. The earth materials that are removed when excavating.

Spokeshave. A hand tool consisting of a blade set between two handles for shaping concave or convex shapes.

Sponge Float. A plasterer's or concrete mason's finishing float that has a sponge surface, for producing a slightly textured finish on the plaster or concrete.

Spool, Insulator. See Insulator Spool.

Spool. A reel for winding rope, cord, cable, or wire.

Spot Grounds. Pieces of wood attached to the plaster base at various intervals for gauging plaster thickness.

Spot Mopping. 1. In roofing, the application of bitumen in small spots or daubs. 2. An applied mopping pattern in which the hot bitumen is applied in circular areas of about 18 inches in diameter, with a grid of unmopped perpendicular bands.

Spot Repair. Preventive maintenance; repainting of small areas.

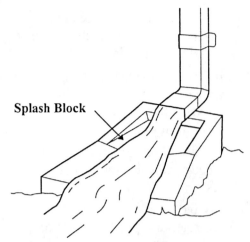

Splash Block

Spot Welding. A resistance-welding process wherein coalescence is produced by the heat obtained from resistance to the flow of electric current through the work parts held together under pressure by electrodes; the size and shape of the individually formed welds are limited primarily by the size and contour of the electrodes.

Spot Zoning. The assigning of a particular zoning classification to a parcel of property that is different to the pattern of classification of surrounding property.

Spotlight. A light or lamp which directs a narrow intense beam of light on a small area.

Spots. Any dark dots on the face of the tile more than 1/64 inch in diameter.

Spotting. In gypsumboard finishing, to cover fastener heads with joint compound.

Spout. 1. A pipe or conductor through which a liquid is discharged or conveyed in a stream. 2. A pipe carrying rainwater from a roof.

SPR. Simplified Practice Recommendation.

Spray Application. The use of mechanical equipment to apply ceiling or wall texture materials.

Spray Booth. An area in a building or structure used for spray painting; blocked off by walls to prevent dust and dirt from work surface.

Spray Cap. Front enclosure of spray gun equipped with atomizing air holes.

Spray Coating. A coat of paint or other material applied to a surface by use of a spray gun.

Spray Cooling. Method of refrigerating by spraying expendable refrigerant or by spraying refrigerated water.

Spray Gun. An apparatus resembling a gun for applying a substance such as paint to a surface.

Spray Head. Combination of needle, tip, and air cap.

Spray Loss. See Drift, 2.

Spray Pattern. Configuration of spray when gun is held steady.

Spray Texture. A mechanically applied material used to produce various decorative finishes; may contain aggregates for different effects.

Spray Transfer. A mode of metal transfer in gas metal-arc welding in which the consumable electrode is propelled axially across the arc in small droplets.

Sprayed Fireproofing. Fire resistant substances sprayed on construction elements.

Sprayed Insulation. A plastic foam of polyurethane sprayed on a surface to insulate.

Spread. The quantity of paint or adhesive per unit area applied to a surface, usually expressed in pounds of adhesive per thousand square feet of area or square feet per gallon of paint; Single Spread refers to application of adhesive to only one adherent; Double Spread refers to application of adhesive to both adherents.

Spreader. Distribution of poolwater inside filter.

Spreading Rate. Area of surface over which unit volume of paint will spread, usually expressed in square feet per gallon.

Spreadsheet. 1. An outsize tabular sheet used by accountants to distribute figures to diverse accounts. 2. A computer program allowing manipulation and flexible retrieval tabulated numerical data.

SPRI. Single-Ply Roofing Institute.

Spring Isolator. An isolating device for setting mechanical equipment using coil springs to prevent vibration transmission into the building structure.

Spring-Board. Flexible board for competition diving.

Springer Points. The starting points from which the under curves of an arch start.

Springer. The brick or stone from which an arch springs.

Spring-Loaded. A device that contains a compressed or stretched spring pressing one part against another.

Springwood. The portion of the annual growth ring that is formed during the early part of the growing season. It is usually less dense and weaker mechanically than latewood; Earlywood.

Sprinkle Mopping. 1. A throwing of heated bitumen in a random pattern onto the roof surface from a mop or broom, the bitumen consisting of bead size granules. 2. In roofing, the installation of insulation by dipping the mop into hot bitumen and sprinkling it onto the deck.

Sprinkler Head. The outlets from which water is sprayed from a fire sprinkler or irrigation system.

Sprocket Wheel. A toothed wheel designed to interlock with a chain to transfer mechanical power.

Sprocket. A tooth on a wheel designed to interlock with a chain.

Sprouting. Temporary condition on new carpets where strands of yarn work loose and project above the pile; can be remedied by careful clipping or spot shearing.

Spud. The removal of roofing gravel and heavy accumulations of bitumen by means of chipping and scraping.

Spur Gears. Meshing gears that are on parallel axes.

Square Cut. A saw cut perpendicular to the length of a member.

Square Edge. A square factory formed edge; not tapered or otherwise deformed.

Square Fixture. A fixture with four sides of equal length.

Square Footage. The area of a building or parcel of land, expressed in square feet.

Square Groove Weld. See Groove Weld.

Square Manhole. A square vertical access shaft from the surface to underground.

Spool Insulator

Square Tub. A bathtub in the shape of a square.

Square. 1. A tool used by tradesmen to obtain accuracy in laying out right angles. 2. An area measuring 10 feet by 10 feet. 3. An area measure of 100 square feet; usually applied to roofing material; sidewall coverings are often packed to cover 100 square feet and are sold on that basis.

Squash Court. A four-walled interior court for playing squash, a game played with balls and racquets.

Squeegee. A rubber edged implement mounted on a handle for spreading, pushing, or wiping liquid material on, across, or off a surface.

Squirrel Cage. Fan which has blades parallel to fan axis and moves air at right angles or perpendicular to fan axis.

SR. Styrene Rubber.

SS Glass. Single strength window glass.

SSD. Saturated Surface Dry

SSPC. Steel Structures Painting Council.

Stability. 1. The ability of a material or structure to remain unchanged. 2. Ability to restore to original condition after being disturbed by some force. 3. The ability of an asphalt paving mixture to resist deformation from imposed loads; stability is dependent upon both internal friction and cohesion.

Stack Bond. A brick bond consisting of all headers or all stretchers with all vertical joints lined up; also called plumb joint bond, straight stack, jack bond, jack on jack. and checkerboard bond.

Stack Bond. See under Bond.

Stack Effect. Pressure-driven airflow produced by convection as heated air rises, creating a positive pressure area at the top of a building and a negative pressure area at the bottom of a building; the stack effect can overpower the mechanical system and disrupt ventilation and circulation in a building.

Stack Vent. 1. The continuation of the soil or waste stack, becoming a vent when it is above the highest horizontal or fixture branch that is connected to the stack; siphonage, aspiration, and back pressure are all relieved by air which enters the system through the stack. 2. A vertical outlet in the interior areas of a built- up roof system to provide relief of water vapor trapped in the insulation.

Stack. 1. Any structure or part thereof which contains a flue or flues for the discharge of gases; a chimney. 2. The vertical main pipe of a soil, waste, or vent pipe system.

Stacking Tile. A method of installation whereby glazed tiles are placed on the wall so that they are in direct contact with the adjacent tiles; the width of the joints is not maintained by the use of string or other means; the tile may be set with either straight or broken joints.

Stadia Rod. A graduated rod used by land surveyors in observing and recording elevations and distances.

Staff Plaster. A kind of stiff plastering held together with fibrous material, now used for precast plaster moldings.

Staff Room. A room used by staff personnel.

Staff. Plaster casts made in moulds and reinforced with fiber; usually wired or nailed into place.

Stage, Band. See Band Stage.

Stage, Theater. See Theater Stage.

Staggered Splices. Splices in steel reinforcing bars which are not made at the same point.

Staggered Studs. Wood wall studs on a plate thicker than the studs with the studs on each side of the wall being staggered, for the purpose of lessening sound transmission from one side of the wall to the other.

Staging. See Scaffold, 1.

Stain Wax, Penetrating. See Penetrating Stain Wax.

Stain, Wood. See Wood Stain.

Stain. 1. A defective discoloration. 2. A discoloration in wood that may be caused by such diverse agencies as micro- organisms, metals, or chemicals. 3. Materials used to impart color to wood.

Stained Glass. 1. Glass colored or stained for use in windows. 2. Colored glass in a lead framework.

Staining. Discoloration caused by a foreign matter chemically affecting the material itself. Application of a paint which lets the grain of wood show through.

Stainless Grab Bar. A length of metal used as a handrail, constructed of stainless steel, attached to a wall in a bathroom, in a shower, above a bathtub or near a water closet; aids people in the use of bathroom fixtures.

Stainless Steel. An alloy of steel with chromium and other metals that is practically immune to rusting and ordinary corrosion.

Stainless Tile. A thin square or rectangular unit used as a finish for walls, floors or roofs, constructed of stainless steel.

Stair Formwork. The support for freshly poured or placed concrete for a stair system.

Stair Horse. See Carriage.

Stair Landing. The level platform between flights of stairs.

Stair Pan. A stair assembly constructed to hold precast or cast-in place masonry or stone in sheet metal pans at the treads.

Stair Rail. A handrail at a stairway.

Stair Reinforcing. The reinforcing steel used in the construction of a concrete stair.

Stair Stringer. One of the structural supporting members of a stairway.

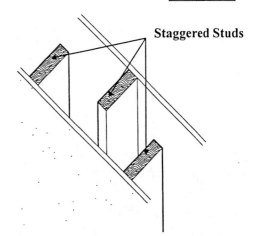

Staggered Studs

Stair Tread, Ceramic Tile. Ceramic tile installed on the horizontal surface of a stairway.

Stair Tread, Metal-Pan Concrete Filled. Section formed from sheet metal to receive concrete fill which will function as a stair tread.

Stair Tread, Steel Grating. Tread fabricated from metal grating.

Stair Tread, Steel Plate. Tread fabricated of floor plate.

Stair Tread. The horizontal surface of a step in a stairway.

Stair, Access. A stair system to provide specific access to roofs and mechanical equipment rooms.

Stair, Concrete. A stair system constructed from concrete.

Stair, Marble. A stair system constructed of or faced with marble.

Stair, Metal. A single or series of metal steps connected by landings.

Stair, Terrazzo. A stairway with terrazzo as the wearing surface.

Stairs, Box. Stairs built between walls, and usually with no support except the wall.

Stairway. One or more flights of stairs including the landings leading from one floor to another.

Stairwork. A unit of horizontal treads, vertical risers, that make up a system of steps from one level to another.

Stake, Side. On a road job, a stake on the line of the outer edge of the proposed pavement; any stake not on the center line.

Stake, Slope. In earthwork, a stake marking the line where a cut or fill meets the original grade.

Stake. 1. A stout stick or post sharpened at one end and driven into the earth as a support or boundary marker. 2. Something that is invested for gain or loss; an interest or share in a financial undertaking. 3. A Mormon territorial jurisdiction comprising a group of wards.

Stall Torque. In an electric motor, the torque developed when starting.

Stall, Shower. A compartment where water is showered on a person.

Stall. 1. A booth or counter in a market or building for the sale of goods. 2. A compartment for one domestic animal in a stable. 3. A compartment for one person in a shower bath; a stall shower. 4. A space marked off for parking one car.

Stanchion. An upright, vertical support or strut.

Stand Oil. Heat-thickened vegetable oil, or combination of vegetable oils, such as linseed and tung.

Stand Pipe System. A system of piping installed for fire protection purposes having a primary water supply constantly or automatically available at each hose outlet.

Standard Air. See Standard Conditions.

Standard Conditions. Used as a basis for air conditioning calculations, temperature of 68° F. (20° C.), pressure of 29.92 inches of mercury and relative humidity of 30 percent; also called Standard Air.

Standard Dimension Ratio (SDR). The ratio of pipe or tubing diameter to wall thickness.

Standard Dimension. The specified dimension of units.

Standard Grade Tile. Highest grade of all types of ceramic tile.

Standard of Living. The degree of comfort available to people.

Standard Operating Procedure. A standard or prescribed method to be followed routinely for the performance of designated operations or in designated situations.

Standard. 1. A stanchion or support. 2. A criterion accepted by custom or general agreement. 3. A level of quality or quantity established by authority for size, weight, value, or extent.

Standardize. To bring into conformity with a standard.

Standing Finish. The finish woodwork of the openings and the base, and all of the interior finish work.

Standing Seam Roof. A sheet metal roof system that has seams that project at right angles to the plane of the roof.

Standpipe. A pipe that extends the full height of a building, with hose connections, used to provide water exclusively for the fighting of fires; see Dry Standpipe and Wet Standpipe.

Stanley Knife. Trademark; a hand tool with removable blades; see Board Knife.

Stannous. Containing or relating to tin.

Staple Fibers. Relatively short natural or synthetic fibers ranging from approximately 1-1/2 to 7 inches in length, which are spun into yarn, as contrasted with continuous filament.

Staple Gun. A spring-loaded or pneumatic devise for driving staples.

Staple. Double pointed, U-shaped metal fastener used for same purposes as nails, but providing additional head holding power.

Star Drill. 1. A drill bit for cutting holes into concrete. 2. A cold chisel, driven by hammer, with a cross-shaped head for cutting holes in concrete.

Starch Coating. Protective coating for surfaces coated with flat paint; it also can be used on wallpapers made with colors that do not smear when wet; the coating is made by soaking ordinary laundry starch in a small quantity of cold water to break up lumps; boiled water is then poured on to cook the starch and make it transparent; mix to consistency of cream, let cool, and apply with large paint or calcimine brush; coating should be stippled while still wet to remove brush marks; it may be removed later by using water and sponge.

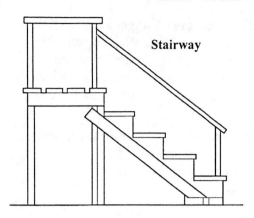

Stairway

Starch. An additive in the core that improves bond between the core and facing paper of gypsumboard.

Starter Strip. A thin strip of wood used to begin the first course of a horizontal siding system.

Starter, Fluorescent Lamp. See Ballast, 1.

Starter, Magnetic. See Magnetic Starter.

Starter. A device that insures that a motor does not receive too high a current when starting up.

Starting Relay. Electrical device which connects or disconnects starting winding of electric motor.

Starting Torque. In an electric motor, the amount of torque available, when at 0 speed, to start and accelerate the load.

Starting Winding. Winding in electric motor used only briefly while motor is starting.

Starved Joint. 1. A glue joint that is poorly bonded because an insufficient quantity of adhesive remained in the joint. 2. See Shrinkage, 1.

Statement of Changes in Financial Position. A financial schedule that shows how funds were obtained and where funds were used; also called Statement of Sources and Application of Funds.

Statement of Sources and Application of Funds. See Statement of Changes in Financial Position.

Stateroom. A private passenger compartment on a ship or train.

Static Bending. Bending under a constant or slowly applied load; flexure.

Static Force. Static Load.

Static Friction. The tangential surface resistance between two bodies in contact which move or tend to move with respect to each other.

Static Head. . Pressure of fluid expressed in terms of height of column of the fluid, such as water or mercury.

Static Load. A non-moving load imposed on a structure.

Static Pressure. Condition that exists when an equal amount of air is supplied to and exhausted from a space; at static pressure, equilibrium has been reached.

Static. 1. Stationary, not acting or changing; passive. 2. Concerned with bodies at rest or forces in equilibrium. 3. In spot, seam and projection welding, the force between the electrodes under welding conditions but with no current flowing and no movement in the welding machine.

Statical Moment. The product of a given area and the distance from its centroid to a point of rotation.

Statically Determinant Structure. A structural frame in which the bending moments and reactions can be determined by the laws of statics alone.

Statically Indeterminant Structure. A structural frame in which the bending moments and reactions cannot be calculated from the equations of statics.

Statics. The science of bodies at rest or forces in equilibrium.

Station Yards of Haul. In hauling excavated earth or fill, the number of cubic yards material multiplied by the number of 100 foot stations through which it is moved.

Stationary Blade Compressor. Rotary pump which uses a non- rotating blade inside pump to separate intake chamber from exhaust chamber.

Stationary. Remaining in one place; not portable; fixed in position.

Stator, Motor. See Motor Stator.

Stator. A stationary part in a machine in or about which a rotor revolves.

Statute of Frauds. A statute that requires certain contracts to be evidenced by a writing.

Statute of Limitations. The period of time after a cause of action arises before the expiration of which a plaintiff must file suit, or lose the right to do so.

Statute of Repose. A statute of limitations, measured from the completion of a work of improvement, protecting builders, contractors, architects, and engineers from liability for defects for an indefinite time in the construction of the improvement.

Stay-Tackling. Temporary tacking of the carpet in stages during a long stretch with the power stretcher or knee kicker, to hold the stretch until the end of the carpet can be kicked over the pins.

STC. 1. Sound Transmission Coefficient. 2. Sound Transmission Class.

Std. Standard.

Steam Bending. The process of forming curved wood members by steaming or boiling the wood and bending it to a desired shape.

Steam Boiler. A boiler for producing steam.

Steam Clean. The act or process of cleaning a structure with a machine that provides pressurized steam through a nozzle.

Steam Curing. Curing of concrete or mortar in water vapor at atmospheric or higher pressures and at temperatures between about 100° F and 420° F (40° C and 215° C).

Steam Distilled Wood Turpentine. Turpentine made from pine tree stumps by treating shredded chips with live steam to produce a distillate which is fractionated to yield turpentine, pine oil, and solid residue.

Steam Heating. Heating system in which steam from a boiler is piped to radiators in space to be heated.

Steam Jet Refrigeration. Refrigerating system which uses a steam venturi to create high vacuum (low pressure) on a water container causing water to evaporate at low temperature.

Steam Kettle. A metallic vessel for creating and holding steam, used in a commercial kitchen.

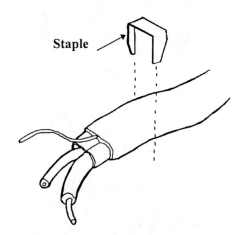

Staple

Steam Meter. An instrument for measuring the passage of steam.

Steam Sterilizer. A machine using steam to sterilize implements.

Steam Table. See Bain-Marie.

Steam Trap. A device that allows the passage of a condensate or air, but prevents the passage of steam.

Steam. The invisible gaseous state of water, produced by boiling.

Steamer, Electric. See Electric Steamer.

Steatite Porcelain. A vitreous ceramic whiteware for technical application in which magnesium metasilicate (MgO, SiO_2) is the essential crystalline phase.

Steatite Talc. Massive talc or the pulverized product, thereof having the general formula 3MgO 4Si O_2 H_2O.

Steel Access Door. Fabricated small steel door and frame, usually flush with adjoining surface to provide access to concealed equipment or system components for inspection and maintenance purposes.

Steel Access Hole Cover. Sheet metal used to cover access hole in deck to facilitate welding operation.

Steel Access Panel. Removable panel, usually flush with adjoining surface to provide access to concealed equipment or system components for inspection and maintenance purposes.

Steel Adjusting Plate. Sheet steel or segments of deck used to fill narrow areas where regular deck units cannot be accommodated.

Steel Anchor Bolt, Foundation. Bolts that tie sill plate and frame of structure to its foundation.

Steel Anchor Bolt, High-Strength. High strength steel bolt used to secure wood construction to concrete or masonry construction.

Steel Anchor Bolt. Steel bolt used to secure wood construction to concrete or masonry construction.

Steel Angle Lintel. Lintel made of structural steel angle.

Steel Angle Sill. Steel angle in concrete or masonry used as a sill.

Steel Angle. An L-shaped member constructed of steel, often used as a lintel or carrying shelf for masonry.

Steel Base Plate. Foundation plate of metal on which heavy piece of machinery or column rests; plate is usually set on masonry or concrete.

Steel Bearing Plate. Flat steel plate attached to supporting building component to uniformly transmit loads to that supporting component.

Steel Bolt, Common Grade. Bolt made of carbon grade steel for use in ordinary bolted connections in contrast to high strength bolted connections.

Steel Bolt, High-Strength. Quenched and tempered or heat-treated steel bolts with higher tensile strengths.

Steel Bolt, Interference Body. High strength bolt with hardened steel ribs on the bolt shank to facilitate driving of bolts in misaligned holes.

Steel Bolt, Load Indicator. High strength bolt with manufactured notch between bolt tip and threads designed to react to opposing rotational torques applied by installation wrench so bolt tip automatically shears off when proper tension of bolt is achieved.

Steel Bolt, Stainless. Stainless steel bolt used in ordinary bolted connections in contrast to high strength bolted connections.

Steel Bolt, Tamperproof. Steel bolt with a head that prevents bolt from being removable.

Steel Bolt, Weathering. Bolt made of steel to be compatible with weathering steel.

Steel Bolt. Metallic pin or rod having a head at one end and threads to receive a nut on the other used for holding members or parts of members together; head may be round, square, flat, oval, hexagonal, or other shapes.

Steel Box Beam. A hollow built-up horizontal structural member.

Steel Cant Strip. Sloped sheet metal strip used as perimeter of roofing membrane to transition membrane from horizontal to vertical surface.

Steel Closure Plate, Deck. Sheet metal plate used to close openings formed between flutes of composite deck and abutting walls or partitions.

Steel Cofferdam. A watertight steel enclosure from which water is pumped to expose the bottom of a body of water and permit foundation construction.

Steel Column Closure. Sheet metal used to close openings around structural steel penetrations of deck.

Steel Column. Long relatively slender vertical load bearing steel member.

Steel Conduit. A pipe, tube, or channel used to enclose electric wires or direct the flow of a fluid.

Steel Control Joint Cover. Exposed linear steel device to cover and conceal a control-joint.

Steel Corner Guard. Steel covering over lower portion of exterior corner of intersection of walls or at corner of column to protect from damage.

Steel Cover Plate and Frame. Sheet steel used to cover non-lapping end joints of deck.

Steel Crane Rail, Overhead, Top Running. Special structural steel shape to support rolling crane.

Steel Deck Institute (SDI). PO Box 3812, St. Louis, Missouri 63122, (314) 965-1741.

Steel Door Institute (SDI). 30200 Detroit Road, Cleveland, Ohio 44145, (216) 899-0010.

Steel Embedment Angle. Steel angle embedded in concrete or masonry; used to anchor or support other materials.

Steel Embedment Plate. Rolled steel plate embedded in concrete or masonry; used to anchor or support other materials.

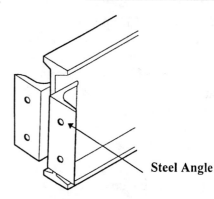

Steel Angle

Steel End Closure. Formed sheet metal at 90 degrees to deck profile to contain wet concrete or gypsum fill during placement at slab edges.

Steel Eye Bolt. Steel bolt comprised of a threaded shank with a lopped head designed to accept a hook, cable, or rope.

Steel Fire Escape. Continuous unobstructed route of escape from building in case of fire constructed of steel components, usually located and mounted on exterior of building and composed of stairs, ladders, and landings.

Steel Float. A concrete finisher's tool used to apply the final finish on a concrete slab, also used to apply stucco or plaster.

Steel Floor Deck, Cellular, Wire Raceway. Steel floor deck fabricated of two sheets to form linear voids and serve as a raceway for conductors.

Steel Floor Deck, Cellular. Steel floor deck which during construction supports wet concrete and construction loads, but after concrete cures does not necessarily perform a structural function in completed construction; deck is fabricated of two sheets to form linear voids which may be used as raceways.

Steel Floor Deck, Composite. Steel floor deck which during construction supports wet concrete and construction loads, but after concrete cures and reaches design strength, acts together with concrete to support dead and live loads; deck and concrete are interlocked by shape of deck.

Steel Floor Deck, Form. Steel floor deck which during construction supports wet concrete and construction loads, but after concrete cures and reaches design strength, does not perform structural function in completed construction.

Steel Floor Deck, Non-Cellular Form. Steel floor deck which during construction supports wet concrete and construction loads, but after concrete cures and reaches design strength, does not perform a structural function in completed construction.

Steel Floor Deck. Sheet steel formed to fluted or ribbed profile to span between supports to support floor system and live loads.

Steel Floor Plate, Patterned. Steel plate with raised pattern wearing surface.

Steel Floor Plate, Raised Tread. Steel plate with raised tread wearing surface.

Steel Floor Plate. Steel plate usually with raised pattern to provide nonslip wearing surface.

Steel Forms. A temporary structure for the support of concrete while setting.

Steel Frame. 1. Frame made of steel. 2. A structural steel framed structure.

Steel Grate. A screen made from sets of parallel steel bars placed at right angles to each other, and which allows water to drain.

Steel Grid Walkway. Walkway fabricated of steel grid placed over roof surface to protect roofing surface from damage from traffic.

Steel Gutter. A shallow channel constructed of galvanized sheet steel, positioned just below and along the eaves of a building for collecting and diverting water from a roof.

Steel Hatch. An opening in a floor or roof having a hinge or completely removable cover constructed of steel.

Steel Joist Bridging. Diagonal or longitudinal structural steel members used to keep joists properly spaced, in lateral position, vertically plumb, in order to distribute load.

Steel Joist Girder. Steel joist of deeper and heavier construction used as primary framing members to support other structural members at points along its span.

Steel Joist Institute (SJI). 1205 48th Avenue North, #A, Myrtle Beach, South Carolina 29577, (803) 449-0487.

Steel Joist, DLH Series. Long span steel joist of deeper and heavier construction for longer spans.

Steel Joist, K Series. Open web, parallel chord, load-carrying members suitable for direct support of floors and roof decks, utilizing hot rolled or cold formed steel.

Steel Joist, LH Series. Steel joist of deeper and heavier construction for longer spans.

Steel Joist. Open web, parallel chord, load-carrying members suitable for direct support of floors and roof decks, utilizing hot rolled or cold formed steel.

Steel Ladder Cage. Open steel framework enclosing ladder on open side to prevent falls from ladder; commonly required by safety codes on high ladders.

Steel Lag Screw. Threaded wood screw with square head, to be turned by wrench; usually longer and heavier than common wood screw with coarser thread and larger head for heavier wood construction.

Steel Lintel. 1. In steel construction, a steel member over an opening to carry loads above the opening. 2. A steel member placed within a masonry wall or partition to support masonry or other construction over the opening.

Steel Lock Washer. Split steel washer which compresses when bolt or screw is tightened to prevent loosening of tightened nut or screw.

Steel Nut, Common Grade. Nut made of carbon grade steel for use with carbon grade bolts in contrast to high strength bolted connections.

Steel Nut, High-Strength. Nut made of alloys for use with high strength bolts.

Steel Nut, Self-Locking. Nut with locking pin that slides along bolt threads. Reversing direction of locking allows nut to be removed without damaging nut or bolt.

Steel Nut, Weathering. Nut made of steel to be compatible with weathering steel.

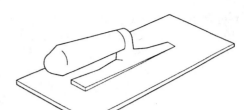

Steel Float

Steel Pile. A long, slender, piece of steel driven into the ground to act as a foundation.

Steel Pin and Roller. Joint used in steel bridge construction to allow for rotation and movement.

Steel Pipe, Seamless. Seamless steel pipe manufactured by a spinning process.

Steel Plate Column Wainscot. Steel covering over lower portion of column to protect it from damage.

Steel Plate. Sheet steel of a heavier thickness.

Steel Pole. A long steel pole installed vertically in the ground.

Steel Rail Standard. Vertical component of railing system supporting handrail.

Steel Repair. The act or process of restoring steel.

Steel Ridge Plate. Sheet metal plate, placed over ridge formed when two differing roof slopes intersect.

Steel Rivet. Metallic pin or rod having head at one end which is inserted through holes in materials to be fastened together; protruding end is flattened (mushroomed) to tie two pieces together.

Steel Roof Deck Diaphragm Action. In-plane action of roof deck system such that columns framing into roof from below and above are maintained in their same position relative to each other.

Steel Roof Deck, Acoustical. Steel roof deck with vertical faces perforated to increase room sound absorption properties and flutes filled with sound absorbing material.

Steel Roof Deck, Cellular. Sheet steel formed to fluted or ribbed profile to span between supports to support roofing system and live loads; deck is fabricated of two sheets to form linear voids which may be used to contain electrical conductors.

Steel Roof Deck, Composite. Steel roof deck which during construction supports wet concrete and construction loads, but after concrete cures, acts together with concrete to support dead and live loads; deck and concrete are interlocked by shape of deck or by mechanical means.

Steel Roof Deck, Form. Steel roof deck which during construction supports wet concrete and construction loads.

Steel Roof Deck, Non-Cellular. Steel roof deck which during construction supports wet concrete loads, but after concrete cures and reaches design strength, does not perform structural function on completed construction.

Steel Roof Deck, Vented. Steel deck with perforations or slots to aid in curing and exiting of moisture from concrete or gypsum fill placed on deck.

Steel Roof Deck. Sheet steel formed to fluted or ribbed profile to span between supports to support roofing system and live loads.

Steel Roof Panel. A rectangular sheet of steel used in roof covering applications.

Steel Roof Walkway. Fabricated steel walkway used over roof surfaces.

Steel Screw, Self-Tapping. Type of screw which drills its own hole through two pieces to be joined, then taps a thread in both pieces while fastening them together.

Steel Screw, Tamperproof. Steel screw with a head that prevents removal.

Steel Screw. Steel pin or rod having head at one end and helically threaded with pointed end on other for penetrating material by being turned used for holding members or parts of members together; head may be round, square, flat, oval, hexagonal, or other shapes.

Steel Sheet Piling. A steel plank in close contact or interlocking with other steel planks to provide a tight wall to resist the lateral pressure of water, adjacent earth, or other materials at excavations.

Steel Shingle. A roof covering or wall finishing unit made of steel applied on roof systems or exterior wall coverings.

Steel Ships Ladder. Ladder fabricated of steel inclined between 50 degrees and 77 degrees with steps in place of rungs and provided with railings.

Steel Siding. Formed and finished steel panels used as exterior wall coverings.

Steel Sill Angle. Steel angle used as a sill.

Steel Square. A carpenter tool which establishes a right angle between two components; the large arm of the square is called the Body or Blade; the smaller arm, at a 90 degree angle to the blade, is called the Tongue; the point where the outside edges of the blade and tongue join is called the Heel; the surface with the manufacturer's name is called the Face; the opposite surface is called the Back.

Steel Stair, Spiral. Stair of steel construction with a closed circular plan, uniform sector shaped treads and supporting center column.

Steel Stair. Flight or series of connected stair flights fabricated of steel members extending between two or more floors within given floor area.

Steel Strap. A steel plate fastened across the inner section of two or more timbers.

Steel Stringer. Steel member upon which stair treads bear.

Steel Structures Painting Council (SSPC). 4516 Henry Street, #301, Pittsburgh, Pennsylvania 15213, (412) 687-1113.

Steel Stud. 1. Steel pin or rod having head at one end for driving into material used for holding members or parts of members together. 2. In light gage construction, a vertical framing member similar to a 2x 4 stud.

Steel Sump Pan. Sheet metal pan forming low point in roof deck to collect water and receive roof drain.

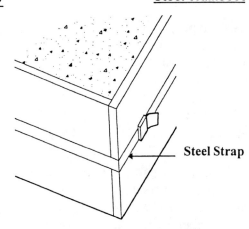

Steel Strap

Steel Tank Institute (STI). 570 Oakwood Road, Lake Zurich, Illinois 60047-1559, (708) 438-8765.

Steel Tank. A receptacle manufactured from steel for holding, transporting or storing liquids.

Steel Toggle Bolt. Steel bolt and nut assembly to fasten objects to hollow construction; assembly from only one side; nut has pivoted wings that close against spring when nut end of assembly is pushed through hole and spring open on other side in void of construction assembly.

Steel Toilet Partition Support Framing. Framing, usually above a ceiling used to support ceiling-hung toilet partitions.

Steel Transom. A steel transverse piece in a structure, lintel; a steel horizontal crossbar in a window, over a door, or between a door and a window or fanlight above it.

Steel Valley Plate. Sheet metal plate, placed over valley formed when two differing roof slopes intersect.

Steel Wainscot. Steel covering over lower portion of wall or partition to protect from damage.

Steel Washer, High-Strength. Flat ring of steel which may be plain split, toothed, or embossed, used in threaded connections to distribute loads, span large openings, relieve friction, or prevent loosening, made of high strength steel; often used with high strength bolts.

Steel Washer, Load Indicator. Compressible washers with direct tension indicator capable of indicating achievement of minimum bolt tension of high strength bolts.

Steel Washer, Weathering. Washer made of steel to be compatible with weathering steel.

Steel Washer, Welding. Washer made of steel to be used for welding in steel fabrication.

Steel Washer. Flat ring of steel which may be plain split, toothed, or embossed, used in threaded connections to distribute loads, span large openings, relieve friction, or prevent loosening.

Steel Welding Washer. Washer for containing arc spot welds on thin sheet metal (thinner than 0.028 inch).

Steel Welding. A localized heating of two pieces of steel to melting until there is a fusing of the material.

Steel Window Institute (SWI). 1300 Summer Avenue, Cleveland, Ohio 44115, (216) 241-7333.

Steel Wood Screw. Steel screw with threads designed for penetration of wood and similar materials.

Steel Wool. An abrasive material consisting of a mass of fine steel shavings, used for polishing and burnishing.

Steel, Structural. Steel that is rolled in a variety of shapes and manufactured for use as structural load-bearing members.

Steel. Iron compounded with other metals to increase strength and wearing or rust resistance.

Steep Asphalt. A roofing asphalt that has a softening point of approximately 190° F (88° C) and that conforms to the requirements of ASTM Standard D 312, Type III.

Steep Slope Roof. Roofing that is over 1-1/2 inches per foot slope or pitch.

Steeple. A spire, as on a church.

Steeplejack. A person who climbs tall structures, usually with safety rigging, for painting and repairs.

Stem. The vertical part of a concrete or masonry retaining wall.

Stenciling. Placing a design on a wall or other surface by applying the finish through a template cut out of thin, flat paper or metal.

Step Flashing. Flashing and counterflashing installed where a sloping roof meets a vertical wall, as at a chimney.

Step Ladder. A short free-standing ladder with flat steps.

Step Return. A bottom step, the nosing of which extends out considerably over the riser; it is frequently wider than the rest of the stairs, and is usually curved at the ends; also called a Bullnose.

Stepped Foundation. A benched foundation on a sloping site with all bottoms flat to prevent sliding.

Stepping Stone. 1. A raised stone used in crossing a stream. 2. Spaced flat stones used as a walk in a garden.

Steri Lamp. Lamp with a high-intensity ultraviolet ray used to kill bacteria; used in food storage cabinets and in air ducts.

Sterilizer, Medical. See Medical Sterilizer.

Sterilizer. A device which destroys or eliminates all forms of bacteria, fungi, viruses, and their spores.

Stethoscope. Instrument used to detect sounds and locate the origin.

STI. Steel Tank Institute.

Stick Shellac. Shellac which comes in sold stick form; used extensively for furniture patching.

Stick, Hook. See Hook Stick.

Stick. 1. A woody piece of twig or branch of a tree. 2. A slender piece of wood. 3. To pierce with a pointed instrument. 4. Adhesive quality.

Stick-Built. Construction of a wood framed building piece by piece on-site as contrasted to prefabrication.

Sticker Stain. A brown or blue stain that develops in seasoning lumber where it has been in contact with the stickers.

Sticker. A piece of metal channel inserted in concrete or masonry walls for the attachment or support of wall furring channels.

Stickers. Strips or boards used to separate the layers of lumber in a pile and thus improve air circulation.

Stickyback. A transparent decal containing technical information that can be adhered to a tracing.

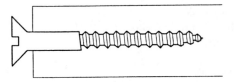

Steel Wood Screw

Stiff Mud Brick. Brick produced by extruding a stiff but plastic clay through a die.

Stiffback. A structural member placed on top of ceiling joists at midspan to line up the joists before plastering or drywall installation.

Stiffener Plate. A steel plate attached to a structural member to support it against heavy localized loadings or stresses.

Stiffener. A horizontal metal shape tied to vertical members (studs or channels) of partitions or walls to brace them.

Stiffness. The quality of resistance to deformation on the part of a material, a component member of a structure, or the whole structure.

Stile. A vertical framing member in a panel door.

Still Water. A part of a stream where the gradient is so gentle that no current is visible.

Stillson Wrench. Pipe wrench.

Stilts. Adjustable extensions worn on the shoes of workers, such as plasterers or painters, to enable reaching higher than normal areas.

Stipple Finishing. Finish obtained by tapping surface with stipple brush before paint is dry.

Stippling. The method of shading by the making of separate marks or points.

Stipulated Sum Contract. A construction contract where the contract sum is a lump sum agreed amount; also called a Lump Sum Contract.

Stirrup Pump. A portable hand pump stabilized with a foot bracket, used especially for fire-fighting.

Stirrup Tie. A stirrup that forms a complete loop, as differentiated from a U-stirrup, which has an open top.

Stirrup. Reinforcing bars used in beams for shear reinforcement; typically bent into a U-shape or box-shape and placed perpendicular to the longitudinal steel.

Stock Dyeing. Dyeing raw fibers before they are carded (combed) or spun.

Stockholder's Equity. Section of the balance sheet also known as Net Worth or Equity; representing the claim of the holders of common stock on the assets of the business.

Stoke. A unit of kinematic viscosity, equal to the viscosity of a fluid in poises divided by the density of the fluid in grams per cubic centimeter.

Stone Facing Guard. Steel components applied to face of vertical stone construction to protect it from damage.

Stone Paver. Blocks of rock processed by shaping, cutting or sizing, used for driveways, patios and walkways.

Stone. 1. Earthy or mineral matter of indeterminate size or shape such as rock. 2. Use of a carborundum stone to eliminate jagged and flaked edges of tile or masonry due to cutting.

Stonemason. A mason who builds in stone.

Stonemasonry. Stonework; the work of a stonemason.

Stoneware. A vitreous or semivitreous ceramic ware of fine texture, made primarily from nonrefractory fire clay.

Stonework. Masonry of stones and mortar.

Stool. A wood molding that provides an interior ledge at the sill of a window.

Stoop. A small unroofed porch.

Stop Box or Curb Box. An adjustable cast iron box that is brought up to grade with a removable iron cover; by inserting a shutoff rod down into the stop box it is possible to turn off the water supply at the curb cock.

Stop Notice. A charge against construction funds in the hands of a property owner or a construction lender for the value of work or materials incorporated into a construction project; this is a lien on funds to be paid whereas a mechanic's lien is a lien on property; this is usually the only form of remedy to one who has done work for a Public Works Agency.

Stop, Door. See Door Stop.

Stop, Gravel. See Gravel Stop.

Stop. A molding covering the crack between a door or window and its frame.

Storage Life. See Shelf Life.

Storage Tank Removal. The act or process of removing an old storage tank.

Storage Tank. A large receptacle used for the holding, transporting, or storage of liquids or gases.

Store Room. A room in which items are stored.

Store. 1. A room or building for retail sales; a shop. 2. A reserve supply of goods.

Storefront. 1. The facade which is constructed on the street side of a building or structure into which persons can enter and transact business. 2. A steel or aluminum tube frame and glass wall.

Storm Drain. A drain used for carrying rain water, cooling water, and sub-surface water, but not waste or sewage.

Storm Sash. An extra window usually placed on the outside of an existing window as additional protection against cold weather and air infiltration; also called a Storm Window.

Storm Sewer. A sewer used for conveying groundwater, rainwater, surface water, or similar nonpollutional wastes.

Storm Water Drainage System. The piping system used for conveying rainwater or other precipitation to the storm sewer or other place of disposal.

Storm Window. See Storm Sash.

Story High. The height between floors or floor joists.

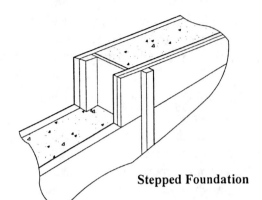

Stepped Foundation

Story Pole. 1. A rod or pole that is the height of the story marked on it from joist to joist and also has the brick courses and any openings marked on it. 2. In ceramic tile installation, a long strip of wood marked at the appropriate joint intervals for the tile to be used; it is used to check the length, width, or height of the tilework; also called an Idiot Stick. 3. Wooden rod used to measure heights and locate rows of siding.

Story. The space within a building between two adjacent floor levels or between a floor and a roof.

Straight Arch. See Jack Arch.

Straight Joint. The usual style of laying tile where all the joints are in alignment.

Straight Peen Hammer. A hammer where the face opposite the flat face is shaped for cutting or grooving metal; the peen is flat and is perpendicular to the handle.

Straight Polarity. The arrangement of direct current arc-welding leads wherein the work is the positive pole and the electrode is the negative pole of the welding arc.

Straight Sheathing. Wood sheathing with the individual boards running at a 90 degree angle to the studs, joists, or rafters.

Straightedge. 1. A straight piece of lumber or metal that is used to rod mortar, align tile, or provide a straight or flush surface. 2. A flat wooden tool or rod, perfectly true, used to straighten the brown coat or screeds.

Straight-Grained Wood. Wood in which the fibers run parallel to the axis of the piece.

Straight-Line Depreciation. A method of calculating depreciation in which the same amount is depreciated each year.

Strain Release. Movement along a fault plane; can be gradual or abrupt.

Strain. 1. Change in shape of a material due to stress. 2. To exert to the utmost. 3. To filter.

Strainer Basket. Device in skimmer and input side of pump used to catch large pieces of debris in pool water.

Strainer. Device such as a screen or filter used to retain solid particles while liquid passes through.

Strandboard. See Oriented Strand Board.

Stranded Conductor. A number of fine wires twisted around a center wire or core, used as a single electric conductor.

Stranded Wire. Fine wires twisted together in a group to create a larger stronger cable or wire.

Strap Channel. A U-shaped or L-shaped iron plate for connecting two or more timbers.

Strap Hinge. A surface hinge of which one or both leaves are of considerable length.

Strap Tie. A metal plate that fastens two parts together as a post, rod, or beam.

Strap, Malleable. See Malleable Strap.

Strap, Steel. See Steel Strap.

S-Trap. A plumbing fitting used to trap gases from a waste system.

Stratification of Air. Condition in which there is little or no air movement in room; air lies in temperature layers.

Strawberry. A small bubble or blister in the flood coating of a gravel-surfaced roof membrane.

Street Ell. A 45 or 90 degree pipe elbow with male threads on one end and female threads on the other end; also called a Service Ell.

Street Removal. The act or process of removing old asphalt from a roadway.

Street Tree. A tree, planted in a specially prepared well, in the sidewalk or parkway adjacent to a public way like a street.

Street. A public way to accommodate automobiles.

Strength Design. A design theory used for most reinforced concrete designs; a design in which members are designed to fail; safety is not provided by limiting stresses, but by using a factored load that is greater than the actual load. The nominal strength multiplied by a strength reduction factor.

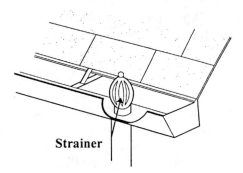

Strainer

Strength of Materials. A branch of mechanics and experimental physics dealing with stresses, strains, and the general behavior or materials and structural elements under the action of forces and moments.

Strength Ratio. In wood, the hypothetical ratio of the strength of a structural member to that which it would have if it contained no strength-reducing characteristics such as knots, slope-of-grain, shakes, or other defects.

Strength Reduction Factor. A factor introduced into the calculation of strength design in reinforced concrete, which increases actual loading, providing an added factor of safety.

Strength, Early. Concrete or mortar strength during the first 72 hours after placement.

Strength, Required. See Required Strength.

Strength. Capacity to resist force.

Stress Diagram. A graphic representation of the value of the stresses in a structural member.

Stress Relief. See Perimeter Relief.

Stress. 1. The internal force of a body that resists external force. 2. The intensity of force per unit area.

Stressed-Skin Panel. A panel consisting of two face sheets of wood or metal bonded to perpendicular spacer strips.

Stress-Grade Lumber. Structural lumber that has been graded and stamped with information to indicate the specific load it will support.

Stress-Relief Heat Treatment. Uniform heating of a structure or portion thereof to a sufficient temperature, below the critical range, to relieve the major portion of the residual stresses, followed by a slow uniform even cooling.

Stretch. The width of the area on which a painter will normally apply paint across a ceiling or down a side wall.

Stretcher, Hospital. See Hospital Stretcher.

Stretcher. A brick that is laid on its flat side and shows its face on the surface of the wall.

Stria. A striped surface effect obtained by loosely twisting two strands of one shade of yarn with one strand of a lighter or darker shade.

Strict Liability. Liability without fault, imposed by the courts for social reasons to distribute the risk of injury to consumers by imposing on the manufacturers of products liability for injury caused by the use of those products even though the manufacturer is not shown to be at fault.

Strike Plate. See Strike, 2.

Strike Slip Fault. A geological fault which runs parallel to the strike of associated strata.

Strike. 1. An organized refusal by employees to work until some grievance is remedied. 2. A metal plate or box that is pierced or recessed to receive the bolt or latch when projected; also called the Keeper or Strike Plate.

Striking In. Materials used in finishing are said to strike in when they soften undercoats and sink into them.

Striking Joints. A process of removing excess grout from the joints by wiping with a sponge or cloth or scraping, compacting or rubbing with a curved instrument.

String Course. A horizontal course of brick or stone on the face of a building; when continuous with a row of window sills or lintels, is referred to as a Sill Course or a Lintel Course; also called a Belt Course.

String Level. A small spirit level attached to a string line for establishing level lines on a construction site.

String Wire. Wire used on open stud construction, placed horizontally around building to support weatherproofing paper.

String. 1. A light cord or twine. 2. See Stringer.

Stringer Bead. A type of weld bead made without appreciable transverse oscillation.

Stringer. 1. The sloping wood or steel member supporting the treads of a stair. 2. A long horizontal timber in a structure supporting a floor.

Stringing Mortar. The procedure of spreading enough mortar on a bed to lay several masonry units.

Stringline. A string used to establish a line or to mark a line on a construction site.

Strip Flooring. Hardwood flooring laid in strips, usually tongued and grooved.

Strip Footing. See Continuous Footing.

Strip Lath. A narrow strip of diamond mesh metal lath sometimes applied as reinforcement over joints between sheets of gypsum lath, at the juncture of two different base materials, and at corners of openings.

Strip Mopping. The mopping of roofing bitumen in parallel bands, generally about 8 inches in width with 4 inch unmopped spaces.

Strip, Cant. See Cant Strip.

Strip, Chamfer. See Chamfer Strip.

Strip. 1. Complete removal of an old finish with paint removers. 2. The removal of wallpaper. 3. To remove forms from set concrete surfaces. 4. To damage the threads of a bolt or screw. 5. A long, narrow, piece of a material.

Striping. Edge painting prior to priming.

Stripping. 1. The technique of sealing the joint between built-up roofing membrane and metal using one or two plies of felt or fabric and hot or cold applied bitumen. 2. The method of taping the joints between insulation board.

Strips. Bands of reinforcing bars in flat slab or flat plate construction.

Stroboscope. An instrument used for determining speeds of rotation by shining a flashing bright light at measured intervals so that a rotating object appears stationary.

Stroke. A single pass with a spray gun in one direction.

Struck Joint. In masonry, a mortar joint which is formed with a recess at the bottom of the joint; it is the reverse of the weathered joint; this joint is used extensively, but chiefly for interior wall surfaces, since it is inferior for outside joints, because of its lack of weather-resisting qualities; the recess at the bottom allows water from rain or snow to seep into the wall.

Structural Bond. The interlocking pattern of masonry units used to tie two or more wythes together in a wall.

Structural Clay Products Institute (SCPI).

Structural Clay. A term applied to various sizes and kinds of hollow and practically solid building units, molded from surface clay, shale, fire clay or a mixture of these materials and laid by masons.

Structural Component. A member of a structural system.

Structural Composite Lumber. Lumber in small pieces glued together to make structurally sound larger members, as in glued laminated beams or columns.

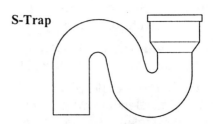

S-Trap

Structural Defects. 1. Defects in a structure that pertain to the structural elements which may cause an unsafe condition. 2. In tilework, cracks or laminations in the body of the tile which detract from the aesthetic appearance or the structural soundness of the tile installation.

Structural Design. The design of the structural system, analysis of support of all dead and live loads, and calculation of sizes and arrangement of all members.

Structural Diaphragms. Structural members, such as floor and roof slabs, which transmit inertial forces to lateral-force-resisting members.

Structural Engineer. A civil engineer who specializes in the design and analysis of structures.

Structural Engineers Association of California (SEAOC).

Structural Fee. The amount of money charged for structural engineering work.

Structural Frame. The columns and girders, trusses, beams and spandrels which directly connect to the columns and all other members necessary to the stability of the building; secondary members are those members of the floor or roof panels which are not connected to the columns and are not part of the structural frame.

Structural Glazed Tile. A hollow clay masonry unit with glazed faces.

Structural Lightweight Concrete. See Concrete, Structural Lightweight.

Structural Lumber Exposed To View. Structural lumber which is exposed to view in finished spaces or at the exterior of the building.

Structural Lumber. 1. Wood members of a structural system which are manufactured by sawing, resawing, passing lengthwise through standard planing machine, crosscutting to length, but without further manufacturing. 2. Lumber that is 2 or more inches thick and 4 or more inches wide; intended for use where strength is required; the grading of structural lumber is based on the strength of the piece and its use.

Structural Pipe. Pipe used in a structure to transfer imposed loads to the ground.

Structural Plywood Diaphragm. Structural plywood sheathing on floors, roofs, or walls which provides shear strength to resist wind and earthquake loads.

Structural Plywood. The highest grade of plywood; plywood of exterior structural grade, secured to top side of floor joists used to create rigidity in building superstructure and also to provide a smooth and even surface to receive finish floor covering.

Structural Shear Wall. Wall constructed of studs and structural-use panel sheathing that in its own plane resists shear forces resulting from applied wind, earthquake, or other transverse loads or provides frame stability.

Structural Steel Door Frame. Door frame fabricated of structural steel shapes, such as channel.

Structural Steel Tubing. Hollow structural tubing, usually circular, square, or rectangular.

Structural Steel, Carbon Grade. Steel having either (1) no specified minimum content of alloying elements; (2) specified minimum copper content not exceeding 0.40 percent, or (3) maximum specified percentage contents as follows. manganese 1.65, silicon 0.60, copper 0.60.

Structural Steel, Corrosion Resistant. Steel having chemical composition specifically developed to impart higher mechanical property values and with resistance to atmospheric corrosion than is four times greater than that of conventional carbon structural steel; this steel weathers (oxidizes).

Structural Steel, High-Strength. Steel having chemical composition specifically developed to impart higher mechanical property values than is obtainable from conventional carbon structural steel.

Structural Steel, Quenched and Tempered. Steel which is quenched (rapidly cooled) and tempered (reheated, quench hardened, or normalized ferrous alloy to temperature below transformation range, followed by cooling of alloy at rate desired); primarily used for welded bridge construction.

Structural Steel. Steel hot-rolled into variety of shapes for use as load-bearing structural members.

Structural Subflooring. Flat sheet material of structural use panel secured to top side of floor joists used to create rigidity in building superstructure and serve as base to receive flooring.

Structural System. All of the supporting parts of a building taken as an integrated whole.

Structural Timbers. Pieces of wood of relatively large size, with a cross section greater than 4 inches x 6 inches, the strength of which is the controlling element in their selection and use.

Structural Trusses. Assemblages of wood, steel, or reinforced concrete members subjected primarily to axial forces.

Structural Tube. A hollow metal product used to carry imposed loads in a building or structure.

Structural Tubing Railing. A metal railing made of structural metal tubing.

Structural Underlayment. Flat sheet material of structural use panel secured to top side of floor joists used to create rigidity in building structures and to provide smooth and even surface to receive finish floor covering.

Structural Use Panel Roof Sheathing. Flat sheet material of structural-use panels secured to exterior side of roof rafters or trusses, used to create rigidity in building superstructure and serve as base to receive roofing.

Structural Use Panel. Flat sheet material of either plywood, oriented strandboard, waferboard, or particleboard.

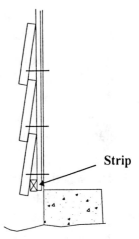

Strip

Structural Wall Sheathing. Flat sheet material of structural-use panel secured to exterior side of exterior wall studs, used to create rigidity in building superstructure and serve as base to receive siding or veneer construction.

Structural Wood Member. Member of a wood structural system.

Structure Borne Sound. A condition in which the sound waves are carried by a solid material.

Structure. That which gives form to something and works to resist changes in the form due to the action of various forces; a building.

Strut. 1. A compression member; a column; it can be placed at any angle, not just vertically. 2. An element of a structural diaphragm, used to provide continuity around an opening in the diaphragm.

Stub Out. A capped pipe or conduit provided for future connection or extension of water, gas, electrical, or other utility.

Stub Tie. A method of attaching metal lath to furring channels, where the wire is twisted and cut off at the twist.

Stub Truss. A truss which has one end truncated by being cut vertically.

Stucco Edge. In gypsumboard, a break in the fold or lapped seal between the face and back papers during the manufacturing process which allows the wet slurry to ooze onto edges and face.

Stucco Lath. Wood or metal lath strips which form the base for the application of a cement plaster on an exterior wall surface.

Stucco Mold. A wood or metal molding to retain the edges of stucco and provide a finish as at doors and windows.

Stucco Netting. 1. Galvanized woven fabric or chicken wire used in the lathing process for exterior stucco over light wood framing construction. 2. See Woven Wire Fabric.

Stucco. 1. A cement plaster used for coating exterior walls and other exterior surfaces of buildings. 2. A plaster used for interior decoration and finish work. 3. A manufacturing term for calcined gypsum prior to addition of water.

Stud Removal. The act or process of demolishing and carrying away old stud walls from a building or structure.

Stud Shear. A hand tool for cutting metal studs.

Stud, Screwable Type. See Screwable Type Stud.

Stud. 1. A vertical wall framing member, wood or steel, may be load bearing or non load bearing. 2. A vertical member to support sheathing of concrete forms. 3. A headed steel device used to anchor steel plates or shapes to concrete members.

Studding. The framework of a partition or wall of a house; usually referred to as 2 by 4's.

Studio Apartment. A small living unit consisting of a combination living-dining-kitchen-sleeping room and a bathroom.

Stuffer Warp. Yarn which runs lengthwise in the carpet but does not intertwine with any filling yarns; serves to give weight, thickness and stability to the fabric; also called Stuffer Yarn.

Stuffer Yarn. See Stuffer Warp.

Style. A distinctive manner in respect to appearance, as in the Greek style or the modern style.

Stylobate. A continuous rectangular base on which a row of architectural columns are placed.

Styrene-Butadiene Resin. Synthetic rubber resin; liquid styrene and butadiene gas, copolymerized to form chemical-resistant product with excellent film-forming properties.

Styrene-Rubber (SR) Pipe & Fittings. Plastics containing at least 50% styrene compounding materials but not more than 15 percent acrylonitrile.

Subbase Course. The course in the asphalt pavement structure immediately below the base course; if the subgrade soil is of adequate quality, it may serve as the subbase.

Sub-bidder. A subcontractor who submits a bid to a prime contractor.

Subcontractor List. A list of the subcontractors a general contractor contemplates using, submitted for the owner's approval, as required by some construction contracts.

Subcontractor. 1. A contractor whose contract is with the general contractor, not the owner. 2. An individual or entity contracting to perform part or all of another's contract.

Subcooling. Cooling of liquid refrigerant below its condensing temperature.

Subdiaphragm. A portion of a larger wood diaphragm designed to anchor and transfer local forces to primary diaphragm struts and the main diaphragm.

Subdivider. One who divides a tract of land into building lots, streets, and open spaces.

Subdivision. The division of a tract of land into building lots, streets, and open spaces.

Subfloor. A wood floor which is laid over the floor joists and on which the finished floor is laid.

Subflooring. Certain material, like plywood, that is installed on the floor joists of a building or structure, onto which the walls and finished flooring is attached.

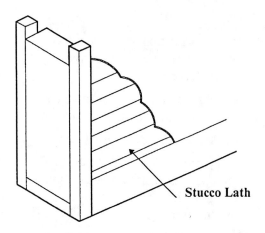

Stucco Lath

Subgrade, Improved. Subgrade, improved as a working platform (1) by the incorporation of granular materials or stabilizers such as asphalt, lime, or portland cement, prepared to support a structure or a pavement system, or (2) any course or courses of select or improved material placed on the subgrade soil below the pavement structure; subgrade improvement does not affect the design thickness of the pavement structure.

Subgrade. The soil prepared to support a structure or a pavement system; it is the foundation for the pavement structure; also called Basement Soil or Foundation Soil.

Subjacent Support. The right of land to be supported by the land adjacent to or below it.

Subject Property. The property that is the subject of a real estate appraisal report.

Sublease. Lease of property by a tenant to a subtenant; also called Subletting.

Subletting. See Sublease.

Sublimation. Conversion of a solid to a gas by heating and from a gas to a solid by cooling, for example, dry ice (carbon dioxide).

Submarginal. Falling below the accepted or required standard of quality.

Submarine. Of or below water.

Submersible Pump. A centrifugal pump, usually driven by electricity or compressed air, that will operate when submerged.

Submission Agreement. An agreement to submit an existing dispute to arbitration.

Submittal Schedule. A time schedule prepared by a contractor showing when each of the specified submittals will be sent to the architect or owner.

Submittals. Various documents submitted by the contractor to the owner or architect, such as shop drawings, material samples, subcontractor list, and insurance certificates.

Subordination Clause. A provision in a mortgage or trust deed that acknowledges that it is subordinate to another loan or will allow a future loan to be a superior lien on the property.

Subpoena Duces Tecum. A subpoena requiring the production of specified documents.

Subpoena. A legal order requiring a person to appear in a court, arbitration hearing, or deposition to give testimony.

Subpurlin. A small roof framing member that spans between joists and purlins.

Subrogation. The assumption by a third party, such as an insurance company, of another's legal right to collect a debt or damages.

Subsidiary Ledger. A supporting ledger consisting of a group of similar accounts, the total of which is in agreement with a controlling account in the general ledger.

Subsoil Drain. A drain used for collecting sub-surface or seepage water and carrying the water to where it can be disposed.

Substance. Any form of matter or material.

Substantial Compliance. A doctrine that may excuse minor violations of licensing statutes.

Substantial Performance. The doctrine that a contract is enforceable by a party who has not strictly complied with all requirements of a contract, but who has substantially complied therewith.

Substitutions. A substitution of materials for those originally specified for the project.

Substrata. Substrate.

Substrate. 1. A lower level, upon which other materials are applied, such as paint, flooring, or roofing. 2. The underlying support for ceramic tile or other finish-type installation. 3. The base or concealed layer of gypsum board in a composite assembly.

Subsubcontractor. An individual or entity contracting to perform part or all of a subcontractor's contract.

Subsurface Investigation. See Geotechnical Investigation.

Subsurface Right. The right to use the subterranean part of a parcel of land, such as for tunneling, extraction of minerals, and pipe lines, usually given in the form of an easement.

Subterranean Termite. See Termite.

Subterranean. Below the surface of the earth.

Subway. 1. An underground railway. 2. A passage under a street for pedestrians or utility lines. 3. An underpass.

Suction Diffuser. A centrifugal pump which forces water out of an area or surface.

Suction Feed. 1. In a sandblast gun, one in which the abrasive is siphoned to the nozzle. 2. In a spray gun, one in which the fluid is siphoned to the spray head.

Suction Line. Tube or pipe used to carry refrigerant gas from evaporator to compressor.

Suction Lines. In a swimming pool, pipework supplying the main circulating pump.

Suction Piping. That portion of the circulation piping located between the pool structure and the inlet side of the pump and usually including the main outlet piping, skimmer piping, vacuum piping, and surge tank piping.

Suction Pressure Control Valve. Device located in the suction line which maintains constant pressure in an evaporator during the running portion of a cycle.

Suction Pressure. Pressure in low-pressure side of a refrigerating system.

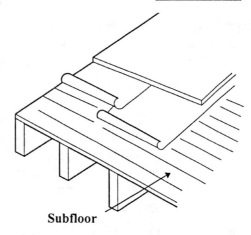

Subfloor

Suction Service Valve. Two-way manually operated valve located at the inlet to compressor; it controls suction gas flow and is used to service unit.

Suction Side. Low-pressure side of the system extending from the refrigerant control through the evaporator to the inlet valve of the compressor.

Suction Spotting. Spotting of paint job caused by oil in new coat being absorbed by spots or porous areas of surface.

Suction Valve. Valve in refrigeration compressor which allows vaporized refrigerant to enter cylinder from suction line and prevents its return.

Suction. The power of absorption possessed by a plastered surface.

Sue and Labor Clause. A clause in a property insurance policy that authorizes the insured to expend funds to protect the insured property.

Sulfate Attack. Deleterious chemical or physical reaction between sulfates in ground water or soil and certain constituents in cement, which result in expansion and disruption of the concrete.

Sulfate Resistance. Ability of cement paste and aggregate to withstand sulfate attack.

Sulphuric Acid. A dense corrosive oily colorless fluid that is a vigorous oxidizing and dehydrating agent; also called Oil of Vitriol.

Summary Judgment. A judgment awarded on the basis of affidavits and legal briefing rather than on the basis of evidence introduced at a trial.

Summer. 1. A horizontal beam such as a lintel. 2. A stone cap on a pier.

Sump Pot. See Main Outlet.

Sump Pump. A small capacity pump that empties pits receiving groundwater, sewage, or liquid waste.

Sump. A sump is a specially made receiving tank to receive wastes or sewage by gravity; from the sump the wastes or sewage is lifted by pump or ejectors to be discharged into the building drain or building sewer.

Sundays. Place skipped when applying finishing materials to a surface.

Sundial. An instrument showing the time by the shadow of a pointer (called a gnomon or stylus) cast by the sun on a graduated disc.

Sunshade. A device that protects from the sun's rays, such as an awning or eyebrow.

Superchlorination. Heavy dose of chlorine added to pool water to burn out nitrogen compounds when bacteria, algae, or ammonia build-up cannot be reduced by normal treatment.

Superconductivity. The property of zero electrical resistance in some substances at very low absolute temperatures.

Superelevation. The additional elevation of the outer side of a curved section of a roadway to overcome centrifugal forces of vehicles.

Superheat. 1. Temperature of vapor above its boiling temperature as a liquid at that pressure. 2. The difference between the temperature at the evaporator outlet and the lower temperature of the refrigerant evaporating in the evaporator.

Superheater. Heat exchanger arranged to take heat from liquid going to evaporator and using it to superheat vapor leaving evaporator.

Superimposed Load. Loads and stresses added to the dead load of the structure; live load.

Superintendent. 1. An individual who is at the top level of a construction team in the field. 2. A person with executive oversight, often of a board or building complex.

Superstructure. 1. The part of a building above the foundation. 2. In bridge construction, the concrete deck or traffic surface of the bridge.

Supervision. A contractor's scheduling, watching, and directing of workers during the execution of work.

Supplementary Conditions. A part of the contract documents that may supplement, delete, add to, or amend the general conditions.

Supplementary Drawings. Additional drawings which augment the contract drawings.

Supplier. One who supplies construction materials to a project.

Supply Air Duct. The duct from a heating, cooling, or ventilation system carrying the treated air to the space of use.

Supply Tank. Separate tank connected directly or by a pump to the oil-burning appliance.

Supports. 1. The structural foundation for essential building elements. 2. Devices for supporting and securing pipe, fixtures, and equipment.

Surcharge. 1. An increase in the lateral earth pressure of a retaining wall, caused by a vertical load behind the wall. 2. A load placed over an area to compact it or change its characteristics.

Surety Bond. A bond guaranteeing the performance of a contract or an obligation; see Performance Bond, Labor and Material Bond, and Completion Bond.

Surety. One who undertakes to guarantee performance by another; usually an insurance company.

Surface Bolt. A door bolt that is mounted on the surface of a door with the bolt sliding into a keeper in the top or side of the door frame or into the sill or floor.

Surface Depression. A mark or indentation on the gypsum board surface.

Surface Drying. Drying of a finishing material on top while the bottom remains more or less soft.

Surface Hinge. One having both leaves applied to the surface of the door.

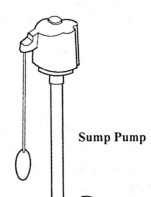

Sump Pump

Surface lumber. Lumber that is dressed by running it through a planer.

Surface Metal Raceway. Similar to a wireway, used for electrical conductors and communication wiring.

Surface Moisture. Free moisture retained on the surfaces of aggregate particles which becomes part of the mixing water in the concrete mix.

Surface Plate. Tool with a very accurate flat surface.

Surface Preparation. All operations necessary to prepare a surface to receive a coating of paint.

Surface Rights. The right to use the surface of real property, usually in the form of an easement.

Surface Skimmer. In a swimming pool, a device designed to continuously remove surface film and water and return it through the filter as part of the recirculation system, usually incorporating a self-adjusting weir, a collection tank, and a means to prevent air lock of the pump; also called a Mechanical Skimmer, an Automatic Skimmer, or a Recirculating Overflow.

Surface Tension. The tension of the surface film (meniscus) of a liquid, tending to minimize its surface area.

Surface Water. Water carried by an aggregate except that held by absorption within the aggregate particles themselves; water in addition to SSD water.

Surface Waters. Rain water collected and running on the surface of the land rather than being confined to drains and water courses.

Surface. The outside layer of a material or body.

Surfacer. A paint used to smooth the surface before finish coats are applied.

Surfacing, ACM. Asbestos-containing material that is sprayed-on, troweled-on or otherwise applied to surfaces, such as acoustical plaster on ceilings and fireproofing materials on structural members, or other materials on surfaces for acoustical, fireproofing, or other purposes.

Surfacing. In welding, the deposition of filler metal on a metal surface to obtain desired properties or dimensions.

Surfactant. Specialized soap used in making paint.

Surficial. Relating to a surface.

Surge Arrestor. A electrical device that minimizes or eliminates the deleterious effect of a sudden rise in voltage or current; also called a Surge Suppressor.

Surge Chamber. In airless spray equipment, a device to eliminate uneven fluid flow.

Surge Suppressor. Surge Arrestor.

Surge Tank. Container connected to the low-pressure side of a refrigerating system which increases gas volume and reduces rate of pressure change.

Surge. 1. To rise or fall rapidly, as electrical current. 2. Regulating action of temperature or pressure before it reaches its final value or setting.

Surgical Chronometer. An instrument used to measure time with great accuracy, used in medical applications.

Surgical Light. A light source used by surgeons for its illumination abilities for close work.

Surgical Scrub. Sterilized clothes worn by physicians in surgery.

Surgical Station. The area in a hospital where surgeons operate and the location of necessary implements for surgeries.

Surgical Table. A surface where surgery is performed.

Surplus. The earnings of a business from previous years which were allowed to accumulate in the business.

Surround. An area or substance that surrounds something, such as the frame around a wall opening.

Surveillance System. An interacting system of devices, such as cameras, that detect presence in and often videotape an area.

Survey. 1. To examine some topic or material to determine its condition, situation, or value. 2. The precise measurement of a parcel of land to determine all its dimensions, both vertical and horizontal.

Surveying. A branch of applied mathematics, including geometry and trigonometry, that aids in determining and delineating the form, extent and position of land.

Suspended Ceiling Removal. The act or process of removing the modular units and skeleton frame of an old suspended ceiling.

Suspended Ceiling. A finish ceiling that is hung on wires from the structure above.

Suspended Structure. A structure supported principally by tension members or carrying its loads principally in tension.

Suspended Unit Heater. A unit heater that is hung from the roof framing of an open space like a factory or workshop.

Suspension Bridge. A bridge with roadway suspended from cables supported by structures at each end.

Suspension of Work. The temporary stopping of the work, a right reserved by owners in some construction contracts.

Suspension System. A system, usually incorporating heavy gage wire, designed to support the finished ceiling membrane in place at a predetermined distance below the structural framing.

Sustainability. 1. Capable of supporting the weight for a long period. 2. Ability to keep up, prolong, keep going for a long time.

Sustainable Architecture. Architectural design based on the principles of sustainable development.

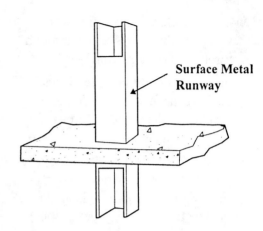

Surface Metal Runway

Sustainable Building. Construction that is based on energy conservation and reduction in use of non-renewable resources.

Sustainable Development. Construction utilizing materials that are replaceable, like wood, and minimizing use of non-replaceable materials and energy resources.

SW. Severe Weather; see Grade SW Brick.

Swage. 1. A tool used for metal forming by holding it on the work or the work on it and striking with a hammer or sledge. 2. To increase or decrease the diameter of a pipe by using a special tool which is forced around the pipe.

Swaging. Enlarging one tube end so end of other tube of same size will fit within.

Swale Excavation. The digging up of low-lying land such as a small meadow, swamp, or marshy depression.

Swale. A low-lying, often depressed and swampy, area of land; an open ended swale can be created for use as a land drainage device.

Swamp Cooler. Evaporative type cooler in which air is drawn through porous mats soaked with water.

Swash Plate-Wobble Plate. Device used to change rotary motion to reciprocating motion; used in some refrigeration compressors.

S-Wave. Shear wave, produced essentially by the shearing or tearing motions of earthquakes at right angles to the direction of wave propagation; see Seismic Wave.

Sway Brace. Diagonal bracing to prevent lateral movement caused by horizontal forces.

Sweat Out. A defective condition occasionally occurring in gypsum plaster; characterized by a soft, damp area remaining after the surrounding area has set hard. Often caused by insufficient ventilation which inhibits normal drying.

Sweat Soldering. Method of soldering in which the parts to be joined are first coated with a thin layer of solder (tinned) then joined while exposed to heat.

Sweating. 1. Condensation of moisture from air on cold surface. 2. Method of soldering in which the parts to be joined are first coated with a thin layer of solder; see Tinning.

Swedish Putty. See Spackling Compound.

Sweet Water. Tap water.

SWI. Steel Window Institute.

Swimming Load. Number of people using pool at a given time.

Swimming Pool, Private. All constructed or prefabricated pools which are used as a swimming pool in connection with a single family residence, and available only to the family of the householder and their private guests.

Swimming Pool, Public. Any constructed or prefabricated pool other than a private swimming pool.

Swimming Pool. A tank of concrete, steel, or plastic suitable for swimming.

Swing Bridge. A moveable bridge, being pivoted in the middle to move out of the way of river traffic.

Swing Gate. The operable member of a fence system that is hinged for opening and closing.

Swing Joint. A joint in a threaded pipe line which permits motion in the line of a plane normal to the direction of one part of the line.

Swirl Texturing. A method of applying gypsumboard texturing material in a decorative circular pattern.

Switch Plate. A cover plate covering an electric switch.

Switch, Dimmer. See Dimmer Switch.

Switch, Fusible. See Fusible Switch.

Switch, Oil. See Oil Switch.

Switch, Safety. See Safety Switch.

Switch, Transfer. See Transfer Switch.

Switch. A device used to continue, disrupt, or redirect an electrical current.

Switchboard. A large dead-front, single panel, frame, or assembly of panels, designed and manufactured as a unit, and which may house bussings, externally operable fused switches, circuit breakers, or other protective or regulating devices and associated instrumentation.

Switchgear. A freestanding assembly including primary (disconnect) switches, secondary (feeder) switches, and overcurrent protection device (fuses and circuit breakers).

Swivel. A device joining two parts so that the moving part can pivot freely.

Sylphon Seal. Corrugated metal tubing used to hold seal ring and provide leakproof connection between seal ring and compressor body or shaft.

Symbol. A mark, sign, or character taken as the conventional representation of some object, idea, function, or process; for example, the letters standing for the chemical elements or mathematical constants.

Symmetrical. Having identical parts on both sides of a center line.

Syndicate. A combination of individuals or commercial firms to promote some common interest.

Synergy. A combined effect or action.

Synthesis. The process of combining a set of component elements into a whole.

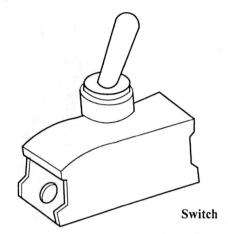

Switch

Synthetic Fiber. A fiber made by a chemical process. for example, nylon, polyester, dacron, rayon, acetate, and acrylic.

Synthetic Resin. A solid or liquid organic compound made by polymerization, the material of which a plastic is made, including epoxy resin, acrylic resin, and many other polymers; chemical compounds react to form synthetic resins.

Synthetic Rubber. Made by polymerizing hydrocarbons; see Neoprene.

Synthetic Wall Mud. An alternative to conventional plaster mortar, contains a lightweight aggregate in place of sand; available in 50 pound bags.

System. An ordered assemblage.

Systematic Errors. Cumulative errors.

T

T Hinge. A surface hinge with the short member attached to the jamb and the long member attached to the door.

T&G Siding. Tongue and groove exterior siding.

T&G. 1. Tongue and Groove. 2. Tongued and grooved.

T&M. Time and Material.

T. Transmittance.

Taber Abrader. An instrument used to test the abrasion resistance of a material.

Table Saw. A power saw in which the saw remains stationary on a table and the material to be cut is passed under it.

Table, Laboratory. See Laboratory Table.

Tableware. All utensils and decorative articles used on the table for meal service.

Tachometer. An instrument for measuring the speed of a rotating shaft or a moving vehicle.

Tack Board. A board, often of cork, for tacking up notices and displays.

Tack Coat. Application of asphaltic material on old surface to insure its bond to new construction.

Tack Hammer. A small hammer for installing tacks.

Tack Rag. A piece of cheesecloth or cotton rag moistened with thinned varnish, used by painters to pick up small particles of abrasive dust.

Tack Strip. A wooden strip containing a series of inclined barbs for grasping the edges when installing carpeting.

Tack Weld. A small weld to hold steel pieces together temporarily.

Tack Welding. Before welding, pipes are properly aligned and then tack welded; special line-up clamps are used to insure correct alignment.

Tack. 1. Short sharp pointed nail with large head used to secure thin or woven materials to wood and similar materials. 2. Degree of stickiness.

Tackboard. Soft panel used for attachment of items with thumbtacks.

Tackiness. 1. Stickiness. 2. When a painting material dries out, gels, or sets up, it loses tackiness or stickiness.

Tacky. 1. Not quite dry; sticky. 2. Shabby; seedy.

Tail Line. Short piece of blast hose smaller than the main hose to permit better maneuverability.

Tail Pipe. Outlet pipe from the evaporator.

Tails. In airless spray painting, finger-like spray pattern.

Take-Off Man. Someone who can read blueprints and is familiar with the specifications; makes notes of special details concerning the project after gathering the necessary information and then estimates the quantities of labor, materials, equipment and special items needed to complete the job; also called a Take-Off Person or a Quantity Surveyor.

Take-Off Person. See Take-Off Man.

Takeout Loan. See Permanent Loan.

Talc. 1. Any crystalline form of magnesium silicate that occurs in soft flat plates, used as a lubricator or dryer. also called Talcum Powder or Soapstone. 2. A fine powder applied to the back of gypsumboard to minimize friction and permit easier sliding of bundles. 3. A hydrous magnesium aluminum silicate used as an extender in paints; helps make paint smooth.

Talcum Powder. See Talc.

Tall Oil. A blend of resin and oil acids obtained as a by- product from the sulfate process for making paper.

Tamp. Pound or press soil to compact it.

Tamper. A tool for compacting soil in spots not accessible to rollers.

Tamping Rod. Used in preparing concrete cylinders for testing, a round, smooth, straight steel rod, 5/8 inch in diameter and approximately 24 inches in length, having the tamping end rounded into a hemispherical tip, the diameter of which is 5/8 inch.

Tandem. A device with two parts, one behind the other.

Tangent. 1. A straight line that touches a curve at one point but continues on without crossing it. 2. The trigonometric ratio of the sides opposite and adjacent to an angle in a right-angled triangle.

Tangential. 1. Acting along or lying in a tangent. 2. Strictly, coincident with a tangent at the circumference of a tree or log, or parallel to such a tangent; in practice, however, it often means roughly coincident with a growth ring; a tangential section is a longitudinal section through a tree or limb perpendicular to a radius; flat-grained lumber is sawed tangentially.

Tangible. Perceptible by touch; palpable; definite; clearly intelligible.

Tank White. Good hiding, self-cleaning white paint for exterior metal converging lines.

Tank, Supply. See Supply Tank.

Tap Box. The electrical box where the public service electrical supply line is connected with a branch to serve a particular building or structure.

Tap-a-Line. See Saddle Valve.

Tape Creaser. In gypsumboard joint finishing, a hand tool used in folding joint tape for use in inside corners.

Tape Recorder. A mechanical device which makes sound recordings on magnetic tape.

Tape Test. A particular type of adhesion test.

Tape. 1. A tape measure, marked off in measuring units. 2. A narrow woven fabric. 3. A strip of narrow transparent or opaque material coated with adhesive for fastening, sticking, or insulating. 4. Gypsumboard joint reinforcing tape. 5. A plastic reinforcing mesh or paper used to reinforce angles and to bridge lath joints in veneer plastering.

Taper. 1. A gradual and uniform decrease in size, as of a round or rectangular piece or hole. 2. The factory edge that is progressively reduced in caliper from the face to the outer edge allowing for the concealment of joint tape below the plane of the gypsum board surface.

Tapered Edge Strip. 1. A tapered insulation strip used to elevate the roof at the perimeter and at curbs that extend through the roof. 2. A tapered insulation strip used to provide a gradual transition from one thickness of insulation to another.

Tapered Edge. See Taper.

Tapered Pattern. Elliptical shaped spray pattern; a spray pattern with converging lines.

Tapered Steel Girder. A fabricated steel plate girder with a sloping top flange.

Tapestry. A looped pile fabric woven on the velvet loom.

Taping Compound. Gypsumboard joint compound especially formulated to embed joint tape.

Taping Tools. See Ames Tools.

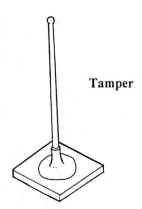

Tamper

Tapping Tile. An inspection technique whereby a coin, key, or other small metallic object is tapped against an installed tile to determine by sound whether the tile is completely bonded to its backing; a hollow sound occurs if the tile is not set properly; tilesetters often tap the tile with a pointing trowel to determine that a good bond has been achieved.

Tapping Valve. A device to open or close a duct, pipe, or other passage, or to regulate flow; it is inserted into an existing pipeline by piercing the wall of the pipe and thus tapping into the flow.

Tapping. 1. The process of forming internal screw threads in a drilled hole. 2. The tapping or pounding of a brick or other masonry unit down into the bed of mortar.

Taproot. A large root that grows downward from the base of a tree.

Tar. A dark thick inflammable liquid distilled from wood or coal and used as a sealant, a wood preservative, and for roofing and road paving.

Tare. The weight of an empty truck or container.

Tarp. See Tarpaulin.

Tarpaulin. A waterproofed canvas or other material used for protecting construction projects, athletic fields, goods, or other exposed objects or areas; also called a Tarp.

Tarred Felt. A felt that has been saturated with refined coal tar.

Taupe. A brownish gray.

Tax Shelter. The offsetting of investment income by non-cash charges, such as reserves for depreciation, for the purpose of reducing income tax.

Tax. A charge imposed by authority upon persons or property for public purposes.

T-Bar Ceiling System. A suspended ceiling system where the members are T-shaped metal that will accept acoustical board panels, lighting fixtures, or air conditioning registers and grilles.

T-Beam. A reinforced concrete beam that contains of a portion of the slab above and which the two act together.

TCA. Tile Council of America Research Center.

Teak Veneer. Thin sheets of teak, a dark wood, used for plywood or other finishes.

Teak. Yellowish-brown wood from the teak tree; used primarily for shipbuilding, furniture, and wood finish for buildings.

Tear Strip. Heavy paper ribbon under bundling tape to facilitate opening of bundles.

Tear-Off. A roofing repair job entailing complete removal of the old roof and insulation down to the deck.

Technical Competence. Knowledge of business and trade practices to successfully accomplish purposes of business.

Technically Infeasible. With respect to an alteration of a building or a facility, that it has little likelihood of being accomplished because existing structural conditions would require removing or altering a load-bearing member which is an essential part of the structural frame; or because other existing physical or site constraints prohibit modification or addition of elements, spaces, or features which are in full and strict compliance with the minimum requirements for new construction and which are necessary to provide accessibility.

Technobabble. Incomprehensible technical jargon.

Technology. 1. Applied science. 2. Technical language.

Tedlar. A registered trademark name of du Pont Co.; an elastomeric membrane consisting of polyvinyl fluoride.

Tee Joint. A joint between two members located approximately at right angles to each other in the form of a T.

Tee Member. A metal or precast concrete member with a cross section resembling a T.

Tee Square. A 48-inch long T shaped metal guide used in scoring gypsumboard to job size requirements; also called a Hanger's Tee.

Tee Turn. Turn Piece.

Tee Weld. Weld in a joint between two members located approximately at right angles to each other in the form of a T.

Tee, Bulb. Rolled steel in the form of a T with a formed bulb at the end of the web.

Tee. A metal or precast concrete member with a cross section resembling a T.

Telamon. A male figure used as a supporting column.

Telescopic. Having parts that slide or pass within one another.

Telescoping Bleacher. A seating structure, usually in a gymnasium, that folds back into itself.

Teller Window. The opening in a wall partition through which transactions in a financial institution take place.

Temblor. Earthquake.

Temper. 1. The condition of metal as regards hardness and elasticity. 2. To mix plaster to a workable consistency. 3. To mix, by the adding of water to make the mortar the right consistency for use.

Tempera. 1. A process of fine art painting in which an albuminous or colloidal medium, as egg yolk, is employed as a vehicle instead of oil; poster color; water color. 2. A water-thinned or water-emulsion paint.

Temperature Bars. See Temperature Reinforcement.

Temperature Control. Temperature-operated thermostatic device which automatically opens or closes a circuit.

Temperature Inversion. The phenomenon of a layer of warmer air trapping a layer of cooler air closer to the ground thus restricting its rise and with it trapping any air pollutants.

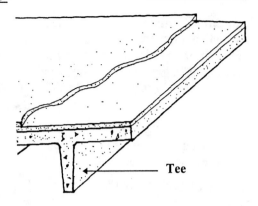

Tee

Temperature Reinforcement. Steel reinforcing bars distributed throughout the concrete to minimize cracks due to temperature changes and concrete shrinkage, also called Temperature Bars or Temperature Steel.

Temperature Rise. The increase of concrete temperature caused by heat of hydration and heat from other sources.

Temperature Steel. See Temperature Reinforcing.

Temperature, Curing. See Curing Temperature.

Temperature. 1. Degree of heat in a body as measured by a thermometer. 2. Measurement of speed of motion of molecules.

Temperature-Humidity Index. Actual temperature and humidity of air sample compared to air at standard conditions.

Tempered Air. See Conditioned Air.

Tempered Glass. Glass that has been treated to increase it toughness and its resistance to breakage.

Template Hardware. Items of hardware that are made to template as to spacing of all holes and dimensions, as in hinges, locks, exit devices, and closers for hollow metal doors and frames.

Template. A gauge, pattern, or mold used as a guide to the form of a piece being made.

Temple. A building devoted to worship.

Temporary Centering. Temporary forming used during the construction process for support of a masonry arch or a concrete slab.

Temporary Closure. Temporary construction used during the construction process to close openings to provide weather protection or to separate construction areas from occupied areas.

Temporary Facility. A structure erected for temporary use.

Temporary Restraining Order. An order of brief duration issued by a court pending a hearing on the merits.

Temporary Services. The utilities (water, electricity, telephone) brought onto a jobsite for the contractor's temporary use during construction.

Tenant at Sufferance. A holdover tenant that remains in possession with the permission of the lessor; also called Tenant at Will.

Tenant at Will. See Tenant at Sufferance.

Tenant. A person who rents property or land from a landlord; a Lessee.

Tenant. Lessee.

Tenants in Common. Two or more persons who own land together without the right of survivorship.

Tender. 1. The person who tends the needs of a mason, such as a hod carrier, pack carrier, or wheelbarrow person. 2. An offer or bid.

Tendon. A concrete prestressing bar, cable, strand or wire.

Tenement. 1. Apartment dwelling. 2. An apartment building of minimum standards occupied by poorer families, usually in a city.

Tenements. The rights and privileges that accompany real property when it is conveyed.

Tennis Court. A level, quadrangular space divided by a low net, upon which tennis is played.

Tennis Net. A low net that divides a tennis court.

Tenon Saw. A handsaw with fine teeth, a thin blade, and a stiffened back, for cutting tenons and other fine work.

Tenon. A tongue-like protrusion on the end of a piece which is tightly fitted into a rectangular slot (mortise) in the side of another other piece.

Tensile Strength. Resistance to elongation; the greatest longitudinal stress a substance can bear without rupture or remaining permanently elongated; the pulling force necessary to break a given specimen divided by the cross sectional area; units are pounds per square inch;

Tensile Stress. A stress caused by stretching of a material.

Tension Ring. A structural element, forming a closed curve in plan, which is in tension because of the action of the rest of the structure; a concrete or masonry dome commonly has a tension ring.

Tension Wood. Abnormal wood found in leaning trees of some hardwood species and characterized by the presence of gelatinous fibers and excessive longitudinal shrinkage; tension wood fibers hold together tenaciously, so that sawed surfaces usually have projecting fibers, and planed surfaces often are torn or have raised grain; tension wood may cause warping.

Tension. Stress which tends to elongate a member.

Tepidarium. The room containing the tepid bath in Roman baths (thermae).

Term Loan. A loan generally obtained from a bank or insurance company with a maturity greater than one year and less than 7 to 10 years.

Term. The time period stated in a lease or mortgage.

Terminal Adapters. Electrical fittings attached to the end of a conductor or to a piece of equipment for taking power from an outlet in a way for which it was not designed.

Terminal Block. A decorative element forming the end of a block structure.

Terminal Box. A metal electrical box, usually with a removable cover, that contains leads from electrical equipment ready for connection to a power source.

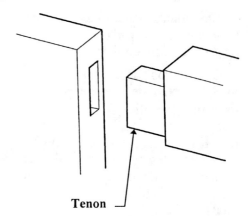

Tenon

Terminal Reheat Handling System. A reheat system that is a modification of a single zone system; this system contains a cooling coil only (unless a preheat coil is needed); air is ducted to all spaces at the same temperature as in a single zone system, but at a temperature low enough to satisfy the temperature and humidity requirements of the zone with the highest cooling load; a terminal reheat box at each zone reheats the air as needed; a constant supply of air is supplied to the space served.

Terminal. 1. A device for fastening and terminating an electrical conductor. 2. A final ending. 3. A terminus for trains or long-distance buses.

Termination for Cause. The right of an owner or contractor to end a construction contract if the other commits certain specified acts.

Termination for Owner's Convenience. The owner's right to end a construction contract for any reason to suit the owner's convenience.

Termite Control. The act or process of spraying chemicals into buildings and structures to control the spread of termites.

Termite Shield. A sheet metal barrier to discourage entry into the structure by subterranean termites.

Termite. A wood destroying social insect; Subterranean Termites live in the earth, and will build mud tunnels from the earth up to the wood structure; Kalotermes, also called Drywood Termites, swarm and will enter the structure where they will bore into the wood and hollow out the structural members.

Terms. Conditions of payment in a sale or purchase of merchandise.

Terne. An alloy of lead and tin, used to coat sheets of carbon steel, stainless steel, or copper for use as metal roofing sheet.

Terra Cotta. Hard baked clayware or tile, of variable color, averaging reddish red-yellow in hue and of high saturation.

Terra Sigillata. A porous red clay ware characterized by embossed decorations of the same color and a satin-like unglazed surface; originated on the Island of Samoa.

Terrazzo Concrete. See Terrazzo.

Terrazzo Divider. A metal strip that separates areas of terrazzo.

Terrazzo Epoxy. A two-part adhesive, employing epoxy resin, an epoxy hardener used for bonding marble or other stone chips set in Portland cement to a backup material.

Terrazzo Floor. A high polished floor made from small marble or stone fragments embedded in a portland cement matrix.

Terrazzo Receptor. A shallow sink of a shower, forming its floor, made of Terrazzo.

Terrazzo Removal. The act or process of demolishing and discarding old terrazzo floors or walls or terrazzo tile.

Terrazzo Tile. A terrazzo surface, on a portland cement and sand body, made by a mixture of marble chips and portland cement and usually ground smooth.

Terrazzo Wainscot. A wainscot made of terrazzo or terrazzo tile wall facing.

Terrazzo. A finish floor material consisting of concrete with an aggregate of marble chips selected for size and color, which is ground and polished smooth after curing; may also be pre-cast and pre-finished.

Terrestrial. Of or relating to the earth.

Tertiary Color. A color produced by mixing two secondary colors or by mixing a primary color with a secondary color.

Tessera (pl. tesserae). A small chip of glass or marble used in mosaic formations.

Test Cut. A sample of the roof membrane, usually 4 inches x 40 inches in size, that is cut from a roof membrane to determine the weight of the membrane, the number of plies, and the average interply bitumen poundages.

Test Light. Light provided with test leads; used to test or probe electrical circuits to determine if they have electricity.

Test Pattern. Spray pattern used in adjusting spray gun.

Test Pile. A foundation pile which is driven before the final design is done to see what bearing is developed and what length of pile is actually needed; a pile which is tested by placing a predetermined load on it, commonly done by erecting a crib on the pile and then filling it with ballast.

Test Plug. A temporary plumbing fitting which is inserted into the end of a plumbing system to facilitate in pressure and leakage testing.

Test Set. Equipment used to check standards of water condition.

Test. A trial, examination, observation, or evaluation used as a means of measuring the physical or chemical characteristics of a material, or a physical characteristic of a structural element or a structure.

Testate. One who dies leaving a will.

Testimony. Oral evidence offered by a witness in the course of a hearing, a trial, or a deposition.

Testing Laboratory. A place equipped to test construction materials to determine if they meet certain specified chemical, physical, and engineering criteria.

Testing Machine. A device for applying test conditions and accurately measuring results.

Testing Weld. The act or process of testing the strength of a weld.

TEV. Thermostatic Expansion Valve.

Texas Forest Service (TFS). PO Box 310, Lufkin, Texas 75902-0310, (409) 639-8180.

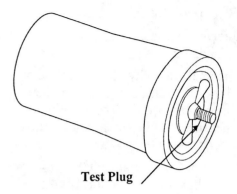

Test Plug

Texture Paint. One which may be manipulated by brush, trowel or other tool to give various patterns.

Texture. 1. Visual or tactile surface characteristics. 2. A surface decoration. 3. In carpet, a surface effect obtained by using different heights of pile or two or more forms of yarn, or by alternating the round and cut pile wires, by brocade engraving, simulated or actual carving or shaving with an electric razor, or other special treatment of the design, to give added interest beyond that provided by the woven design or tones.

TFS. Texas Forest Service.

Theater Equipment. Props, lighting, curtains, stage platforms and other devices and objects used in a theater.

Theater Stage. A raised platform where plays, musicals or other theatrical events are performed.

Theodolite. A surveyor's instrument for measuring vertical and horizontal angles; a transit.

Theoretical. In spot, seam and projection welding, the force, neglecting friction and inertia, available at the electrodes of a resistance-welding machine by virtue of the initial force application and the theoretical mechanical advantage of the system.

Therm. A heat unit; 100,000 BTUs.

Thermae. Roman public baths; see Calidarium.

Thermal Capacity. 1. A measure of how much energy is needed to heat up a substance. 2. The quantity of heat needed to warm a collector up to its operating temperature.

Thermal Conductance (C). A unit of heat flow that is used for specific thicknesses of material or for materials of combination construction, such as laminated insulation. The formula for thermal conductance is. $C = k$ times the thickness in inches

Thermal Conductivity (k). The heat energy that will be transmitted by conduction through one square foot of one inch thick homogeneous material in one hour when there is a difference of 1 degree Fahrenheit perpendicularly across the two surfaces of the material. The formula for thermal conductivity is $k = BTU/SQUARE FOOT/INCH/HOUR/DEGREE FAHRENHEIT$.

Thermal Cutout. An overcurrent protective device which contains a heater element in addition to and affecting a renewable fusible member which opens the circuit; it is not designed to interrupt short circuit currents.

Thermal Fracture. A compression crack caused by expansion of peripheral building components.

Thermal Inertia. The tendency of a building to remain at the same temperature or to fluctuate only very slowly.

Thermal Insulation. Material used to retard the flow of heat through an enclosing surface.

Thermal Insulator. An insulation material that lowers the thermal conductivity of a part of a building system, such as a floor, wall, ceiling, or roof assembly.

Thermal Lag Factor. The time interval between the time heat enters a material at one side and exits at the other side.

Thermal Movement. The expansion and contraction of any material caused by temperature differences.

Thermal Protector. An electrical device that automatically opens a circuit in the event of overheating, thereby protecting the appliance from possible damage.

Thermal Radiation. Electromagnetic radiation emitted by a warm body.

Thermal Relay. Heat operated electrical control used to open or close a refrigeration system electrical circuit; this system uses a resistance wire to convert electrical energy into heat energy; also called Hot Wire Relay.

Thermal Resistance (R). An index of a material's resistance to heat flow; it is the reciprocal of thermal conductivity (k) or thermal conductance (C). The formula for thermal resistance is.

Thermal Shock. 1. Stress built up by sudden and appreciable changes in temperature. 2. The shock produced and also the stress produced resulting from sudden temperature changes in a roof membrane; for example, when a rain shower follows brilliant sunshine. 3. A stress created by an extreme change in temperature that may result in cracking of the plaster which has not yet attained its ultimate strength.

T

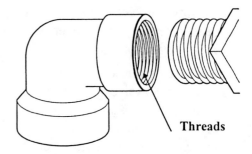

Threads

Thermal System Insulation. Insulative material applied to pipes, fittings, boilers, tanks, ducts or other interior structural components to prevent heat loss or gain or water condensation.

Thermal Transmission Value. The resistance factor to the conductance of heat.

Thermal. Pertaining to heat.

Thermistor. A semiconductor which has electrical resistance that varies with temperature.

Thermocouple Thermometer. Electrical instrument using thermocouple as source of electrical flow, connected to a milliammeter calibrated in temperature degrees.

Thermocouple. A pair of dissimilar metals create a thermoelectromotive force that can be measured and converted to temperature differences.

Thermodisk Defrost Control. Electrical switch with bimetal disk controlled by temperature changes.

Thermodynamics. The science dealing with the relationship between heat and other forms of energy.

Thermoelectric Refrigeration. Refrigerator mechanism that depends on Peltier effect; direct current flowing through electrical junction between unlike metals provides heating or cooling effect depending on direction of flow of current.

Thermometer. Device for measuring temperatures.

Thermomodule. A number of thermocouples used in parallel to achieve low temperatures.

Thermopane Window. Tradename; a type of glass constructed in a manner to protect against sound, heat, heat loss or moisture; double or triple glazed glass.

Thermopile. Number of thermocouples used in series to create a higher voltage.

Thermoplastic Glues and Resins. Glues and resins that are capable of being repeatedly softened by heat and hardened by cooling.

Thermoplastic. Having the property of softening when heated and rehardening when cooled.

Thermosetting Glues and Resins. Glues and resins that are cured with heat and do not soften when subjected later to high temperatures.

Thermosetting. Material which undergoes irreversible chemical reaction when heated and molded; once formed, it cannot be reheated and reshaped.

Thermostat Cable. A specific electrical system cable which operates an automatic device for regulating the temperature in a room, space, or area.

Thermostat. Device which senses ambient temperature conditions and, in turn, acts to control a circuit.

Thermostatic Control. Device which operates system or part of system based on temperature change.

Thermostatic Expansion Valve (TEV). Control valve operated by temperature and pressure within evaporator. It controls flow of refrigerant; control bulb is attached to outlet of evaporator.

Thermostatic Motor Control. Device used to control cycling of unit through use of control bulb; bulb reacts to temperature changes.

Thermostatic Valve. Valve controlled by temperature change response elements.

Thermostatic Water Valve. Valve used to control flow of water through system, actuated by temperature difference; used in units such as water-cooled compressor and/or condenser.

Thickness. 1. The smallest of the three dimensions of a solid object. 2. The quality of being thick. 3. A layer of material. 4. The dimension at right angles to the face of the wall, floor, or other assembly in which masonry units are used.

Thin Coat. A one coat plaster system over gypsum board.

Thin Set. The bonding of tile with suitable materials, applied approximately 1/8" thick.

Thin Shell. A three dimensional spatial structure made up of one or more curved concrete slabs or folded plates whose thicknesses are small compared to their other dimensions; thin shells are characterized by their three dimensional load carrying behavior which is determined by the geometry of their forms, by the manner in which they are supported, and by the nature of the applied load.

Thin Wall Conduit. See EMT Conduit.

Thincoat-High Strength Plaster. See Veneer Plaster.

Thingamabob. See Thingamajig.

Thingamajig. Something that is difficult to classify or whose proper name is unknown or forgotten; a Thingamabob.

Thinner. A volatile liquid added to finishing material to make it flow more easily and smoothly; also called a Diluent.

Thinners. Volatile liquids used to reduce the viscosity of paint and varnish; also called Diluents.

Third Party Beneficiary. A person who is not a party to a contract but who is an intended beneficiary thereof and may therefore enforce it.

Thixotropic Paint. Property exhibited by some paint of becoming fluid when shaken or disturbed; after cessation of mechanical disturbance, such as stirring or putting brush into paint, rigidity develops again.

Thixotropy. The property of various gels of becoming fluid when disturbed, as by shaking or stirring.

Thomas Jefferson Ideas. The advocacy of grid pattern planning where frequent public squares are left open between building blocks.

T

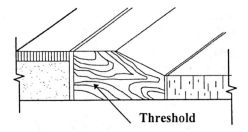

Threshold

Threaded Coupling. A fitting for joining two lengths of threaded pipe.

Threading. The process of forming external screw threads on a rod or pipe.

Threadless Coupling. A fitting for joining two lengths of pipe fastened by soldering or cementing.

Threads. Spiral ridges on a member, like a pipe, by which parts can be screwed together.

Three Coat Plastering. The application of plaster in three successive coats, leaving time between coats for setting and drying of the plaster.

Three Dimensional. Having, or appearing to have, three dimensions, height, width, and depth.

Three Point Lock. A device to lock a door to the jamb at three points.

Three Quarter. A brick with one end broken off; a three quarter brick.

Three-Phase. Operating by means of combination of three alternating current circuits which differ in phase by one-third of a cycle.

Three-Quarter Bath. A room containing a toilet, lavatory, and shower.

Three-Way Switch. A switching circuit enabling a light fixture to be controlled from two points.

Three-Way Valve. Multi-orifice flow control valve with three fluid flow openings.

Threshold Limit Value (TLV). Maximum concentration of solvent vapor in parts per million parts of air in which a worker may work eight consecutive hours without an air fed mask; the lower the TLV number the more toxic the solvent.

Threshold, Door. A beveled piece of floor trim over which a door swings.

Threshold. 1. The wood or metal beveled floor piece at door openings which commonly separates non-continuous floor types. 2. A strip of wood or metal beveled on each edge and used above the finished floor under outside doors; also called a Sill or a Saddle.

Throat. The opening at the top of a fireplace, through which the smoke passes into the smoke chamber and chimney; the damper is placed in the throat.

Throne. A chair of state for a sovereign or bishop.

Throttling. Expansion of gas through an orifice or controlled opening without gas performing any work as it expands.

Through Dry. Ability of film to show no loosening, detachment or evidence of distortion when the thumb, placed on film with maximum arm pressure, is turned through 90 degrees in the plane of the film.

Through-Wall Flashing. A sheet metal sleeve through a parapet wall, integrated with the abutting counter-flashing and roofing to prevent water from intruding.

Throw. The distance that a door latch or bolt may be fully extended.

Thrust Block. A concrete mass around water piping to prevent movement caused by the sudden change in velocity of water flow.

Thrust Fault. A geological fault caused by compression in which the upper older layers of rock are thrust over the lower younger ones at a very low angle.

Thrust. A lateral or inclined force resulting from the structural action of an arch, vault, dome, suspension structure, or rigid frame.

Thumb Screw. A screw with a knurled or flat-sided head that may be grasped between the thumb and forefinger for turning.

Thumb Tack. A tack with a broad flat head so it may be pressed into a surface with the thumb.

Thumbnail Proof. Checking hardness of a finish by pressing thumbnail against it.

Thyristor. An electrical semiconductor device for converting alternating current into direct current.

Ticky-Tacky. 1. Sleazy or shoddy material. 2. Cheap, repetitive housing built of shoddy materials.

Tie Bars. 1. Steel reinforcing bars at right angles and tied to main reinforcement to keep it in place. 2. Bars extending across a construction joint.

Tie Beam. A reinforced concrete beam cast as part of a masonry wall, whose primary purpose is to hold the wall together, especially against seismic loads, or cast between a number of isolated foundation elements to maintain their relative positions.

Tie Beam. See Collar Beam.

Tie-Down. See Anchorage, 1.

Tie Elements. Elements which serve to transmit inertia forces and prevent separation of such building components as footings and walls.

Tie Rod. A steel rod that acts in tension and commonly holds together wall forms while concrete is being poured.

Tie Strap. A metal plate that fastens two parts together as a post, rod, or beam.

Tie Wire. 1. Light gauge wire used as temporary fasteners in construction work. 2. Soft annealed steel wire used to join lath supports, attach lath to supports, and attach accessories. 3. Wire (generally No. 16, No. 15 or No. 14 gauge) used to secure intersections of reinforcing bars for the purpose of holding them in place until concreting is completed.

Tie, Masonry. See Masonry Tie.

Tie, Mesh. See Mesh Tie.

Tie. 1. A device for holding components together, a structural device that acts in tension. 2. A stirrup tie.

Tied Column. A square or rectangular reinforced concrete column containing longitudinal reinforcing and lateral ties.

Tie-In. The point or line of interface of new materials with old in alterations or repairs.

Tier. 1. A row or rank or unit of a structure, as one of several placed one above another. 2. One of the vertical layers in the thickness of a masonry wall.

Ties. The attaching of metal lath to furring channels by use of light gauge malleable wire; see Butterfly Tie, Saddle Tie, and Stub Tie.

TIG Welding. See Gas Tungsten-Arc Welding.

Tile Adhesive. Organic adhesive used for bonding tile to a surface; rubber solvents and resin-based and rubber emulsions can be used as adhesives.

Tile Assemblies. Tile assembled into units or sheets by suitable material to facilitate handling and installation; tile may be face-mounted, back-mounted or edge-mounted.

Tile Council of America Research Center (TCA). PO Box 326, Princeton, New Jersey 08542, (609) 921-7050.

Tile Cutter. Any device that cuts tile; one of the most efficient and economical tools is the adjustable hand-drawn tile cutting board.

Tile Mounted. Tile fastened together in sheets to facilitate handling and setting.

Tile Setting. The skilled trade of installing ceramic tile on floor, wall, stairs, or counter surfaces.

Tile, Acoustical. See Acoustical Tile.

Tile, Back Mounted. See Back Mounted Tile.

Tile, Clay. See Clay Tile.

Tile, Conductive. See Conductive Tile.

Tile, Face Mounted. See Face Mounted Tile.

Tile, Mounted. See Mounted Tile.

Tile, Quarry. See Quarry Tile.

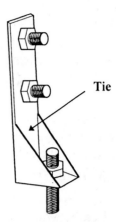

Tie

Tile, Resilient. See Resilient Flooring.

Tile, Sewer. See Sewer Tile.

Tile, Terrazzo. See Terrazzo Tile.

Tile. 1. To install ceramic tiles, flooring tiles, roofing tiles, or drainage tiles. 2. A ceramic surfacing unit, usually relatively thin in relation to facial area, made from clay or a mixture of clay and other ceramic materials, called the body of the tile, having either a glazed or unglazed face and fired above red heat in the course of manufacture to a temperature sufficiently high to produce specific physical properties and characteristics. 3. A fired clay product that is thin in cross section as compared to a brick; either a thin, flat element (ceramic tile or quarry tile), a thin, curved element (roofing tile), a hollow element with thin walls (masonry tile), or a pipe-like shape (drainage tile).

Tilt Angle. The angle that a flat solar collector surface forms with a horizontal.

Tilt-Up Construction. A method of constructing concrete walls in which panels are cast and cured flat on the floor slab, then tilted up into their final positions.

Tilt-Up Wall. See Tilt-Up Construction.

Timber Framing. Structural framing by the use of timber for load carrying applications.

Timber Parking Barrier. A barrier made from thick wood to prevent the encroachment of vehicles.

Timber, Standing. Timber still on the stump.

Timber. Lumber with cross-section over 4 by 6 inches, such as posts, sills, and girders.

Time and Material. A method of paying a contractor for work, consisting of reimbursement for time expended and cost of materials.

Time Clock. A clock that stamps an employee's starting and ending times on a time card.

Timekeeper. A clerk who keeps records of the time worked by employees.

Timer. Clock-operated mechanism used to control opening and closing of an electrical circuit.

Timer-Thermostat. Thermostat control which includes a clock mechanism; unit automatically controls room temperature and changes temperature range depending on time of day.

Time-Sharing Ownership. A method of sharing in the ownership, usually of recreational properties, in which each co-owner has the right to occupancy at specified times only.

Time-Weighted Average. In air sampling, this refers to the average air concentration of contaminants during a particular sampling period.

Tin Shingle. A small piece of tin used in flashing and repairing a shingle roof.

Tin Snip. A hand tool for shearing tin.

Tin. A soft faintly bluish white lustrous low-melting crystalline metallic element that is malleable and ductile at ordinary temperatures and that is used as a protective coating, in tinfoil, and in soft solders and alloys.

Tinge. Slight trace of color.

Tinner. A tinsmith.

Tinning. In soldering, coating the metals to be joined with a thin layer of solder.

Tinsmith. A worker in tin.

Tint. A hue lightened by addition of white; a color produced by adding white pigment or paint to a colored pigment or paint, with the amount of white greater than the amount of colored pigment.

Tinted Glass. Glass that has been treated to reduce transmitted glare.

Tissue Dispenser. A container with a slim opening that holds tissues and allows one to be drawn at a time.

Titania Porcelain. A vitreous ceramic whiteware for technical application in which titania, TiO_2, is the essential crystalline phase.

Titanium Calcium. Paint pigment made by combining titanium dioxide of rutile type and calcium sulphate.

T

Titanium Dioxide. White pigment used extensively in paint making; comes in two forms, rutile and anatase; it is chemically inactive and is not affected by dilute acids, heat or light.

Titanium. A metal which is the basis for the pigment, titanium dioxide.

Title Deed. Evidence if ownership of property.

Title Insurance. A guarantee of title issued by an insurance company.

Title Search. A research of the county records to determine the condition of title of a parcel of real property.

Title. The right to ownership of property.

TL. Transmission Loss.

TLV. Threshold Limit Value.

TMS. The Masonry Society.

To the Weather. A term applied to any part of the structure which faces the elements; a shingled roof is to the weather, but the framing system is not.

Toe Nail. Nailing at an angle, as from the bottom of a stud to the sole plate.

Toe Plate. 1. A metal bar or plate fastened to the outer edge of a floor grating. 2. A metal plate fastened to the rear of a stair tread. 3. A protective metal plate fastened to the bottom rail of a door.

Toe Space. A recess at the bottom of a cabinet supporting a work counter, to provide space for the worker's toes.

Toeboard. Wood or metal vertical metal member at floor or landing level placed at floor or landing edge to prevent objects from falling or persons from stepping off the edge.

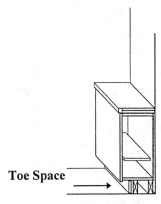

Toe Space

Toggle Bolt. A bolt and nut assembly to fasten objects to hollow construction assembly from only one side. Nut has pivoted wings that close against spring when nut end of assembly is pushed through hole and spring open on other side in void of construction assembly.

Toilet Bowl. The oval part of a toilet which receives the waste and fills with water after flushing the toilet tank.

Toilet Carrier. A metal structure attached to a bathroom wall to support and mount a wall-hung toilet.

Toilet Partition. Privacy panels in a toilet enclosure.

Toilet Room Accessories. Toilet room equipment such as toilet paper dispensers, paper towel dispensers, hand dryers, sanitary napkin dispensers, grab bars, and grab rails.

Toilet Room. A room containing one or more toilets and usually one or more lavatories.

Toilet Seat and Cover. A hinged seat for a toilet and its hinged cover.

Toilet Tank. The reservoir located in the back of a wate. Tool used to cut internal threads.

Toilet. A bathroom or washroom fixture that serves as a receptacle for human waste; also called a Water Closet.

Tolerance. 1. The allowable deviation from a standard. 2. Extra space to allow for dimensional differences.

Toluene. Lacquer diluent made normally by coal tar distillation; also called Toluol.

Toluidine Red. Brilliant non-bleeding red pigment made from coal tar product.

Toluol. See Toluene.

Tomb. 1. A grave; place of entombment. 2. A house, chamber, or vault for the dead.

Ton of Refrigeration. Refrigerating effect equal to the melting of 1 ton of ice in 24 hours; this may be expressed as. 288,000 Btu/24 hours or 12,000 Btu/hour or 200 Btu/minute.

Ton. 1. A measure of weight equaling 2000 pounds. 2. The metric ton is 1000 kilograms or 2200 pounds.

Tone Down. The process of reducing visual prominence of an installation by the application of external coatings; blending of overall color scheme with the surrounding environment.

Tone. 1. A sound of only a single frequency. 2. A graduation of color, either a hue, a tint, or a shade, as a gray tone.

Tone-on-Tone. Carpet pattern made by using two or more shades of the same hue.

Toner. A color modifier.

Tongue and Groove (T&G). A type of lumber, metal, or precast concrete having matching or mated edges to provide a tight fit.

Tongue. 1. A projecting edge, as on a board, that fits into a groove of another piece, as in a tongue and groove joint. 2. See Spline. 3. See Steel Square.

Too Much Drag. Refers to paint that has excessive pull or drag in its application.

Tool Belt. A webbed or leather belt worn by carpenters, electricians, and other artisans from which frequently used hand tools are slung.

Tooled Joint. A masonry joint where the surface has been tooled to compress the mortar; the exposed surface can be flat, concave, or v-shaped, depending on the configuration of the tool.

Tooling. Compressing and shaping the face of a mortar joint with a special tool other than a trowel; also called jointing. See Jointing, 3.

Tooth. 1. Profile; mechanical anchorage; surface roughness. 2. Roughened or absorbent quality of a surface which affects adhesion and application of a coating.

Toothed Ring. A shear-resisting timber connector used in the manufacturing of large member wood trusses.

Toother. A brick projecting from the end of a wall against which another wall will be built.

Toothing In. The joining of a new masonry wall on to the old toothed wall.

Toothing. The system of the construction at the end of a wall in which every other course projects one half of a brick length; then another wall may be tied into this staggered brick end.

Top Bars. Steel reinforcing bars near the top of reinforced concrete.

Top Chord. The top flat or sloping member of a truss.

Top Coating. Finish coat.

Top Colors. Colors of the yarn used to form the design, as distinguished from ground color.

Top Dip. The highest water or waste point in the bottom section of a trap.

Top Mopping. The finish coat of hot bitumen on a built up roof.

Top Plate. The horizontal member at the top of a stud wall, usually supporting rafters.

Top Set Base. Vinyl or rubber base with an integral cove, cemented to the wall, set on top of resilient floor coverings.

Top Weir. The top part of the outlet of a trap.

Topographic Map. See Topographic Survey.

Topographic Survey. A land survey that shows topography and all other physical features; also called a Topographic Map or a Contour Map.

Topography. 1. The form of the terrain. 2. The description of the surface features in graphic terms, depicted by contour lines showing ground elevation.

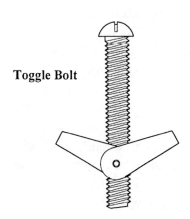

Toggle Bolt

Topology. The study of the geometric properties and spatial relations of flat and solid shapes that are unchanged by squeezing, stretching, twisting, or changing of size.

Topping Compound. A specially formulated gypsumboard joint compound designed for the final joint finishing coat; not to be used to embed tape.

Topping, Concrete. See Concrete Topping.

Topping, Granolithic. See Granolithic Topping.

Topping. Fine material forming a surface layer or dressing for a road or grade.

Topsoil. Surface soil at and including the average plow depth, soil which is used as a planting or growing medium.

Torch Brazing. A brazing process wherein coalescence produced by heating with a gas flame and by using a nonferrous filler metal having a melting point above 800° F. but below that of the base metal; the filler metal is distributed in the joint by capillary attraction.

Torching. Application of direct flame to a membrane for the purpose of heating, melting, or adhering.

Tornado. A violent storm of whirling winds of high speed, over a narrow path often accompanied by a funnel shaped cloud.

Torque Wrench. Wrench which may be used to measure torque or pressure applied to a nut or bolt.

Torque, Full Load. See Full Load Torque.

Torque, Stall. See Stall Torque.

Torque, Starting. See Starting Torque.

Torque. 1. Turning or twisting action. 2. Moment.

Torsion. The rotation of a diaphragm caused by lateral forces and whose center of mass does not coincide with the center of rigidity.

Tort. A negligent or intentional wrongful act that damages the person or property of another, the wrongful nature of which is independent of any contractual relationship.

Torus. A geometric solid shaped like a doughnut.

Total Absorption. The amount of water a masonry unit will absorb when immersed in water.

Total Alkalinity. Actual amount of alkali salts present in swimming pool water.

Total Cost. A method of computing damages sustained by a contractor because of breaches of contract causing the contractor to operate in an inefficient or unproductive manner.

Total Design Displacement. The design-basis of earthquake lateral displacement, including additional displacement due to actual and accidental torsion, required for design of the isolation system, or an element thereof.

Total Heat. Sum of both the sensible and latent heat.

Total Maximum Displacement. The maximum capable earthquake lateral displacement, including additional displacement due to actual and accidental torsion, required for verification of the stability of the isolation system, or elements thereof, design of building separations, and vertical load testing of isolator unit prototype.

Total Static Head. Static head from the surface of the supply source to the free discharge surface.

Touch Up Primer. Field touch-up of prime painting for steel members after erection.

Touch Up Painting. Spot repair painting usually conducted after initial painting.

Toughness. A quality of a material which permits it to absorb a relatively large amount of energy, to withstand repeated shocks, and to undergo considerable deformation before yielding or breaking.

Towel Bar. A horizontal rod, usually secured to a wall, on which a towel can be hung or laid.

Towel Dispenser. A container that holds paper towels for future use, and which has an opening allowing one towel to be removed at a time.

Tower, Cooling. See Cooling Tower.

Tower. A tall structure, constructed of frames, braces, and accessories rising to a greater height than the surrounding area.

Townhouse. A single-family house of two or three stories, often connected to similar houses in a row.

Township. In land ownership and legal description, a township is a square 6 miles on a side, totaling 36 square miles; each square mile is called a section.

Toxic. Poisonous.

Toxicity. Degree of poisonousness or harmfulness.

Toxins. A poison produced by a living organism.

TPI. Truss Plate Institute.

TPO. Thermoplastic Polyolefin.

Trace. 1. A very small, barely perceptible, amount or vestige of a material. 2. To delineate the outline of an object. 3. To copy a drawing by use of an overlay of thin paper.

Tracer Gases. Compounds, such as sulfur hexafluoride, which are used to identify suspected pollutant pathways and to quantify ventilation rates; tracer gases may be detected qualitatively by their odor or quantitatively by air monitoring equipment.

Tracery. 1. Ornamental stone openwork in the upper part of a gothic window. 2. A fine decorative pattern.

Tracheid. The elongated cells that constitute the greater part of the structure of the softwoods (frequently referred to as fibers); also present in some hardwoods.

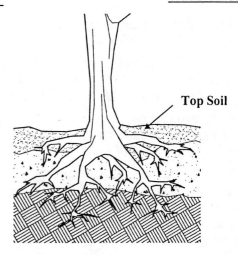

Top Soil

Tracing Paper. Transparent drafting medium from which photosensitive prints can be made of the drawing.

Tracing. A reproducible drawing made on semitransparent paper called tracing paper; such a drawing on any kind of semitransparent drawing medium.

Track Lighting. An assembly with a U-shaped electrified member attached to a ceiling or wall, acting as a channel for sliding light fixtures.

Track. 1. A metal way serving as a guide for a movable lighting fixture. 2. Metal channel; see Runner. 3. A specially laid out course, as for racing.

Tract. A defined area of land.

Traction. The friction grip of a wheel or track on the surface on which it moves.

Trade Discount. A discount or deduction from a catalog or list price extended to the trade or quantity buyer.

Trade Union. An organized association of workers in a trade, group of trades, formed to protect and further their rights and interests.

Traffic Deck. A walking surface placed on top of a roof membrane.

Traffic Paint. A paint, usually white, red or yellow, used to designate traffic lanes, safety zones, and intersections.

Traffic Sign. A posted command, warning, or direction for motor vehicle drivers or pedestrians.

Traffic. 1. The movement of people or vehicles through an area. 2. The passing to and fro of persons with special reference to carpet wear resulting therefrom.

Trailer. A vehicle designed to be hauled or to serve parked as a dwelling or place of business.

Trammel Bar. A tile layout tool used to erect perpendicular lines and to bisect angles.

Trammel. A Beam Compass.

Transducer. Device turned on by change of power from one source for purpose of supplying power in another form to second system.

Transept. Either arm of the part of a cross-shaped church at right angles to the nave.

Transfer Switch. A switch in an electrical system that transfers the load to another circuit when the voltage drops below a predetermined level.

Transfer. That act of transferring stress in prestressing tendons from jacks or pretensioning bed to concrete member.

Transformer Pad. A precast concrete block or stone placed under a transformer to spread and support its weight.

Transformer. A device that changes, or transforms, alternating current from one voltage to another.

Transformer-Rectifier. Combination transformer and rectifier in which input a-c current may be varied and then rectified into d-c current.

Transistor. An electronic component that switches or amplifies a current; a semiconductor device with three connections.

Transit Mixed Concrete. Concrete produced from a central batching plant, where the materials are proportioned and placed in truck mixers for mixing enroute to the job or after arrival there.

Transit. See Theodolite.

Transition Lot. A lot in a subdivision that consists of both cut and fill.

Transition Point. In a swimming pool, the point at which the floor slope changes from shallow to deep area.

Transition Primer. Coating compatible with primer and with a finish coat, though the latter is not compatible with the primer; also called Barrier Coating or Block Coating.

Transition. 1. A passing or change from one place, state, or condition to another. 2. The physical part needed to effect the change, such as a piping reducer.

Translate. 1. To move from one place to another, without rotation. 2. Motion of a body along a line without rotation or turning.

Translation. Motion of a body along a line without rotation, or turning.

Translucent Panel. A building panel that permits the passage of light but not vision.

Translucent. Permitting the passage of light.

Transmission Factor. See Transmittance.

Transmission Loss (TL). The reduction of airborne sound power that is caused by placing a barrier, wall or material, between the reverberant sound field of a source and its receiver; transmitted loss is a property of the barrier.

Transmittance. The ratio of light transmitted through a material to the total incident light falling on it; also called Transmission Factor or T.

Transom Bar. The part of a door or window frame that separates the top of a door or window from the bottom of a transom light or sash.

Transom Catch. A fastener applied to a transom sash and having a ring by which the latch bolt is retracted.

Transom Chain. A short length of chain used to limit the opening of a transom sash, usually with a plate provided at each end for attachment.

Transom Lift. A vertically operated device attached to a door frame and transom sash by which the transom sash may be opened or closed.

Transom Sash. The framework of a glass window over a door or window; may be hinged for ventilation.

Transom. 1. A transverse piece in a structure, lintel. 2. A horizontal crossbar in a window, over a door, or between a door and a window or a fanlight above it. 3. A window above a door or other window built on and commonly hinged to a transom.

Transparent Mirror. A mirror with a transparent coating that allows vision from the space on the darker side to the space on the more brightly lit side; from the brighter side, it reflects like a mirror; also called One-Way Glass.

Transom

Transudate. A material that has transuded or seeped.

Transude. To pass through a membrane or permeable substance; to seep or exude.

Transverse Strength. A standard measurement of the relative flexural strength of gypsum board products conducted in accordance to ASTM C 473; expressed as pounds force or Newtons.

Transverse. At right angles to the long direction of the member; crosswise; also referred to as lateral.

Trap Seal. The vertical distance between the crown weir and the top dip of the trap.

Trap Vents. In all traps it is necessary that a back vent connection be made to the drain or waste pipe on the sewer side of the trap; if the back vent is not provided, the trap could be emptied by siphonage.

Trap, Float. See Float Trap.

Trap, Grease. See Grease Trap.

Trap. Located at a plumbing fixture waste line, designed to hold a quantity of water that provides a seal to prevent gasses in the sewer system from entering a room.

Trapdoor. A door or hatch in a floor, ceiling, or roof.

Trash Chute. A device either constructed on the inside of a building or structure or hung outside for the removal of waste materials from upper floors.

Trash Compactor. A device that compresses waste materials to conserve storage space.

Trash Handling. The act or process of transporting or removing waste materials.

Traveling Crane. A tower crane mounted on tires, rails, or crawlers.

Traverse Rod. A device that draperies or curtains are hooked on so they can be opened or closed.

Travertine Floor. A floor constructed of richly patterned, marble-like limestone.

Travertine. A richly patterned, marble-like form of limestone.

Tray, Cable. See Cable Tray.

Tread, Grating. See Grating Tread.

Tread. The horizontal surface of a step in a stair.

Treated Lumber. Lumber infused or coated with stain or chemicals to retard fire, decay, insect damage, or deterioration due to weather.

Treated Pole. Wooden pole infused or coated with stain or chemicals to retard fire, decay, insect damage, or deterioration due to weather.

Treated Post. Vertical post infused or coated with stain or chemicals to retard fire, decay, insect damage, or deterioration due to weather.

Tree Ball. See Root Ball.

Tree Cutting. The trimming of leaves, branches or larger portions of a tree, or the chopping down of a tree.

Tree Maintenance. The pruning or trimming of certain parts of a tree for aesthetic or practical purposes.

Tree Moving. The digging out of a tree with its root system intact, to relocate it to a different place.

Tree Removal. The chopping down, carrying off, or disposal of a tree or tree segments.

Tree Trimming. The pruning or light cutting of branches, leaves or other portions of a tree for practical or aesthetic purposes.

Tree Well. A planting space for one or more trees.

Treenail. A wooden pin, peg, or spike used chiefly for fastening planking and ceiling to a framework.

Trefoil. A three cusped element in Gothic tracery.

Trellis. A light latticework on which trees or shrubs may be trained.

Tremie. A large funnel with a tube attached, used to deliver concrete into deep forms or beneath water or slurry; a tremie slows down the concrete and resists segregation of the aggregates.

Trench Frame and Cover, Cast Iron. Frame and cover of cast iron components to cover recessed lineal element designed to collect water from surface and deliver to drainage pipe or pit.

Trench. A long, narrow excavation; a ditch.

Trencher. A mechanical device used to dig narrow channels in the ground.

Trenching. The act or process of digging narrow channels in the ground.

Trespass. Unauthorized entry upon the real property of another.

Trestle. 1. A braced framework bridge supporting a railroad track. 2. A table frame consisting of a horizontal member supported by a pair of divergent legs at each end.

Triad Color Harmony. Harmony obtained by using colors from three equidistant points of the color wheel; red, yellow, and blue make up a triad.

Trial Batch. A batch of concrete prepared to establish or check proportions of the constituents.

Trial. A hearing conducted by a judge at which litigants are entitled to introduce evidence and arguments supporting, and defending against, claims.

Triangle. 1. Three sided plane figure; a polygon of three sides. 2. A transparent plastic drafting tool in the shape of a triangle, usually either 45 degrees, 30-60 degrees, or adjustable.

Tribology. The science dealing with friction, wear, and lubrication of interacting surfaces.

Trichloromonofluoromethane. Low pressure, synthetic chemical refrigerant which is also used as a cleaning fluid.

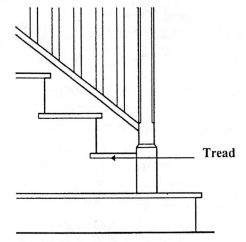

Tread

Trichloromonofluoroethane. Complete name of R-113, a synthetic, nontoxoic and nonflammable Group 1 refrigerant in rather common use; chemical components which make up this refrigerant are chlorine, fluorine, and ethane.

Trig. The bricks laid in the middle of a wall between the two main leads to overcome the sag in the line and also to keep the center plumb in case there is a wind bearing upon the line.

Triglyphs and Metopes. Parts of a Doric frieze; the triglyphs are vertical lines and the metopes are the spaces between.

Trigonometry. The branch of mathematics dealing with the relations of the sides and angles of right triangles and with the relevant functions of any angles.

Trilinear. Involving three lines.

Trim Enamel Paint. Surface coating differing from ordinary house paint by faster drying, by having more gloss and showing fewer brush marks; used mostly on trim, shutters, and screens.

Trim Painting. The act or process of applying paint to trim work.

Trim. Finish materials, such as moldings applied around openings (window trim, door trim) or where walls join the floor or ceiling of a room (baseboard, cornice, and other moldings).

Trimmer Arch. An arch, usually of brickwork and of low rise used for supporting the fireplace hearth.

Trimmer. 1. A beam or floor joist into which joists are framed. 2. A beam or joist to which a header is nailed in framing, as for a chimney or a stairway.

Trimmers. Tile units of various shapes consisting of such items as bases, caps, corners, moldings, and angles necessary or desirable to make a complete installation and to achieve sanitary purposes as well as architectural design for all types of tile work.

Trimming. Putting the inside and outside finish and hardware on a building.

Triple Glazing. An assembly of three panes of glass separated by air spaces; triple glass.

Triple Oven. A mechanical cooking device that has three separate chambers with doors.

Triple Point. The condition of temperature and pressure under which the gaseous, liquid, and solid phases of a substance can exist in equilibrium.

Triplex Apartment. An apartment having rooms on three floors.

Triplex. 1. A three-family house. 2. A Triplex Apartment.

Tripod. A three legged support for something, such as a transit or camera.

Tripoli. See Rottenstone.

Triptych. A picture or carving in three panels.

Trisect. 1. To divide into three parts. 2. To divide into three equal parts.

Trisodium Phosphate (TSP). A crystalline compound that forms the base for cleaning compounds; especially good for washing walls prior to repainting.

Troffer. A channel-like enclosure for light sources.

Trombe Wall. A masonry wall, usually glazed on the outside, used to trap solar heat, as a part of a passive solar heating system.

Troposphere. Part of the atmosphere immediately above the earth's surface in which occur most weather disturbances.

Troughing. Making repeated dozer pushes in one track, so that ridges of spilled material hold dirt in front of the blade.

Trowel Finish. The smooth finish coat surface of concrete, plaster, or stucco produced by troweling.

Trowel. A hand tool used by masons and plasterers to apply, shape, spread, or smooth plastic materials such as plaster and mortar.

Trowelling. Applying, spreading, or smoothing plastic materials such as mortar, concrete, or plaster, using a hand trowel.

Truck Crane. A mechanical device mounted on the bed of a truck for hoisting or lifting materials.

Truck Mixer. A concrete mixer capable of mixing concrete in transit when mounted on a truck chassis.

Truck Shelter. An enclosure or structure for the storage of motorized trucks.

Truck Well. A ramped-down depressed area adjacent to a loading door to allow trucks to be loaded or unloaded with the truck bed level with the building floor.

Truck, Refrigerated. See Refrigerated Truck.

Truck. 1. A wheeled vehicle for moving heavy articles. 2. A strong automotive vehicle for hauling. 3. A small wheelbarrow consisting of a rectangular frame having at one end a pair of handles and at the other end a pair of small, heavy wheels and a projecting edge to slide under a load.

True Complement Color Harmony. Two colors directly across the color wheel from each other are true complements; examples are red and green, and orange and blue.

Truncate. To shorten by cutting off the top of an object.

Truss Panel. A chord segment between two successive joints of a truss.

Truss Plate Institute (TPI). 583 D'Onofrio Drive, #200, Madison, Wisconsin 53719, (608)833-5900.

Truss Reinforcing. The addition of cross bracing or other bracing members to stiffen and strengthen a framework of trusses.

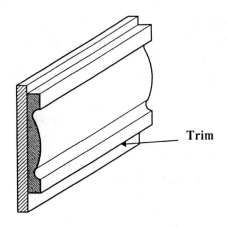

Trim

Truss Steel. Steel framework of triangular units for supporting loads over long spans.

Truss, Wood. See Wood Truss.

Truss. 1. A triangular arrangement of structural members that reduces nonaxial forces on the truss to a set of axial forces in the members. 2. Structural framework of triangular units for supporting loads over long spans.

Trussed Bars. Steel reinforcing bars bent up to act as both top and bottom reinforcement.

Trust Deed. A document evidencing a security interest in real property; used in some states in place of a mortgage.

Trustee. In a trust deed, the entity to whom title to the property is legally entrusted for the benefit of the beneficiary (the lender).

Trustor. In a trust deed, the entity that has pledged property to guarantee repayment of a loan.

TSG. Tapered Steel Girder.

TSI. Thermal System Insulation.

TSP. Trisodium Phosphate.

T-Square. A drafting tool that, used with a drawing board, can be used to produce parallel lines on a drawing.

T-Stat. Thermostat.

Tsunami. A sea wave produced by a large displacement of the ocean bottom, the result of earthquakes or volcanic activity.

Tub Enclosure. An enclosure of glass or plastic mounted on a bathtub rim with sliding, folding, or swinging door panels, to keep shower spray in the compartment.

Tub, Bath. A basin with a drainage unit and water supply used primarily for bathing or washing purposes.

Tube, Constricted. Tubing reduced in diameter.

Tube, Structural. See Structural Tube.

Tube. 1. A duct or pipe to convey liquids. 2. A structural load-bearing member in the form of a square or round pipe. 3. A tunnel. 4. Electron tube. 5. Vacuum tube. 6. Television Tube.

Tube-In-Plate Absorber. An aluminum or copper sheet metal absorber plate in which the heat transfer fluid flows through passages formed in the plate itself.

Tube-Within-A-Tube. Water-cooled condensing unit in which a small tube is placed inside large unit; refrigerant passes through outer tube, water through the inner tube.

Tubing, Non-Metallic. A round sheath product, of round cross-section, fabricated from a moisture-resistant, flame-retardant material.

Tubing. Any thin-walled pipe which can be bent easily.

Tubular Lock. See Cylindrical Lock.

Tubular. Having the form of a round or polygonal pipe or tube.

Tuck Pointing. The process of removing deteriorated mortar from the surface of an existing brick wall, and inserting fresh mortar.

Tufted Carpet. Carpet made by inserting the pile yarns through a pre-woven fabric backing on a machine with hundreds of needles, similar to a huge sewing machine.

Tufts. The cut loops of a carpet pile fabric; applies to both woven and tufted carpets.

Tumbler Switch. A lever-operated snap switch, powered by electric current.

Tumbler. A lock mechanism that holds a bolt until operated by a key.

Tumblestone Aggregate. A random collection of aggregate.

Tumbling. Method of finishing or polishing by using a tumbling barrel; articles to be finished and finishing material are put into barrel which is turned or tumbled.

Tung Oil. A yellow drying oil obtained from the seed pods of tung trees and widely used in water-resistant varnishes, lacquers, and high-gloss paints.

Tungsten. 1. A gray-white dense high-melting point ductile hard metallic element that resembles chromium and molybdenum in many of its properties and is used for electric light filaments and in hardening alloys.

Tunnel. A horizontal passageway through or under an obstruction; also called Wolfram.

Tunneling. The act or process of digging a horizontal passageway through or under an obstruction.

Turbidity. Degree to which pool water is visually obscured by suspended particles.

Turbine. A rotary engine actuated by the reaction of a current of fluid, as water, steam, or air, subject to pressure and usually made with a series of curved vanes on a central rotating spindle.

Turf. The layer of grass, matted roots, and earth that comprise the surface of a grassland.

Turn Piece. A small knob, lever, or tee turn with spindle attached for operating the deadbolt of a lock or a mortise bolt.

Turnbuckle. A mechanical device for tightening tie rods by turning a threaded link.

Turner. One who operates a lathe.

Turning Vanes. Thin curved blades placed at bends in air ducts to direct air flow with a minimum of eddying and pressure loss.

Turnkey. A project delivery system where the owner pays an all-inclusive price for the land, financing, design, and construction.

Turnout. 1. A widened space in a highway for vehicles to pass or park. 2. A railroad siding.

Turnover Rate. The number of times a quantity of water, which is equal to the total capacity of water in the swimming pool, passes through the filters in a given period; this is usually expressed in turnovers per day.

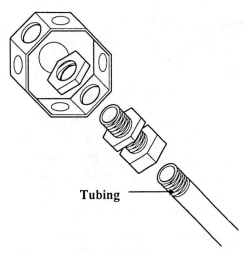

Tubing

Turnstile. A gate for admission or exit with revolving arms allowing people to pass singly.

Turntable. 1. A circular revolving platform for turning a vehicle like a locomotive, truck, bus, or automobile. 2. A lazy susan installed in a corner cabinet.

Turpentine. 1. An oleoresin secreted by various conifers. 2. A volatile pungent oil distilled from turpentine, used in mixing paints and varnishes; also called Turps.

Turps. Turpentine.

Turret. A small tower, often projecting from a building's wall as an ornament.

Turtleback. See Blistering, 2.

Tuscan Red. A red pigment consisting of a combination of iron oxides and a lake.

TV. Television.

TVOC. Total Volatile Organic Compound.

TWA. Time-Weighted Average.

Twist. A distortion caused by the turning or winding of the edges of a board so that the four corners of any face are no longer in the same plane.

Two Coat Work. See Double-Up.

Two-Coat System. Two-coat paint application for initial painting.

Two-Component Gun. One having two separate fluid sources leading to spray head, for spraying a coating and a catalyst simultaneously.

Two-Family Dwelling. A dwelling containing two dwelling units.

Two-Point Latch. A device used to lock an inactive door leaf to the frame head and to the floor.

Two-Pole Motor. 3600 rpm, 60 Hz electric synchronous speed motor.

Two-Temperature Valve. Pressure-opened valve used in suction line on multiple refrigerator installations which maintains evaporators in system at different temperatures.

Two-Way Action. Bending of a slab or deck in which bending stresses are approximately equal in the two principal directions of the structure.

Two-Way Concrete Joist System. A reinforced concrete framing system in which columns directly support an orthogonal grid of intersecting joists.

Two-Way Flat Plate. A reinforced concrete framing system in which columns directly support a two-way slab that is planar on both its surfaces.

Two-Way Flat Slab. A reinforced concrete framing system in which columns with mushroom capitals and/or drop panels directly support a two-way slab that is planar on both its surfaces.

Two-Way Valve. Valve with one inlet port and one outlet port.

Two-Way. Construction with steel reinforcing running in two perpendicular directions.

Tyloses. Masses of parenchyma cells appearing somewhat like froth in the pores of some hardwoods, notably the white oaks and black locust; tyloses are formed by the extension of the cell wall of the living cells surrounding vessels of hardwood.

Tympanum. The recessed triangular face of a pediment.

Typ. Typical.

Type X Gypsum Board. A gypsumboard used where increased fire resistance is required.

Type X Lath. See Gypsum Lath.

Typical. Exhibiting the essential characteristics of a group; a representative example.

U Bolt. A U-shaped, bent iron bar that has bolts and threads at both ends.

U Stirrup. An open-top, U-shaped loop of steel bar used as reinforcing against diagonal tension in a beam.

U.S. Department of Justice (DOJ). 10th & Constitution Avenue, NW, Washington, DC 20530, (202) 514-2000.

UBC. Uniform Building Code.

UCC. Uniform Commercial Code.

UF Cable. Underground Feeder Cable.

UHF Cable. Cable that is designed for ultra-high frequency.

UL. Underwriters Laboratories.

ULC. Underwriters Laboratories of Canada.

Ultimate Compressive Strength. The stress at which a material crushes.

Ultimate Load. The absolute maximum magnitude of load which a structure can sustain, limited only by ultimate failure.

Ultimate Strength. Maximum strength that can be developed in a material.

Ultimate-Strength Design. Method of structural analysis of continuous concrete structures based on calculating the loadings which will cause actual failure of sections, rather than the loadings which will cause stresses to reach allowable, safe, values.

Ultramarine Blue. Blue pigment made by heating a mixture of China clay, sodium carbonate, carbon and sulphur; it is not affected by alkalis, but is easily affected by acids.

Ultramarine. A vivid blue pigment, made from powdered lapiz lazuli.

Ultramicroscopic. Too small to be seen with a normal microscope.

Ultrasonic Transmitter. A mechanism that converts vibrations with the same physical nature as sound into equivalent waves above the range of human hearing.

Ultra-Violet Light. Light of short wave length which is invisible but has active chemical effect on finishing materials.

Ultraviolet. 1. The invisible rays of the spectrum at the violet end. 2. Invisible radiation waves with frequencies shorter than wave lengths of visible light and longer than X-ray.

Umber. A natural earth pigment like ochre but darker and browner.

Unadulterated. Not contaminated; pure.

Unbalanced Wall. See Unsymmetrical.

Unbalancing the Schedule of Values. See Front end loading.

Unbonded Construction. 1. Post-tensioned concrete construction in which the tendons are not grouted to the surrounding concrete. 2. A construction contract in which the contractor has not posted a surety bond.

Unbuffed End. An untrimmed, serrated factory cut end.

Uncertainty. The doctrine that contracts are unenforceable if unintelligible

Unconscionability. A doctrine that courts will not enforce contractual provisions that put one party at the mercy of another.

Unctuous. Oily; of minerals, having a greasy or soapy feel.

Uncut loop. Carpet pile yarns that are continuous from tuft to tuft, forming visible loops.

Underatomized. Not dispersed or broken up fine enough.

Underbead Crack. In welding, a crack in the heat-affected zone not extending to the surface of the base metal.

Underbid. To bid less than a competing bidder.

Undercapitalized. Having too little capital for efficient operation.

Undercoat. A coating applied prior to the final or top coat of a paint job; second coat in three-coat work or first coat in repainting.

Undercourse. A course of shingles laid beneath an exposed course of shingles at the lower edge of a wall or roof, in order to provide a waterproof layer behind the joints in the exposed course.

Undercroft. A subterranean room, such as a crypt under a church.

Undercut. A groove at the base metal adjacent to the toe of a weld and left unfilled by weld metal.

Underdrain. 1. Perforated pipe drain installed in crushed stone under a slab to intercept ground water and drain it away from the structure. 2. The distribution system at the bottom of a sand filter to collect the filtered water during a filter run and to distribute the backwash water during backwash.

Underestimate. An estimate that is too low.

Underfloor Duct. A round or rectangular metal pipe placed under a wood floor construction or in a concrete floor to distribute warm air from a heating or air conditioning system.

Underglaze Decoration. A ceramic decoration applied directly on the surface of ceramic ware and subsequently covered with a transparent glaze.

Underground Feeder Cable. Insulated electrical cable suitable for underground use.

Underground Tank Removal. The excavation and removal process of a storage tank buried in the ground.

Underground Tank. A container that holds various liquid or solid matter and is found underground.

Underground. Subterranean.

Underground Paper. Wallpaper without a basic background color.

Underlay. 1. One or more layers of felt applied as required for a base sheet over which finish roofing is applied. 2. See Carpet Padding.

Underlayment. A material placed under finish coverings, such as flooring or shingles, to provide a smooth, even surface for applying the finish.

Underpinning. The process of placing new foundations beneath an existing structure.

Underslab Drainage. The process of continuous interception and removal of ground water from under a concrete slab with the installation of perforated pipe.

Undertone. A color covered up by other colors but when viewed by transmitted light, shows through the other colors modifying the effect.

Underwater Light. A lighting fixture designed to illuminate a pool from beneath the water surface; it can be wet-niche, located in the pool water, or dry-niche, located in the pool wall and serviced from outside the pool.

Underwriters Laboratories (UL). An organization which classifies roof assemblies for their fire characteristics and wind-uplift resistance for insurance companies in the United States.

Underwriters Laboratories of Canada (ULC). 7 Crouse Road, Scarborough, Ontario, Canada M1R 3A9, (416) 757-3611.

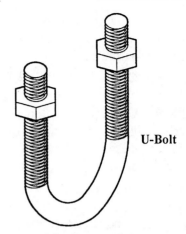

U-Bolt

Underwriters Laboratories, Inc (UL). 333 Pfingsten Road, Northbrook, Illinois 60062, (708) 272-8800.

Undivided Interest. Joint ownership of an interest in the whole property rather than in a specific identifiable part.

Unemployment Tax. A program funded by employers based on the wages of employees, paid on a quarterly basis to the Unemployment Insurance Fund from which benefits are paid to qualifying ex-employees.

Unencumbered Property. Real property that has no liens, encumbrances, easements, or debt; free and clear.

Unequal Angle. A steel angle in which the two legs are of different lengths.

Unfaced Insulation. Insulation which has been manufactured without a vapor barrier.

Unfinished Bolt. An ordinary carbon steel bolt.

Unforeseen Conditions. Unexpected or unanticipated events or situations on the jobsite or locality above or below grade that affect the contract price or time.

Unglazed Floor Tile. A hard, dense floor tile deriving color and texture from the materials used and the method of manufacture, with uniform composition throughout.

Unglazed Floor. A hard ceramic floor surface deriving its color or texture from the materials used with uniform composition throughout.

Unglazed Tile. A hard, dense tile of homogeneous composition throughout, deriving color and texture from the materials of which the body is made; the colors and characteristics of the tile are determined by the materials used in the body, the method of manufacture, and the thermal treatment.

Ungrounded Cable. Two-wire non-metallic sheathed cable that contains one neutral wire and one hot wire.

Uniform Building Code (UBC). A model building code written by the International Building Officials Conference and adopted by many cities and counties as their official building code.

Uniform Commercial Code (UCC). A statute governing commercial transactions, sales, and commercial paper that has been adopted in substantially the same form in all 50 states.

Uniform Sand. A mixture of sand in which most of the granules are the same size.

Union Hub. A pipe fitting used to join two pipes without turning either pipe, with one end of the fitting enlarged into a bell or socket.

Union T. A pipe tee with a fitting on one end that joins two pipes without turning either pipe.

Union, Brass. A pipe fitting constructed of brass, used to join two pipes without turning either pipe.

Union, PVC. A pipe fitting used to join two pipes without turning either pipe, manufactured from polyvinyl chloride and used mainly in drain lines because of its resistance to chemicals.

Union, Stainless. A pipe fitting used to join two pipes without turning either pipe, manufactured from stainless steel.

Union. A type of pipe fitting used to join two pipes in line without turning either pipe.

Unit and Mullion System. A curtain wall system consisting of prefabricated panel units secured with site-applied mullions.

Unit Cost. 1. Cost per unit. 2. Cost per unit of length, volume or area.

Unit Heater. A device for heating a space without the use of ductwork.

Unit Masonry. Manufactured or natural building units of concrete, burned clay, glass, stone, gypsum, etc.

Unit of Work. Calorie, erg, foot-pound, joule, poundal; unit of electrical power, amp, amperage, ampere, kilowatt, megawatt, ohm, voltage, volt, wattage, watt.

Unit Price. 1. The price charged for a single unit or minimum unit of production. 2. The price charged for a unit of length, volume, or area.

U

Union Hub

Unit Stress. The force on a member divided by its cross-sectional area, such as pounds per square inch.

Unit Substation. A freestanding assembly including a transformer, switchgear and meter.

Unit Vent. A vent pipe which services two or more traps.

Unit Ventilator. A fan coil unit with an opening through an exterior wall to admit outside air.

Unit Water Content. The quantity of water per unit volume of freshly mixed concrete, often expressed as gallons or pounds per cubic yard; this is the quantity of water on which the water- cement ratio is based and does not include water absorbed by the aggregate.

Unit Weight. The weight of a unit volume of a substance; its density.

Unitary Package Unit. A system of conditioning air in an outdoor package unit, similar to a rooftop unit, except that the supply and return ductwork passes horizontally through an outside wall.

Unitary System. A heating/cooling system factory assembled in one package and usually designed for conditioning one space or room.

United States Customary Systems. The measuring units used in the U.S. consisting of the mile, foot, inch, gallon, second, and pound; also called Imperial Units; these units are gradually being replaced by SI units.

United States Standard. Sheet metal gauge.

Universal Motor. Electric motor which will operate on either a-c or d-c.

Universal. Description of a lock, door closer, or other device that can be used on doors of any hand without change.

Unjust Enrichment. A situation in which one party is unjustly enriched at the expense of another party.

Unlawful Detainer Action. A lawsuit brought to evict a defaulting tenant from real property.

Unlined Stack. A chimney built of brick only, possibly lined with a coat of cement plaster, but with no terra cotta flue lining.

Unplasticized Polyvinyl Chloride (uPVC). A stiff plastic material of which doors, windows, corrugated roofing and other building components are made.

Unrated Door. A door which has not been rated for any ability to withstand the spread of fire.

Unreasonable Hardship. This exists when the enforcing agency finds that compliance with the building standard would make the specific work of the project affected by the building standard unfeasible, based on an overall evaluation of all relevant factors, including cost, value, structural stability, safety, conservation, and disabled access.

Unreinforced. Concrete members constructed without steel reinforcing bars or welded wire fabric.

Unsaponifiable Water. Substance in resins and fats which does not unite with caustic alkali to form a soap.

Unstable. A structure that lacks stability.

Unsymmetrical Wall. An assembly of multiple layers of gypsumboard where one side has more layers or a different construction than the other; also called an Unbalanced Wall.

Untreated Joint. A gypsumboard joint without joint compound, tape, or battens.

Uplift. An upward force.

Upset Price. A guaranteed maximum price in a construction contract.

Upside Down Roof. A membrane roof assembly in which the thermal insulation lies above the membrane.

uPVC. Unplasticized Polyvinyl Chloride.

Urban Renewal. The redevelopment of a deteriorated inner city area.

Urea. A crystalline colorless nitrogenous compound found primarily in the urine of mammalian animals; can be synthesized from carbon dioxide and ammonia for use in fertilizers, resins, and plastics.

Urea-Formaldehyde Foam. An aerated foam of a thermosetting synthetic resin composed of urea and formaldehyde; used for thermal insulation.

Urea-Formaldehyde Resin. Product obtained by chemical reaction between urea and formaldehyde in presence of catalyst.

Urea-Melamine Resin. Melamine-formaldehyde resins which produce tough finish that approaches porcelain.

Urethane Board. Rigid form of plastic foam of polyurethane.

Urethane Foam. Type of insulation which is foamed in between inner and outer walls of a container.

Urethane Insulation. Board form or sprayed on plastic foam of rigid polyurethane.

Urethane Padding. Plastic foam of polyurethane used under carpeting for cushioning.

Urethane Resins. A particular group of film formers; Isocyanate resins.

Urethane Roof. A roof with plastic foam of rigid polyurethane; made into board form or sprayed on.

Urethane, Rigid. Sheet or board form of plastic foam of rigid polyurethane.

Urinal Carrier. A metal structure attached to a bathroom wall to support and mount a urinal.

Urinal Screen. A privacy panel that separates urinals from each other; the screens are either floor, wall or ceiling mounted.

Urinal. A water flushed plumbing fixture designed to receive urine directly from the user of the fixture.

Urn, Coffee. See Coffee Urn.

USC&GS. United States Coast and Geodetic Survey.

USC. United States Code.

USCS. United States Customary Systems.

USDA. United States Department of Agriculture.

Useful Life. The length of time a coating is expected to remain in service.

USGS. United States Geological Survey.

USP. Letters affixed to name of material to indicate that it conforms in grade to specifications of United States Pharmacopoeia and that it is approved for use in medicinal preparations. The material does not necessarily have to be chemically pure.

USS. United States Standard.

Usury. Charging more than the maximum legally allowable rate of interest.

Utensil Washer, Medical. See Medical Utensil Washer.

Utility Excavation. The act or process of either digging up existing cable buried in the ground, or trenching to lay new cable.

Utility Knife. 1. A commonly used razor-type blade and holder. 2. See Board Knife.

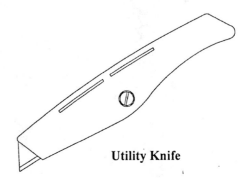

Utility Knife

Utility Pole. A vertical pole with cross arms, where electrical and telephone lines are carried.

Utility Room. A room housing domestic appliances such as a laundry.

Utility. A service provided by a public utility, such as electrical power or water.

Utilization Equipment. Equipment which utilizes electric energy for mechanical, chemical, heating, lighting, or similar useful purposes.

UV. Ultraviolet.

U-Value. See Coefficient of Heat Transmission.

V Beam Roof. A roof with corrugated sheeting with flat, V-angled surfaces.

V. A symbol for vertical shear.

VA. Veterans Administration.

Vacations. The uncoated portion of a finished object; also called Skips and Holidays.

Vacuum Breaker. 1. An electrical breaker with a space that contains reduced air pressure. 2. A device which prevents the formation of a vacuum in a water supply pipe; installed to prevent backflow.

Vacuum Cleaner. One of several types of suction devices used to collect dirt from the bottom of a swimming pool; these units work in a variety of ways - some force collected dirt to the filter, others to the waste drain, or into a porous container.

Vacuum Control System. Intake manifold vacuum is used to operate dampers and controls in some automobile systems.

Vacuum Gauge. Instrument used to measure pressures below atmospheric pressure.

Vacuum Piping. The pipe from the suction side of a pump connected to a vacuum fitting located at the pool and below the water level to which underwater cleaning equipment may be attached.

Vacuum Pump. Special high efficiency device used for creating high vacuums for testing or drying purposes.

Vacuum. 1. A space absolutely devoid of matter. 2. Pressure lower than atmospheric pressure.

Valley Flashing. Thin sheet metal used to line the valley of a roof.

Valley Rafter. A diagonal rafter that supports a roof valley.

Valley. 1. An elongated depression in the earth's surface between two raised areas. 2. A trough or internal angle formed by the intersection of two roof slopes.

Valuation. The opinion of a real estate appraiser as to the value of a parcel of real property.

Value Engineering. The process of analyzing the cost versus the value and alternative materials, equipment, and systems.

Value of the Assets. The assets of a company in terms of dollar value.

Value. 1. The worth of something measured monetarily. 2. Term used to distinguish dark colors from light ones; dark values are known as shades; light values as tints.

Valve Plate. Part of a compressor located between the top of the compressor body and the head, containing compressor valves and ports.

Valve Suction. See Suction Valve.

Valve, Expansion. See Expansion Valve.

Valve, Solenoid. See Solenoid Valve.

Valve, Suction. See Suction Valve.

Valve, Tapping. See Tapping Valve.

Valve, Water. See Water Valve.

Valve. Numerous mechanical devices by which the flow of liquid, gas, or loose material in bulk may be started, stopped, or regulated by a movable part that opens, shuts, or partially obstructs one or more passageways.

Van Dyke Brown. A brown pigment which consists of decomposed vegetable matter that has almost approached the coal state; it is weak in hiding power compared to umber and sienna.

Vanadium. 1. A hard gray malleable ductile metallic element found in several ores and used as an alloy in small quantities for strengthening some steels. 2. Vanadium steel.

Vane. A revolving pointer mounted on a high place to show the direction of the wind.

Vanity Cabinet. Case, box, or piece of furniture which rests on the floor and receives a lavatory, commonly has shelves and doors and is primarily used as storage for below lavatory.

Vanity, Bath. See Bath Vanity.

Vapor Barrier. 1. A material with a high resistance to vapor movement, such as foil, plastic film, or specially coated paper, that is used in combination with insulation to control condensation. 2. Any material that has a water vapor permeance (perm) rating of one or less. 3. A type of plastic sheeting that both eliminates drafts and keeps moisture from damaging a building or structure. 4. A waterproof membrane placed under concrete floor slabs that are on grade. 5. Material used to retard the movement of water vapor into walls, and prevent condensation in them; applied separately over the warm side of exposed walls or as a part of batt or blanket insulation. 6. Thin plastic or metal foil sheet used in air-conditioned structures to prevent water vapor from penetrating insulating material.

Vapor Compression Refrigeration. A cooling process that transforms heat from one location to another; heat is forced to flow from a region of lower temperature to a region of higher temperature; similar to a refrigerator for cooling food.

Vapor Degreasing. A cleaning process utilizing condensing solvent as the cleaning agent.

Vapor Lock. Condition where liquid is trapped in line because of bend or improper installation; such vapor prevents liquid flow.

V

Vapor Migration. The movement of water vapor from a region of high vapor pressure to a region of lower vapor pressure.

Vapor Pressure Curve. Graphic presentation of various pressures produced by refrigerant under various temperatures.

Vapor Pressure. Pressure imposed by either a vapor or gas.

Vapor Retarder Felt. Sealing material made of asphalt-saturated felt.

Vapor Retarder. A layer of material intended to obstruct the passage of water vapor through a building assembly; also called a Vapor Barrier; sealing material placed between building components or materials to arrest movement of moisture within building enclosure assembly.

Vapor, Saturated. See Saturated Vapor.

Vapor. The gaseous state of any substance.

Vaporization. Conversion from liquid or solid to a gaseous state.

Variable Air Volume (VAV). An air handling system, unlike a single zone, terminal reheat, multi-zone, or dual duct, supplies air at a constant temperature and varies the air quantity supplied to each zone.

Variable Pitch Pulley. Pulley which can be adjusted to provide different pulley drive ratios.

Variable Rate Mortgage. A mortgage that allows the lender to raise or lower the interest rate from time to time as some specified index rises or falls.

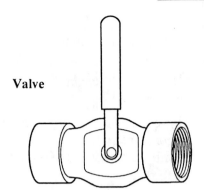

Valve

Variable. Changing; subject to change, not constant; indeterminate.

Variance. A deviation in a building or zoning ordinance granted by an appeal authority upon relevant grounds being proven.

Varnish. A thickened preparation of drying oil or drying oil and resin; when applied to a surface, it leaves a hard, glossy, transparent coating; may also be mixed with pigments to make enamels; clear varnish is a slightly yellow, semitransparent liquid.

Varnishing. Application of varnish to a surface using a brush.

Varnish-Stain. Interior varnish tinted with pigments or dyes.

Vat Dyes. Dyes formed in fabrics by oxidation and precipitation of the original dye liquor, for example, indigo; vat dyeing refers to a kind of dye rather than a method of dyeing; raw stock dyeing, skein dyeing, or solution dyeing can be performed with vat dyes.

Vault Door. A hinged, pivoted, or sliding member that permits passage to an enclosure built for safety or security.

Vault. 1. An arched surface. 2. An arch translated along an axis normal to the plane of its centerline curve. 3. A room to store valuable items.

VAV. Variable Air Volume.

V-Belt. Type of belt commonly used in refrigeration work; it has a contact surface with the pulley which is in the shape of the letter V.

VCP. Vitreous Clay Pipe.

VDU. Video Display Unit.

Vector. 1. A quantity having direction as well as magnitude; usually represented by a line in space with an arrowhead indicating direction and the scale of the line representing magnitude. 2. A carrier of disease, such as a roach, rat, or mouse.

Vegetable Oils. Oils obtained from the seeds or nuts of vegetable growth, including linseed, soybean, perilla, tung, or castor.

Vegetation Control. The act or process of spraying chemicals or placing powders on unwanted vegetation.

Vehicle Lift. 1. An apparatus for lifting vehicles in order to have access to the carriage underneath. 2. A mechanical device used to move vehicles up and down in a parking structure.

Vehicle. 1. Any conveyance for transporting people or goods. 2. The liquid portion of paint and other finishing materials, consisting of the binder (nonvolatile) and thinners (volatile).

Vehicular Way. A route used for vehicular traffic, such as a street, driveway or parking lot.

Veining. The characteristic stretch marks that develop during the aging process of soft bitumen.

Vellum Glaze. A semi-mat glaze having a satin-like appearance.

Vellum. Fine parchment-like paper used for drawings for reproduction.

Velocimeter. Instrument used to measure air speeds using a direct-reading air speed indicating scale.

Velocity of Sound. A product of frequency times wavelength.

Velocity. Speed, usually expressed in length units per time unit, like miles per hour or feet per second.

Velvet Carpet. A pile fabric woven on a velvet loom; it is the simplest of all carpet weaves and is used mostly for solid colors; woven with wires, a looped pile is created when the wires are withdrawn; a cut pile results when knife blades on the ends of the wires cut the loops; tightly twisted yarns in cut pile provide a frieze surface; varied heights of looped pile can be achieved by using shaped wires; see Woven Carpet.

Veneer Plaster Base. A gypsumboard used as the base for application of a gypsum veneer plaster.

Veneer Plaster. A wall finish system in which a thin layer of plaster is applied over a special gypsumboard base.

Veneer, Ashlar. See Ashlar Veneer.

Veneer, Brick. See Brick Veneer.

Veneer, Granite. See Granite Veneer.

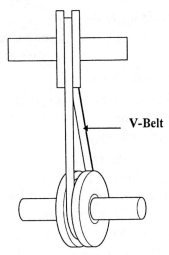

V-Belt

Veneer. 1. A thin layer, sheet or facing. 2. A thin layer or sheet of wood; usually one that has beauty or value and is intended to be overlaid on an inferior surface. 3. Nonstructural facing of brick, concrete, stone, tile, metal, plastic or other similar approved material attached to a backing for the purpose of ornamentation, protection or insulation.

Venetian Blind. A window blind which has numerous horizontal slats that may be set at several different angles in order to vary the amount of light admitted.

Venetian Red. Pigment with brick-red color made synthetically by calcining copperas and whiting.

Venetian Terrazzo. A mosaic type of terrazzo topping in which large chips of stone are incorporated.

Vent Cap. The top piece installed over the terminus of a ventilation pipe.

Vent Screed. A sheet metal plaster screed incorporating a vented section used chiefly in roof overhang soffits.

Vent Stack. A plumbing vent pipe in a multistory building, a separate pipe used for venting, that either connects with a stack vent above the highest fixture, or extends through the roof.

Vent, Foundation. See Foundation Vent.

Vent, Masonry. See Masonry Vent.

Vent, Roof. See Roof Vent.

Vent, Smoke. See Smoke Vent.

Vent, Soffit. See Soffit Vent.

Vent. 1. A vertical pipe connected to a waste or soil distribution system that prevents a back pressure or a vacuum that might siphon the water out of a trap. 2. Vertical pipe to provide passageway for expulsion of water vapor and vent gases from gas-burning equipment to outside air. 3. A free opening to provide air intake, expulsion, or circulation in such areas as underfloor crawl spaces and attics.

Ventilated. Provided with a means to permit circulation of air sufficient to remove an excess of heat, fumes or vapors.

Ventilation Air. 1. The total volume of air in a system, which is a combination of the air brought into the system from the outdoors and the air that is being recirculated within the building. 2. The fresh air brought into the system from the outdoors.

Ventilator, Gravity. See Gravity Ventilator.

Ventilator, Relief. See Relief Ventilator.

Venting Clip. Clips to provide means to aid in curing and exiting of moisture from concrete or gypsum fill placed on a deck.

Venting of Roof Assembly. The necessary venting to eliminate moisture and vapor in roof systems.

Venturi. A short tube with a tapering constriction in the middle that causes an increase in the velocity of flow of a fluid and a corresponding decrease in fluid pressure.

Verdigris. 1. A green or bluish-green crystallized substance formed on copper by the action of acetic acid, used as a pigment. 2. Green oxidation on copper, bronze, or brass.

Verge Boards. The boards which serve as the eaves finish on the gable end of a building.

Vermiculite Insulation. A loose mineral fill for thermal insulating applications.

Vermiculite Plaster. 1. A covering that is put on steel beams, concrete slabs, and other heavy construction materials, as a fire-retardant. 2. An insulating and soundproofing plaster.

Vermiculite. A mineral closely related to mica, with the faculty of expanding on heating to form lightweight material with insulation quality; used as bulk insulation and also as aggregate in insulating and acoustical plaster and in insulating concrete floors.

Vermilion. Sulphide of mercury used as a pigment.

Vermont Slate. A fine grained thin-layered rock used for roofing, and paving.

Vernacular. The common building style of a period or place; concerned with ordinary rather than monumental buildings; vernacular architecture.

Vernier. A supplementary scale used on an instrument for obtaining fine adjustment.

Vertical Angle. An angle in a vertical plane.

Vertical Application of Roofing. Roll roofing laid parallel to the roof slope.

Vertical Application. Of gypsum-board, see Parallel Application.

Vertical Bar. An upright reinforcing bar in a reinforced concrete shape.

Vertical Broken Joint. Style of laying tile with each vertical row of tile offset for half its length.

Vertical Check Valve. A device, mounted vertically, which allows fluid or air to pass through in only one direction.

Vertical Curve. The meeting of different gradients in a road or pipe.

Vertical Drains. Usually columns of sand used to vent water squeezed out of humus by weight of fill.

Vertical Forces. Loads imposed on a structure caused by gravity or the vertical component of wind or earthquake forces.

Vertical Joint. See Butt Joint, 4.

Vertical Lift Door. A door assembly on a lifting or hoisting device that is opened or closed, bottom to top or top to bottom.

Vertical Louver. A louver equipped with slats that are mounted vertically; see Louver.

Vertical Pattern. A spray pattern whose longest dimension is vertical.

Vertical Pipe. Any pipe or fitting which makes an angle at 45 degrees or less with the vertical.

Vertical Position. In pipe welding, the position of a pipe joint wherein welding is performed in the horizontal position and the pipe may or may not be rotated.

Vertical Reinforcing. Steel reinforcing bars in concrete, running vertically, perpendicular to the horizontal bars.

Vertical Shear. The sum of the algebraic forces that are one side of a given cross-section of a member.

Vertical Siding. Exterior wall covering attached vertically to the wood frame of a building or structure.

Vertical Slotted Shelf Standard. Metal strip with slots attached to vertical surfaces as the support part of adjustable shelf hardware.

Vertical Transportation. A general term for elevators, lifts, escalators, and dumbwaiters.

Vertical Wood Board and Batten Siding. Linear vertical wood boards with wood strips covering vertical joints used as an exterior cladding for a framed wall.

Vertical Wood Board Siding. Linear vertical wood material used as exterior surface or cladding for exterior framed wall.

Vertical. Perpendicular to the plane of the horizon; upright.

Vertical-Grained Wood. Edge-grained lumber.

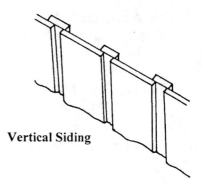

Vertical Siding

Vertically Laminated Wood. Laminated wood in which the laminations are so arranged that the wider dimension of each lamination is approximately parallel to the direction of load.

Vessels. Wood cells of comparatively large diameter that have open ends and are set one above the other to form continuous tubes; the openings of the vessels on the surface of a piece of wood are usually referred to as pores.

Vestibule. An entrance to a house; usually enclosed. *Entrance Hall; HALL; FOYER; LOBBY; LARGE ENTRANCE OR RECEPTION ROOM/AREA*

VG. Vertical Grain.

Vibration Aresstors. Soft or flexible substance or device which will reduce the transmission of a vibration.

Vibration Detector. A part of a burglar alarm system.

Vibration Isolation. A method of isolating equipment from a structure to reduce the vibration transmission from the equipment, thus reducing the sound transmitted to the structure.

Vibration Isolator. A flexible device supporting vibrating mechanisms or machinery and reducing the vibrations transmitted to the rest of the building or structure.

Vibration. 1. A periodic motion which repeats itself after a definite interval of time. 2. Energetic agitation of concrete to assist in its consolidation, produced by mechanical oscillating devices at moderately high frequencies.

Vibrator, External. See External Vibrator.

Vibrator, Internal. See Internal Vibrator.

Vibrator. Device for agitating wet concrete to eliminate air pockets.

Vicat Apparatus. A penetration device used to determine the setting characteristics of hydraulic cements.

Video Display Unit (VDU). 1. Cathode Ray Tube. 2. A television screen. 3. A computer monitor.

Vierendeel Truss. A rigid frame, used as a beam, assembled from parallel top and bottom chords tied together by vertical members.

Vinyl Asbestos Tile. Resilient floor tile consisting of vinyl reinforced with asbestos fibers.

Vinyl Coated Fence. A fence or enclosure built with posts and rails and covered with chain-link which has been factory coated with plastic.

Vinyl Coating. One in which the major portion of the binder is of a vinyl resin.

Vinyl Copolymer. Resins produced by copolymerizing vinyl monomers such as vinyl acetate and vinyl chloride.

Vinyl Faced Panel. Sheathing that has been covered by a flexible film or a liquid bonded by heat.

Vinyl Floor Tile. A resilient floor tile made of vinyl.

Vinyl Foam Cushioning. Carpet cushioning made from a combination of foamed synthetic materials.

Vinyl Gasket. A plastic seal between glass and aluminum.

Vinyl Resins. Synthetic resins made from vinyl compounds such as vinyl acetate.

Vinyl Sheet Floor. A thin sheet of vinyl used for the finish of floor surfaces.

Vinyl Sheet. The rolled form of vinyl.

Vinyl Sheetrock. Gypsumboard with a thin layer of vinyl as the finished surface.

Vinyl Siding. Exterior wall coverings made from a thermoplastic compound.

Vinyl Tile. A semi-flexible, resilient floor tile made from polymerized vinyl chloride, vinylide chloride, or vinyl acetate.

Vinyl Trim. Extruded vinyl moldings used for concealing edges, ends, joints, and corners of various sheet materials.

Vinyl Wall Covering. A tough, flexible and shiny film or liquid bonded by heat to a paper or fabric backing material.

Vinyl. A plastic made by polymerization, used in flooring, wall coverings, gaskets, and miscellaneous parts.

Vinyl-Covered Gypsum Board. See Predecorated Wallboard.

Virgin Growth. The growth of mature trees in the original forests; to be distinguished from Second Growth.

Viscoelasticity. The ability of a material to simultaneously exhibit viscous and elastic responses to deformation.

Viscometer. An instrument for the measurement of Viscosity.

Viscosity Cup. A device for measuring viscosity.

Viscosity Grading. A classification system of asphalt cements based on viscosity ranges at 140° F (60° C); a minimum viscosity at 275° F (135° C) is also usually specified; the purpose is to prescribe limiting values of consistency at these two temperatures; 140° F approximates the maximum temperature of asphalt pavement surface in service in the U.S.; 275° F approximates the mixing and laydown temperatures for hot asphalt pavements; there are five grades of asphalt cement based on the viscosity of the original asphalt at 140° F.

Viscosity. Internal friction in a fluid; resistance to flow; the ratio of the shear stress existing between laminae of moving fluid and the rate of shear between these laminae.

Vise Grip. A pliers-like hand tool that has great gripping power through a simple system of highly efficient leverage.

Vise. Bench tool with two jaws that can be closed by turning a screw, for holding work.

Vision Panel. Glass placed in an opening of a door.

Visual Aid Board. A display that has information posted on it with writing, diagrams or pictures to dispense helpful facts.

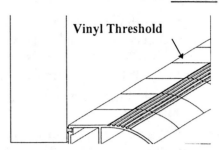

Vinyl Threshold

Vitreous China. The material of a china plumbing fixture with a finish resembling glass, in color, composition, brittleness, and low porosity, used for toilets, urinals, and lavatories.

Vitreous Enamel. A fired-on opaque glassy coating on metal, such as steel.

Vitreous Slip. A slip coating matured on a ceramic body, producing a vitrified surface.

Vitreous Tile. Tile with water absorption of more than 0.5%, but not more than 3.0%.

Vitreous. That degree of vitrification evidenced by low water absorption; vitreous generally signifies less than 0.5% absorption, except for floor and wall tile and low-voltage electrical porcelain which are considered vitreous up to 3.0% water absorption.

Vitrification Range. The maturing range of a vitreous body, producing a vitrified surface.

Vitrification. The progressive reduction in porosity of a ceramic composition as a result of heat treatment, or the process involved.

Vitrified Clay Pipe. Pipes used especially for underground drainage, that are made of hard baked clay.

Vitrified. That which is fused by heat; well burned to a greater degree of hardness.

VM&P Naphtha. Varnish and paint manufacturers naphtha; a low power flammable hydrocarbon solvent.

VOCs. Volatile Organic Compounds.

Void Volume. Total empty spaces in a compacted mix.

Void. 1. An unfilled space in a material, of trapped air or other gas. 2. In gypsumboard, a hollow space in the core caused by the entrapment of air during the manufacturing process. 3. In concrete, air spaces resulting from segregation and improper consolidation during placement. 4. In asphalt paving, empty spaces in a compacted mix surrounded by asphalt coated particles.

Voidable. A valid provision of a contract that can be made void by following some specified procedure.

Volatile Content. Those materials which evaporate; usually expressed as a percentage.

Volatile Flammable Content. A flammable liquid having a flash point below 100° F (37.8° C) or a liquid whose temperature is above its flash point.

Volatile Organic Compounds (VOCs). Compounds that evaporate from the many housekeeping, maintenance, and building products made with organic chemicals; these compounds are released from products that are being used and that are in storage; in sufficient quantities, VOCs can cause eye, nose, and throat irritations, headaches, dizziness, visual disorders, memory impairment; some are known to cause cancer in animals; some are suspected of causing, or are known to cause, cancer in humans. At present, not much is known about what health effects occur at the levels of VOCs typically found in public and commercial buildings.

Volatile Thinner. A liquid that evaporates readily and is used to thin or reduce the consistency of finishes without altering the relative volumes of pigments and nonvolatile vehicles.

Volatiles. Fluids which evaporate rapidly.

Volcanic Ash. A fine powder similar to diatomite but lighter in weight, it is used as a filter medium or filter aid in diatomite filters.

Volt. A unit of electrical potential difference; the SI unit of electromotive force; the difference of potential that would carry one ampere of current against one ohm resistance.

Voltage Control. Device used to provide some electrical circuits with uniform or constant voltage.

Voltage Drop. Loss of voltage due to length of run or resistance.

Voltage Regulator. An automatic electrical control device for maintaining a constant voltage supply to the primary of a welding transformer.

Voltage to Ground. In grounded circuits the voltage between the given conductor and that point or conductor of the circuit which is grounded; in ungrounded circuits, the greatest voltage between the given conductor and any other conductor of the circuit.

Voltage. 1. Term used to indicate the electrical potential or electromotive force in an electrical circuit. 2. Voltage or electrical pressure which causes current to flow. 3. Electromotive force.

Voltmeter. An instrument for measuring in volts the differences of potential between different points of an electrical circuit.

Volume Ceiling. Any ceiling higher than the normal.

Volume. The amount of space occupied by a three dimensional solid or gas, measured in cubic units like cubic feet or cubic centimeters.

Volumetric Efficiency. The relationship between the actual performance of a compressor or of a vacuum pump and calculated performance of the pump based on its displacement.

Volute. 1. A spiral scroll characteristic of Ionic capitals, also used in Corinthian and composite capitals. 2. A spiral or scroll-shaped form, as in volute pumps.

Vomitory. One of the passages for entrance and exit in a theater or amphitheater.

Vortex Tube Refrigeration. Refrigerating or cooling device using principle of vortex tube, as in mining suits.

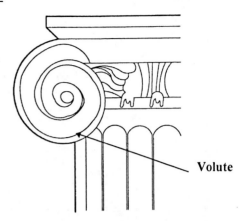

Volute

Vortex Tube. Mechanism for cooling or refrigerating which accomplishes cooling effect by releasing compressed air through a specially designed tube.

Voussoir. One of the wedge-shaped pieces of stone or brick forming an arch.

VRM. Variable Rate Mortgage.

VTR. Vent through roof.

Vulcanizing. Treating rubber with sulphur at a high temperature to harden and toughen it.

W. The total dead load used in earthquake design.

W/C Ratio. Water Cement Ratio.

Wadding. The act of hanging staff by fastening wads made of Plaster of Paris and excelsior or fiber to the casts and winding them around the framing.

Wading Pool. Any constructed or prefabricated pool used for wading which is less than twenty-four (24) inches in depth.

Waferboard Sheathing. Exterior wall or roof covering panels of waferboard.

Waferboard. A building panel made by bonding together large, flat flakes of wood.

Waffle Slab. 1. A two-way concrete joist system. 2. Two-way slab or flat slab made up of a double system of narrow ribs or joists, usually at right angles to each other, forming a pattern of waffle-like coffers.

Wagner Fineness. The fineness of materials such as portland cement expressed as total surface area in centimeters per gram as determined by the Wagner turbidimeter apparatus and procedure.

Wainscot, Terrazzo. See Terrazzo Wainscot.

Wainscot. The lower 3 or 4 feet of an interior wall when it is finished differently from the remainder of the wall.

Waiver. The intentional relinquishment of a known right.

Wale. A horizontal beam.

Waler. A horizontal structural member in a concrete forming or excavation shoring system.

Walk. A path specially arranged or paved for walking.

Walk-In Cooler. Larger commercially refrigerated space kept below room temperature; usually installed in supermarkets, restaurants, food processing plants, and wholesale meat distribution centers.

Walk-Up. A two or more story apartment building or living unit without an elevator.

Walkway, Roof. See Roof Walkway.

Wall Angle. 1. An L-shaped aluminum angle used as the wall termination of T-bar suspended ceiling systems to support the perimeter acoustic tiles. 2. A structural steel angle attached to a masonry wall.

Wall Blocking. Framing lumber cut in short lengths and installed horizontally between wall studs as filler pieces to stabilize the framing, to act as fire blocking, or to provide a backing for fastening a finish item.

Wall Bracket. 1. A wall-mounted support for shelving or other object. 2. A wall-mounted lighting fixture.

Wall Cabinet. Case, box, or piece of furniture which mounts on a wall, commonly with shelves and doors, used for storage.

Wall Cladding. Exterior wood or metal building siding.

Wall Cleanout. An opening in a wall for removal of refuse.

Wall Covering. Any of a variety of final applications to finish a wall surface.

Wall Expansion Joint. A break or space in wall construction to allow for thermal expansion and contraction of materials.

Wall Fabric. Wallpaper or other fabric used to cover or finish off an existing wall.

Wall Finish. The final planing, sanding, staining, varnishing, waxing, or painting of a wall.

Wall Flange. A ridge on a wall that prevents movement. A supporting rim on a wall for attachments.

Wall Footing. A continuous spread footing that supports a uniform load from a wall.

Wall Formwork. The system of wood support for a freshly placed concrete wall left in place until the concrete has set.

Wall Framing. Building construction where exterior and other bearing walls are made of wood.

Wall Furring. Strips of wood applied to make a wall surface level, form an air space, or provide fastening surfaces.

Wall Grille. A wall grating used to cover an opening as protection or as an ornament.

Wall Heater. A heating unit installed in or on a wall.

Wall Hung Sink. A lavatory mounted on brackets attached to a wall.

Wall Hydrant. A connection to a water main cut through and mounted on a wall; see Hose Bibb.

Wall Insulation. Material placed in wall cavities for the reduction of fire hazard or for protection from heat and cold.

Wall Lath. Expanded metal, gypsum sheets, or thin strips of wood attached to stud walls, acting as a foundation for plastering.

Wall Louver. Openings in walls for ventilating spaces.

Wall Mounted Flagpole. A pole to raise or display a flag, mounted or attached to a wall.

Wall Mounted Heater. A heating unit mounted on or attached to a wall.

Wall Mounted Oven. A domestic oven for cooking purposes designed for mounting in or on a wall or other surface.

Wall Panel. Form sheathing, constructed from plywood, boards, or metal sheets, that are installed as a unit.

Wall Pier. A wall segment with a horizontal length to thickness ratio between 2.5 and 6 and whose clear height is at least two times its horizontal length.

Wall Plaster. A paste-like composition that hardens on drying and is used for coating walls.

Wall Railing. A band of ornamental wood installed horizontally.

Wall Reinforcing, Masonry. Steel reinforcing rods or mesh used in masonry walls between courses.

Wall Reinforcing. To strengthen a wall by the addition of new or extra materials.

Wall Scupper. See Scupper, 2.

Wall Sheathing. The first layer of covering on an exterior wall, fastened to the wall studs.

Wall Size. Solution such as glue, starch, casein, shellac, varnish or lacquer, used to seal or fill pores of wall surface to stop suction, counteract chemicals or stains and prepare surface for paint, paper or fabric.

Wall Spreader. An accessory, usually fabricated from reinforcing bar to a Z or U shape, used to separate and hold apart two faces of curtains of steel reinforcements in a concrete wall.

Wall System. A modular set of shelves and cabinets that can be arranged along a wall in various combinations.

Wall Tie. A mechanical metal fastener which connects wythes of masonry to each other or to other materials.

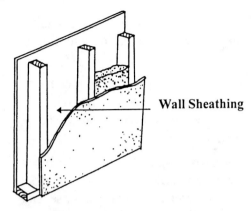

Wall Sheathing

Wall Tile. A glazed tile with a body that is suitable for interior use and which is usually non-vitreous, and is not required nor expected to withstand excessive impact.

Wall Z Tie. A Z-shaped reinforcing strip used as a support bracket from the structural wall to the masonry veneer.

Wall, Cavity. See Cavity Wall.

Wall, Retaining. See Retaining Wall.

Wall. A member, usually vertical, used to enclose or separate spaces.

Wallboard. Large, rigid sheets of wood pulp, gypsum, pressed cellulose fibers, gypsumboard, plywood, or similar materials, used in place of plaster in interior surfaces.

Wall-Hung. Anything hung from a wall, such as cabinets or plumbing fixtures.

Wallpaper Paste. An adhesive used for attaching wallpaper.

Wallpaper. A decorative paper for applying to the walls of a room.

Wall-to-Wall. Covering the entire floor area.

Walnut Door. A door which has a veneer of walnut.

Walnut Veneer. An overlay of a thin layer of walnut wood for outer finish or decoration.

Walnut. The richly grained moderate reddish-brown wood of the walnut tree, used for veneers, cabinetmaking, and moldings.

Wandering Block Sequence. A block sequence wherein successive blocks are completed at random after several starting blocks have been completed.

Wane. Bark or lack of wood from any cause on edge or corner of a piece of lumber except for eased edges.

Ward. 1. A large room in a hospital, that will accommodate several patient beds. 2. The inner court of a castle or fortress. 3. A local Mormon congregation. 4. A projecting ridge of metal on a lock to prevent insertion of a key that does not have a matching notch.

Wardrobe Locker. A clothes storage cabinet with a locking door.

Wardrobe. A room or freestanding closet where clothes are kept.

Warehouse. A storage room or building.

Warm Air Heating. Any heating system which depends upon the circulation of warm air.

Warm Colors. Colors in which red-orange predominates; so termed because of the association with fire, heat, and sunshine.

Warming Colors. Any color except green may be warmed by adding red; green is warmed by adding yellow.

Warp. 1. Any variation from a true or plane surface. 2. Warp includes bow, crook, cup, and twist, or any combination thereof. 3. In carpet, the backing yarns running lengthwise.

Warpage. A concave or convex curvature of a material that was intended to be perfectly flat.

Warranty. A promise that a fact is true.

Warren Truss. A truss in which the top and bottom chords are parallel or nearly so, the top chord sometimes sloping for roof drainage.

Wash Coat. A very thin coat of finishing material, usually shellac.

Wash Fountain. A waist high sink which supplies a steady stream of water to cleanse the hands.

Wash Primer. A thin inhibiting paint usually chromate pigmented with a polyvinyl butyrate binder.

Wash. The slant of a sill, wall, parapet, or capping, to allow the water to run off easily.

Washboarding. In asphalt paving, see Corrugations.

Washer, Coin. See Coin Washer.

Washer, Flat. See Flat Washer.

Washer, Steel Welding. See Steel Welding Washer.

Washer. A flat thin ring or a perforated plate used in joints or assemblies to ensure tightness or relieve friction.

Washer-Extractor. A clothes washer that includes a high speed centrifugal drying cycle that removes all the free water except for dampness.

Washing. 1. Rapid dissolution or emulsification of a paint film when wet with water. 2. Erosion of a paint film after rapid chalking.

Wash-Out. Lack of proper coverage and texture build-up in machine-dash textured plaster caused by the mortar being too soupy.

Washroom Faucet. A device that dispenses hot and cold water, mounted above a sink.

Waste and Overflow Fitting. A bathtub drain fitting which provides both the outlet for the bathtub drain and an overflow to drain excess water from the tub.

Waste Handling. The act or process of transporting or removing rubbish.

Waster Mold. An unreusable precast plaster mold made for the forming of decorative monolithic or cast-in-place concrete; the mold cannot be removed without being destroyed.

Waste Pipe. A plumbing line that carries waste water from fixtures except water closets and urinals.

Waste Piping. See Filter Waste Discharge Piping.

Waste Receptacle. A container for the temporary storage of rubbish.

Waste Stack. A vertical line of piping that extends one or more floors and receives the discharge of fixtures other than water closets and urinals.

Waste Water. In a swimming pool, the water from any filter, perimeter overflow, pool emptying line, or similar apparatus or appurtenance.

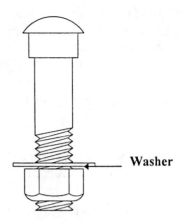

Washer

Waste. 1. Valueless material. 2. Digging, hauling, and dumping of valueless material to get it out of the way. 3. Liquid discharged from any plumbing fixture, except water closets and urinals.

Watchman. A guard who keeps watch over a certain area.

Water Absorption. 1. The amount of water absorbed by a material under specified test conditions; expressed as weight percent of the test specimen. 2. The amount of water ingested into the core and surface papers of gypsumboard; expressed as a percent of water added over dry weight.

Water Blasting. Blast cleaning using high velocity water.

Water Closet. See Toilet.

Water Conditioner. A device used to dissolve minerals from water; normally after the minerals are removed, the water tastes better and reduces the likelihood that mineral deposits will build up within the plumbing system; the water is considered to be soft after the minerals have been removed.

Water Cooled Chiller. A piece of equipment that produces chilled water for circulation through a building and used for cooling.

Water Cooled Condenser. A condenser in which water is normally circulated through a cooling tower through which heat is dispersed to the atmosphere.

Water Cooler. An apparatus that cools, holds, and dispenses cold water.

Water Defrosting. Use of water to melt ice and frost from evaporator during off-cycle.

Water Gain. Of concrete, see Bleeding, 2.

Water Gauge. An instrument to measure the depth of water, or to indicate the height of its surface, as in a steam boiler.

Water Glass. 1. An instrument consisting of an open box or tube with a glass bottom, used for examining objects in or under water. 2. A Water Gauge. 3. A viscous, syrupy solution of sodium or potassium silicate that is used especially as a cement, as a protective coating and fireproofing agent, and in preserving eggs.

Water Hammer. A banging sound in water supply lines caused by the sudden stopping of the water flow.

Water Heater. An apparatus for heating and storing water.

Water Level. A length of clear plastic hose, 3/8 to 1/2 inch in diameter and approximately 50 feet long, filled with water and used as a leveling device to check level in walls or structures.

Water Loss. A measurement of the amount of free water evaporated from gypsumboard products during the drying stage in the manufacturing process; expressed in pounds per 1000 S.F.

Water Main. Water supply pipe generally located at the street which may supply a number of buildings.

Water Meter. An instrument for measuring water consumption.

Water of Hydration. The water required to replace the water lost during the calcination process.

Water Paint. A paint in which the vehicle is a water emulsion and in which water is used as thinner.

Water Pipe Restoration. To install new linings in pipes to improve the condition of existing water pipes; usually includes the cleaning and removal of built-up scale or debris out of the existing pipe and the relining with a compatible material.

Water Piping Removal. The act or process of tearing out and carrying away old water piping systems.

Water Proofing. Treatment of a surface or structure to prevent the passage of water under hydrostatic pressure.

Water Reducing Admixture. Material added to cement or a concrete mix to cut down on its water content.

Water Repellent Paper. A special paper treated to minimize wetting of the surface.

Water Repellent. 1. A finish that is resistant but not impervious to penetration by water. 2. A liquid that penetrates wood which, after drying, materially retards changes in moisture content and in dimensions without adversely altering the desirable properties of wood.

Water Resistant Core. A special gypsumboard core formulation with additives to reduce water absorption; water resistant gypsum backing board is recommended for use as a base for ceramic tile in bathrooms and other wet areas.

Water Resistant Gypsum Backing Board. A gypsumboard designed for use on walls as a base for the application of ceramic or plastic tile.

Water Resistant Gypsumboard. A gypsumboard designed for use in locations where it may be exposed to occasional dampness; plasterboard that has had a chemical treatment to make it resistant to moisture, but not necessarily waterproof.

Water Retentivity. That property of a mortar which prevents the rapid loss of water to masonry units of high suction; it prevents bleeding or water gain when mortar is in contact with relatively impervious units.

Water Service Pipe. The pipe from the water main or other source of water supply to the building serviced.

Water Slide. A sloping trough down which water is carried by gravity.

Water Softener. A device attached to a water system to remove unwanted minerals and substances.

Water Spotting. Spotty changes in the color or gloss of a paint film; may be caused by various factors, such as emulsification or the solution of water soluble components.

Water Stain. A colored dye that is soluble in water.

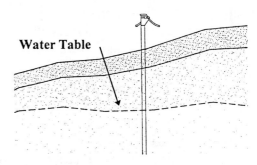

Water Table

Water Supply System. The water supply system of a building is composed of the water service pipe, the water distributing pipes and the various connecting pipes, control valves and fittings in or on private property.

Water Table. 1. The level at which the pressure of water in the soil is equal to atmospheric pressure; effectively, the level to which ground water will fill an excavation. 2. The finish at the bottom of a house which carries water away from the foundation. 3. A projection on the bottom of an exterior wall to prevent rain or water from seeping through to the wall below.

Water Valve. A device to regulate the flow of water in a pipe or other passage.

Water Vapor Transmission (WVT). See Permeability.

Water Vapor. Moisture existing as gas in the air.

Water White. Transparent and colorless like water.

Water. A colorless, transparent, odorless compound of hydrogen and oxygen, H_2O.

Waterbar. A rubber or plastic strip cast into concrete at joints to prevent water leakage; also called Waterstop.

Water-Cement Ratio. The proportion of water to portland cement; the number of gallons of water per 94 pound sack of cement; only a small amount of water is needed to hydrate the cement and complete the chemical reaction; all additional water is for workability only and too much of which will weaken the concrete or mortar.

Water-Cooled Condenser. Condensing unit which is cooled through use of water flow.

Waterproof Cement. Portland cement to which waterproofing agents, such as surface repellents, have been added at time of blending materials at the mill.

Waterproof Membrane. An impermeable layer of plastic film, roofing material, bitumen, or other material placed to stop transfer of water or water vapor.

Waterproof. So constructed or protected that moisture will not interfere with its successful operation.

Waterproofing. 1. The act or process of making something waterproof. 2. A coating capable of stopping penetration of water or moisture.

Water-Reducing Admixture. Material placed in concrete mix to increase slump or maintain workability with a reduced amount of water.

Water-Repellent Paper. Gypsumboard paper surfacing which has been formulated or treated to resist water penetration.

Water-Resistant Core. A gypsumboard specially formulated to resist water penetration.

Waterstop. See Waterbar.

Water-Struck Brick. See Soft Mud Brick.

Water-Thinned Paint. A paint whose thinner is mainly water; the binder may be a material that (1) requires water for setting, e.g., portland cement; (2) which is soluble in water, e.g., casein; (3) which is emulsifiable in water, e.g., flat wall paint binders.

Watertight Manhole. A cover for a vertical access shaft that prevents the elements from coming in.

Watertight. So constructed that moisture will not enter the enclosing case.

Water-Tube Level. A leveling device consisting of a water-filled tube with a transparent section at each end; also called Water Level.

Watt. 1. A unit of electrical power. 2. The absolute meter-kilogram-second unit of power equal to the work done at the rate of one absolute joule per second. 3. The rate of work represented by a current of one ampere under a pressure of one volt and taken as the standard in the U.S. of 1/746 horsepower.

Watt-Hour. A unit of work or energy equivalent to the power of one watt operating for one hour.

Wattmeter. An instrument for measuring electric power in watts.

Wave Length. 1. The distance between successive similar points on two wave cycles. 2. In color, the computed distance between vibrations of light that produce visible color sensation on the eye; in the visible spectrum, red-orange has the longest wave length; violet the shortest; wave lengths are measured in millimicrons; wave lengths shorter than violet are called ultraviolet; wave lengths longer than red-orange are called infrared.

Wave. Any disturbance that advances through a medium with a speed that is completely determined by properties of that medium, such as sound or light.

Wavy-Grained Wood. Wood in which the fibers collectively take the form of waves or undulations.

Wax, Floor. See Floor Wax.

Wax. 1. A fatty material obtained from the honeycombs of bees or from similar plant, animal, or mineral substances; used for providing an attractive, protective coating, as for wood; waxes may be used by themselves or combined with other ingredients to make certain polishes, paints, varnishes, and paint removers. 2. Ingredient in many lubricating oils which may separate from the oil if cooled enough.

WC. Water closet.

WCLIB. West Coast Lumber Inspection Bureau.

Wearing Course. A topping or surface treatment to increase the resistance of a pavement or slab to abrasion.

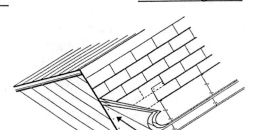

Weatherproof

Weather Delay. Lack of construction progress caused by inclement weather.

Weather Joint. A masonry joint where the mortar has been formed at a 45 degree angle, sloping down and out, exposing approximately 1/4 to 1/2 inch of the brick above, for the purpose of shedding water.

Weather Seal. A flanged channel installed on the edges of an exterior door.

Weather Vane. A vane; a moveable device for showing the direction of the wind.

Weatherability. Capability of withstanding the weathering process.

Weathered Joint. A mortar joint finished in a sloping profile that tends to shed water to the outside of the wall.

Weathered. In masonry, stonework which has been cut with sloped surfaces so it will shed water from rain or snow.

Weatherhead. A weather resistant fitting for feeding aerial electrical conductors into a building.

Weathering Steel. A steel alloy that forms a tenacious, self-protecting rust layer when exposed to the atmosphere.

Weathering. 1. Changes in color, texture, strength, chemical composition or other properties of a natural or artificial material due to the action of the weather. 2. The mechanical or chemical disintegration and discoloration of a wood surface resulting from exposure to light, action of dust and sand carried by winds, alternate shrinking and swelling brought about by changes in the weather or a combination of these causes; weathering does not include decay.

Weatherometer. A testing device intended to simulate atmospheric weathering.

Weatherproof Box. An electrical outlet or switch box which has been manufactured to withstand the outside elements.

Weatherproof Cover. A protective layer constructed so that exposure to the weather will not interfere with successful operation.

Weatherproof Duct. An enclosure for wires or cables constructed so that exposure to weather will not interfere with successful operation.

Weatherproof Insulation. A material used to reduce heat transfer, constructed so that exposure to the weather will not interfere with its successful operation.

Weatherproof Switch. A device designed to be used outdoors that opens, closes or changes the connection of an electric circuit and is weatherproofed against moisture.

Weatherproof. So constructed that exposure to the weather will not cause damage or loss of function.

Weatherstrip. A ribbon of resilient material used to reduce air infiltration through the crack around a sash or door.

Weather-stripping. The process of reducing air or rain infiltration by covering joints of doors or windows with strips of resilient material.

Weave Bead. A type of weld bead made with transverse oscillation.

Weaving. Process of forming carpet on a loom by interlacing the warp and weft yarns.

Web Belt. Commonly a used military type belt adapted to carry hand tools at the jobsite.

Web Members. The vertical and angular members between the top and bottom chords of a truss.

Web. 1. The vertical plate connecting the top and bottom flanges of a metal beam. 2. An interior solid portion of a hollow concrete block.

Wedge. A piece of material, such as wood or metal, tapering to a sharp edge that is driven between two objects or parts of an object to secure or separate them.

Weed Control. The act or process of spraying chemicals or placing powders to control the spread of weeds.

Weep Hole. 1. A small opening, the purpose of which is to permit drainage of water that accumulates inside a building component. 2. Openings placed in mortar joints of facing material at the level of flashing to permit the escape of moisture. 3. Openings in retaining walls to permit water to escape.

Weep Screed. A plaster screed at the bottom of an exterior stucco wall that is designed to allow moisture to seep out.

Weeping Joints. A masonry joint treatment in which mortar extruding from the joint in laying is not cut off, but is allowed to harden; gives informal rustic appearance but difficult to waterproof.

Weft. Backing yarns which run across the width of the carpet; in woven carpets, the weft shot (filling) yarns and the warp chain (binder) yarns interlock and bind the pile tufts to the backing; in tufted carpets, pile yarns which run across the carpet are also considered weft yarns.

Weight. The force experienced by a body as a result of the earth's gravitation.

Weir of a Trap. The highest portion of the inside channel through a plumbing trap; compare with Dip of a Trap; the depth of the water seal is the distance from the dip to the crown weir.

Weir. 1. A dam in a stream over which water flows; see Crest. 2. A dam in a stream to raise the water level or to divert the stream. 3. An orifice through which the liquid only partially flows.

Weld Bead. A weld deposit resulting from a pass.

Weld Crack. A crack in weld metal.

Weld Inspection. Methods to determine existence and extent of defects and discontinuities in welds.

Weld Plate. A steel plate anchored into the surface of concrete, to which another steel element can be welded.

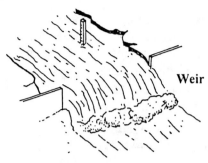

Weir

Weld Test. Welding examination or inspection; the loading of welds to determine load capacity of welds is not normal practice.

Weld X-Ray. To examine, treat, or photograph the connection of surfaces that have been welded together.

Weld, Destructive Test. Methods to determine existence and extent of defects and discontinuities in welds which affect capabilities of weld and may require repairs after testing.

Weld, Liquid Penetrant Examined. Nondestructive method of inspection to determine existence and extent of discontinuities that are open to surface in weld being inspected; indications are made visible through use of dye or fluorescent chemical in liquid employed as inspection medium.

Weld, Nondestructive Examined. Nondestructive method of inspection to determine existence and extent of discontinuities in welds; finely divided magnetic particles applied to magnetized part are attracted to any magnetic leakage created by discontinuities in mass of sample; methods to determine existence and extent of defects and discontinuities in welds which do not affect capabilities of weld.

Weld, Radiographic Examined. Nondestructive method of inspection to determine existence and extent of discontinuities in welds; X rays are used to penetrate the weld to produce radiograph to detect defect.

Weld, Shielded Metal-Arc. See Shielded Metal-Arc Weld.

Weld, Ultrasonic Examined. Nondestructive method of inspection to determine existence and extent of discontinuities in welds; ultrasonic waves (frequency of mechanical vibrations above 20,000 vibrations per second) are used to penetrate weld to detect defects.

Weld, Visual Examined. Nondestructive method of inspection to determine existence and extent of discontinuities in welds; exposed weld surfaces are visually inspected (without auxiliary equipment) for evidence of existence and extent of discontinuities in welds.

Weld. 1. A joint between two pieces of metal formed by fusing the pieces together, usually with the aid of additional metal melted from a rod or electrode. 2. To join two pieces of metal together by heating until fusion of material either with or without filler metal.

Weldability. The capacity of a metal to be welded under the fabrication conditions imposed into a specific, suitably designed structure and to perform satisfactorily in the intended service.

Welded Joint. A union of two or more members produced by the application of a welding process.

Welded Pipe. Piping where connections and fittings are welded.

Welded Railing. Railing sections with the components fastened with welds.

Welded Truss. Trusses with components fastened together with welds.

Welded Wire Fabric. A series of longitudinal and transverse steel wires arranged substantially at right angles to each other and welded together at the intersections, in sheets or rolls, used to reinforce mortar and concrete.

Welder. One who is capable of performing manual or semiautomatic welding operations based on training, experience, testing, or certification, or any combination of these.

Welding Electrodes and Rod. The electrode and rod are the components of the welding circuit through which current is conducted between the electrode holder and the arc.

Welding Goggles. Goggles with tinted lenses, used during welding or oxygen cutting, which protect the eyes from harmful radiation and flying particles.

Welding Inspector. One who is capable of inspection of welds based on training, experience, testing, or certification, or any combination of these.

Welding Leads. The work lead and electrode lead of an arc- welding circuit.

Welding Process. A metal-joining process wherein coalescence is produced by heating to suitable temperatures, with or without the application of pressure, and with or without the use of filler metal.

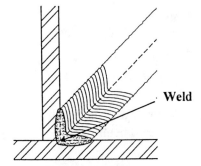

Weld

Welding Rod. Filler metal, in wire or rod form, used in gas welding and brazing processes, and in those arc-welding processes wherein the electrode does not furnish the filler metal.

Welding Technique. The details of a manual, machine or semi- automatic welding operation which within the limitations of the prescribed joint welding procedure are controlled by the welder.

Welding Test. The act or process of testing the strength of a weld.

Welding Tip. A welding-torch tip designed for welding.

Welding Torch. A device used in gas welding or torch brazing for mixing and controlling the flow of gases.

Welding Transformer. A transformer used to supply current for welding.

Welding. Fusing metallic parts by heating and allowing the metals to flow together.

Weldment Connection. The assembling together of pieces by welding to create a unit.

Weldor. See Welder.

Well Graded Aggregate. Aggregate having a particle size distribution which will produce maximum density of concrete or asphalt, that is, minimum void space.

Well. 1. A pit or hole sunk into the earth to reach a supply of water. 2. An open space extending vertically through floors of a structure.

Wellpoint System. A series of vertical pipes in the ground connected to a header and pump to drain marshy areas or to control ground seepage.

Wellpoint. A perforated pipe surrounded by sand to permit the pumping of ground water.

West Coast Lumber Inspection Bureau (WCLIB). PO Box 23145, Portland, Oregon 97281, (503) 639-0651.

Western Frame. See Platform Frame.

Western Red Cedar Lumber Association (WRCLA). 1500 Yeon Building, 522 SW Avenue, Portland, Oregon 97204, (503) 224-3930.

Western Wood Products Association (WWPA). 1500 Yeon Building, 522 SW Avenue, Portland, Oregon 97204, (503) 224-3930.

Wet and Dry Bulb Hygrometer. See Hygrometer.

Wet Areas. Interior or exterior tiled areas subject to periodic or constant wetting; examples are showers; sunken tubs; pools; exterior walls; roofs; exterior paving, and interior floors.

Wet Cell Battery. Cell or connected group of cells that converts chemical energy into electrical energy by reversible chemical reactions.

Wet Edge Time. The length of time before a stretch of paint sets up without showing lap marks when the painter applies the next stretch.

W

Wet Film Gauge. In painting, a device for measuring wet film thickness.

Wet Film Thickness. Thickness of liquid film of paint immediately after application.

Wet Heat. Heating system using hot water heat or steam heat; see Hydronic.

Wet Location Fluorescent. A water-tight fluorescent fixture that is sealed to protect against moisture.

Wet Location. A location subject to saturation with water or other liquids, such as exposed to weather, washrooms in garages and like location; installations underground or on concrete slabs or masonry in direct contact with the earth are wet locations.

Wet Mixing Period. The interval of time between the beginning of application of asphalt material and the opening of the mixer gate.

Wet Niche. Underwater swimming pool light with a water cooled sealed beam unit.

Wet Sand. To smooth a finished joint of sheetrock with a coarse wet sponge. A preferred method to reduce dust created in the dry sanding method.

Wet Spray. Paint spraying so that surface is covered with paint that has not started to dry.

Wet Sprinkler System. A sprinkler system that is filled with water at design pressure for immediate use upon activation.

Wet Standpipe. A fire fighting pipeline in a building that is always full of water under pressure and ready for use.

Wet Vent. A pipe that acts as a drain or waste pipe and also as a vent when not carrying any liquids; this wet vent usually serves as a waste pipe for the fixture it is closest to; when this fixture is not in use, but other fixtures farther down the line are in use the wet vent acts as a vent pipe for these fixtures farther down the line.

Wet-bulb Temperature. The temperature indicated by the wet-bulb thermometer of a psychrometer.

Wet-Edge. Fluid paint boundary.

Wettability. A condition of a surface that determines how fast a liquid will wet and spread on the surface or if it will be repelled and not spread on the surface.

Wetting Agent. A substance capable of lowering the surface tension of liquids, facilitating the wetting of solid surfaces and permitting the penetration of liquids into capillaries.

Wetting Oils. Products used to promote adhesion of applied coatings when all mill scale and rust cannot be removed.

Wetting. The process in which a liquid spontaneously adheres to and spreads on a solid surface; the more viscous a fluid, and the higher its surface tension, the more difficult it is for the liquid to wet a material; certain additives, for example, water softeners, reduce surface tension, or viscosity and improve wetting properties, allowing the material to better flow.

WF Beam. Wide Flange Beam.

WFI. Wood Flooring Institute.

WH. 1. Wall Hydrant. 2. Water Heater.

Wharf. A structure that provides berthing space for vessels, to facilitate loading and discharge of cargo.

Wheel Barrow. A small vehicle with handles and one or more wheels used for carrying small loads.

Wheel Chair Partition. A dividing wall in a bathroom or bathing room which forms the perimeter of a private area that has been made accessible to the disabled.

Wheel Stop. A concrete or wood bumper, approximately 4 to 6 feet (1.20 to 1.80 meters) long, at the end of a parking space to stop the car.

Wheelchair Occupant. A person who, due to a physical impairment or disability, utilizes a wheelchair for mobility; also called Wheelchair User.

Wheelchair User. See Wheelchair Occupant.

Wheelchair. A chair mounted on wheels to be propelled by its occupant manually or with the aid of electric power, of a size and configuration conforming to the recognized standard models of the trade.

Wheeled Extinguisher. A fire extinguisher mounted on a wheeled cart that can be pushed or pulled by a person.

Whiskering. See Efflorescence.

White Blast. Blast cleaning to white metal.

White Cedar Shingle. A light colored weather-resistant cedar wood used for roofing and siding.

White Cement. Cement made from materials with low iron content to produce mortar or concrete that is white in color.

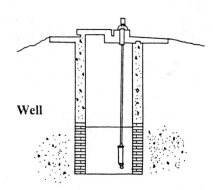

Well

White Coat. A term denoting a gauged lime putty trowel finish.

White Lead. 1. Basic carbonate white lead is a compound of lead, carbon dioxide and water; lead is melted and cast into disks or buckles, which are about six inches in diameter; the buckles are placed in porcelain pots containing dilute acetic acid; the pots are covered with boards and layers of tanbark; heat and carbonic acid generated by fermentation of the tanbark, with the acid vapors, combine to transform the lead into basic carbonate white lead. 2. Basic sulphate white pigment is obtained from lead sulphide ore by a process of fuming or burning. 3. Carter process white lead is made by blowing melted lead in fine granules by a jet of air or superheated steam; the powdered lead is placed in revolving drums or cylinders and subjected to the action of air and carbon dioxide gas from burning coal.

White Metal. Any of several lead-base or tin-base alloys, such as babbitt metal, used for bearings and fusible plugs.

White-Rot. In wood, any decay or rot attacking both the cellulose and the lignin, producing a generally whitish residue that may be spongy or stringy rot, or occur as pocket rot.

Whitewash. A solution of quicklime or of whiting and size for whitening walls.

Whiting. Calcium carbonate, limestone, or chalk in pigment form; used extensively for making putty and as an extender in paints.

WHO. World Health Organization.

WI. Wrought Iron.

WIC. Woodwork Institute of California.

Wicking. The absorption of water by capillary action into the core of gypsumboard.

Wide Flange Section. Any of a wide range of steel sections rolled in the shape of a letter T or H.

Wide Stile Door. Wider than normal vertical members forming the outside framework of a door.

Wide-Selvage Asphalt Roll Roofing Surfaced with Mineral Granules. See Nineteen-Inch Selvage.

Width and Length. In a swimming pool, the actual water dimension taken from wall to wall at the maximum operating water level.

Wilton Carpet. A fabric woven on a loom controlled by a Jacquard pattern device, which raises one of from 2 to 6 surface yarns over a bladed pile wire that is then withdrawn to cut the tufts and give a plushlike face; the other yarns run dormant through the center and back of the fabric in the warp direction.

Winch. A hoist or windlass for lifting or moving heavy loads.

Wind Brace. A diagonal structural member whose function is to stabilize a frame against lateral forces.

Wind Drift. Horizontal deflection of a frame caused by wind forces.

Wind farm. An area of land with a cluster of wind turbines for producing electrical power.

Wind Load. A load on a building caused by wind pressure.

Wind Pressure. The horizontal live load on the side of a structure imposed by the wind; it increases in magnitude at higher elevations and causes a suction on the leeward side of buildings and roofs.

Wind Restraint System. The collection of structural elements which provide restraint of the seismic-isolated structure for wind loads; the wind-restraint system may be either an integral part of isolator units or may be a separate device.

Wind Shelf. In a fireplace, the concave ledge just back of the damper at the bottom of the smoke chamber; its function is to direct the cold air downdraft in the flue to join the hot air in rising in the flue.

Wind Tunnel. A tunnel-like passage through which wind is blown at a known velocity past a model of an object to determine the effect of wind forces on the object.

Wind Uplift. Upward forces on a structure caused by negative aerodynamic pressures that result from certain wind conditions.

Wind. The surface of a board when twisted (winding) or when resting upon two diagonally opposite corners, if laid upon a perfectly flat surface; wind rhymes with kind.

Windchill Factor. See Windchill.

Windchill Index. See Windchill.

Windchill. The still-air temperature that has the same cooling effect on exposed human flesh as a given combination of temperature and wind speed; also called Chill Factor, Windchill Factor, or Windchill Index.

Winder. A stair tread that is wider at one end than at the other, located at a turn in the stairway.

Window Blind. An adjustable window covering.

Window Box. A box on an outside window sill for growing flowers.

Window Covering. Any window treatment such as curtains, draperies, or blinds, for regulating light, for decoration, or for privacy.

Window Frame. The structure which holds a window assembly in place.

Window Guard Lock. Tamperproof hasp and padlock for window guards.

Window Guard, Diamond Mesh. See Diamond Mesh Window Guard.

Window Guard, Steel Bar Grille. Guard fabricated from a steel bar grille.

Window Guard, Woven Wire. Guard fabricated of woven wire.

Window Head. The assembly of parts at the top of a window frame including the frame, stops, casing, shims, and flashing.

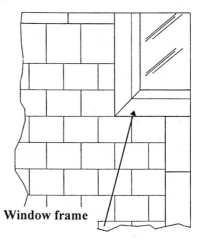

Window frame

Window Header. A horizontal structural member placed across the top of a window opening to support the load above.

Window Pane. A pane of glass in a window.

Window Screen. A fine mesh screen fitted in a window frame to exclude insects.

Window Seat. A seat built into a window recess.

Window Shade. Any adjustable window covering to regulate light or provide privacy.

Window Sill. The horizontal member at the bottom of a window.

Window Stool. Wood, ceramic tile, or masonry installed on the plate at a window sill on the inside of the window, fitted against the bottom rail of the lower sash.

Window Treatment. The addition of hanging fabrics, curtains, or blinds to the interior of a window.

Window Unit. Air conditioner which is placed in a window.

Window Well. Recess located at or below grade to allow for natural light to reach a ground level or basement window, often created by use of corrugated metal in half-round shape.

Window, Drive-Up. See Drive-Up Window.

Window, Service. See Service Window.

Window. An opening in the wall of a building or structure for the admission of light and air, closed by casements or sashes containing glass panes.

Windowwall. The opening in a wall surface which contains a window assembly or wall of assemblies.

Wiping Stain. See Pigment Oil Stain.

Wire Brush. 1. A hand cleaning tool comprised of wires for bristles. 2. The act of cleaning a surface with a wire brush, including power brushes.

Wire Cloth Lath. A plaster reinforcement of wire not lighter than No. 19 gauge, 2-1/2 meshes per inch and coated with zinc or rust-inhibitive paint; not to be used as reinforcement of exterior portland cement plaster.

Wire Cut Brick. A brick having its surfaces formed by wires cutting the clay before it is baked.

Wire Edge Joint. In wallpaper, a joint made by trimming both selvedges and lapping one edge slightly over the other.

Wire Glass. Glass in which a wire mesh was embedded during manufacture.

Wire Guard. Flexible strands of metal that have been manufactured into a unit to act as an enclosure around moving parts of machinery, around an excavation, equipment, or materials to prevent injury to the operator.

Wire Hanger. A wire that supports or connects material.

Wire Hook Hanger. Flexible strand of metal in the shape of a hook to hold a construction member in place.

Wire Mesh Partition Adjustable Cap. Top capping channel that is tightly fitted over top of entire partition.

Wire Mesh Partition Adjustable Floor Shoe. Adjustable floor socket to permit level installation of wire mesh partition channel.

Wire Mesh Partition. Dividing wall constructed of metal framing and wire mesh.

Wire Mesh. A series of longitudinal and transverse wires arranged substantially at right angles to each other sheets or rolls, used to reinforce mortar and concrete.

Wire Nut. A device for connecting electrical wire conductors together by twisting.

Wire, Aluminum. See Aluminum Wire.

Wire, Chicken. See Chicken Wire.

Wire, Stranded. See Stranded Wire.

Wire. A metal drawn out into the form of a thread or thin flexible rod, used for fencing, binding, or to conduct an electrical current.

Wire-Cut Brick. Brick formed by forcing plastic clay through a rectangular opening designed for the purpose, and shaping of clay into bars; before burning, wires pressed through the plastic mass cut the bars into brick units.

Wires. In carpet making, metal strips inserted in the weaving shed under the surface yarns to form loops when the yarns are bound by the weft shuttle in the Velvet and Wilton weaves; a round wire will withdraw, leaving uncut pile loops for round-wire Velvet and Wiltons, while bladed knife-ends on flat wires will cut these loops, forming a plushlike surface of tufts; the term has come to indicate the number of rows of tufts per inch of warp; thus 11 wires means 11 rows of tufts to each inch of length.

Wireway. A sheet metal trough with hinge or removable cover to carry several electrical cables.

Wiring, Flexible. See Flexible Wiring.

With Framing. Of gypsumboard application, see Parallel Application.

Withdrawing Room. Drawing Room.

Withe. 1. A single tier of brick or stone in a wall. 2. A masonry separation or partition between flues in a chimney with more than one flue.

Withering. Withering or loss of gloss is often caused by varnishing open-pore woods without filling pores, use of improper undercoating and applying topcoat before undercoat dried.

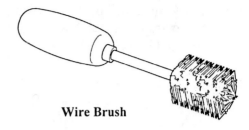

Wire Brush

Witness. A person who observed events and is called to testify concerning those events at a hearing.

Wobble Friction. In prestressed concrete, friction caused by unintended deviation of prestressing sheath or duct from its specified profile.

Wobble Plate-Swash Plate. Type of compressor designed to compress gas, with piston motion parallel to crankshaft.

Wolfram. See Tungsten.

Wolmanized Lumber. Lumber treated with Wolman salts to resist decay.

Wood Alcohol. Poisonous alcohol obtained by destructive distillation of wood.

Wood Anchor. A bolt or fastening device which attaches wood to wood or wood to another material.

Wood and Plastics. A category of the CSI Masterformat which is represented in Division 6 of the format. Commonly called just Wood or Carpentry.

Wood Backing. See Backing, 2.

Wood Batten. Wood strips covering vertical joints on boards used as exterior siding.

Wood Beam. Horizontal wood structural member that supports uniform and concentrated loads.

Wood Blocking. Small pieces of wood used to secure, join, or reinforce members, or to fill spaces between members.

Wood Board Roof Sheathing. Wood board material placed diagonally and secured to exterior side of roof rafters or trusses used to create rigidity in building superstructure and serve as base to receive roofing.

Wood Board Subflooring. Wood board material placed diagonally and secured to top side of floor joists used to create rigidity in building super-structure and serve as base to receive flooring.

Wood Board Wall Sheathing. Wood board material placed diagonally and secured to exterior side of exterior wall studs used to create rigidity in building superstructure and serve as base to receive siding or veneer construction.

Wood Bottom Plate. 1. In Western or Platform Framing construction, a horizontal wood lumber member which supports wall studs and ties them together; rests on the floor sheathing. 2. In Balloon Framing construction, a flat horizontal member, also called a mudsill, that supports the vertical wall studs and posts.

Wood Bridging. Diagonal or longitudinal wood members used to keep wood joist members properly spaced, in lateral position, vertically plumb, and to distribute load.

Wood Buck. Wood frame typically built into concrete or masonry wall to accommodate finish door frame.

Wood Bumper. Wood component used to absorb impact and prevent damage to other surfaces, such as at a loading dock.

Wood Cant Strip. Sloped wood strip used at perimeter of roofing membrane to transition membrane from horizontal to vertical surface.

Wood Cap. Wood member used on top of an assembly to provide termination or finish.

Wood Carriage. Sloping beam installed between stringers to support steps of wood stair.

Wood Chip Mulch. Wood chips, spread on the ground to prevent erosion, control weeds, minimize evaporation, and improve the soil.

Wood Chisel. A sharp ended hand tool used to carve special shapes into wood.

Wood Column. Vertical wood structural member, usually supporting a beam.

Wood Cornice. Horizontal wood molding that may be combination of several shaped pieces.

Wood Decking. Plywood, lumber, or glued laminated member placed over roof or floor structural members for structural rigidity of building frame and to provide a surface for traffic or substrate for roofing or flooring system.

Wood Diagonal Bracing. Diagonal wood member used to prevent buckling or rotation of wood studs.

Wood Failure. The rupturing of wood fibers in strength tests of bonded joints usually expressed as the percentage of the total area involved which shows such failure.

Wood Fascia. Flat vertical wood member of cornice, eaves, or gable or other finish, generally that part of the assembly to which the gutter is secured.

Wood Fiber. 1. A wood cell, comparatively long (1/25 or less to 1/3 inch) (1 to 8 millimeters), narrow, tapering, and closed at both ends. 2. Ground or shredded, non-staining wood used as an aggregate with gypsum plaster.

Wood Filler. A heavily pigmented preparation used for filling and leveling off the pores in open-grained woods.

Wood Finish Concrete. The act or process of using a wood float to smooth irregularities left in curing concrete; work the surface or compact the concrete.

Wood Float. A flat wooden trowel used for floating mortar, normally used just before the pure coat.

Wood Floor Joist. Horizontal structural member of a framed floor.

Wood Floor Removal. The act or process of tearing up old wood floors.

Wood Flooring Institute (WFI). 1800 Pickwick Avenue, Glenview, Illinois 60025, (312) 724-7700.

Wood Flooring. Floor coverings consisting of dressed and finished boards.

Wood Flour. Wood reduced to finely divided particles, approximately those of cereal flours in size, appearance, and texture, and passing a 40-100 mesh screen.

Wood Frame. Floors, roofs, exterior and bearing walls of a building or structure constructed with wood.

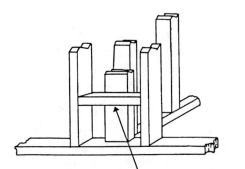

Wood Blocking

Wood Furring. Strips of wood applied to surfaces (usually concrete or masonry) to provide a planar surface and to provide a fastening base for finish material.

Wood Girder. Large horizontal wood beam which supports concentrated loads at isolated points along its length such as the support of joists or rafters.

Wood Ground. Narrow wood strips used around openings and at the perimeter to provide a guide for strike off of plaster to provide a straight and level or plumb line for plaster.

Wood Header. Wood member placed across joist ends or at openings in a wall to support joists or studs at openings in a framing system.

Wood Joist. Horizontal framing member of a floor, ceiling or flat roof.

Wood Lath. Strips of wood, 1-1/4 inches wide by 3/8 inch thick, formerly used as a plaster base; now entirely replaced by gypsum lath.

Wood Lintel. Wood header over openings in wood frame wall construction.

Wood Nailer. Strip of wood attached to steel or concrete to facilitate making nailed connections.

Wood Pile. A long, slender wooden pole driven into the ground to act as a foundation; a member embedded into the ground that supports vertical loads.

Wood Plate. Horizontal wood lumber member on top or bottom of wall studs which ties them together and supports studs, joists, or rafters.

Wood Pole. A long piece of wood used to carry utility lines.

Wood Railing Newel. Principal wood post at foot of stairway or central support of a winding flight of stairs.

Wood Railing Post. Large vertical wood member to support railing.

Wood Rays. Strips of cells extending radially within a tree and varying in height from a few cells in some species to 4 or more inches in oak; the rays serve primarily to store food and transport it horizontally in the tree; on quartersawed oak, the rays form a conspicuous figure, sometimes referred to as Flecks.

Wood Ridge Rafter. Horizontal wood supporting member at top of sloping roof immediately beneath sheathing.

Wood Riser. The vertical wood board under the tread in a stairway system.

Wood Roof Cricket. Relatively small elevated area of wood roof constructed to divert water around chimney, curb, or other projection.

Wood Roof Curb. Wood member elevated above plane of roof surface used for mounting of equipment or other elements.

Wood Roof Decking. Plywood, lumber, or glued laminated member placed over roof structural members for structural rigidity of building frame and to proved surface for traffic or substrate for roofing system.

Wood Roof Edge Strip. Wood strip (usually in plane of roof insulation) at perimeter of roof secured to structural roof deck used for securing roofing membrane.

Wood Roof Nailer. Wood strip (usually in plane of roof insulation) secured to structural roof deck used for securing roofing membrane.

Wood Saddle. Short horizontal wood member set on top of wood column to serve as seat for a girder.

Wood Saw. See Handsaw.

Wood Screw. A screw for fastening objects in wood.

Wood Shake. A hand-split shingle.

Wood Shingle. Factory cut and shaped roof covering of wood (usually of cedar), cut into modular lengths, widths, and rectangular profile.

Wood Siding. Wood used as exterior surface or cladding for exterior framed wall to provide protection from the elements.

Wood Sill Plate. Horizontal wood lumber member on bottom of wall studs which ties them together and rests on concrete or masonry; horizontal timbers of a house which either rest upon the masonry foundations or, in the absence of such, form the foundation.

Wood Sleeper. Wood member laid on concrete floor to support and receive fastening of wood subfloor or finish flooring.

Wood Stain. Finish for wood containing a dye or pigment; stain sinks into fibers of the wood to a certain extent while paint and lacquer ordinarily do not penetrate the wood.

Wood Stair Framing. Wood structural members supporting stairs or stair openings.

Wood Structural Panel. A structural panel product composed primarily of wood; wood structural panels include all-veneer plywood, composite panels containing a combination of veneer and wood-based material, and mat-formed panels such as oriented strand board and waferboard.

Wood Substance. The solid material of which wood is composed.

Wood Threshold. Wood strip fastened to floor beneath door.

Wood Treatment. The act or process of applying a variety of stains or chemicals to retard fire, decay, insect damage or deterioration, due to the elements.

Wood Trim. Wood millwork, primarily moldings, used to finish off and cover joints and openings.

Wood Truss Joist. Joists of rigid open framework construction with top and bottom chords, fabricated of wood web and chord members or combination wood chord members with metal web members.

Wood Truss Rafter. Truss where chord members also serve as rafters and ceiling joists.

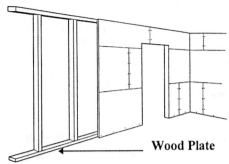

Wood Plate

Wood Truss. A structural component formed of wood members in a triangular arrangement, often used to support a roof.

Wood Wool. Long, curly, slender strands of wood used as an aggregate component for some particleboards.

Wood. The hard fibrous material that makes up most of the trunk and branches of trees.

Wooden Brick. Piece of seasoned wood, made the size of a brick, and laid where it is necessary to provide a nailing space in masonry walls.

Wood-Fibered Plaster. Neat gypsum basecoat plaster containing wood fiber as an aggregate and designed for use either with or without addition of other aggregates.

Woodwork Institute of California (WIC). 1833 Broadway, PO Box 11428, Fresno, California 93773, (209) 233-9035.

Woodwork, Architectural. A higher than average quality feature of finish work using wood for ornamental design.

Wool, Mineral. Any of various lightweight vitreous fibrous materials used for heat and sound insulation.

Wool. 1. Fine soft wavy hair from the fleece of sheep, goats, and other animals. 2. A yarn produced from this hair. 3. Fabric made from it. 4. Any of various wool-like substances like steel wool, lead wool, or mineral wool.

Woolen Yarn. Soft, bulky yarn spun from both long and short wool fibers which are not combed straight but lie in all directions so they will interlock to produce a felt-like texture.

Work Load. The electric conductor between the source of arc- welding current and the work.

Work Product Exclusion. A provision in a liability insurance policy that withholds coverage for damages to work performed by or on behalf of the insured.

Work Station. An area defined by equipment or work surfaces intended for use by employees only, and usually for one or a small number of employees at a time; examples include ticket booths, the employee side of grocery store checkstands, the bartender area behind a bar, the employee side of snack bars, sales counters and public counters, guardhouses, toll booths, kiosk vending stands, lifeguard stations, maintenance equipment closets, counter and equipment areas in restaurant kitchens, file rooms, and storage areas.

Work Top. A counter top.

Workability. 1. The ease with which material can be worked or smoothly cut and shaped with hand or machine tools. 2. In masonry, the texture of mortar such that it behaves properly under the trowel. 3. In painting, the texture and consistency of the paint such that it spreads properly. 4. In woodworking, the degree of ease and smoothness of cut obtainable with hand or machine tools. 5. The property of freshly mixed concrete or mortar which determines the ease and homogeneity with which it can be mixed, placed, compacted, and finished. 6. A property of plaster mortar closely related to plasticity which determines the ease and speed with which the mortar can be applied and finished. 7. The ease with which paving mixtures may be placed and compacted.

Workbench. Table at which work is accomplished.

Worker's Compensation. A system established by statute under which employers are responsible for medical expenses and disabilities of workers injured while on the job; compensation is payable even if the employer is not at fault and even if the carelessness of the worker contributed to the injury, but the employer is not necessarily liable for damages for pain and suffering.

Working Capital. The amount of capital available for current use in the operations of a business measured by the excess current assets (cash and assets readily converted into cash) after current liabilities have been subtracted.

Working Day. A day on which construction work may be done, eliminating Saturdays, Sundays, and Holidays.

Working Drawings. Drawings of the project that are used in the construction of structure, they are part of the contract documents.

Working Life. The period of time during which an adhesive, sealant, or other material, after mixing with catalyst, solvent, or other compounding ingredients, remains suitable for use; also called Pot Life.

Working Properties. The properties of an adhesive that affect or dictate the manner of application to the adherends to be bonded and the assembly of the joint before pressure application, such as viscosity, pot life, assembly time, and setting time.

Working Stress Design. A design theory that is used in the design of concrete and masonry members; safety is provided by limiting allowable stresses.

Working Stress. The maximum permissible stress used in the working stress design of a member.

Workout. A procedure whereby a borrower negotiates a restructuring of a mortgage or trust deed in lieu of a foreclosure.

Worm Gear. Gear teeth spirally cut into a shaft meshing with a worm wheel; the worm gear, or worm, and the worm wheel are on perpendicular axes.

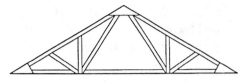

Wood Truss

Worsted Yarn. Strong, dense yarn made from long staple fibers which are combed to align the fibers and remove extremely short fibers.

Woven Carpet. Carpet made by simultaneously interweaving backing and pile yarns on one of several types of looms from which the carpets derive their names.

Woven Wire Fabric. A plaster reinforcement of zinc-coated wire, not lighter than No. 18 gauge when woven into 1 inch openings, or not lighter than No. 17 gauge when woven into 1-1/2 inch openings; lath may be paper-backed, flat or self-furring.

WP. Water proofing.

Wraparound Deed of Trust or Mortgage. See All-Inclusive Deed of Trust or Mortgage.

WRCLA. Western Red Cedar Lumber Association.

Wrench. A hand or power tool, some adjustable, for gripping and turning bolts and nuts.

Wrinkle Finish. A varnish or enamel film which exhibits fine wrinkles or ridges. Used extensively as a novelty finish.

Wrinkle. 1. A pattern of ridges usually caused by roofing plies not being flat in the bitumen intermopping. 2. See Cockle.

Wrinkled Sheets. Pertaining to ceramic mosaics mounted on paper. Due primarily to rough handling in shipment.

Wrinkling. A surface defect resembling the skin of a prune.

Wrist Action. In using a spray gun, swiveling of wrist without arcing forearm.

Wrongful Death. Unlawful homicide, whether by negligence or intent.

Wrought Iron. A tough malleable, relatively soft, form of iron suitable for forging or rolling; not cast.

WWF. Welded Wire Fabric.

WWM. Welded Wire Mesh.

WWPA. Western Wood Products Association.

Wye. A Y-shaped pipe fitting which joins 3 pipes.

Wythe. 1. Each continuous vertical section or layer of masonry one unit in thickness; also called a Tier. 2. The thickness of masonry separating flues in a chimney.

X, Y, Z

X Bracing. That form of bracing wherein a pair of diagonal braces cross near mid-length of the bracing members.

Xanthic. Containing yellow or pertaining to yellow color.

X-Axis. One of the axes in a three-dimensional system, that is not the y- or z-axis.

Xenon. A heavy colorless odorless inert gaseous element occurring in traces in the atmosphere and used in fluorescent lamps.

Xeric. Requiring only a small amount of moisture, as a xeric habitat.

Xerography. A dry electrostatic document copying process.

Xerothermic. A hot dry environment.

XO, XOO, XOOX. Designations of the arrangement of sliding and fixed panels in sliding glass doors and windows; X indicates a sliding panel; O indicates a fixed panel; see OX, OXO, OXXO.

X-Ray, Weld. See Weld X-Ray.

X-Ray. Electromagnetic radiations of a short wavelength that can penetrate various thicknesses of all solids.

Xylem. The portion of the tree trunk, branches, and roots that lies between the pith and the cambium.

Xylene. A solvent obtained from wood tar, coal tar, or petroleum.

Xylography. The art of making engravings on wood.

Xylol. Xylene.

Xylonite. A thermoplastic material like celluloid, used in model building.

Xyst. In Greek and Roman architecture, a long covered portico for exercise in bad weather.

Y

Y Strainer. A device in the shape of a Y for withholding foreign matter from a flowing liquid or gas.

Yard lumber. Run of the mill lumber in a lumber yard, including all sizes and patterns.

Yard. 1. A unit of linear measure equal to 3 feet or 36 inches; equal to 0.9144 meters. 2. A square yard, as in yards of carpeting. 3. A cubic yard, as in yards of earth cut or fill. 4. A piece of enclosed ground. 5. The space around a residence; see Curtilage. 6. The space between buildings.

Yarn Size. Carpet yarn is measured by the number of yards in length to the ounce of weight in a single ply; two to four plies of this single strand will be twisted to form the best balance for any particular weave and quality of fabric desired; it is necessary to know yarn sizes and plies in order to make fair comparisons.

Y-Axis. One of the axes in a three-dimensional system, that is not the x- or y-axis.

Yellow Fir. Douglas Fir.

Yellow Metal. Muntz metal.

Yellow Ochre. An earthy pigment made from impure iron ore, used in paint.

Yellow Ware. A yellow semi-vitreous ware or an earthenware with a colorless clear glaze.

Yellowing. Development of yellow color or cast, in whites, on aging.

Yield Point. The point at which a material deforms with no increase in load; the stress at which a material ceases to deform in a fully elastic manner.

Yield Strength. The specified minimum yield strength or yield point of reinforcement in psi.

Yield. 1. The ratio of the net income of a real estate or other investment compared to the acquisition and operating costs. 2. The amount of concrete produced by a given combination of materials; the total weight of ingredients divided by the unit weight of the freshly mixed concrete. 3. The cubic feet of concrete produced per sack of cement. 4. The number of product units, such as block, produced per batch of concrete or sack of cement.

Yoke Vent. A pipe connecting upward from a soil or waste stack to a vent stack for the purpose of preventing pressure changes in the stacks.

Young's Modulus. The ratio of the unit stress to the unit strain below the proportional limit.

Z

Z Bar. A Z-shaped steel or sheet metal strip used for light framing or furring.

Z Tie, Wall. See Wall Z Tie.

Z. A numerical coefficient used in the design of earthquake forces and that is dependent upon site location.

Zaffer. An impure oxide of cobalt used as a blue ceramic coloring.

Zax. An implement used for splitting and installing slates; a kind of hatchet with a sharp point on the head for perforating the slate to receive a nail or pin.

Z-Axis. One of the axes in a three-dimensional system, that is not the x- or y-axis.

Zebrawood. The mottled or striped wood of several trees, such as various leguminous African timber trees with pale golden heartwood uniformly striped with dark brown or black.

Zenith. The highest or culminating point.

Zero Ice. Trade name for dry ice. See Dry Ice.

Zero Slump Concrete. A concrete mixed with so little water that it has a slump of zero when tested.

Zeta. A closed or small chamber; a room over a church porch where documents were kept.

Z-Furring Channel. A Z-formed metal channel for mechanically attaching gypsum board and insulation material on masonry walls.

Ziggurat. A stepped pyramid, as in the sacred architecture of Western Asia in antiquity.

Zigzag. Making short and sharp turns; in architecture, especially in the moldings in arched door heads of Romanesque style.

Zinc Chromate. Metal priming pigment with important rust-inhibitive properties; Zinc Yellow.

Zinc Dust. Finely divided zinc metal, gray in color; used primarily in metal primers.

Zinc Oxide. A compound of zinc used as a white pigment in many types of paint.

Zinc Phosphate Coating. Treatment used on steel to improve adhesion of coatings.

Zinc Silicate. Inorganic zinc coating.

Zinc Sulphide. Compound of zinc used as white pigment in paints.

Zinc Yellow. Commercial zinc chromate pigment.

Zinc, Leaded. See Leaded Zinc.

Zinc. 1. A bluish white crystalline metallic element of low to intermediate hardness that is ductile when pure but in the commercial form is brittle at ordinary temperatures and becomes ductile on slight heating. 2. Zinc, as galvanizing, is widely used as a protective coating for iron and steel. 3. Zinc is used extensively as paint pigment.

Zincky. Containing or resembling zinc.

Zincoid. Relating to or resembling zinc.

Zip Tape. In gypsumboard bundles, a reinforcement paper strip to facilitate the removal of end bundling tapes.

Zippers. See Paper Rollers.

Zircon Porcelain. A vitreous ceramic whiteware for technical application in which zircon (ZrO_2 SiO_2) is the essential crystalline phase.

Zone. 1. In space heating or air conditioning, a specific area in a building that is kept at the same temperature. 2. A defined area of land that is limited in use to some particular purpose, such as residential, commercial, or industrial. See Zoning, 2.

Zoning. 1. Government regulation of the use of privately owned land. 2. The official designation of parts of a municipality or other governmental territory to be used only for certain specified land uses.

Zonolite. A lightweight insulating concrete composed of portland cement, vermiculite aggregate, and water.

Zoomorph. A lifelike form.

Zoophoric Column. A pillar supporting the figure of an animal.